STO

TED L. HANES is professor of biology at California State University, Fullerton. He has taught courses in botany, biology, and plant ecology since 1955 and since 1969 has taught ecology at the National Audubon Society Workshop of the West. JEAN MACQUEEN is a well-known British biologist.

Jean Macqueen / Ted L. Hanes

The Living World

exploring modern biology

PRENTICE-HALL, INC., Englewood Cliffs, New Jersey 07632

Library of Congress Cataloging in Publication Data

Macqueen, Jean.
The living world

(A Spectrum Book)
A revision of Success in Biology.
Bibliography: p.
Includes index.
1. Biology. I. Hanes, Ted L., joint author.
II. Success in biology. III. Title.
QH308.2.M36 574 78-2601
ISBN 0-13-538975-5
ISBN 0-13-538967-4 pbk.

A Spectrum Book

10 9 8 7 6 5 4 3 2 1

Printed in the United States of America

PRENTICE-HALL INTERNATIONAL, INC., *London*
PRENTICE-HALL of AUSTRALIA PTY. LIMITED, *Sydney*
PRENTICE-HALL of CANADA, LTD., *Toronto*
PRENTICE-HALL of INDIA PRIVATE LIMITED, *New Delhi*
PRENTICE-HALL of JAPAN, INC., *Tokyo*
PRENTICE-HALL of SOUTHEAST ASIA PTE. LTD., *Singapore*
WHITEHALL BOOKS LIMITED, *Wellington, New Zealand*

Contents

Preface xi

Unit 1 **The Nature of Living Organisms** 1

1.1 The Characteristics of Living Organisms 1.2 Differences between Plants and Animals 1.3 The Varieties of Living Organisms 1.4 Questions

Unit 2 **Cells** 7

2.1 The Characteristics of Cells 2.2 Parts of the Cell 2.3 Differences between Plant and Animal Cells 2.4 Relation of Cells to the Organism as a Whole 2.5 Specialization 2.6 Questions

Unit 3 **Diffusion and Osmosis** 17

3.1 Introduction 3.2 Diffusion 3.3 Osmosis 3.4 Osmosis in Plants 3.5 Osmosis in Animals 3.6 Questions

Unit 4 **Respiration** 35

4.1 Introduction 4.2 Aerobic Respiration 4.3 Anaerobic Respiration 4.4 Fermentation 4.5 Metabolism 4.6 Questions

Unit 5 **Structure of the Flowering Plant** 46

5.1 Introduction 5.2 Stems 5.3 Leaves 5.4 Buds 5.5 Winter Twigs 5.6 Roots 5.7 Questions

Unit 6 **Storage Organs and Vegetative Reproduction** 71

6.1 Introduction 6.2 Bulbs 6.3 Corms 6.4 Rhizomes 6.5 Stem Tubers 6.6 Advantages of Food Storage 6.7 Advantages of Vegetative Reproduction 6.8 Artificial Propagation 6.9 Questions

Unit 7 **Sexual Reproduction in Flowering Plants** 84

7.1 Introduction 7.2 Flower Structure 7.3 Pollination 7.4 Fertilization 7.5 Fruit Formation 7.6 Dispersal of Fruits and Seeds 7.7 Questions

Unit 8 **Seeds, Germination, and Tropisms** 106

8.1 Structure of seeds 8.2 Germination 8.3 Experiments on Germination 8.4 Experiments on the Sensitivity of Plants 8.5 The Auxin Theory of Tropistic Response 8.6 Additional Experiments 8.7 Questions

Unit 9 **Translocation and Transpiration** 129

9.1 Introduction 9.2 Translocation 9.3 Transpiration 9.4 Experiments on Transpiration 9.5 Questions

Unit 10 **The Nutrition of Green Plants** 143

10.1 Types of Nutrition 10.2 Photosynthesis 10.3 Experiments on Photosynthesis 10.4 Direct Evidence for Photosynthesis 10.5 Limiting Factors 10.6 The Elements Essential for Plant Growth 10.7 The Source of Salts in the Soil 10.8 Questions

Unit 11 **Bacteria, Blue-Green Algae, and Virus** 161

11.1 Classification 11.2 Structure of Bacteria 11.3 Bacteria in Nature 11.4 Harmful Bacteria 11.5 Prevention of Infection 11.6 Culturing Bacteria 11.7 Structure of Blue-Green Algae 11.8 Blue-Green Algae in Nature 11.9 Viruses 11.10 Questions

Unit 12 **Fungi** 172

12.1 The Characteristics of Fungi 12.2 Mucor 12.3 Antibiotics 12.4 Parasitic Fungi 12.5 Yeast 12.6 Questions

Unit 13 **Algae** 180

13.1 Introduction 13.2 Green Algae 13.3 Diatoms 13.4 Brown Algae 13.5 Red Algae 13.6 Questions

Unit 14 **The Animal Kingdom** 184

14.1 Introduction 14.2 Protozoans 14.3 Sponges 14.4 Coelenterates

14.5 Flatworms 14.6 Roundworms 14.7 Segmented Worms
14.8 Crustaceans 14.9 Insects 14.10 Arachnids 14.11 Mollusks
14.12 Echinoderms 14.13 Fish 14.14 Amphibians 14.15 Reptiles
14.16 Birds 14.17 Mammals 14.18 Questions

Unit 15 **Insects** **197**

15.1 Introduction 15.2 Life History 15.3 Cuticle and Ecdysis
15.4 Breathing 15.5 Blood System 15.6 Sensory System
15.7 Locomotion 15.8 Feeding Methods 15.9 Insects as Disease Carriers
15.10 The Large White Butterfly 15.11 Questions

Unit 16 **Fish** **225**

16.1 Introduction 16.2 External Features 16.3 Swimming
16.4 Breathing 16.5 The Three-spined Stickleback

Unit 17 **Frogs** **235**

17.1 Amphibia 17.2 External Features 17.3 Locomotion
17.4 Breathing 17.5 Feeding 17.6 Skin and Color 17.7 Habitat
17.8 Life History

Unit 18 **Birds** **248**

18.1 Introduction 18.2 Locomotion 18.3 Reproduction 18.4 Questions

Unit 19 **Food and Diet in Man** **259**

19.1 Introduction 19.2 Energy Value of Food 19.3 Nutrients
19.4 Water 19.5 Roughage 19.6 Milk 19.7 Practical Work: Food
Tests 19.8 Questions

Unit 20 **Digestion, Absorption, and Metabolism of Food** **269**

20.1 Introduction 20.2 Movement of Food through the Alimentary Canal
20.3 Digestion in the Mouth 20.4 Digestion in the Stomach
20.5 Digestion in the Duodenum 20.6 Digestion in the Ileum
20.7 Absorption in the Ileum 20.8 The Cecum and Appendix 20.9 The
Large Intestine 20.10 How Digested Food is Used 20.11 Storage of
Digested Food 20.12 The Liver 20.13 Homeostasis 20.14 Practical
Work 20.15 Questions

Unit 21 **Blood and the Circulatory System** **291**

21.1 The Composition of Blood 21.2 The Functions of Blood 21.3 The Circulatory System 21.4 Questions

Unit 22 **Breathing** **310**

22.1 Introduction 22.2 The Lungs 22.3 Gaseous Exchange in the Lungs 22.4 Ventilation of the Lungs 22.5 The Nose 22.6 Voice 22.7 Practical Work 22.8 Questions

Unit 23 **Excretion** **322**

23.1 Introduction 23.2 Structure of the Kidneys 23.3 Mechanism of Kidney Excretion 23.4 The Bladder 23.5 Water Balance and Osmoregulation 23.6 Questions

Unit 24 **Skin and Temperature Control** **330**

24.1 Skin Function and Structure 24.2 Temperature Control 24.3 Questions

Unit 25 **Sexual Reproduction** **339**

25.1 Introduction 25.2 The Reproductive Organs 25.3 Production of Gametes 25.4 Fertilization 25.5 Pregnancy and Development 25.6 Birth 25.7 Parental Care 25.8 Secondary Sexual Characters 25.9 Menstruation 25.10 Birth Control 25.11 World Population 25.12 Questions

Unit 26 **The Skeleton, Muscles, and Movement** **359**

26.1 Introduction 26.2 Functions of the Skeleton 26.3 Joints 26.4 Muscles 26.5 Girdles 26.6 Locomotion 26.7 Questions

Unit 27 **Teeth** **372**

27.1 Introduction 27.2 Tooth Structure 27.3 Specialization of Teeth 27.4 Jaw Action 27.5 Teeth in Man 27.6 Questions

Unit 28 **The Sensory Organs** **379**

28.1 The Sensory System 28.2 Taste 28.3 Smell 28.4 Structure of the Eye 28.5 Sight 28.6 Structure of the Ear 28.7 Hearing 28.8 The Semicircular Canals and the Sense of Balance 28.9 Questions

Unit 29 **Coordination** 403

29.1 Introduction 29.2 The Nervous System 29.3 Reflex Action 29.4 The Endocrine System 29.5 Interaction and "Feedback" 29.6 Questions

Unit 30 **Heredity: Chromosomes** 425

30.1 Introduction 30.2 Cell Division 30.3 Mitosis 30.4 Chromosomes 30.5 Genes and Their Function 30.6 How Genes Work 30.7 Variation 30.8 Meiosis 30.9 New Combinations of Genes in the Gametes 30.10 Fertilization 30.11 Mutations 30.12 Questions

Unit 31 **Heredity: Genetics** 456

31.1 Genes and Inheritance 31.2 Human Genetics 31.3 Discontinuous and Continuous Variation 31.4 Heredity and Environment 31.5 Applications of Genetics to Human Problems 31.6 Intelligence 31.7 Questions

Unit 32 **Evolution and Natural Selection** 473

32.1 The Theory of Evolution 32.2 The Theory of Natural Selection 32.3 Heritable Variation 32.4 Balanced Polymorphism 32.5 Isolation and the Formation of New Species 32.6 "The Survival of the Fittest" 32.7 Natural Selection in Man

Unit 33 **Ecology—Interrelationships In Nature** 489

33.1 Introduction 33.2 The Environment 33.3 Resource Allocations 33.4 Adaptation 33.5 Significance of Reproduction 33.6 Succession 33.7 The "Balance of Nature" 33.8 Applied Ecology 33.9 Questions

Unit 34 **Soil** 508

34.1 Components of Soil 34.2 Types of Soil 34.3 Experiments on Soils 34.4 Soil Erosion 34.5 Questions

Further Reading 517

Index 519

Preface

The Living World presents the factors of biology without fanfare. Many contemporary textbooks in biology include so much of the research and speculative aspects of the subject that the student often cannot readily find what is known about biology. Other beginning biology books are filled with color plates and "window dressings," but there is little factual content. *The Living World* is written to alleviate both of these problems. It is a straightforward presentation of the principles and facts of biology.

The Living World should be especially useful as a textbook in introductory courses both for the major and non-major student. It is ideal also for students working independently in extension courses since experiments and chapter questions are included in the book. *The Living World* provides reference information for high school students, amateur naturalists, and all readers with a general interest in the world of living things. This book should give the reader a full understanding of the diversity as well as the unity of life as we know it today.

The Living World is a revision of *Success in Biology* published in England. In turn, that book was adapted from the textbook *Introduction to Biology* by D. G. Mackean. *The Living World* is written to be self-explanatory and covers all the concepts of modern biology in a clear and factual manner.

Acknowledgments

Most of the diagrams are by D. G. Mackean, and any requests for permission to reproduce them should be addressed to D. G. Mackean, care of the publishers. The sources of other diagrams and of photographs are acknowledged where each occurs.

Unit 1
The Nature Of Living Organisms

1.1 THE CHARACTERISTICS OF LIVING ORGANISMS

Biology is the study of life (Greek *bios* = life, *logos* = knowledge); in practice, this means the study of living things.

The characteristics by which we know that an animal is alive are usually self-evident: it moves about, it feeds, it produces young, and it responds to changes in its surroundings.

These features are less obvious in plants and certain small animals; and when dealing with organisms like bacteria and viruses the distinction between living and non-living material can often be drawn only by a trained scientist with the appropriate apparatus and techniques at his disposal. The main differences between living organisms and non-living objects can be summarized as follows:

(a) **Respiration.** This is the process of obtaining energy as a result of chemical changes within the organism, the commonest of which is the chemical decomposition of food by reaction with oxygen. It is not always a particularly obvious occurrence in plants and animals; but it is fairly easy to demonstrate that living creatures take in air, remove some of the oxygen from it and increase the amount of carbon dioxide in it. More simply, it can be said that living organisms take in oxygen and give out carbon dioxide. Sometimes this takes place with obvious breathing movements. Respiration also results in a rise of temperature, which is more easily detectable in animals than in plants.

(b) **Nourishment.** All living things need substances to repair and build their bodies from an outside source of energy. Some organisms, like green plants, take in simple nutrients and light energy and then manufacture their own food. Other organisms, like animals, must obtain their nutrients and energy from complex chemical compounds, namely organic substances. But for both types of organisms, the food they manufacture or the food

they feed upon becomes the source of energy (by respiration) upon which all other life processes depend.

***(c)* Excretion.** Living involves a vast number of chemical processes, including respiration, many of which produce substances that are poisonous as they accumulate. The elimination of these from the body of the organism is called excretion. Excretion should not be confused with *egestion,* or the removal of undigested food from the body. This undigested food has not taken part in any of the organism's vital chemistry and most of it has been unaffected by digestive fluids. Thus its removal, although essential, is not usually included in the term excretion.

***(d)* Growth.** Feeding results in the growth of the organism. Strictly, growth is simply an increase in size, but it usually implies also that the organism is producing more cells and becoming more complicated and more efficient. For example, a tadpole grows into a frog, and a caterpillar into a butterfly.

***(e)* Movement.** Living creatures are usually capable of some kind of movement, although in most plants this is very restricted. But within all living cells there is continuous movement and activity.

***(f)* Reproduction.** Very few organisms have, even in theory, a limitless life, but although individuals must die sooner or later, their life is handed on to new individuals of the same kind by reproduction, so that the existence of this particular form of life is continued.

***(g)* Irritability.** Irritability, or sensitivity, is the ability to respond to a stimulus. Obvious signs of sensitivity are the movements made by animals as a result of noises, or on being touched by or seeing an enemy, and the upward growth of the young shoots of seedlings toward the light, and the growth of their roots downward in response to gravity.

1.2 DIFFERENCES BETWEEN PLANTS AND ANIMALS

Most living creatures can be assigned to one of the two great categories of *plants* and *animals,* although some simple, microscopically small organisms, described later have certain characteristics of both. Plants and animals have in common, to a greater or lesser extent, all the features discussed in Section 1.1, but there are some fundamental differences between them.

***(a)* Method of feeding.** Animals take in food that is chemically very complicated (i.e. composed of large molecules); it consists either of plant products or of other animals. This food is reduced to simpler material by the process of digestion, and in this form it can be taken up by the body.

Plants, in general, take in very simple substances that are composed of small molecules, namely carbon dioxide from the air, and water and dissolved mineral salts from the soil. In their leaves they combine this carbon dioxide and water into glucose (a simple sugar), using the energy of sunlight to effect the change— a process called *photosynthesis.* From the glucose so produced, together with the mineral salts taken in from the soil, green plants can make any of the substances needed for their existence.

(b) **Chlorophyll.** The green color of most plants is due to the presence of a compound called chlorophyll, which is not found in any animal. This substance is important in the absorption of sunlight, and its presence indicates the fundamental difference between the nourishing processes of plants and animals. However, fungi and most bacteria lack chlorophyll and therefore must obtain both their nourishment and energy from the foods they consume, much as animals do.

(c) **Cellulose.** A considerable proportion of a plant body consists of this substance. It is not present in animal structures.

(d) **Movement.** An animal can generally move its whole body; but the movement of higher plants is usually restricted to certain parts—for example, the opening and closing of the petals of some flowers—or to movements of parts as a result of growth. However, certain microscopic plants move as actively as microscopic animals.

(e) **Sensitivity.** While both plants and animals respond to stimuli, the response of an animal usually follows almost immediately after the application of even a very brief stimulus; in general, the more complex and highly evolved the animal, the greater is the range of its sensitivity to stimuli like heat, light and chemicals. In most plants, on the other hand, a response may take place over a period of hours or days, and then only if the stimulus persists for a relatively long time. A few plants are sensitive to touch and respond in seconds. Insect-eating plants use this rapid response in catching their prey.

1.3 THE VARIETIES OF LIVING ORGANISMS

Plants and animals can be divided into groups, the members of any of which show strong likenesses to one another. These similarities are not always immediately obvious, but soon become apparent when the distinctive features of the group are studied. A bee and a butterfly, for example, differ considerably in appearance, size, color and habits, but they belong to the same group because they both have hard outer skeletons, three divisions to their bodies, six legs and two pairs of wings. Similarly, a rose does not greatly resemble a sunflower, yet because their leaves are broad rather than narrow and strap-shaped, and their seeds contain two cotyledons (leaf-like structures; see Section 8.1) rather than one, they can reasonably be grouped together.

The groups of animals and animals named in Tables 1.1 and 1.2 are called *phyla* (singular: *phylum*). Eight other phyla containing less familiar animals are not listed, for simplicity's sake. While the classification of the vertebrates and plants that follows is convenient, it is somewhat oversimplified.

Table 1.1 *Living organisms: animals*

A: ANIMALS WITHOUT VERTEBRAL COLUMNS: INVERTEBRATES

1 **Protozoans.** Very abundant, these minute animals all live in water and can usually be seen only with a microscope. They are one-celled animals.
2 **Sponges.** Porous, "headless" animals found growing on solid objects in fresh and salt water environments. Some are soft, others are hard and brittle.
3 **Coelenterates.** Examples: sea anemones, jelly-fish, *Hydra,* coral-building "anemones," all of which (except *Hydra*) live in the sea.
4 **Flatworms.** Mostly small fresh-water animals, flatworms are often found under stones and floating leaves in streams. The group also includes the parasitic tapeworms and liver flukes.
5 **Roundworms.** Mostly small, white, cylindrical, parasitic worms. Live in soil and in host animals.
6 **Segmented Worms.** These include the earthworm, many little worms that live in ponds, and the bristle-worms of the sandy coasts.
7 **Crustacea.** These animals have hard outer shells. Examples: crabs, lobsters, crayfish, prawns, shrimps, many small fresh-water creatures such as the fresh-water shrimp and water-flea.
8 **Insects** possess six legs, three body parts and usually wings. Examples: butterflies, ants, bees, grasshoppers, flies, mosquitoes and beetles.
9 **Arachnids** have eight walking legs. Examples: spiders, scorpions, ticks, mites.
10 **Mollusks.** Examples: snails and slugs, whelks, oysters and other "shellfish," squid and octopus.
11 **Echinoderms.** These are marine animals; they include the familiar sea star (starfish) and sea urchin.

B: ANIMALS WITH VERTEBRAL COLUMNS: VERTEBRATES

I POIKILOTHERMIC (with variable body temperature)

1 **Fish** breathe by means of gills and have bodies covered with scales. Examples: shark, herring, pike, stickleback, bass, trout.

Table 1.1 *Living organisms: animals* (Cont'd)

2 Amphibia have no scales on their bodies; they spend much of their lives on land. Examples: frogs, newts, toads.
3 Reptiles are land-dwelling animals with scaly bodies. Examples: lizards, snakes, tortoises, crocodiles, alligators.

II HOMOIOTHERMIC (with constant body temperature)

4 Birds have bodies that are covered with feathers. Examples: sparrow, duck, penguin.
5 Mammals have bodies that are covered with fur; their young are born alive and suckle milk. Examples: cows, dogs, cats, whales, seals, apes and humans.

Table 1.2 *Living organisms: plants*

A: PLANTS THAT DO NOT HAVE FLOWERS

1 Bacteria are extremely small organisms lacking an organized nucleus, living in and on most things, and reproducing rapidly.
2 Blue-Green Algae are much larger then bacteria but also lack an organized nucleus. Live in chains or groups in water, moist soil, snow, hot springs.
3 Fungi include molds and mushrooms. They are classified as plants even though they differ considerably from most other plants in their method of nourishment.
4 Algae include the green slimy filaments in ponds and also seaweeds.
5 Liverworts are small, flat, green, leaf-like plants found in clusters in damp places, stream banks, and in cellars and caves to which light has access.
6 Mosses are small green plants growing in dense clusters in damp, shady places.
7 Ferns are leafy plants of many kinds, and include bracken.
8 Coniferous trees. Examples: spruce, pine, cypress, fir, redwood.

B: FLOWERING PLANTS

1 Monocotyledons. Narrow-leaved plants with parallel veins and only one cotyledon in their seeds. Examples: cereals and other grasses, reeds, rushes, iris, daffodil, palms, yuccas.
2 Dicotyledons. Broad-leaved plants with net veins and two cotyledons in their seeds.
(*i*) *Herbaceous plants*. Examples: daisy, buttercup, dandelion, begonia.
(*ii*) *Shrubs*. Woody, bushy plants. Examples: privet, rose, azalea.
(*iii*) *Deciduous trees*. Examples: oak, ash, hazel, beech, maple.
(*iv*) *Evergreen trees*. Examples: southern magnolia, orange, live oak.

The smallest natural group of animals or plants is the *species*. For example, robins form a single species; the term *birds* comprises many species. Generally speaking, all the members of a species look and behave alike in every important respect, and can breed among themselves. Breeding between members of different species does not usually occur in nature, but when it does occur the resulting individuals are called *hybrids*. Hybrids show some of the traits of each of their parents.

1.4 QUESTIONS

1. Try to classify the following animals and plants according to the scheme in Tables 1.1 and 1.2: Python, chestnut tree, antelope, mussel, cowslip, lugworm, tadpole, Douglas fir, whale, caterpillar, turtle, puffball, jackdaw, alligator, flea, trout, bluebell, mildew, earwig, cedar, otter, barley. [Initially it is good practice to set out the answers as follows:
   Cow. Animal; vertebrate; homoiothermic; mammal.
   Mushroom. Plant; non-flowering; fungus.]
2. An automobile takes in oxygen and gives out carbon dioxide, consumes fuel but nevertheless is not a living creature. In what ways does it not "qualify" as a living organism?
3. A sponge-like organism is found adhering to a rock in a marine pool. How would a microscopic examination help to decide whether it was a plant or an animal?

UNIT 2
Cells

2.1 THE CHARACTERISTICS OF CELLS

If a structure from a plant or animal is examined under a microscope, it can be seen to be made up of numbers of more or less distinct units, or *cells.* Cells are usually too small to be seen with the naked eye, but vast numbers of them go to make up a structure like a leaf or a lung.

Most cells are specially adapted in their size, shape and chemistry to carry out one particular function (for example, muscle cells are adapted for contracting), so that strictly speaking there is no such thing as a "typical" plant or animal cell. Nevertheless, certain features are common to most cells.

2.2 PARTS OF THE CELL (Figs. 2.1 to 2.5)

Protoplasm. Protoplasm is the material inside the cell which is truly alive. There are two principal kinds of protoplasm; the protoplasm which constitutes the nucleus (see below) is called *nucleoplasm.* All other forms of protoplasm are referred to as *cytoplasm.*

Cytoplasm. Cytoplasm is jelly-like and transparent; it may be fluid or semi-solid. Sometimes it contains structures like starch grains or chloroplasts, the minute bodies in which the green pigment chlorophyll is found. In the cytoplasm the chemical reactions essential to life are carried on. The boundaries of the cytoplasm are *selectively permeable,* that is, they allow some substances to pass freely into or out of the cytoplasm, and prevent others from doing so. This selectivity helps to maintain the best conditions for chemical reactions in the protoplasm.

Nucleus. The nucleus consists of nucleoplasm bounded by a nuclear membrane. It is always embedded in the cytoplasm, is frequently ovoid in

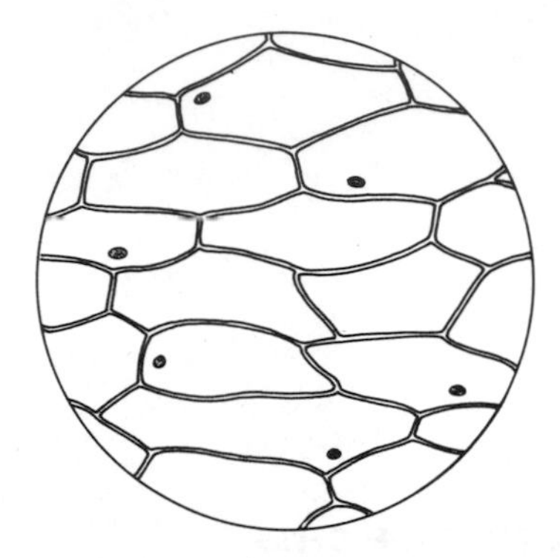

Fig. 2.1 *Epidermis ("skin") from an onion scale seen under the microscope*

The epidermis is only one cell thick, so that under the microscope the transparent cells can be seen clearly. The shape of each cell is partly determined by the pressure of the other cells around it. If a cell were isolated, it would be rounded or oval.

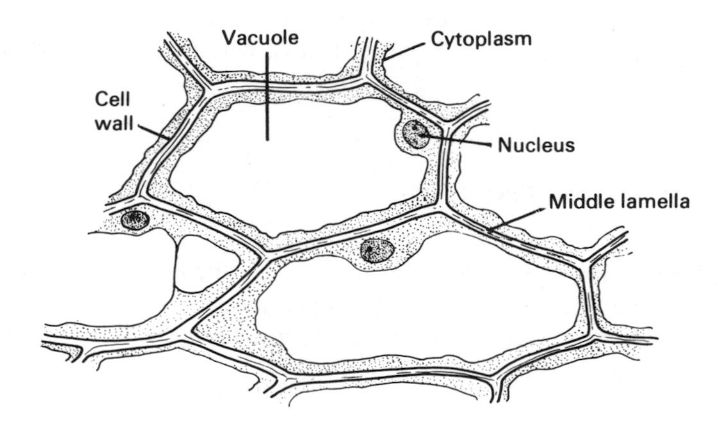

Fig. 2.2 *A group of similar cells highly magnified to show the structures of individual cells*

shape and lighter in color than the cytoplasm. (In diagrams it is often shaded darker because most microscopical preparations are stained with dyes to show it up clearly. It is less easily seen in the unstained cell.) The nucleus is thought to be a center of chemical activity, playing a part in determining the shape, size and function of the cell and controlling most of the physiological processes within it.

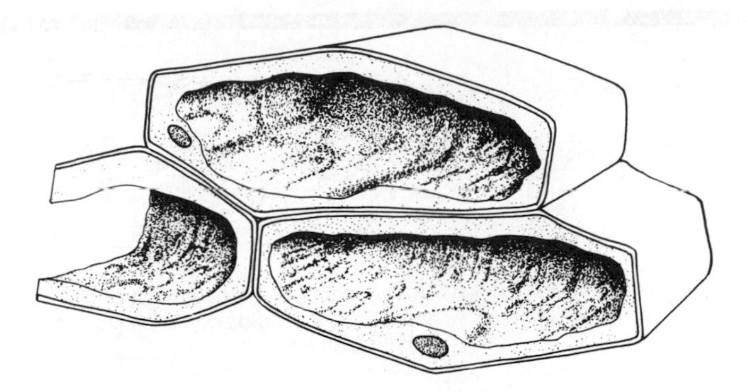

Fig. 2.3 *Stereogram of plant cells*

It is important to remember that, although cells look flat in sections or strips of epidermis, they are in fact three-dimensional and may seem to have different shapes according to the direction in which the section is cut.

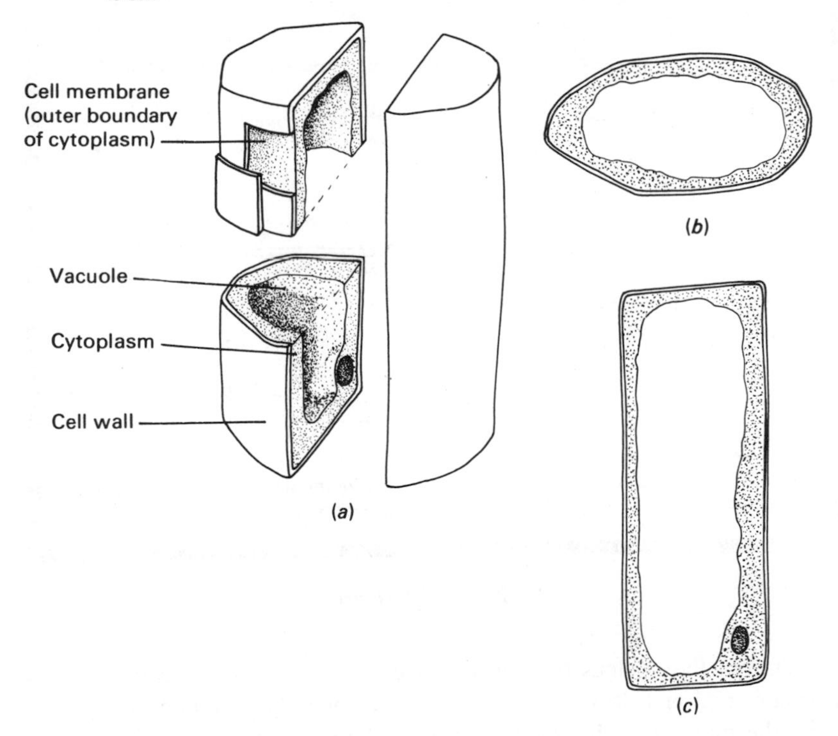

Fig. 2.4 *A single plant cell: (a) stereogram; (b) transverse section; (c) longitudinal section*

If the cell at (a) is cut across, it will look like (b) under the microscope; if cut longitudinally, it will look like (c).

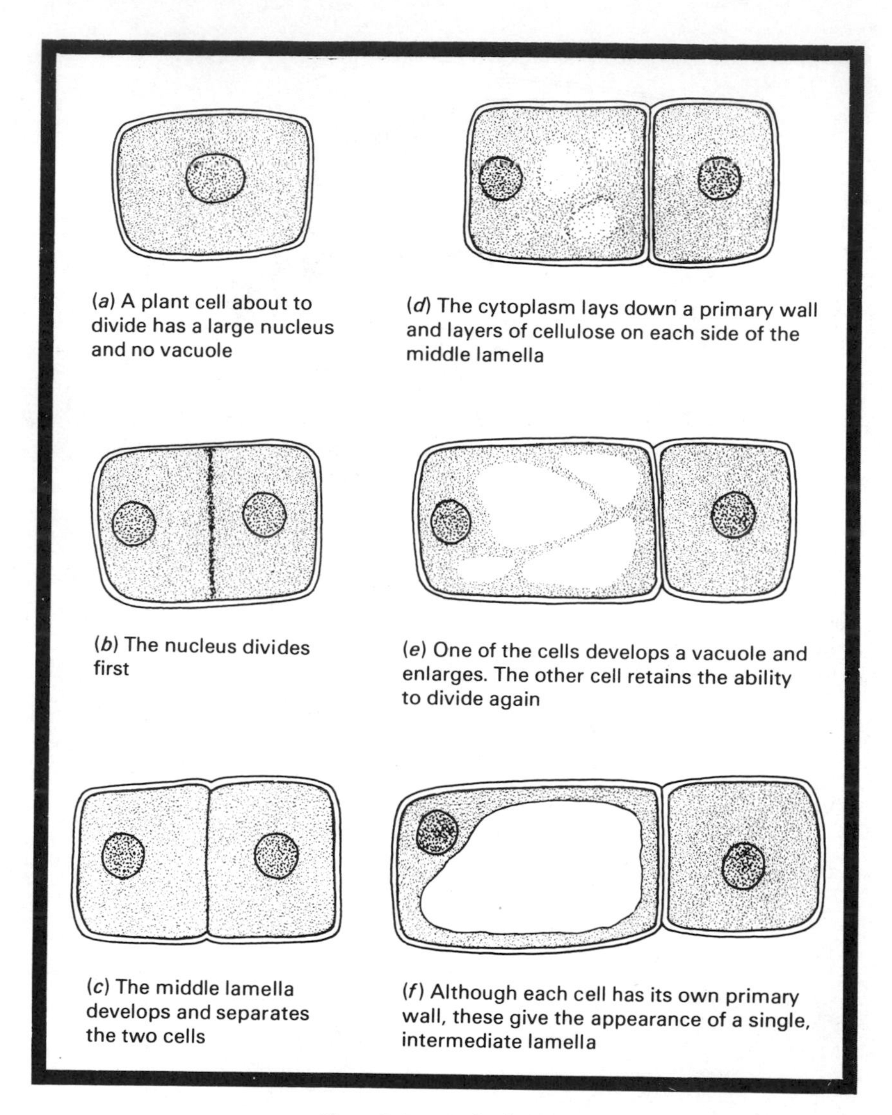

Fig. 2.5 *Cell division*

Without the nucleus the cell is not capable of its normal functions or of division, although it may continue to live for a time. When cell division occurs, the nucleus initiates and controls the process (Fig. 2.5).

Vacuoles. Vacuoles are fluid-filled cavities in the cytoplasm. They can sometimes be seen in animal cells, and in these they are small and variable in size and position. In plant cells they are usually permanent and occupy the greater part of the cell; the fluid they contain is called *cell sap,* and may contain salts, sugars and pigments dissolved in water. Usually it exerts a fairly high pressure, pushing out on all sides, flattening the cytoplasm against the cell wall and, in turn, squeezing the cells closely together, making a firm structure.

Cell membrane. All cells are bounded by a very thin, flexible membrane which retains the cell contents and controls the transfer of food and waste substances into and out of the cell (also called the *plasma membrane*).

Cell wall. Plant cells, but not animal cells, have a wall lying just outside the cell membrane. This cell wall confers shape and, to some extent, rigidity to the cell. While the cell is growing, the cell wall is fairly plastic and extensible, but once the cell has reached full size, the wall becomes tough and resists stretching.

Unless impregnated with chemicals, as in the cells of corky tree bark, the cell wall is freely permeable to gases and water, that is, it allows them to pass through in either direction. The cell wall is made by the cytoplasm but is not a part of it; it is non-living and is made of a transparent substance called *cellulose.* Because of the presence of cell walls, plant cells are often easier to demonstrate under a microscope than are animal cells.

Cell division. Some, but not all, cells are able to reproduce themselves by dividing into two (Fig. 2.5). The growth of an organism takes place by the repeated division and subsequent enlargement of many of its constituent cells. When a plant cell divides, the nucleus first splits into two, and then the cytoplasm forms a layer, or primary wall, between the two new nuclei, separating the two daughter cells. This layer is then thickened by the deposition of cellulose, like the rest of the cell wall. However, it often remains visible in microscopical preparations (compare Fig. 2.2) and is called the *middle lamella.*

2.3 DIFFERENCES BETWEEN PLANT AND ANIMAL CELLS

(*a*) Plant cells are usually larger than animal cells.

(*b*) Plant cells have cell walls made of cellulose, which give them a well-defined outline. Animal cells contain no cellulose and have no wall, they are bounded only by a cell membrane.

(*c*) Plant cells generally have only a thin lining of cytoplasm, with a large central vacuole. Animal cells consist almost entirely of cytoplasm;

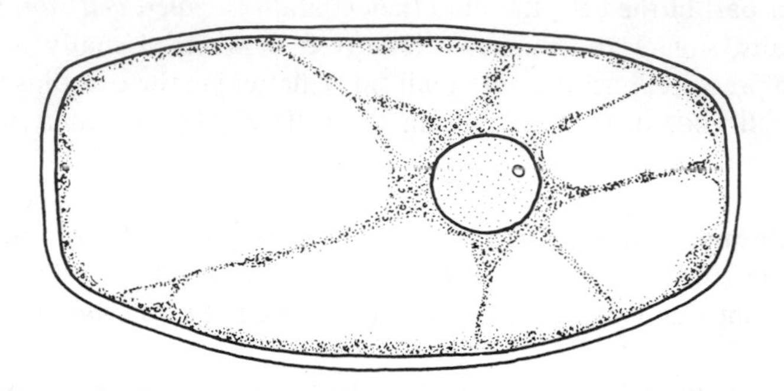

Fig. 2.6 *A plant cell*

Sometimes the nucleus appears in the center of the cell, but it is always surrounded by cytoplasm connected by strands to that lining the wall.

if any vacuoles are present they are usually temporary and small, concerned with excretion or secretion (Figs. 2.7 and 2.8).

(*d*) Animal cells never contain chloroplasts (see Section 3.3(*c*)), whereas these are present in a great many plant cells.

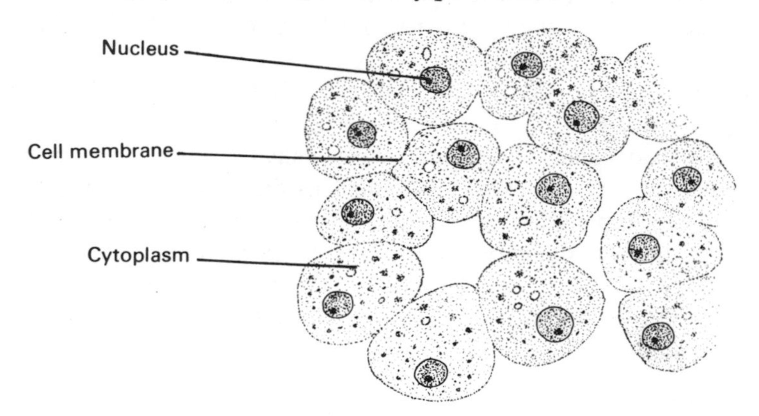

Fig. 2.7 *A group of animal cells (similar to those that line the mouth cavity)*

2.4 RELATION OF CELLS TO THE ORGANISM AS A WHOLE

Although each cell of a complex organism can carry on the vital chemistry of living, it is not capable of existence on its own. The contractile cells in a muscle, for example, cannot obtain their own food or oxygen; other specialized cells in the muscle are necessary for the purpose. Unless individual cells are grouped together in large numbers and made to work

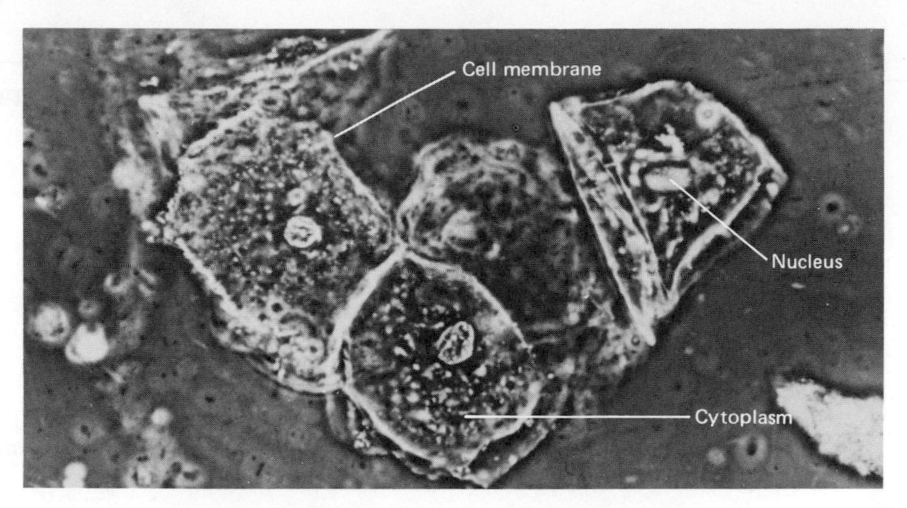

Fig. 2.8 *Cells from the lining of the cheek*
(Brian Bracegirdle)

together by the coordinating mechanisms of the body, they cannot live for long.

Tissue. A tissue is made up of many hundreds of cells of onc or a few types, in which the cells of each type have a more or less identical structure or function: for example, the bones, muscle and brain of an animal, the phloem and pith of a plant stem and the flesh of a fruit like an apple. The tissue itself can also be said to have a specific function: thus nervous tissue in animals conducts impulses, and the phloem of plants (see Section 3.2(*c*)) carries food.

Organs. An organ is a functional unit consisting of several tissues grouped together: for example, a muscle is an organ containing long cells held together with connective tissue and permeated with blood vessels and nerve fibers. The arrival of a nerve impulse causes the muscle fibers to contract, using the food and oxygen brought by the blood vessels to provide the necessary energy. Some of the organs of a plant are its roots, stems and leaves.

Systems. This term usually refers to a group of organs whose functions are coordinated to produce effective action in the organism. For example, the heart and blood vessels constitute the circulatory system; the brain, spinal cord and nerves make up the nervous system; the veins and green tissues of a leaf are integrated with the tissues of the stem and roots.

Organism. The efficient coordination of the organs and systems enables an individual to exist and be able to perpetuate its own kind independent of other organisms.

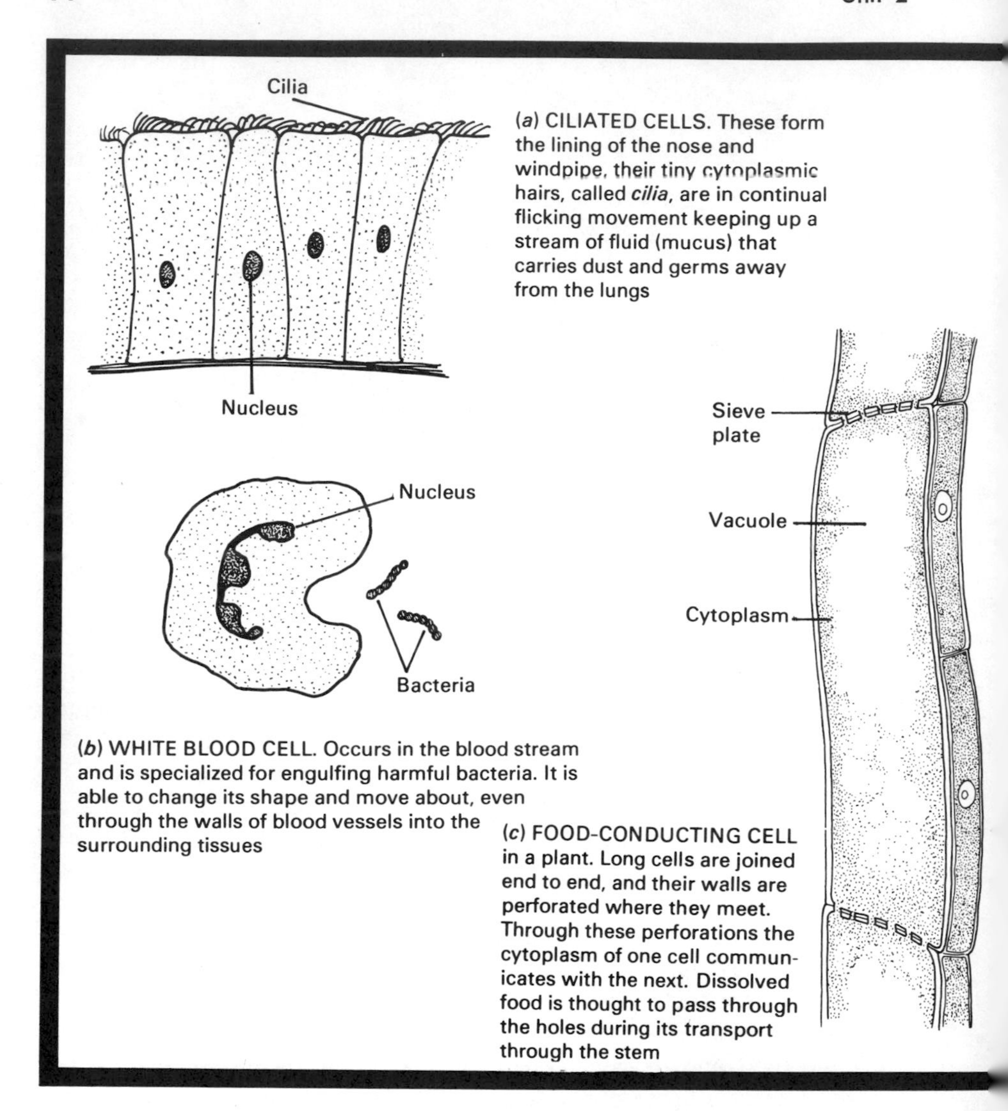

(*a*) CILIATED CELLS. These form the lining of the nose and windpipe, their tiny cytoplasmic hairs, called *cilia*, are in continual flicking movement keeping up a stream of fluid (mucus) that carries dust and germs away from the lungs

(*b*) WHITE BLOOD CELL. Occurs in the blood stream and is specialized for engulfing harmful bacteria. It is able to change its shape and move about, even through the walls of blood vessels into the surrounding tissues

(*c*) FOOD-CONDUCTING CELL in a plant. Long cells are joined end to end, and their walls are perforated where they meet. Through these perforations the cytoplasm of one cell communicates with the next. Dissolved food is thought to pass through the holes during its transport through the stem

Fig. 2.9 *Specialized cells*

2.5 SPECIALIZATION

The cells of higher animals and plants often have a structure and physiology (that is, an internal chemistry) which is specially adapted to the particular function of the organ of which they form a part. Fig. 2.9 illustrates a few examples from the enormous range of specialization which exists.

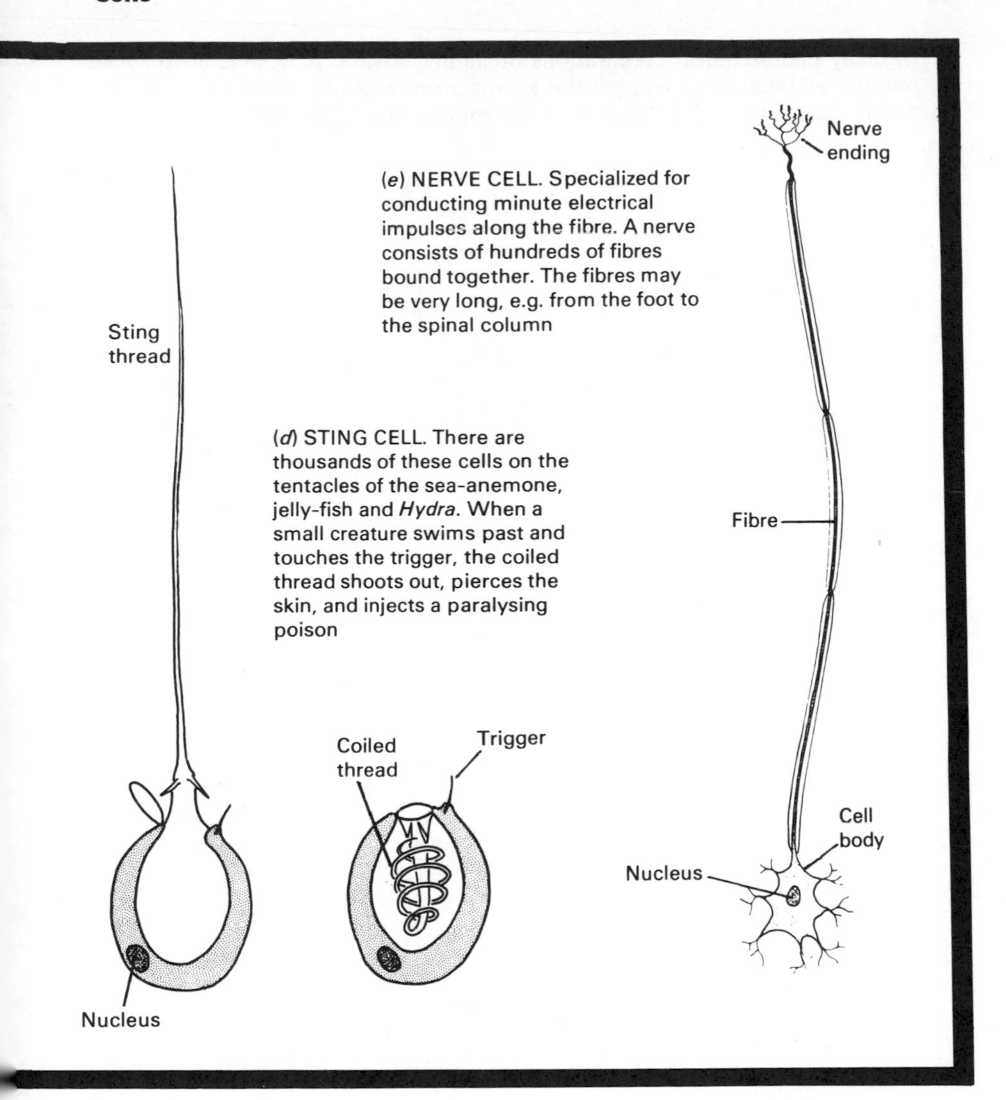

2.6 QUESTIONS

1. What features are (*a*) possessed by both plant and animal cells, (*b*) possessed by plant cells only, (*c*) possessed by animal cells only?
2. With what materials must cells be supplied if they are to survive?
3. In what ways would you say that the white blood cell (Fig. 2.9(*b*)) is less specialized than the nerve cell (Fig. 2.9(*e*))?

4. In many microscopical preparations of animal tissues, it is difficult to make out the cell boundaries and yet the arrangement and the number of cells can usually be determined. Which cell structure makes this possible?

Unit 3
Diffusion And Osmosis

3.1 INTRODUCTION

Diffusion and osmosis are vitally important processes in the metabolism of plants and animals, although they do not *only* take place in living things. Indeed, several of the experiments described in this unit do not involve any biological material at all. Some knowledge of diffusion and osmosis is, however, essential to an understanding of the ways by which living things take up their food materials and oxygen, by which water and other substances move from one part of the organism to another, and by which unwanted substances are eliminated from the organism.

3.2 DIFFUSION

All chemical substances are made up of minute particles called *molecules*. For example, the smallest possible particle of water is a molecule consisting of one atom of oxygen joined to two atoms of hydrogen.

The molecules in a solid, like an ice crystal, are packed closely together. However, they are in constant motion, but can only vibrate about a more or less fixed position; they have little or no freedom to move in relation to each other. In a liquid, like water, the molecules are spaced farther apart and move about at random within the volume of the liquid. In a gas, like steam, the molecules are so widely spaced that they can move constantly in all directions and at very high speeds, limited only by the walls of the vessel that contains them. Because of this constant random movement the molecules of a gas tend to distribute themselves evenly throughout any space in which they are confined. The same principle holds true for substances which dissolve in a liquid; for example, if sugar is placed at the bottom of a beaker of water the sweetness of the solid will eventually spread throughout the water as the sugar dissolves. This movement of the molecules which tends to result in their uniform distribution in a liquid or a gas is called *diffusion*.

The following experiments illustrate the process of diffusion in air and water.

Experiment 3.1 *To demonstrate the diffusion of a gas (ammonia) in another gas (air)*

Red litmus paper is used in this experiment to detect the presence of the alkaline gas, ammonia, which changes the color of the paper to blue.

Squares of wetted red litmus paper are pushed with a glass rod or wire into a wide glass tube, corked at one end, so that they stick to the side and are evenly spaced out (Fig. 3.1). The open end of the tube is closed with a cork carrying a plug of cotton saturated with a strong solution of ammonia. The alkaline ammonia vapor diffuses along inside the tube at a rate which can be determined by observing how long it takes for each square of litmus paper to turn completely blue. If the experiment is repeated using a more dilute solution of ammonia the rate of diffusion will be found to be slower.

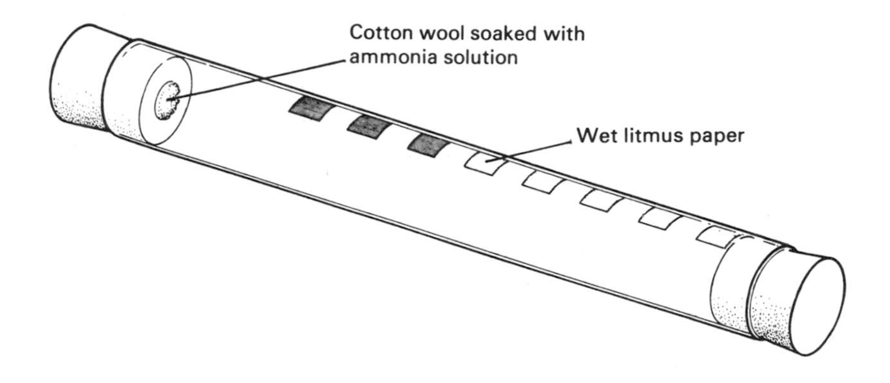

Fig. 3.1 *Diffusion of ammonia in air*

Experiment 3.2 *To demonstrate diffusion of a soluble solid in a liquid*

Diffusion of a soluble salt in water can be demonstrated, as we have mentioned, by placing a crystal of a colored salt like copper sulphate in a beaker of water: after a time the characteristic color spreads throughout the liquid. However, the distribution of the blue copper ions in this experiment is helped by every slight disturbance of the water, including convection currents: diffusion in liquids is actually quite slow. But the water can be

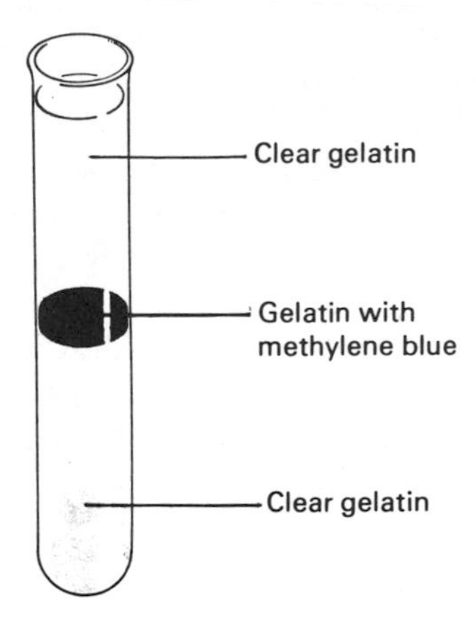

Fig. 3.2 *Diffusion in a liquid*

"held still" with gelatin; although the *gel* which is formed appears nearly solid, on the molecular scale it can be pictured as an open network of gelatin molecules, through the meshes of which water and other small molecules can move unhindered. Cytoplasm is a gel.

10 g gelatin is dissolved in 100 g very hot water. A test-tube is half-filled with the solution, which is allowed to cool and set while the rest of the solution is kept just warm. Some of the solution is colored with a little of the dye methylene blue, and as soon as the gelatin in the test-tube has set firmly a thin layer of blue gelatin is poured on top. When the blue layer is also cold and firm the tube is filled with the cooled but still liquid gelatin solution and chilled quickly, so that the blue gelatin is sandwiched between two layers of clear gelatin (Fig. 3.2). After a week, the blue dye can be seen to have diffused into the clear gelatin, upward and downward to equal extents, showing that diffusion is not affected by gravity.

The rate of diffusion. The rate of diffusion of a substance depends on the size of its molecule, and is also affected by temperature and concentration. Large molecules diffuse more slowly than small ones, and all substances diffuse more quickly when warm than when cold. Experiment 3.1 has already shown that the direction of diffusion is from a region of high concentration of a substance to a region of low concentration of that substance, and also that a gas diffuses more rapidly from a highly concentrated source than from a more dilute one. The difference in concentration which results in diffusion is called the *concentration gradient* or the *diffusion gradient,* and the more marked the gradient, the more rapid is the diffusion produced.

For example, when rapid photosynthesis is going on in a leaf cell, the production of carbohydrate uses up carbon dioxide from the cytoplasm. As the concentration of carbon dioxide in the cytoplasm falls, a diffusion gradient is set up between the cytoplasm and the intercellular spaces resulting in more carbon dioxide diffusing into the cell from intercellular spaces. At the same time, the cell is producing oxygen as a by-product of its photosynthesis, so creating a diffusion gradient with respect to oxygen in the opposite direction to that for carbon dioxide. Consequently oxygen diffuses out of the cell into the intercellular spaces. Diffusion, therefore, refers to the movement of each substance independent of other substances.

Diffusion in Living Organisms

Diffusion is important whenever an organism takes up material from, or loses material to, its environment or when substances move from one part of the organism to another. In many microscopic single-celled organisms, the uptake of oxygen and the loss of waste products like carbon dioxide take place over the entire surface of the cell. Diffusion is rapid enough to account for the whole process in these very small creatures, in which the ratio of surface area to total volume is very large. But diffusion alone is too slow to meet the needs of large organisms, and is supplemented by processes like the circulation of blood, and by mechanisms of transferring material into, across or out of cells by chemical activity, referred to as *active transport.*

Specific examples of the role of diffusion in biological processes are mentioned in various parts of this book, and can be located by reference to the index. The key to understanding diffusion in living organisms is that substances are put into solution (made soluble) where they can move as "free agents" in response to their relative concentrations (numbers) from one place to another.

3.3 OSMOSIS

In Experiment 3.2 the presence of the gelatin can be disregarded and the experiment can be considered simply as a study of a strong solution of dye in contact with water. After diffusion has proceeded for a little while it is more accurately considered in terms of a strong dye solution in contact with a weaker solution. A unit volume—say one cubic centimeter (1 cm^3) —of the weak solution contains less dye and more water than the same volume of the strong solution. Thus *for the dye* there is a diffusion gradient from the strong to the weak solution, but *for the water* the diffusion gradient is in the reverse direction. If the experiment could be left for several weeks and protected from bacterial spoilage, the dye would diffuse throughout the test-tube and the color of the gel would become a uniform blue: that is, dye molecules would have diffused from the stronger solution to the weaker, and water molecules from the weaker to the stronger, until both kinds of molecule are evenly distributed.

Suppose now that the two solutions are separated by a membrane which allows the water, but not the dye, to diffuse through it. (Membranes like these are said to be *differentially permeable* or *semipermeable*; cellophane and some kinds of parchment have this property of semipermeability.) The strong solution would then become more and more dilute as water enters it by diffusion from the weak solution. The diffusion of water—strictly speaking, of any solvent—from a weak to a strong solution across a semipermeable membrane is called *osmosis*.

Experiment 3.3 *To demonstrate osmosis*

A length of cellophane tubing (sold as "dialysis tubing") is closed with a knot at one end and filled with a strong syrup or sugar solution. The open end is fitted over the end of a capillary tube and held with a rubber band (Fig. 3.3). The tube is clamped vertically so that the cellophane section is immersed in a beaker of water.

Result. After a few minutes the level of liquid in the capillary tube begins to move upwards, and may continue to rise up to a height of a meter or so if the tube is long enough.

Interpretation. The most plausible interpretation is that water molecules have passed through the cellophane tubing into the sugar solution, increasing its pressure and hence its volume, thus forcing it up the capillary tube. Theoretically, this movement should continue until the downward pressure of the column of liquid in the capillary tube just equals the pressure exerted by the water diffusing into the dialysis tubing.

In practice, cellophane is not a perfectly semipermeable membrane: it is permeable to water, but not quite impermeable to sugar, so that sugar molecules are not completely prevented from passing from the syrup into the water. If the experiment were left to stand for some time, the concentrations of sugar in the beaker and in the cellophane tubing would eventually become the same, and the liquid in the capillary tube would drop gradually back to its level at the beginning of the experiment.

In this experiment the liquid is forced up the capillary tube by a

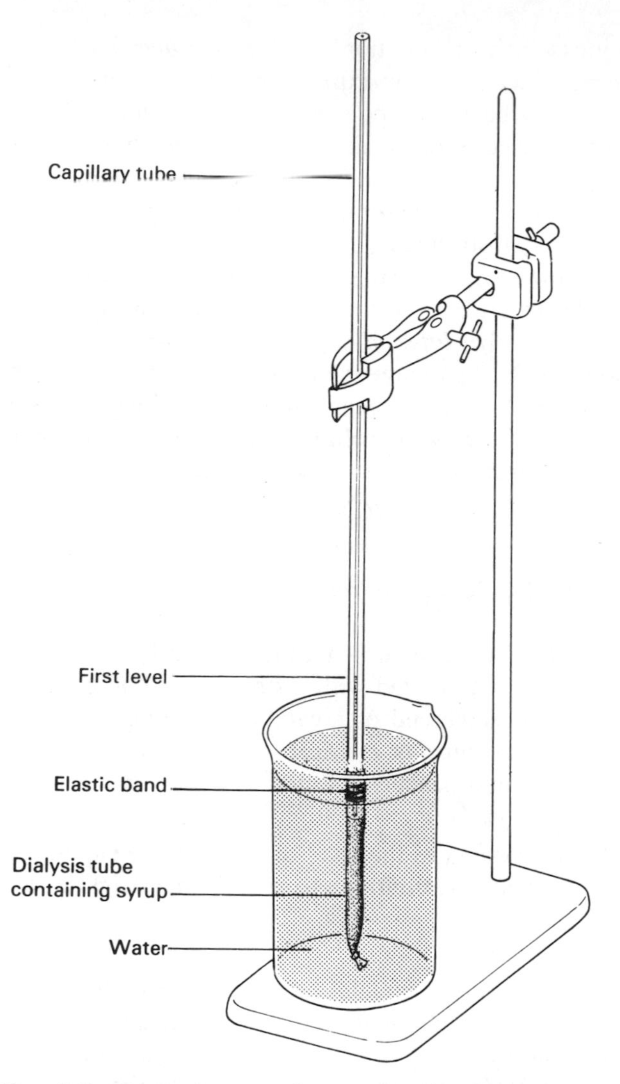

Fig. 3.3 *Demonstration of osmosis and osmotic pressure*

pressure, *osmotic pressure,* built up in the sugar solution in the dialysis tube; this pressure can be large enough to burst the dialysis tubing if the capillary tube is blocked.

The *osmotic potential* of a solution is a measure of the tendency for water molecules to diffuse out of it. A very concentrated solution, having relatively few water molecules, has a low osmotic potential while a dilute solution with a larger proportion of water molecules has a high osmotic potential.

Semipermeable Membranes

There are several theories as to why some membranes have the particular property of semipermeability. One supposes that the membrane acts as a kind of sieve, having tiny pores in it that are too small to allow large molecules like sugar ($C_{12}H_{22}O_{11}$: molecular weight = 342) to pass, but large enough to admit freely small molecules like water (H_2O: molecular weight = 18). This theory considers molecular size differences to explain diffusion, but it does not account for the discriminating ability of cell membranes or the movement of some substances against a concentration gradient (i.e. from an area of low concentration to an area of high concentration).

The currently accepted theory of cell membrane structure is illustrated in Fig. 3.4 (page 24). The cell membrane has minute pores or holes of a given size much like a sieve or window screen. Certain small molecules or charged atoms (ions) may pass freely, while certain large molecules cannot pass through. The membrane itself is described as a lipoprotein sandwich, composed of an inner and outer layer of protein, and sandwiched between them are two fatty (lipid) layers. The total thickness of the cell membrane is about 1/1000 mm, but it varies in thickness in spots. The thinner areas lack protein, and fat-soluble substances can pass through even though they may be too large to pass through the pores. Since the cell membrane is a part of the cytoplasm, and hence alive, it can expend energy to transport scarce commodities in the environment across the cell membrane against normal diffusion gradients. This is called *active transport.*

3.4 OSMOSIS IN PLANTS

The cell membrane in every living cell has the property of semipermeability, so that whenever cells are surrounded by fluids which are either weaker or stronger than those which they contain, osmotic forces are set up.

(a) Turgor

The cellulose wall of a plant cell is freely permeable to water and dissolved substances, but the membrane immediately inside the wall and the layer of cytoplasm between the membrane and the vacuole are both semipermeable. The cell sap in the vacuole contains dissolved salts and sugars, so that an isolated plant cell surrounded by water will take in water by osmosis. The vacuole expands, pushing the cytoplasm outward against

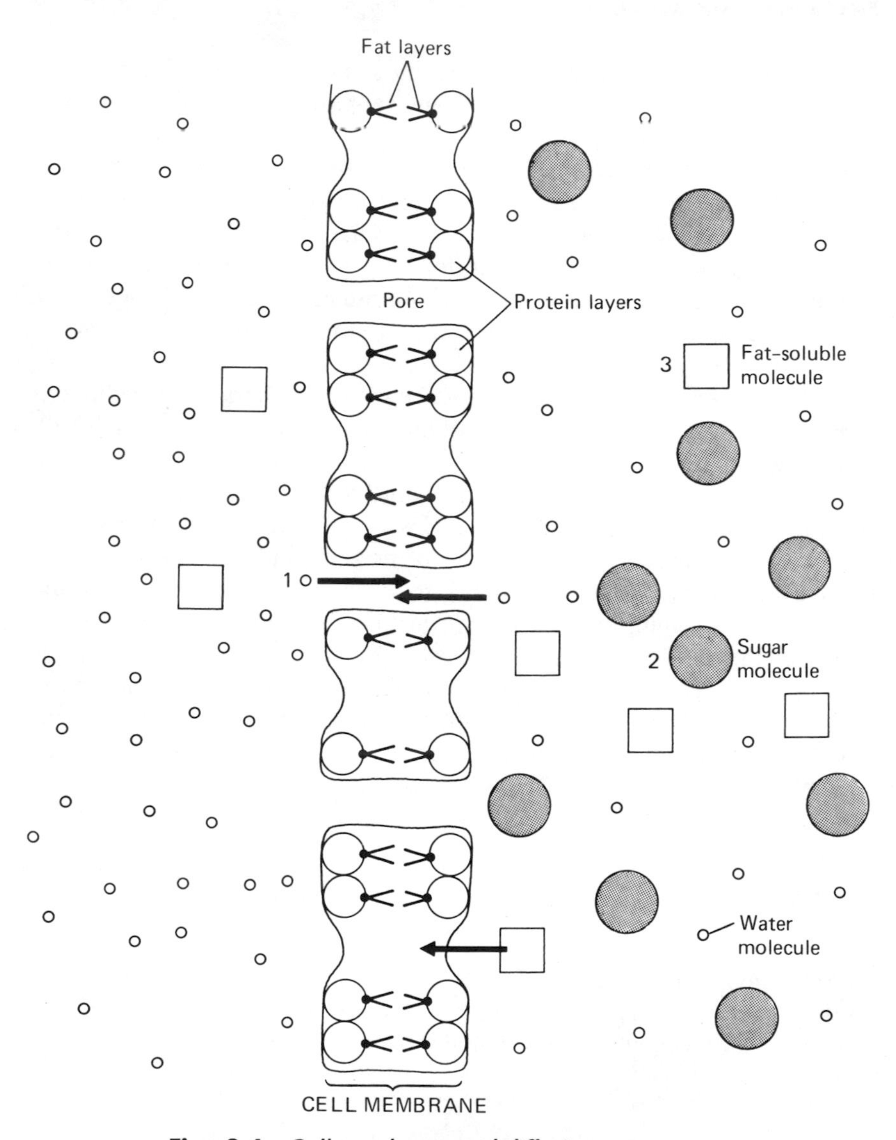

Fig. 3.4 *Cell membrane and diffusion*

Water molecules (1) move through membrane pores in both directions but mainly from area of high concentration (left) to area of lower concentration (right). Sugar molecules (2) are too large to move through the membrane pores. Fat soluble molecules (3), even though too large to pass through the membrane pores, pass directly through the lipoprotein layers to the left.

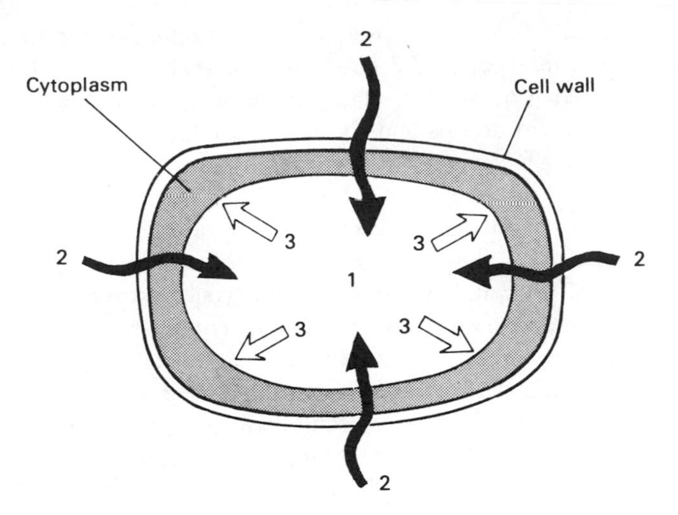

Fig. 3.5 *Turgor in a plant cell*

(1) Dissolved salts and sugars in cell sap give it a low osmotic potential; (2) water enters by osmosis, passing through the cell wall and the semipermeable cytoplasm; (3) the cell-sap volume increases, creating an outward pressure on the cell wall and making the cell turgid.

the cell wall (Fig. 3.5). Water intake continues until the outward pressure of the vacuole is just balanced by the resistance of the relatively inelastic cell wall: the cell can then take in no more water. It is said to be *turgid* and the vacuole is exerting *turgor pressure*.

A plant structure made of turgid cells is resilient and strong; the plant stem stands upright and the leaves are held out firmly. Many young plants depend entirely upon this turgidity for their support, but in older plants woody and fibrous tissues take over this function so that they do not collapse even though they are dead.

(i) **Growth.** The walls of newly formed cells are fairly plastic and the vacuoles they contain are quite small (see Fig. 2.5). Water enters the young cell by osmosis, and its vacuoles enlarge and join up. The outward pressure of the growing vacuole against the plastic cell wall extends the cell. As hundreds of young cells extend in this way—for example, in the region near the growing point of a stem—the plant grows rapidly by *extension* or *expansion growth* even though the number of new cells may not be increasing at the time.

(ii) **Wilting.** When plants are exposed to conditions in which they lose water to the atmosphere faster than they can obtain it from the soil, water

is lost from the vacuoles. The turgor pressure in the vacuoles decreases and they no longer push out against the cell wall. The cell becomes limp or *flaccid* (like a deflated football). A plant structure made of such cells is weak and flabby, the stem droops and the leaves are limp; in other words, the plant is *wilting,* such as a limp stalk of celery.

(iii) **Plasmolysis.** If a plant cell is surrounded by a solution more concentrated than the cell sap, water passes out of the vacuole by osmosis to the solution outside of the cell. To understand this you must remember that the presence of substances in water makes the concentration of *water* molecules *less* than they are in pure water. If the concentration of water molecules in the concentrated solution is less than their concentration in the cell sap, net movement of water molecules will be out of the cell. The vacuole therefore shrinks, and pulls the cytoplasm inward away from the comparatively rigid cell wall, drawing fluid into the cell from the surrounding solution (Fig. 3.6). There is now no outward pressure on the cell wall

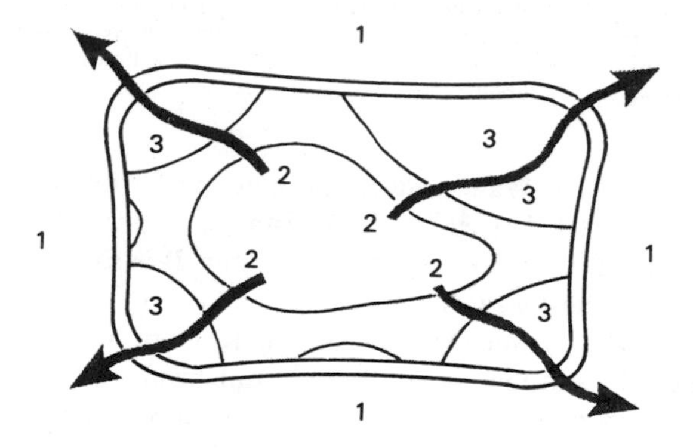

Fig. 3.6 *Plasmolysis in a plant cell*

(1) The solution outside the cell is more concentrated than the cell sap; (2) water passes out of the vacuole by osmosis; (3) the vacuole shrinks, pulling the cytoplasm away from the cell wall and leaving the cell flaccid.

and the cell is flaccid. This condition, called *plasmolysis,* can be induced experimentally in living cells without necessarily harming them. Sometimes the beginning gardener will over-fertilize a plant and bring about plasmolysis. The plant wilts and will die unless abundant water is used to flush out the excess fertilizer (and raise the water concentration) around the roots of the plant. Plasmolysis in nature is an extreme condition and rarely occurs.

(iv) **Control of gas exchange.** The leaves and tender portions of plants have minute openings on their surfaces called *stomata* (*stoma*, singular). These pores are bounded by a pair of specialized cells called *guard cells*. The stomata are opened and closed by the guard cells in response to light primarily. But the expansion and contraction of the guard cells are osmotic phenomena. Although the details of the process have not been fully worked out, it seems that in daylight when photosynthesis is rapid the consequent fall in carbon dioxide concentration causes the guard cells to convert part of their stored starch into sugar. Because of the increase in sugar concentration in their cell sap, water moves by osmosis into the guard cells and their turgor pressure increases. Their inner walls, which bound the stoma, are thicker than the outer, and cannot stretch as easily (Fig. 3.7); under the increased turgor pressure the outer walls extend more than the inner, so that the guard cells develop a curvature and the stoma between them opens.

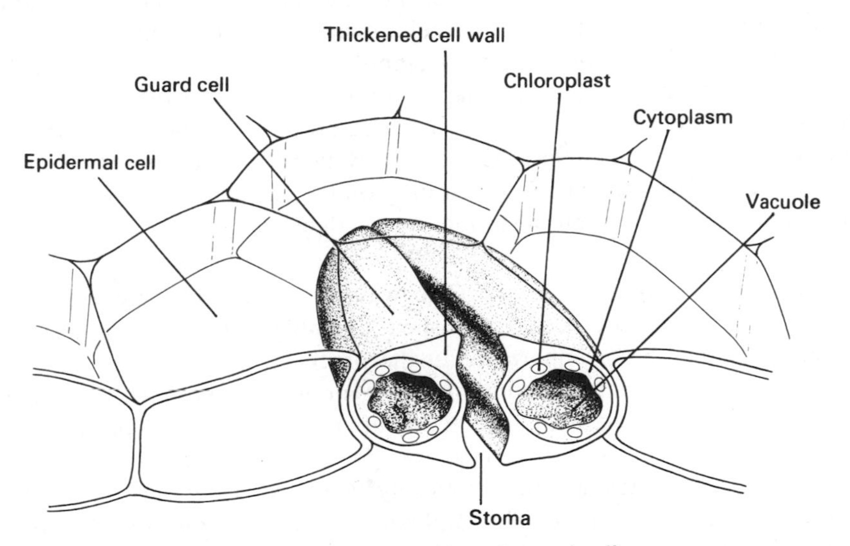

Fig. 3.7 *Structure of guard cells*

This mechanism ensures that the stomata open in daylight, admitting carbon dioxide to the chlorophyll-containing cells at a time when it is needed for photosynthesis, and close in the dark when photosynthesis stops.

(b) Movement of Water in a Plant

(i) **Water movement from cell to cell.** Osmosis is responsible for the passage of water from one cell in a plant to another. Imagine two adjacent

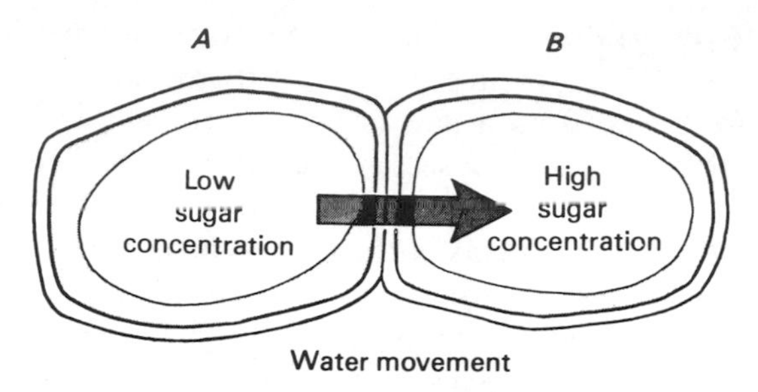

Fig. 3.8 *Water movement from cell to cell*

cells, *A* and *B*; suppose that *B* has more sugar in its cell sap than *A*. The more dilute cell sap in *A* will force water by osmosis into *B* (Fig. 3.8). This water dilutes the cell sap in *B*, raising its turgor pressure. The diluted cell sap in *B* now tends to force water into the next cell in line in the same way. Water thus passes from cell to cell down an osmotic gradient.

However, if *B* is fully turgid it cannot expand and so can take in no more water, even though its cell sap is more concentrated than that of *A*. Indeed, if *A* is not fully turgid and can therefore still expand, water will be forced out of *B* into *A* against the osmotic gradient. The movement of water between cells depends, therefore, not only on the concentration of salts and sugars in their cell sap but also how turgid they are.

(ii) **Water absorption by roots.** The way in which roots absorb water from the soil has not been established unambiguously, but osmosis is generally held to be of importance in the process.

Fig. 3.9 represents a section of a root along its radius, showing some of the cells concerned with the taking up of water. The root hair, an extension of the cell R 1, has grown out from the root, pushing between the soil particles which adhere to it closely. A film of water surrounds both the root hair and the soil particles, and contains small amounts of dissolved salts. The cell sap in the root hair is more concentrated than the soil water, so that water passes by osmosis from the soil through the thin layer of cytoplasm lining the root hair, and thence into the vacuole. The extra water dilutes the cell sap in R 1 so that it is now weaker than that in R 2. It also increases the volume of the vacuole and hence its turgor pressure. Water is thus forced from R 1 to R 2, and this is the first step in a chain of similar osmotic movements by which water passes from cell to cell from the root hair toward the center of the root. Eventually it reaches the vessels in the xylem and is carried up the root to the stem and thence to the rest of the plant in the *transpiration stream* (Section 9.3(*a*)), the flow of water through the plant resulting from the loss of water by evaporation (Fig. 3.10).

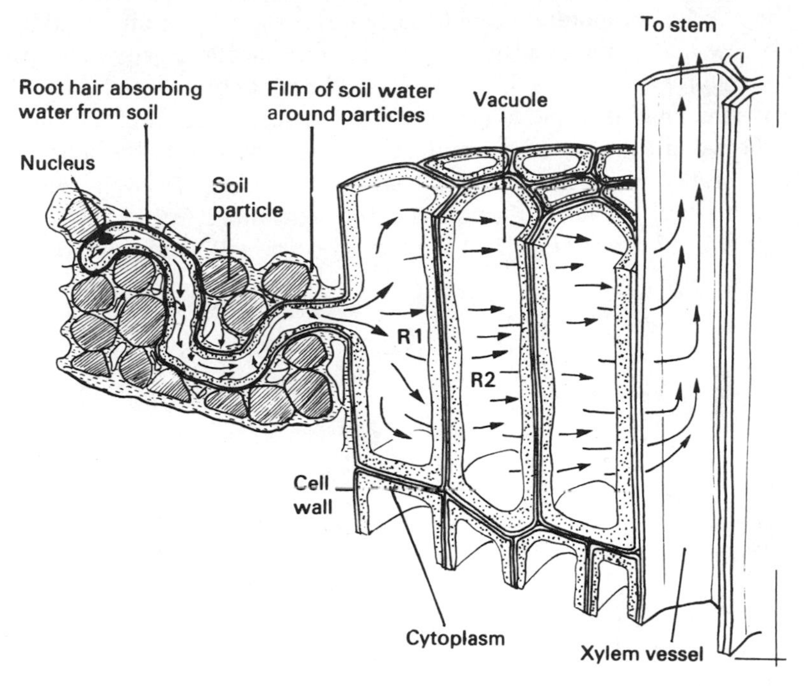

Fig. 3.9 *Diagram to show one theory of the passage of water from soil to xylem vessels in a root*

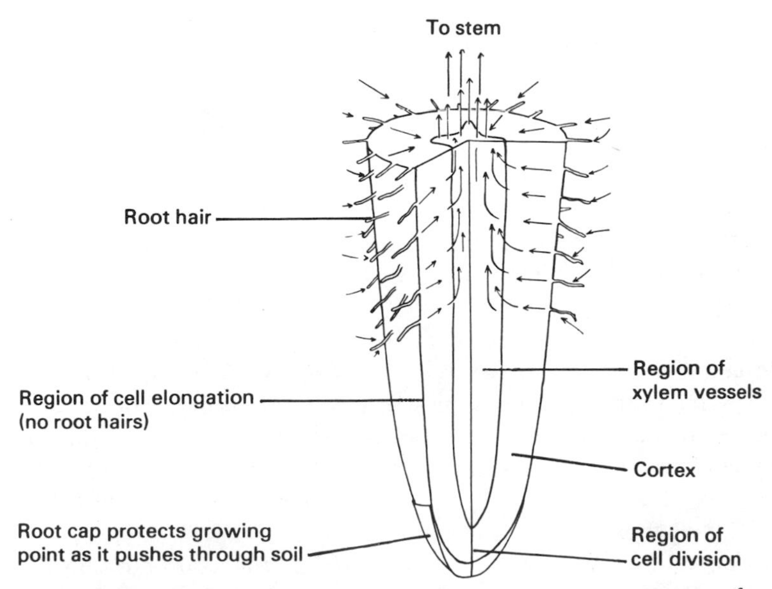

Fig. 3.10 *Diagrammatic section of root to snow passage of water from the soil*

The effect of evaporation can actually lower the pressure in the xylem vessels below that of the atmosphere—that is, produce a negative pressure —so that water is drawn into the xylem from the root cortex.

It is sometimes possible to demonstrate a positive pressure, *root pressure,* in the wood (xylem) for example, sap often oozes from the stump of a tree or a branch which has been cut back in the spring. If a well-watered potted plant is cut off above the soil and the stem fitted with a glass tube containing a little water (Fig. 3.11), root pressure will produce a rise of

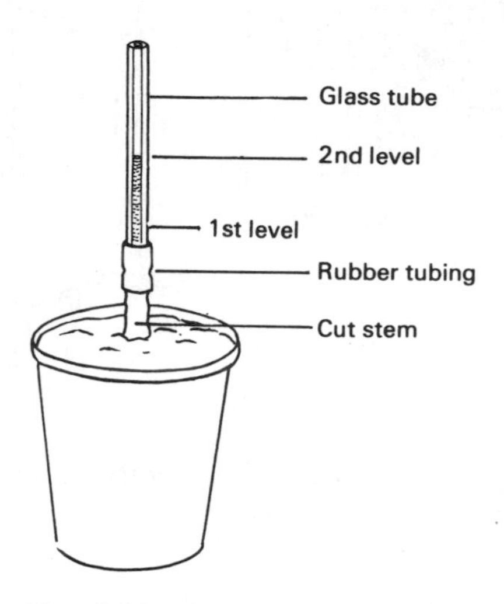

Fig. 3.11 *Experiment to show root pressure*

several centimeters in the level of water in the tube. Root pressure is also believed to be responsible for *guttation,* the exudation of drops of water from the tips of the leaves of certain plants. Although there is some evidence that root pressure arises from osmotic effects, the mechanism is not completely certain.

(iii) **Water movement in leaves.** Fig. 3.12 shows a few adjacent cells in a leaf. The palisade cell *A* is losing water by evaporation from its surface into the intercellular spaces. This loss of water results in a reduction of turgor pressure and an increase in the concentration of the sugars and salts in its vacuole. The cell sap in the vacuole becomes more concentrated than that of its neighboring cell, and water flows into it by osmosis from the neighboring cell. Such a loss of water increases the concentration of cell sap in the vacuole of cell *B,* and also reduces the volume of the vacuole

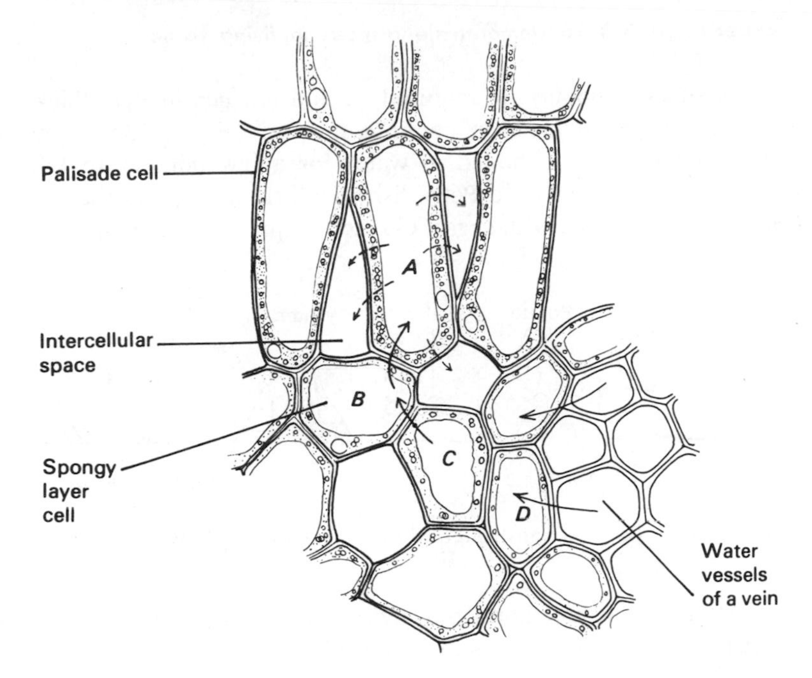

Fig. 3.12 *Diagram of a few leaf cells to show how water movement could occur by osmosis*

and hence its outward pressure on the cell wall. The effect of these two changes is that *B* absorbs water from its neighbor *C*. Cell *D* is next to a water vessel running in one of the veins of the leaf, and it will take water from it, again, by osmosis. In this way water can flow from a vein to the cells of a leaf.

The account of water movement between cells given in this section describes the passage of water by osmosis from the vacuole of one cell to the vacuole of another. However, there are other possible explanations: water and dissolved substances could perhaps move from cell to cell through the cell walls without passing through the cytoplasm at all, or from the cytoplasm of one cell into the cytoplasm of another without entering the vacuoles. Again, the intercellular spaces in some tissues may provide an alternative pathway. The relative importance of the different mechanisms is not known.

Osmosis is concerned with the movement of water only, and cannot directly affect the movement of dissolved substances into the roots and through the plant. This is discussed more fully in Unit 9.

Osmosis can only take place in cells when they are alive. If the cytoplasm dies it loses its semipermeability.

Experiment 3.4 *To demonstrate osmosis in living tissue*

Three slices of potato are required, each a few centimeters thick and having a cup-shaped depression cut into the top (Fig. 3.13): they are labeled *A*, *B* and *C*. *A* is boiled in water for a few minutes to kill the cytoplasm. Each is placed on a Petri dish in a little water, and the cavities in *A* and *B* are half-filled with sugar, *C* is left empty, as a control.

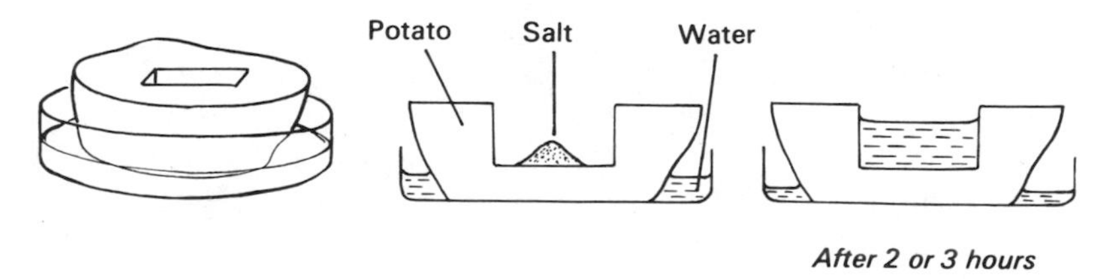

Fig. 3.13 *Demonstration of osmosis in living plant tissue*

Result. After two or three hours the cavity in *B* is nearly full of liquid. No liquid accumulates in the cavities in *A* and *C*.

Interpretation. The sugar in *B* dissolves the sap exuding from the damaged cells, and the solution so formed withdraws water from the surrounding tissues by osmosis. No water movement can take place in *A*, for semipermeability is a property of *living* cytoplasm. The control *C* shows that liquid does not collect in the cavity without the presence of the sugar.

3.5 OSMOSIS IN ANIMALS

(a) Fresh-water Animals

The body fluids of fresh-water animals like minnows, frogs or water-fleas are more concentrated than the pond or stream water in which they live. As their skins are more or less semipermeable, water tends to enter their bodies by osmosis. Unless this water is removed continuously their blood would be diluted, their bodies would swell and become water-logged, their metabolism would be seriously disturbed and they would soon die. Different animals eliminate the excess water—that is, *osmoregulate*—in different ways. The kidneys of fish and frogs, for example, extract the excess water from the blood as it circulates through them, and it leaves the body as a dilute urine. The osmoregulation of the *Ameba,* a single-celled animal, is discussed in Unit 14. Some animals, like crayfish or water-beetles, have impermeable outer coverings which greatly reduce the proportion of the body's surface which can absorb water. However, the ways in which some fresh-water animals are affected by osmosis and by which they are protected are still not clearly understood.

(b) Salt-water Animals

Sea water is a more concentrated solution than the body fluids of most marine animals, so that water tends to be withdrawn from their bodies by osmosis. Salt-water fish replace this by swallowing water and absorbing it through their alimentary canals; there is evidence that they have a special elimination mechanism to deal with their consequent considerable intake of excess salts.

Animals living in estuaries are surrounded by salt water when the tide comes in, and by fresh river water when it goes out. These animals must have very specialized osmoregulatory mechanisms to withstand such extreme osmotic changes.

(c) Land Animals

Land animals lose water from their body surfaces by evaporation, and take in water in their food and drink, rather than by osmosis. Osmotic effects are nevertheless important *within* their bodies (see, for example, Section 21.3(*c*)). These animals too must keep the concentration of their body fluids very uniform and this is a part of the role of the kidneys (see Sections 23.3 and 23.5), although other organs are concerned as well.

Loss of water by evaporation is reduced to some extent in many land animals by body coverings: for example, the hair of mammals, the feathers of birds and the impermeable outer covering of insects, adaptations which contribute to the successful survival of these groups on land.

3.6 QUESTIONS

1. A solution of salt is separated by a semipermeable membrane from a solution of sugar or equal concentration (*equimolecular*).* Discuss whether osmosis would take place and justify your conclusions.
2. Pieces of fresh beetroot are washed and left in water. Similar pieces of cooked beetroot are washed and left in water. After an hour or two, the

* That is, having the same number of moles of solute per liter of solution.

water round the cooked beetroot is colored red while that around the fresh beetroot is not. Explain this result.

3. What features in a fish gill would help to maintain a steep diffusion gradient of oxygen between it and its immediate environment?
4. In an actively photosynthesizing cell, the sugars being formed are quickly converted to starch. What is the biological advantage of this in terms of osmosis?
5. What activities in man are likely to increase and decrease the osmotic potential of blood and body fluids?
6. What osmotic problems confront the salmon, which is hatched in a river, grows to maturity in the sea and returns to the river to lay its eggs?

UNIT 4
RESPIRATION

4.1 INTRODUCTION

Respiration was the first-named in the list of the vital characteristics of living organisms in Section 1.1, and there it was defined as the series of chemical changes which release energy from food material in both plants and animals. All living things respire, although different respiration mechanisms are known: non-living things do not respire.

The energy so produced is used for the wide range of activities which characterize living things. These may include muscular contraction (for example, for locomotion or for breathing), the conduction of impulses along nerves, and the driving of the cell reactions which are part of the vital chemistry of life, and which may lead to the build-up of new cells or the maintenance of older ones, or to the secretion of products such as the digestive juices of animals or the nectar of flowers. A detailed discussion of all the reactions comprised in the term "respiration" is beyond the scope of this book, and for our purposes we may consider the process in its simplest terms as energy production by the breakdown of *carbohydrates*—foodstuffs like sugar and starch, which are made up of carbon, hydrogen and oxygen only [see Section 19.3(*a*)]—and in particular of the fairly simple and very common carbohydrate *glucose*. The chemical formula of glucose is $C_6H_{12}O_6$, and the products of its complete breakdown are carbon dioxide (CO_2) and water (H_2O).

If some glucose is heated over a bunsen burner in the laboratory it catches fire and burns vigorously, releasing energy as heat and light and producing carbon dioxide and water as the products of the reaction. Obviously the mechanism by which a plant or an animal obtains its energy from glucose is very different from this. In fact every stage of the complex chain of reactions involved is, like all the reactions taking place in a living cell, very precisely controlled by its specific *enzyme* or biological catalyst. An enzyme controls the *rate* of a chemical reaction in living material, and generally that of one particular reaction; they are discussed in more detail in Section 20.1.

The term "respiration" is often used loosely in reference to breathing, as in "artificial respiration" or "pulse and respiration rate," or in connection with gaseous exchange, "the respiratory surface of a gill," "organs of respiration" and so on. For this reason, the "respiration" described in this chapter is sometimes called *cellular respiration* or *internal respiration* to distinguish it from either the breathing movements (ventilation) or the intake of oxygen and output of carbon dioxide (gaseous exchange).

A distinction is usually made between two forms of, or stages in, respiration. *Aerobic respiration* involves the release of energy by the reaction of oxygen with carbohydrates, which are completely broken down to carbon dioxide and water. *Anaerobic respiration,* on the other hand, releases energy from carbohydrates by reactions which do not involve oxygen from the air.

4.2 AEROBIC RESPIRATION

The following equation summarizes the overall process of complete oxidation of glucose which takes place in aerobic respiration:

$$\underset{\text{glucose}}{C_6H_{12}O_6} + \underset{\text{oxygen}}{6O_2} \rightarrow \underset{\text{carbon dioxide}}{6CO_2} + \underset{\text{water}}{6H_2O} + \text{energy}$$

Thus for aerobic respiration to take place the organism must take in food and oxygen, and the end-products of their reaction, carbon dioxide and water, must constantly be removed.

Methods of Demonstrating Respiration

If respiration can be demonstrated in a sample of material, this is an indication that the material is living.

The equation (Fig. 4.1) suggests that if an organism is respiring it will

(*a*) use up carbohydrate,
(*b*) take in oxygen,
(*c*) give out carbon dioxide,
(*d*) produce water or water vapor, and
(*e*) release energy.

Since non-living matter may give off water vapor simply by evaporation, (*d*) is not a reliable criterion of respiration. The experiments which follow illustrate the other four tests; in most of them germinating seeds are used as convenient demonstration material.

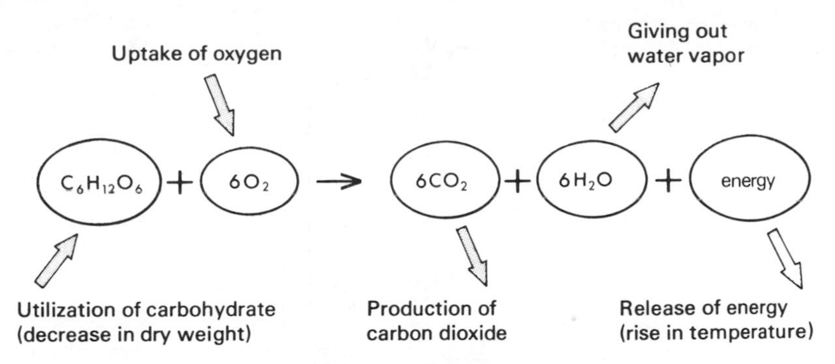

Fig. 4.1 *Ways of detecting respiration*

Experiment 4.1 *To demonstrate the decrease in weight of respiring material*

If living material is converting part of its carbohydrate into carbon dioxide and water, and if these end-products are allowed to escape into the surrounding air, there will be an overall loss in weight of the material. However, any water-containing substance may lose weight by evaporation over a period of time: hence the *dry weight* of the material must be measured.

One hundred seeds—peas or beans are suitable—are soaked in water for 12 hours. Half of them are killed by boiling them in water for a few minutes: these are the controls. Each group is placed on moist cotton on a shallow dish, and the two dishes are then left in identical conditions of temperature and lighting. The cotton must be kept moist.

Each day ten seeds or seedlings are taken from each dish, heated for 12 hours in an oven at 120 °C to drive off all their water, and then cooled and weighed.

Result. The average dry weight of the living seeds shows a steady daily decrease as the solids in the food reserves in the endosperm or the cotyledons are used up in the process of respiration. The dry weight of the killed seeds shows no such change.

Experiment 4.2 *To demonstrate the uptake of oxygen during respiration*

The apparatus for this experiment, shown in Fig. 4.2, is designed for the comparison of changes in volume of the gases in the two tubes. Temperature variations can produce volume changes, but these are minimized by the presence of the water bath (a large beaker of water) which also ensures that both tubes are kept at the same temperature. The soda lime

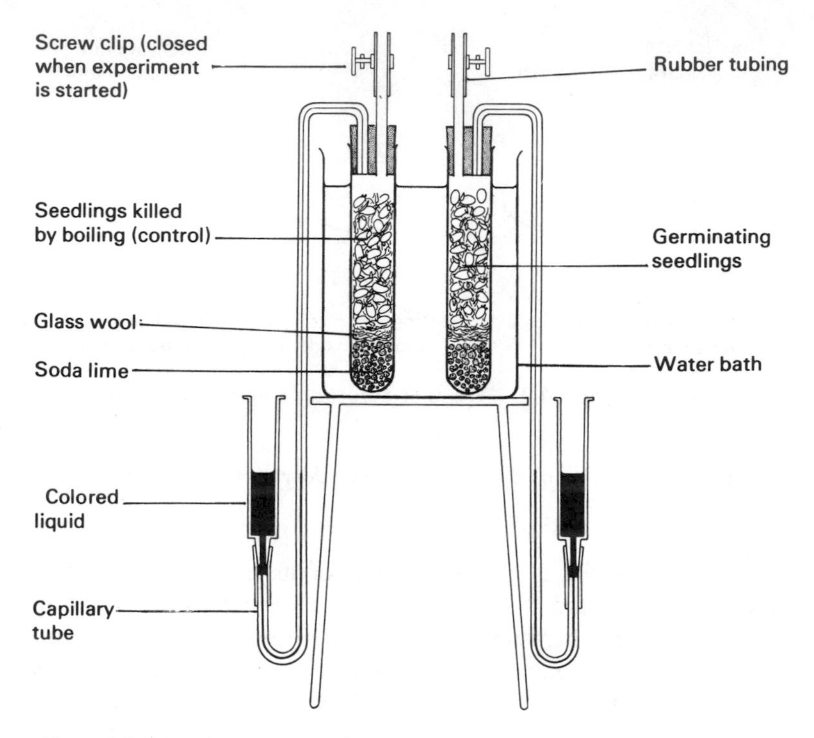

Fig. 4.2 *To demonstrate the uptake of oxygen during respiration*

in the bottom of both tubes rapidly absorbs any carbon dioxide which may be produced by the seeds, so that the changes in volume observed can be assumed to be due only to uptake of oxygen. (Nitrogen, the other main constituent of air, is chemically very inactive and is unlikely to be involved in any reaction under these conditions.) Seeds which have been killed by boiling are put into one tube, and seeds which have been germinating for a day or two into the other.

The apparatus is left to stand for five or ten minutes so that the tubes can acquire the temperature of the water. The screw clips are then closed and the levels of liquid in the capillary tubes are observed.

Result. At the beginning of the experiment the liquid levels in the two capillaries are the same. After 20 minutes, or so, however, a difference develops, showing a decrease in volume in the tube containing the germinating seeds compared with the control. This difference cannot be due to temperature variations or physical absorption by the seeds, either of which would affect both tubes equally, and therefore arises from oxygen uptake by the respiring seeds.

Experiment 4.3 *To demonstrate carbon dioxide production during respiration*

About 40 seeds (peas are suitable) are soaked in water for 24 hours, and half of them are killed by boiling them for a few minutes. Both lots of peas are then soaked in a solution of sodium hypochlorite (e.g. commercial hypochlorite bleach diluted 1 : 4) for 15 minutes or so: this antiseptic will kill fungi or bacteria, the respiration of which might interfere with the experiment. The seeds are then rinsed with tap water.

Wet cotton is put into two flasks, *A* and *B*. The living seeds are put into flask *A* and the killed seeds into *B*; the flasks are tightly corked and left in the same conditions of light and temperature until the seeds in *A* have germinated. The seeds in *B* will not germinate.

The gases in both flasks are then tested for the presence of carbon dioxide. This gas gives a milky appearance to lime water, normally a clear colorless liquid. Since carbon dioxide is considerably heavier than air, the simplest way of testing the contents of a flask is to tilt it over a test-tube of lime water so as to "pour" the gases into the tube (Fig. 4.3); the tube is then shaken up.

Result. The gases from flask *A* turn the lime water milky, whereas those from *B* have no effect. This shows that the carbon dioxide has been produced by the germinating seeds in *A*. The control experiment with flask *B*

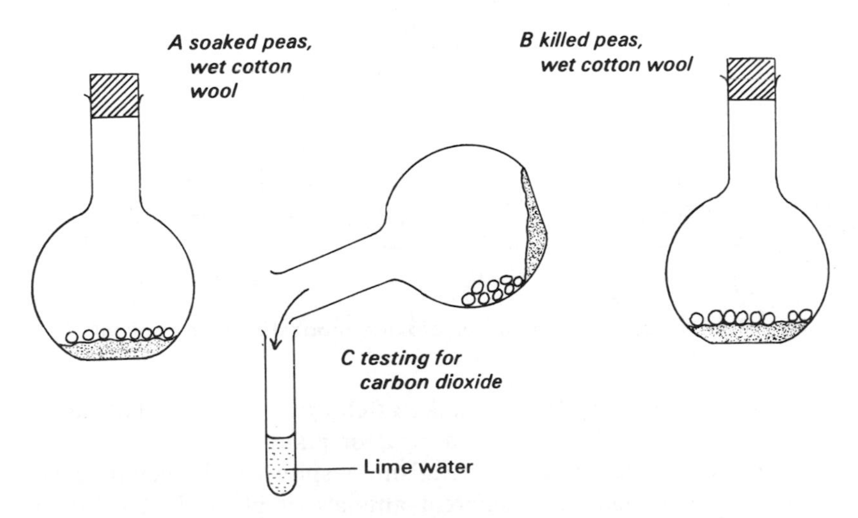

Fig. 4.3 *To demonstrate carbon dioxide production in germinating seeds*

proves that it is not the cotton or anything other than the germinating seeds that produces carbon dioxide.

Experiment 4.4 *To demonstrate the production of carbon dioxide by small animals or plants*

The apparatus in Fig. 4.4 is suitable for use with small animals. A stream of air is drawn slowly through the apparatus by means of a filter pump at *E*. Soda lime in *A* absorbs the small amount of carbon dioxide always present in the atmosphere; the lime water in *B* stays clear, proving that the air reaching the animal in *C* is free from carbon dioxide.

The same apparatus can be used to demonstrate carbon dioxide production by living plants, but the flask *C* must be "blacked out" to avoid complications from carbon dioxide *uptake* by photosynthesizing leaves. If a potted plant is used the pot must be completely enclosed in a polythene bag so that the respiration of organisms in the soil cannot affect the result.

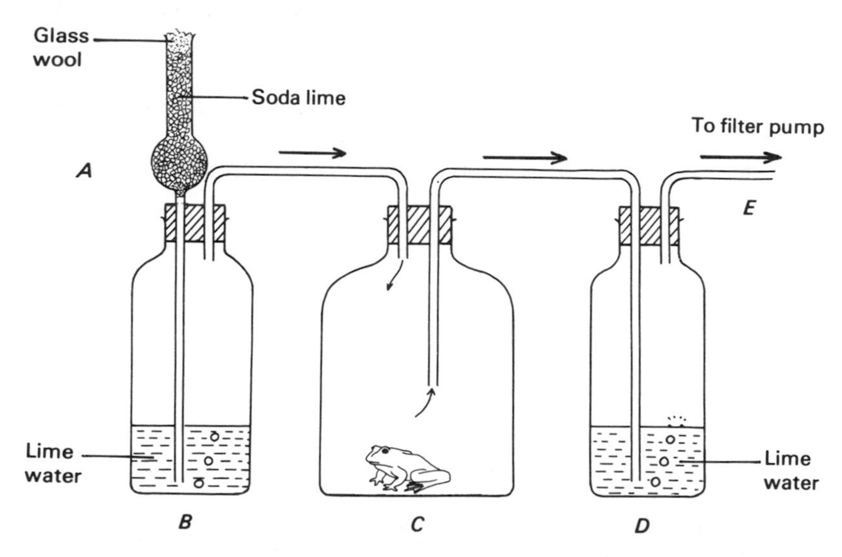

Fig. 4.4 *To demonstrate carbon dioxide production by an animal*

Result. The lime water in *D* turns milky after a time, proving that carbon dioxide is produced by the respiring animal or plant in *C*.

The *rates* at which different organisms respire can be compared by repeating the experiment using different animals or plants in *C*—for example, a mouse, a frog or a young sunflower seedling—and comparing the times taken for the lime water to turn milky.

Experiment 4.5 *To demonstrate carbon dioxide production in man*

The apparatus is shown in Fig. 4.5. Tube *T* is placed in the mouth and normal breathing is maintained: air is thus breathed in (inhaled) through *A* and breathed out (exhaled) through *B*.

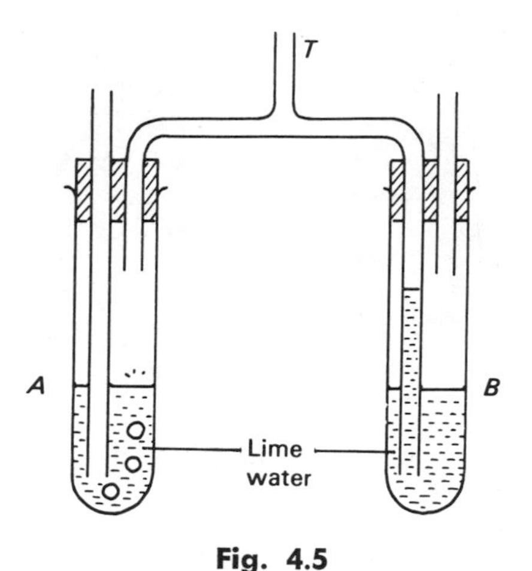

Fig. 4.5

Result. After a few breaths the lime water in tube *B* turns milky while that in *A* remains clear, demonstrating the greatly increased concentration of carbon dioxide in exhaled air as compared with normal air.

Experiment 4.6 *To demonstrate the release of energy by germinating seeds*

Energy release by germinating seeds can be demonstrated by proving the production of heat. Sufficient wheat grains to fill two small vacuum flasks are soaked in water for 24 hours and half of them killed by boiling for ten minutes. Both lots of wheat are soaked for fifteen minutes in a solution of sodium hypochlorite to kill fungal spores on the fruit walls (see Experiment 4.3). The grains are rinsed with tap water; the living grains are placed in one flask, the dead grains in the other. Thermometers are inserted into the flasks so that the bulbs are plunged into the mass of the grains, and the mouths of the flasks plugged with cotton.

Result. After a few days the temperature in the flask containing living wheat will be considerably higher than in the control, demonstrating that heat energy is released by germinating seeds but not by killed seeds. The

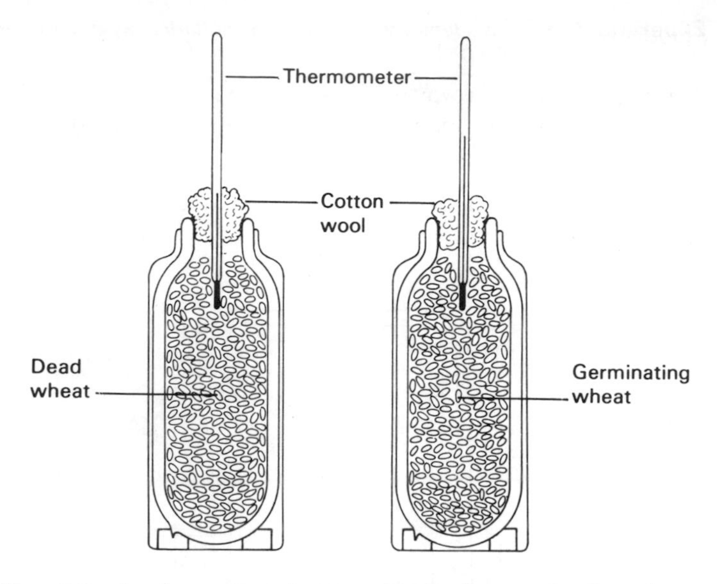

Fig. 4.6 *To demonstrate heat production by germinating wheat*

results, however, do not justify the conclusion that the heat is the result of respiration rather than any other process.

4.3 ANAEROBIC RESPIRATION

Anaerobic respiration is the release of energy from food material by a process of chemical breakdown which does *not* require oxygen. In such a process carbohydrates are not broken down completely into carbon dioxide and water but to intermediate products only: thus glucose, with six carbon atoms in its molecule, is broken down only to lactic acid or to ethanol (the substance called "alcohol" in everyday speech), compounds having three and two carbon atoms respectively. Less energy is available from this partial breakdown of food than from the complete oxidation that takes place in aerobic respiration: for example, the complete oxidation of glucose to carbon dioxide and water releases about 24 times as much energy as its partial breakdown to carbon dioxide and ethanol only.

The two kinds of respiration are not mutually exclusive alternatives; both may well be taking place in a cell at the same time. Indeed the first steps in the breakdown of glucose are normally anaerobic and may be followed by further aerobic steps, depending on the organism and the conditions:

$$\text{glucose} \xrightarrow[\text{respiration}]{\text{anaerobic}} \text{lactic acid} \xrightarrow[\text{respiration}]{\text{aerobic}} \text{carbon dioxide} + \text{water}$$

During vigorous activity like running, the energy demands of an animal's body are very high and the oxygen available to its cells may not be sufficient to oxidize food rapidly enough to meet its needs. The energy therefore must be released anaerobically and products like lactic acid may accumulate in the tissues. When the need for vigorous activity has passed these products are oxidized, or converted back into carbohydrate, so that rapid oxygen uptake continues for some time even though the animal may be at rest: it is said to have incurred an *oxygen debt* as a result of the excess of anaerobic respiration.

4.4 FERMENTATION

Some single-celled organisms, including certain yeasts and bacteria, derive all or most of their energy from anaerobic respiration, and in organisms like these the term *fermentation* is usually applied to the process. The products of fermentation are often carbon dioxide and ethanol, but other compounds are also produced by various species, including acetic acid (the characteristic acid of vinegar), oxalic, citric and butyric acids, and fermentation methods are used for the commercial production of these compounds.

Yeasts (see Section 12.5) and bacteria (Unit Eleven) which bring about fermentation contain enzymes that promote the anaerobic release of energy from carbohydrates. For example, in wine-making the wild yeasts which live on the skins of the grapes ferment the natural sugars in the fruit, while in brewing ethanol is produced by the action of yeasts on the malt sugar prepared from germinating barley. The reactions involved can be summarized by the equation:

$$\underset{\text{glucose}}{C_6H_{12}O_6} \rightarrow \underset{\text{carbon dioxide}}{2CO_2} + \underset{\text{ethanol}}{2C_2H_5OH} + \text{energy}$$

Unless the products of fermentation are removed they accumulate up to a concentration at which the organisms producing them are killed, and fermentation ceases.

Experiment 4.7 *To demonstrate carbon dioxide production by yeast during anaerobic respiration (fermentation)*

Some water is boiled for ten minutes or so in order to expel all the oxygen dissolved in it and it is allowed to cool in a closed flask. It is then used to make up an approximately five per cent solution of glucose and a ten per cent suspension of dried yeast. 5 cm^3 of the glucose solution and 1 cm^3 of the yeast suspension are placed in a test-tube and covered with a thin layer of liquid paraffin to exclude atmospheric oxygen from the mixture. The tube is fitted with a stopper and a delivery tube dipping into clear lime water (Fig. 4.7). The tube is gently heated by standing it in a beaker of warm (*not* boiling) water.

A control experiment can be set up in exactly the same way, but using a yeast suspension which has been boiled to kill the cells: such a suspension will not ferment.

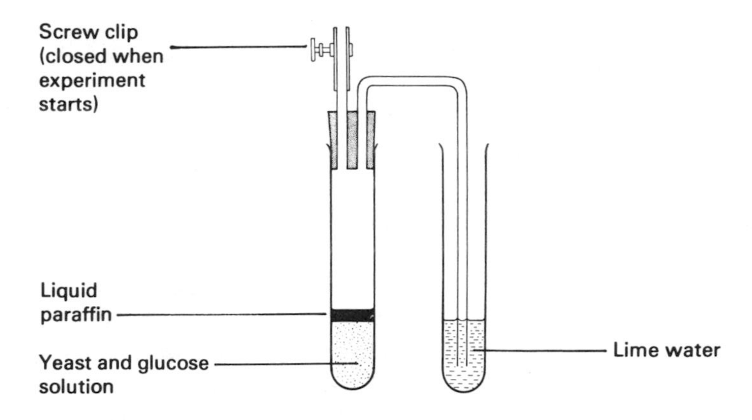

Fig. 4.7 *To demonstrate anaerobic respiration in yeast*

Result. After ten or fifteen minutes, fermentation begins in the mixture containing living yeast cells: the liquid becomes frothy and a stream of gas bubbles escapes through the lime water, which presently turns milky. The gas therefore contains carbon dioxide. Some gas bubbles appear in the control experiment as the tube warms up but the stream soon stops and the lime water remains clear.

4.5 METABOLISM

The thousands of chemical changes taking place in living cells, each controlled by its specific enzyme, are collectively referred to as *metabolism*. These reactions are of many kinds, including changing one compound into another more useful or more reactive, building up small and simple molecules into larger ones which can be incorporated into the tissues of the plant or animal, breaking down complex compounds to make them easily transportable or, as discussed in this unit, to release their energy. Respiration is thus one aspect of the enormously wide range of processes comprised by the term *metabolism*.

4.6 QUESTIONS

1. (*a*) Where does respiration occur?
   (*b*) What is the importance of respiration?
   (*c*) What materials does respiration (*i*) need, (*ii*) produce?
2. List the differences between aerobic and anaerobic respiration. Are these two forms of respiration mutually exclusive? Explain.
3. Which aspects of respiration can be measured or demonstrated?
4. An organism in the course of respiration takes in 50 cm^3 oxygen. It is quite likely to give out 50 cm^3 carbon dioxide in the same period so that there is no volume change in the gas surrounding it. In such a case how can one, in principle, design an experiment to show that oxygen is being taken up?

Unit 5
Structure Of the Flowering Plant

5.1 INTRODUCTION

The next nine units are concerned with *plants;* Table 1.2 gives some indication of the enormous range of organisms comprised by this term. In this and the following three units the structures and functions of flowering plants are discussed, including the way in which they reproduce their kind. The mechanisms by which water and food materials move within a plant are described in Units Three and Nine. Unit Four dealt with respiration, and Unit Ten deals with the way in which green plants obtain their food. Units Eleven and Twelve give an account of bacteria and fungi—which although they are classified as plants contain no chlorophyll and feed in a different way from flowering plants—and Unit Thirteen discusses some very simple plants.

A typical flowering plant consists of two main parts: the *shoot* above the ground, and the *root* below it. This does not mean, however, that any part of a plant below ground is a root; underground stems, leaves and buds are not uncommon.

The shoot is usually made up of a stem, bearing leaves, buds and flowers.

5.2 STEMS

(a) General Characteristics of the Stem (Fig. 5.1)

A stem carries leaves and buds at more or less regular intervals. Most of its growth takes place at the tip, or *growing point,* where there is a *terminal bud.* The part of the stem from which a leaf springs is called a *node,* and a region of the stem between two nodes is an *internode.*

Commonly the stem is erect, but its form varies greatly. The "runners" of the strawberry plant are horizontal stems; rhizomes (Section 6.4) of

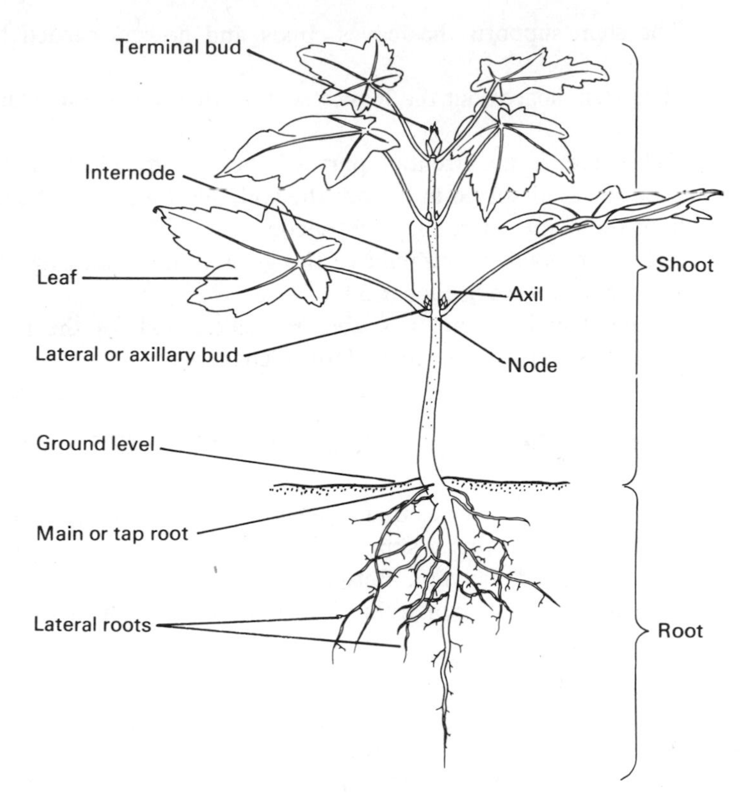

Fig. 5.1 *Structure of a typical flowering plant*

plants like iris are underground stems; in bulbs and corms (Sections 6.2 and 6.3) like hyacinth and crocus the stem is very short indeed and never shows above the ground (it is the flower stalk that is visible). The stems of climbing plants like convolvulus are very long, thin and weak, while the stems of trees are often immensely thick and strong. Young stems are usually green, and contain chlorophyll.

The cells of young stems are living; they obtain the oxygen they need from air which enters the stem through minute pores called *stomata* or, in older stems, through openings called *lenticels* (see Section 5.2(*c*)). Running through the stems are tubes which conduct water from the soil up to leaves, and food manufactured in the leaves to the other parts of the plants. Young stems depend for their rigidity on the pressure of the sap in their cells (see Section 2.2), on the arrangement of their tough conducting tissues and on the opposing stresses of the inner and outer parts of the stem. Older stems are supported by woody and fibrous tissues, layers of which are added year by year.

(b) Functions of the Stem

(*i*) The stem supports the leaves, buds and flowers carried by the shoot.

(*ii*) The stem spaces out the leaves so that they receive adequate air and sunlight.

(*iii*) The stem is an essential part of the system which transports water and substances dissolved in it from the soil, and food from the leaves, to the parts of the plant where they are needed.

(*iv*) The stem holds flowers above the ground, thus assisting pollination by insects or wind (see Section 7.3).

(*v*) If the stem is green, it is able to make food for the plant by photosynthesis (discussed in detail in Unit Ten).

(c) Structure of the Stem (Figs. 5.2, 5.3, 5.4, 5.5, and 5.6)

A fairly typical stem, such as that of a sunflower, is cylindrical in form. The outer layer of cells forms a skin, the *epidermis*, while the inner cells make up the *cortex* and the *pith*. Between the cortex and the pith lie a number of *vascular bundles* containing specialized cells which carry food and water.

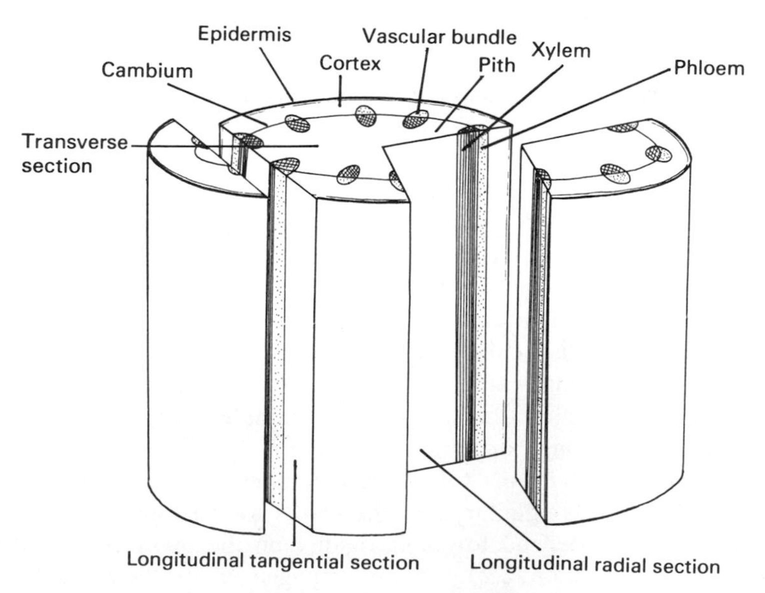

Fig. 5.2 *Stereogram of a plant stem*

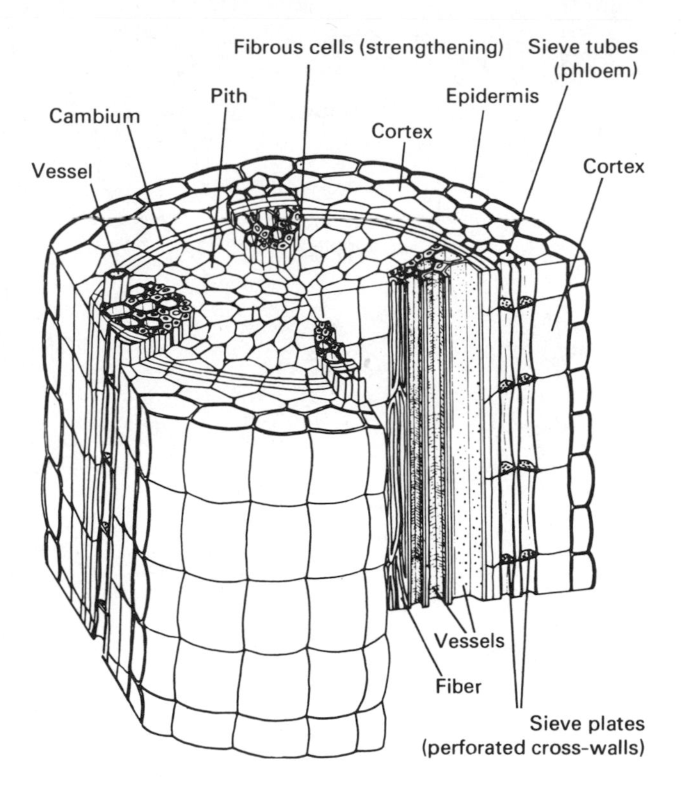

Fig. 5.3 *Stereogram of a plant stem sectioned, showing cells (cell contents not shown)*

Epidermis. This single layer of closely fitting cells is effective in holding the inner cells in shape, preventing loss of water, affording protection from damage and preventing the entry of dust and micro-organisms like fungi and bacteria. This layer is relatively impermeable to liquids and gases, and in young stems oxygen can enter and carbon dioxide escape only through the stomata. These resemble the stomata of leaves, which are described in detail in Section 5.3(*c*). In older, woody stems the epidermis is replaced by a dead, corky layer, *bark,* made by the layer of cells lying just beneath it. Bark is impervious to oxygen and carbon dioxide, and in these older stems the gases pass into and out of the tissues through *lenticels*. These are small gaps in the bark, usually circular or oval and slightly raised on the bark surface. In them, the cells of the bark fit together loosely, leaving air gaps which communicate with the air spaces in the cortex (Fig. 5.7).

Cortex and pith. These are tissues consisting of fairly large, thin-walled cells with air spaces between them. This air-space system is continuous

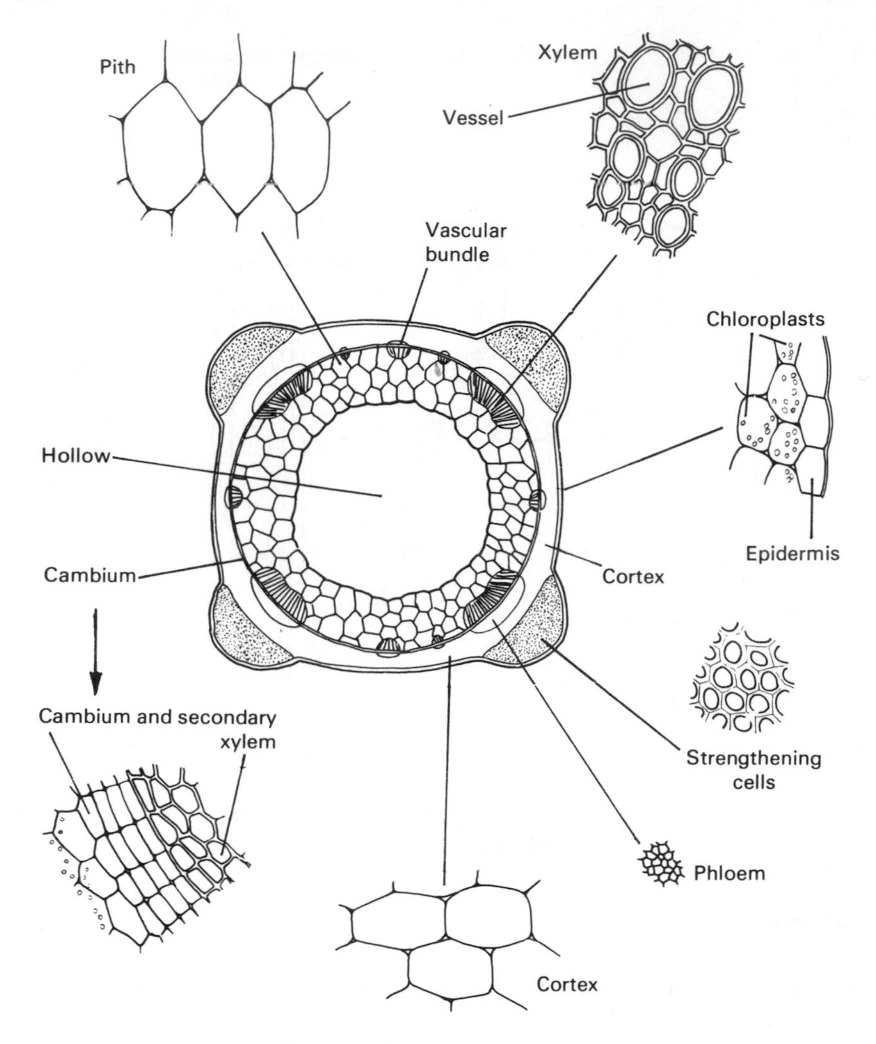

Fig. 5.4 *Section through stem of white deadnettle*

throughout the living tissues and allows air to circulate from the stomata or lenticels to all the living regions of the stem. The cortex and pith contribute to the rigidity of the stem by pressing out against the epidermis and by tending to increase in length against the shrinking tendency of the epidermis. These tissues also space out the vascular bundles and have a general value as packing. Many stems, however, are hollow with only a narrow band of pith within the cortex.

Vascular bundles. These are sometimes called *veins*. They are made up of vessels and sieve tubes, with fibrous and packing tissue between and

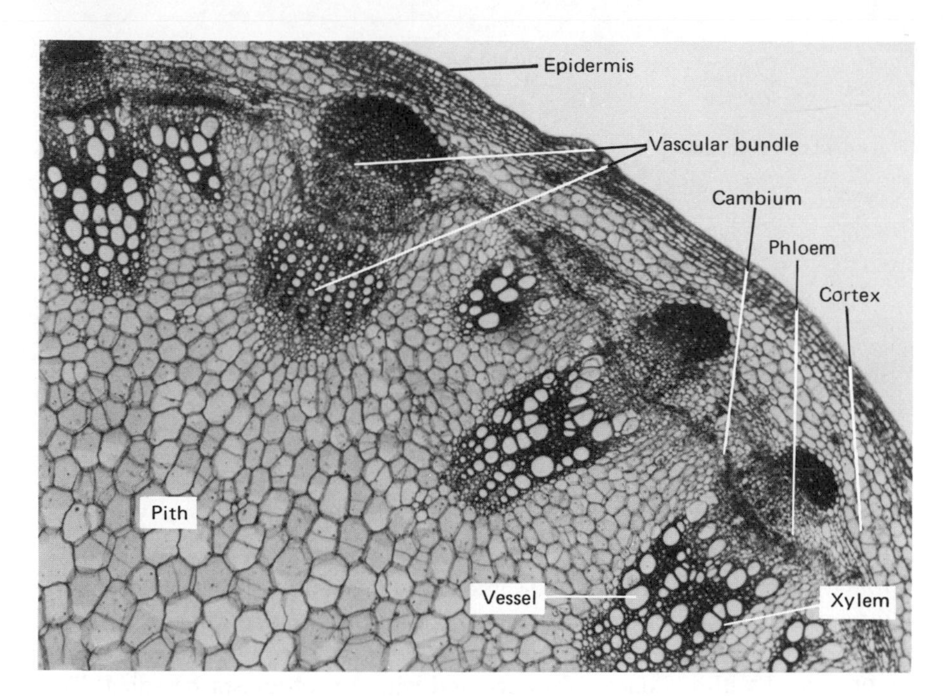

Fig. 5.5 *Transverse section through sunflower stem*
(GBI Laboratories Ltd.)

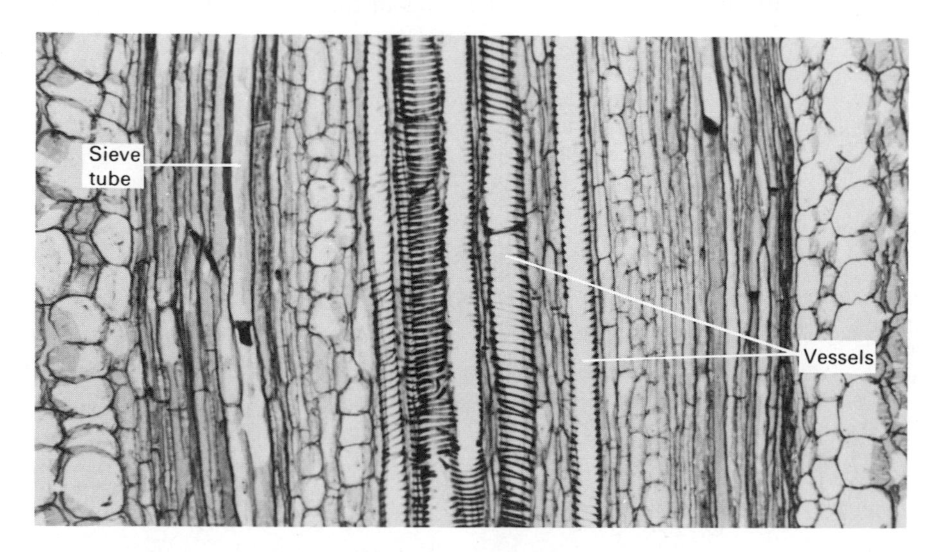

Fig. 5.6 *Longitudinal section through sunflower stem, showing vessels thickened internally with bands of woody material*
(Brian Bracegirdle)

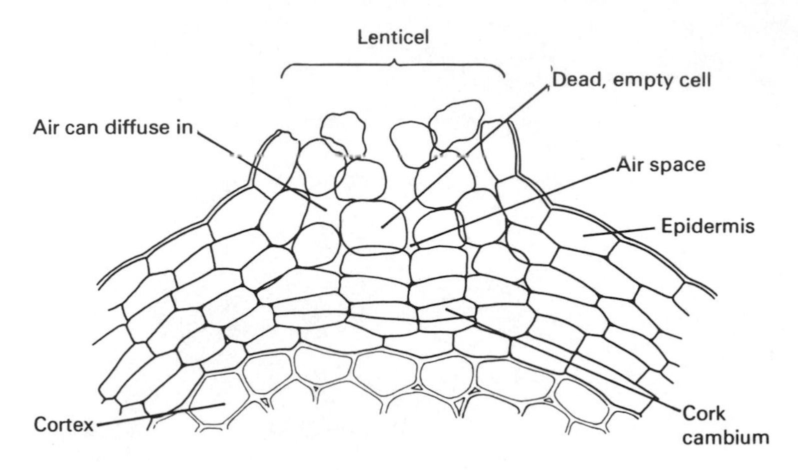

Fig. 5.7 *Section through stem showing lenticel*

around them (the word *vascular* is derived from a Latin word meaning "little vessel"). Vessels are long tubes; they may be up to a meter or so in length, and are formed from columns of cells whose walls have become impregnated with a woody substance and whose protoplasm has died. The horizontal cross-walls of these cells have broken down, so that a long continuous tube is formed (Fig. 5.8). In these vessels water is carried from the roots, through the stem and to the veins in the leaves (Fig. 5.9). *Sieve*

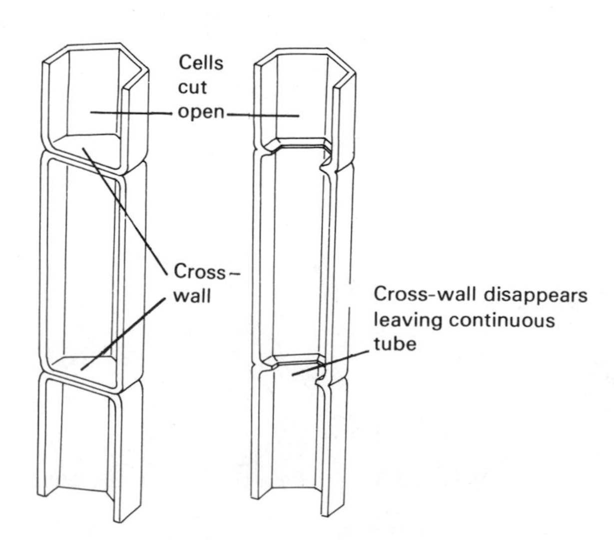

Fig. 5.8 *Diagram to show how vertical columns of cells give rise to vessels*

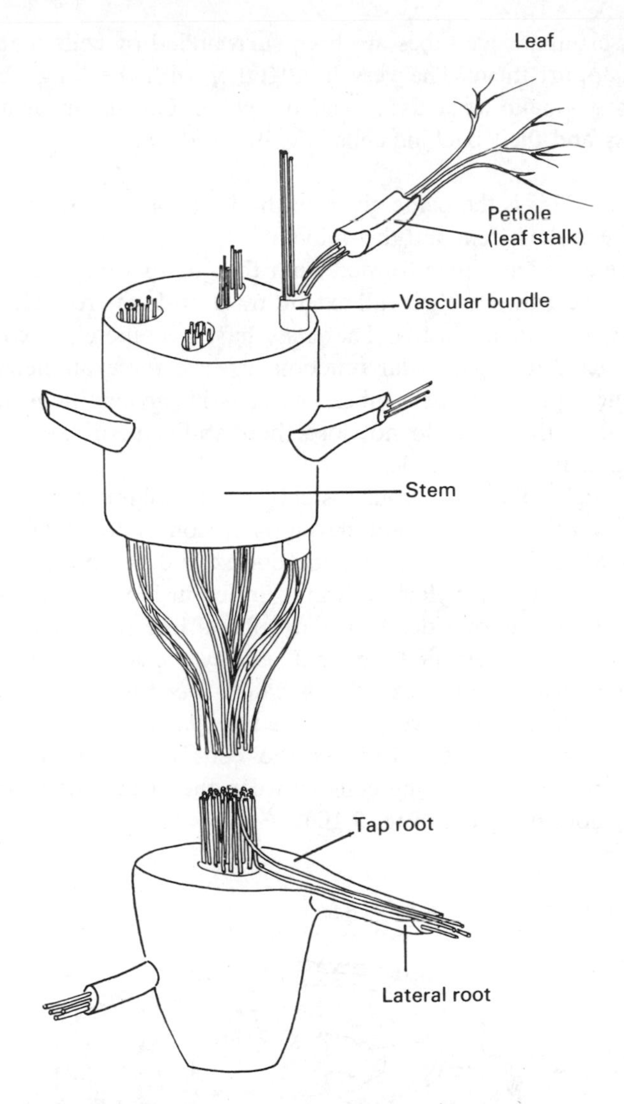

Fig. 5.9 *Diagram showing the distribution of veins from lateral root to leaf*

tubes, unlike vessels, are columns of living cells. The horizontal walls of these cells are perforated (see Fig. 2.9(*c*)) so that dissolved substances can flow from one cell to the next. In this way food made in the leaves can be carried to other parts of the plant, for example, to the ripening fruit, growing points or underground storage organs, according to the species of plant and the time of year.

Vessels and sieve tubes are both surrounded by cells that space them out and support them. The vessels, together with the long fiber-like cells among them, make up a tissue called *xylem*. The tissue made up of the sieve tubes and their packing cells is called *phloem*.

Cambium. This is the name given to the layer of narrow thin-walled cells lying between the xylem and the phloem.

Once cells have been formed from the growing point at the tip of the stem and have grown to their full extent, most of them are no longer capable of dividing to make new cells. They may have become changed in structure and specialized to a particular function, as has, for example, a cell which has become a part of a sieve tube. The cells in the cambium, however, are exceptional in that they do not lose their ability to divide; they can still multiply and make new cells.

Although in a very young stem the cambium is restricted to the vascular bundles, as the plant develops it comes to form a continuous cylinder within the stem between the cortex and the pith. Its cells divide in such a way that new xylem cells are formed inside the cambium and new phloem cells on the outside. In woody plants like trees, this process continues throughout their lifetime, and as new cells are constantly added the stem increases in thickness; this is called *secondary thickening*. In such woody stems a separate layer of *cork cambium* develops just beneath the epidermis, and this layer produces the cells of the bark. The phloem becomes a thin layer of living cells between the dead cells of the bark and the woody core of xylem (Fig. 5.10).

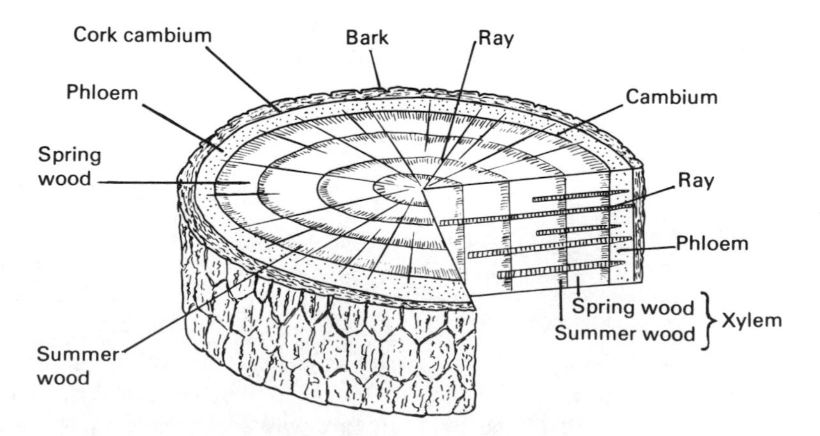

Fig. 5.10 *Diagrammatic section through a woody stem showing four seasons' growth*

(d) Strength of Stems

Vertical stems often experience quite large sideways forces when the wind blows against them. The turgor pressure of the cell sap in the stem cells (discussed in detail in Section 3.4(*a*)) and the opposing forces exerted by the tendency of the epidermis to shrink and that of the pith to expand, all contribute to the stem's resilience. The vascular bundles usually contain the toughest structures in the stem, the woody vessels and often long stringy fibrous cells running alongside them, and they are frequently found to be arranged in a cylinder near the outside of the stem; in such an arrangement they add to its strength, since a hollow cylindrical structure is much more resistant to bending than a solid structure of the same weight. There are other ways too in which the strengthening tissue can be distributed so as to increase the stem's resistance to bending stresses; for example, in plants of the deadnettle family the stem is square in section, and carries the vascular bundles and strengthening strands of cells in the four corners (see Fig. 5.4).

5.3 LEAVES

(a) General Characteristics of the Leaf (Fig. 5.11)

A typical leaf is a flat, green *lamina* or blade consisting of a soft tissue of thin-walled cells supported by a stronger network of *veins*. Leaves are sometimes joined to the stem by a leaf stalk or *petiole* which continues into the *midrib* or main vein of the leaf. In some plants there is no leaf stalk. The angle made by the leaf (or by the petiole, if one is present) with the stem is called the *axil* of the leaf; it always contains a bud.

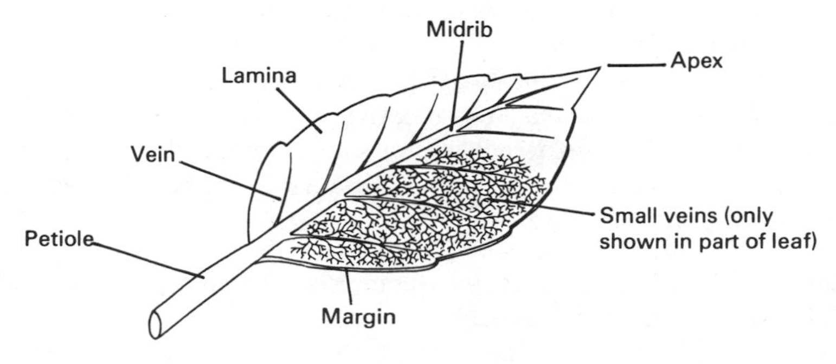

Fig. 5.11 *A typical leaf*

(b) Functions of the Leaf

The important function of leaves is the manufacture of food substances, such as carbohydrates like sugars and starches, from simple, readily available constituents: water and carbon dioxide. This is the process called *photosynthesis:* it requires a supply of energy, which is absorbed from sunlight by the green pigment chlorophyll. The water necessary for the process is obtained from the soil by the roots of the plant (see Section 5.6), and is conveyed to the leaf through the vessels which run in the vascular bundles branching from the stem. These run through the petiole and midrib and then divide repeatedly to form a network of tiny veins extending throughout the lamina. The components of air can diffuse into and out of the leaf through the tiny pores or *stomata* (singular, *stoma*) described in the following section. In this way the plant can obtain directly from the air both the carbon dioxide it needs for photosynthesis and the oxygen which is essential for its respiration: air contains about 0.03 per cent of carbon dioxide (by volume) and about 20 per cent of oxygen.

Most leaves are broad, flat and thin in form; this shape offers the best possible conditions for photosynthesis. The large surface area of the leaf not only allows the maximum amount of sunlight to reach it, but also promotes rapid absorption of carbon dioxide and oxygen. Because most leaves are thin, there is only a short distance from the stomata, through which the gases enter the leaf, to the cells inside it where they are used. However, these characteristics of permeability and of large surface area are conducive to a rapid loss of water vapor by evaporation, especially in hot dry weather (see Section 9.3).

(c) Structure of the Leaf

Epidermis. This is a single layer of cells fitting closely together and covering the entire surface of the leaf. There are no air spaces between them, except for the stomata (see below). The epidermis helps to maintain the shape of the leaf, protects its inner cells from infection by bacteria or fungi and from mechanical damage, and reduces the loss of water by evaporation; in some plants, the epidermis secretes a continuous waxy layer called *cuticle,* which is particularly effective in keeping down water loss. Most of the cells of the epidermis contain no chlorophyll, and are transparent. In consequence, sunlight can pass through to the cells below, which do contain chlorophyll.

Stomata (Fig. 5.13). These minute openings in the epidermis are usually more abundant on the lower side of the leaf. Each is formed between two *guard cells*; the pressure of the fluid in the vacuoles of these cells controls the opening and closing of the stoma. The conditions which affect this pressure are thought to be related principally to the light intensity, and in some cases also to the loss of water. If the concentration of sugars in the cell sap of the guard cells increases, they withdraw water from neighboring cells; the amount of fluid within them is increased and its pressure (turgor pressure) rises. The guard cells begin to swell, but as the cell wall is particularly thick along the side which forms the stoma, most of the stretching takes place in the outer walls. The effect of this is to cause the inner walls of the guard cells to curve away from each other, and the stoma opens. When the turgor pressure in the guard cells falls, they become more flaccid, the inner walls straighten and the stoma closes again. (The process is discussed in more detail in Section 3.4(*a*).) In very general terms, the stomata are open in the daytime and closed at night. There are exceptions found among certain desert plants, which open their stomates at night but keep them closed during the day to conserve vital water.

Palisade layer. Immediately below the epidermis lie one or more layers of tall cylindrical cells, with narrow air spaces between them. These are the cells of the palisade layer, where most of the photosynthesis occurs. The cytoplasm lining the walls of these cells contains large numbers of minute bodies made of protein, called *chloroplasts*. Usually discus-shaped, these contain the green substance chlorophyll which is essential for photosynthesis and which gives plants their characteristic color; chlorophyll is not found throughout the cytoplasm, but only within the chloroplasts. It is believed that the chemical processes of photosynthesis actually take place in the chloroplast when light is shining on it. As the palisade cells lie so near the upper surface of the leaf, they receive and absorb most of the sunlight; the chloroplasts can move up or down the cells according to the intensity of the light. Because of the elongated shape of the palisade cells, little light is wasted by absorption in horizontal cross-walls before it reaches the chloroplasts. In addition, the chloroplasts are never far from supplies of carbon dioxide, which reach them through the air spaces between the palisade cells.

Spongy layer. The cells in this region do not fit closely together, and there are large air spaces between them. The air spaces communicate with each other and with those in the palisade layer and, through the stomata, with the atmosphere, thus allowing air to circulate in them and reach most of the internal cells of the leaf. The cells of the spongy layer can photosynthesize, but they receive less sunlight than do the palisade cells, and contain fewer chloroplasts.

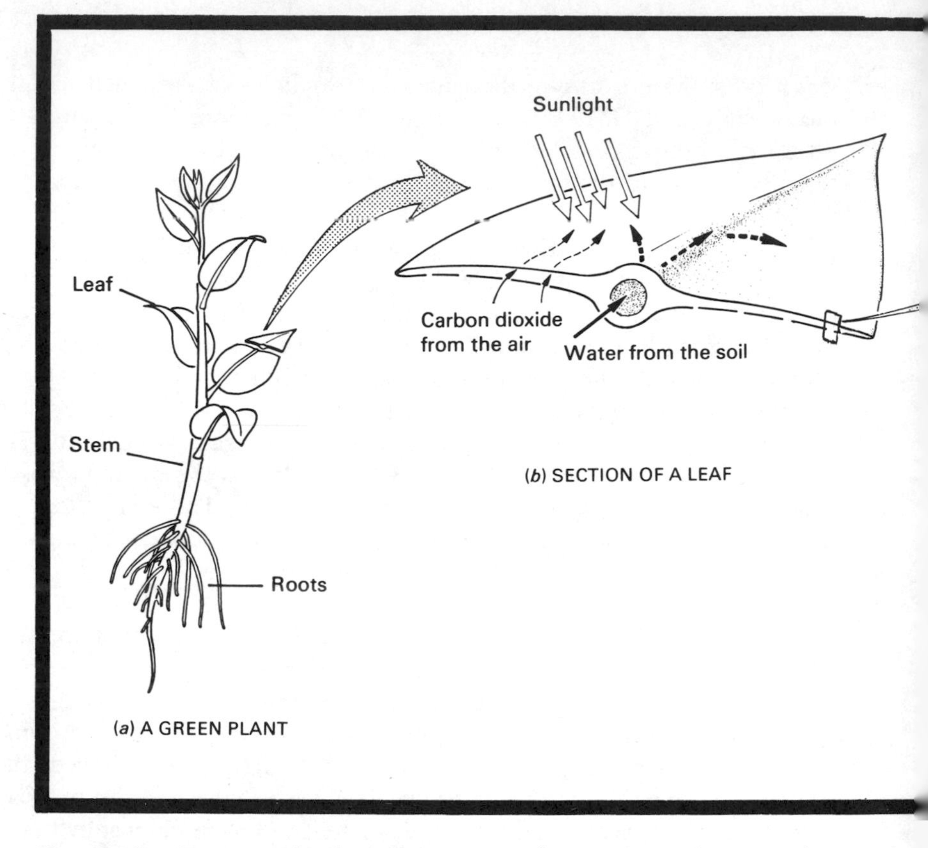

Fig. 5.12 *Structure of a leaf, showing its adaption for its function of photosynthesis*

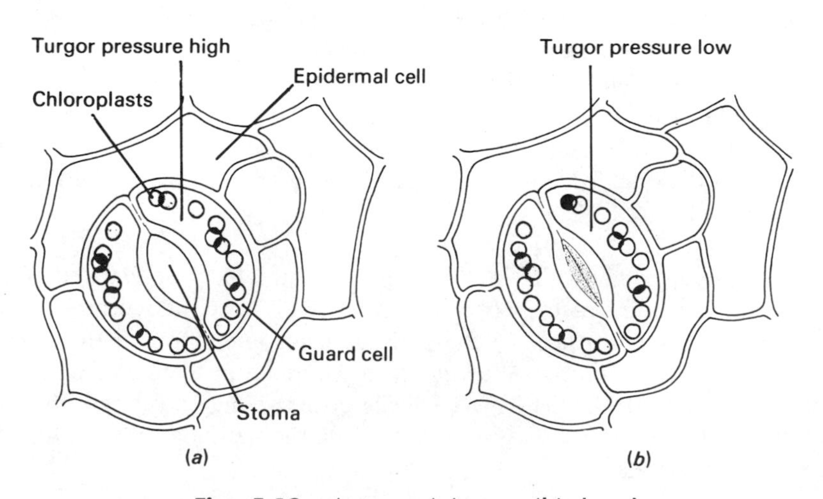

Fig. 5.13 *A stoma: (a) open; (b) closed*

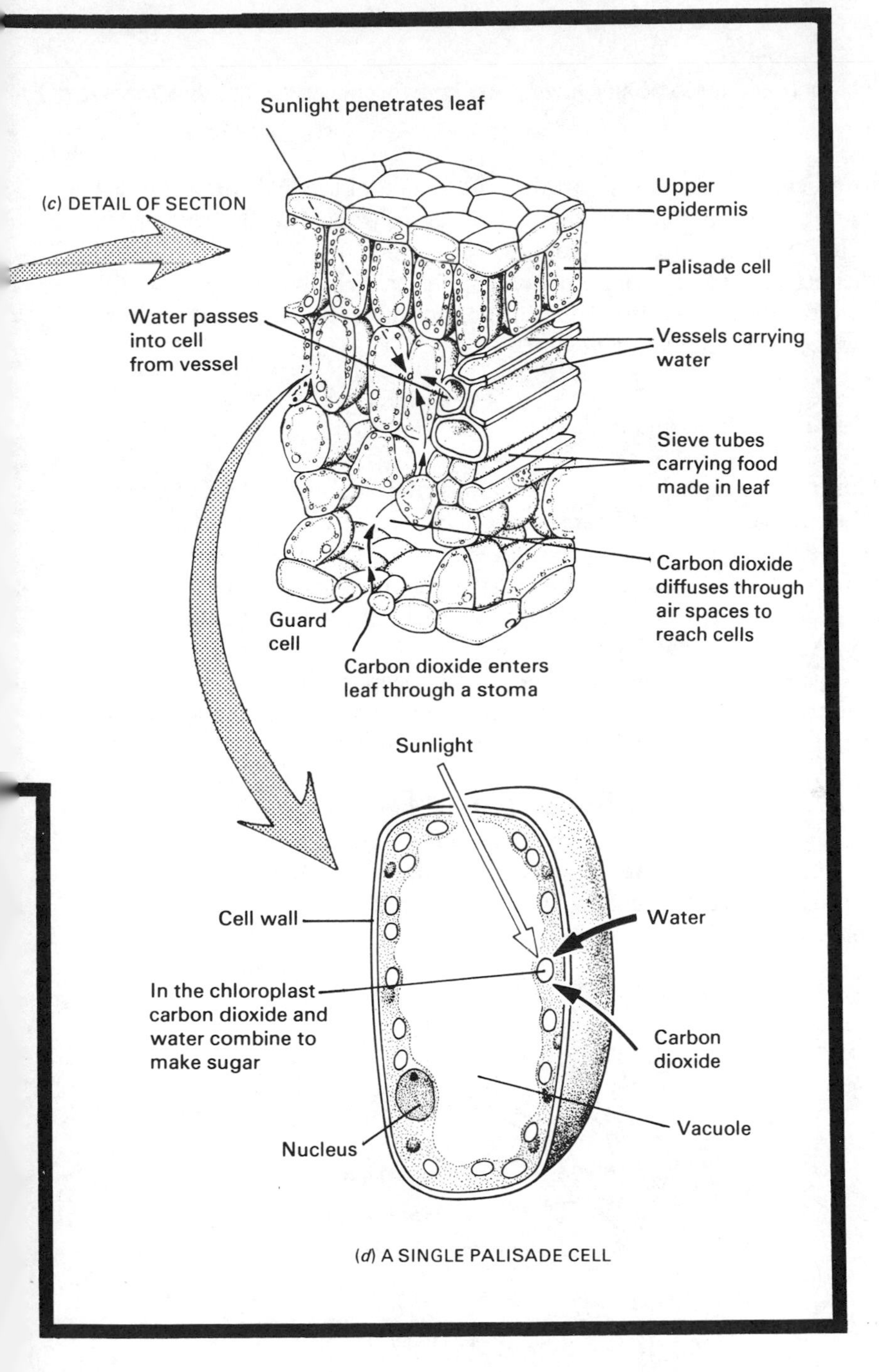

(d) A SINGLE PALISADE CELL

The palisade and spongy layers are known collectively as the *mesophyll.*

Midrib and veins. These support the soft lamina of the leaf; they conduct water into it, and food manufactured there away from it to the rest of the plant. They contain vascular bundles surrounded by other fibrous and strengthening cells. The veins do not supply each individual cell of the leaf, but the network of veins is very fine, and no cell in the leaf is more than a few cells distant from a vein.

(d) **Leaf-fall** (Figs. 5.14 and 5.15)

Most trees shed their leaves. *Deciduous trees* like beech or apple shed them all at the same time of year; *evergreens* like holly shed them in small numbers all the year round, and never stand leafless. The details of the mechanism of leaf-fall, or *abscission,* vary in different species, but in certain trees the process takes place as follows: the cells at the base of the petiole begin to divide, and new layers of cells form across the region where it joins the stem. The contents of the leaf cells begin to break down chemically, the effect of which is to produce the characteristic red and yellow autumn tints. The soluble products of this breakdown are absorbed back into the tree. Meanwhile, the cells of the new tissues lying nearest to the stem become corky, and the vessels become blocked so that the leaf is deprived of water. The cells beyond the corky layer degenerate, and the dried-up leaf falls away, leaving a scar on the stem. However, the stem remains protected from infection by bacteria or fungi by the layer of impermeable cork.

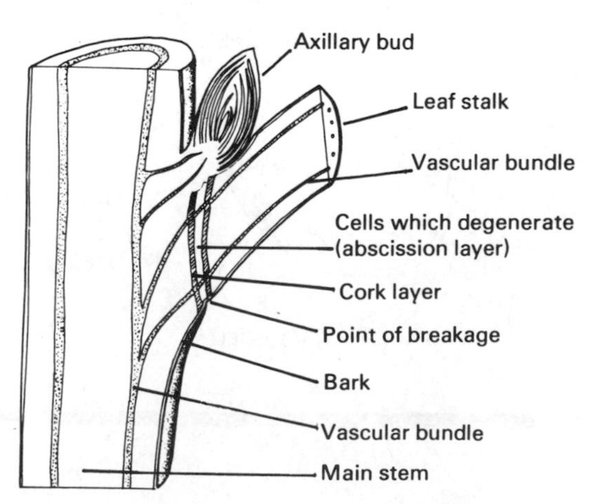

Fig. 5.14 *Diagram to show leaf-fall*

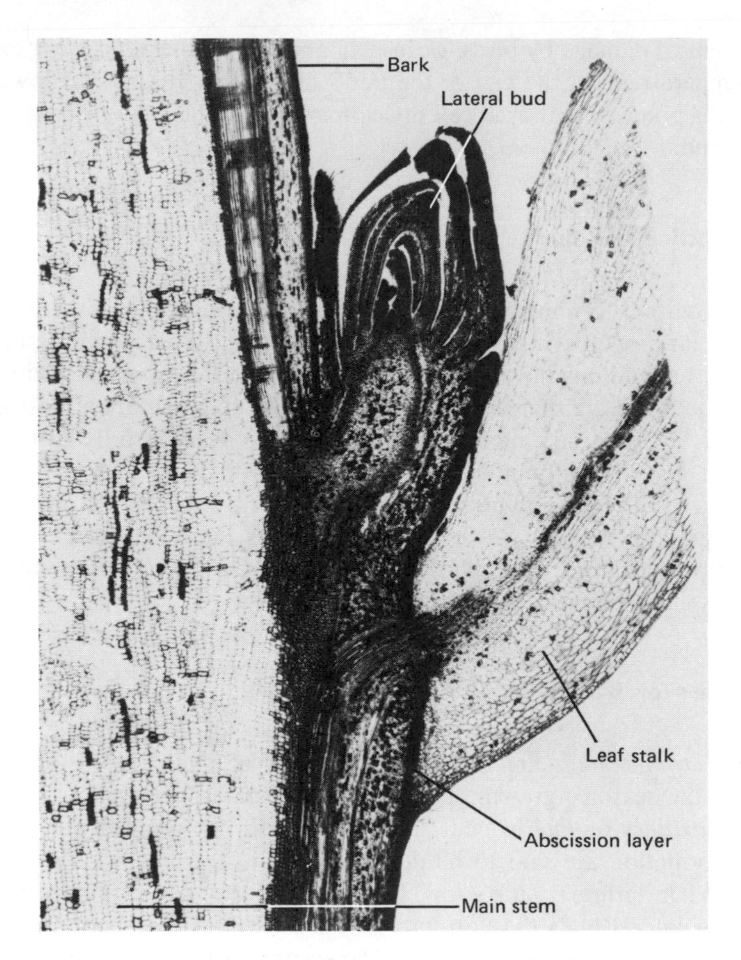

Fig. 5.15 *Leaf-fall in sycamore—longitudinal section*
(Brian Bracegirdle)

5.4 BUDS

(a) Structure of a Bud (Fig. 5.16(*a*))

A bud is a "condensed" shoot. Its stem is very short and its leaves are so close that they overlap, each one wrapping around the next above it. The inner leaves are crinkled and folded, since a large surface area is packed into a small space. The outermost leaves are often thicker and tougher, and sometimes black or brown. These are the *bud scales* and they protect the more delicate inner foliage leaves from drying up, from mechanical damage by birds, or insects and, to some extent, from extremes of temperature. At the end of the bud's short stem is either a flower, or a growing point where rapid cell division will take place later on to form the next bud.

(b) **Functions of Buds**

The bud-forming habit gives the plant the advantages of being able to present a photosynthesizing surface to the atmosphere very rapidly after seasonal conditions have become favorable. The leaves are already formed in the bud, and are therefore available for food manufacture almost at once, whereas it would take many weeks for them to develop fully from the few cells at the growing point.

In most plants the buds are formed when conditions are favorable for plant growth, usually in the summer months, and the close packing of the leaves and the thick bud scales protect the leaves from desiccation and low temperatures during the winter.

(c) **Types of Bud**

Terminal buds are formed at the ends of main shoots or branches during the season's growth. There are also buds in the axils of the leaves, called *axillary* or *lateral* buds. Buds which do not grow in the year after their formation are said to be *dormant.*

When terminal buds grow, the length of the main shoot is increased, whereas lateral buds develop into new branches of the shoot. Either kind of bud may produce a flower or inflorescence (a group of flowers growing from a single main stalk) instead of, or in addition to, a leafy shoot. If this happens in a terminal bud, the growth of the shoot must be continued in the following season by one or more lateral buds, for when the flower or inflorescence fades and falls away no growing point is left.

(d) **Growth of Buds** (Figs. 5.16 and 5.17)

In the spring, the stem of the bud begins to elongate and the bud scales are pushed apart. As the stem grows in length it spaces out the leaves, which unfold and spread out their surface. The bud scales often curl back and fall off after a few weeks. As soon as the leaves are exposed to light their chlorophyll develops fully, and photosynthesis begins soon after.

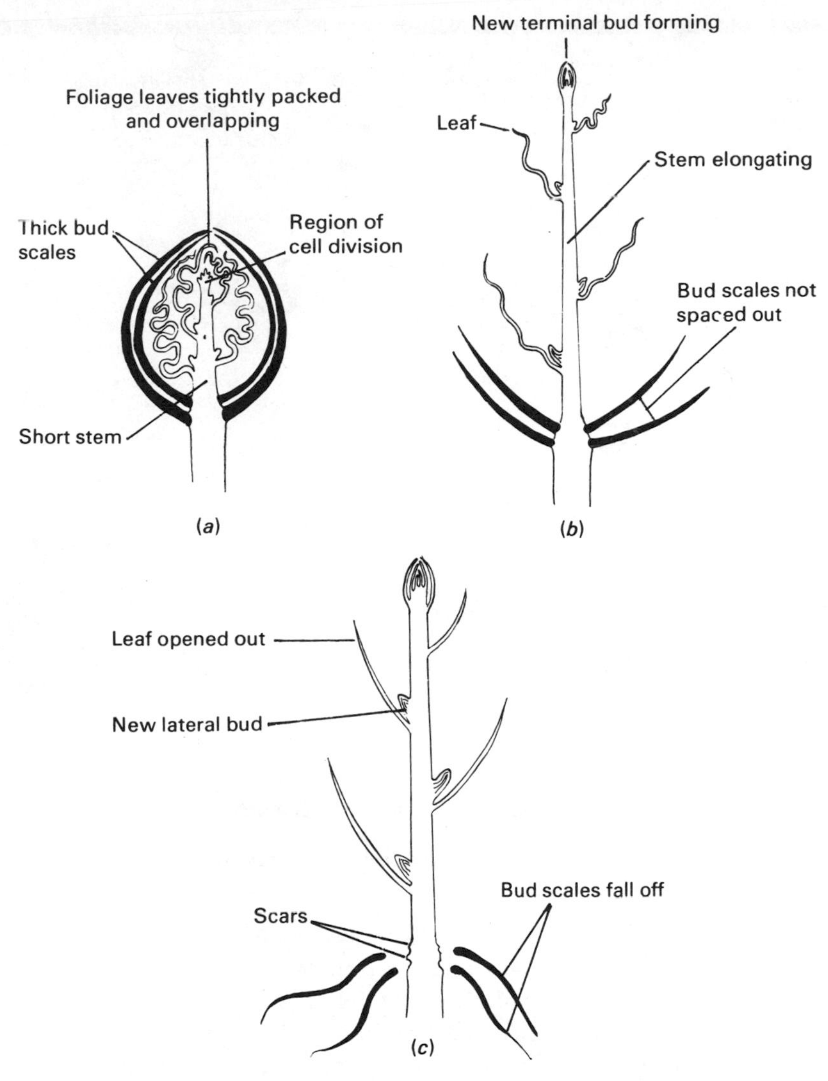

Fig. 5.16 *Bud growth*

5.5 WINTER TWIGS

In winter, the leafless twigs of deciduous trees still have an appearance that is characteristic of each species. In part, this is because they carry the scars where the previous season's leaves and bud scales had grown.

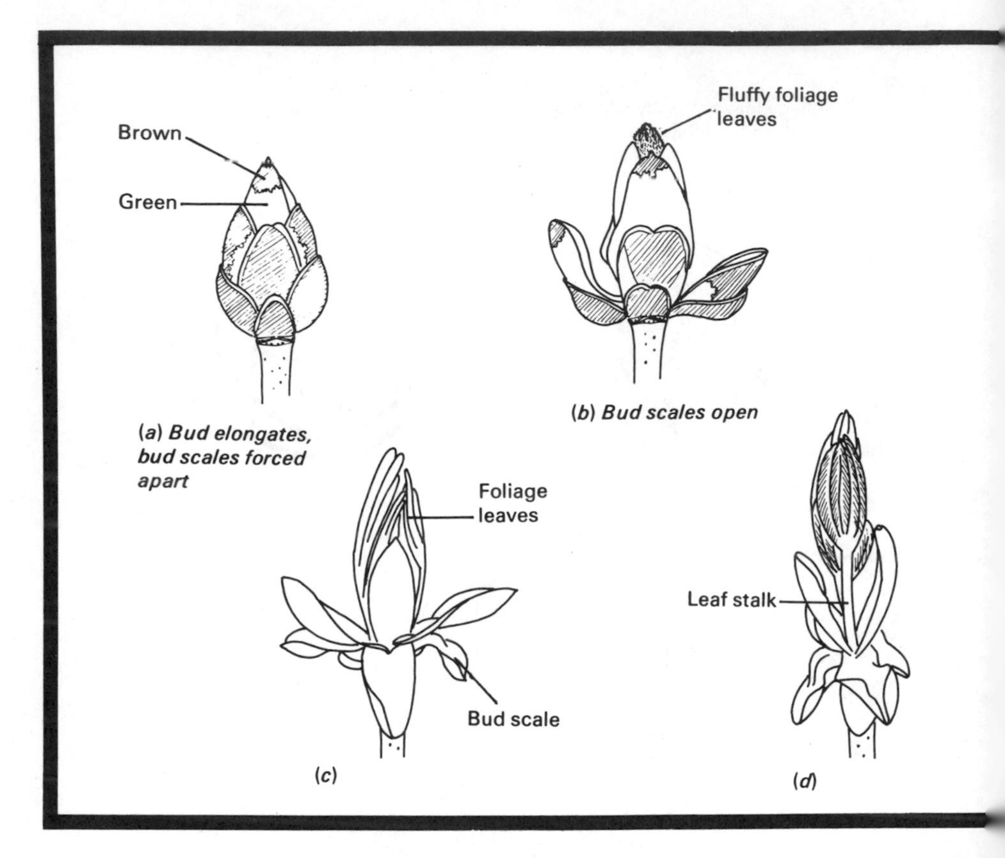

Fig. 5.17 *Stages in growth of horse-chestnut terminal bud*

The leaf scars on twigs are a characteristic shape for every species, the sealed vascular bundles making a pattern of dots in them. Since each leaf usually had a bud in its axil, there is a lateral bud above each leaf scar.

Earlier in the year, when the terminal bud was sprouting, the bud scales, unlike the foliage leaves, were not spaced out on the stem, and when they fell off they left narrow scars, close together, extending from a quarter to half-way around the stem. These are commonly called *girdle* or *terminal bud scale scars* and, since they mark the position of each year's terminal bud, the length of stem between each set of girdle scars represents one year's growth (Figs. 5.18, 5.19 and 5.20).

5.6 ROOTS

Unlike shoots, roots never bear leaves or buds. They are usually white in color, and cannot develop chlorophyll. They have two main functions:

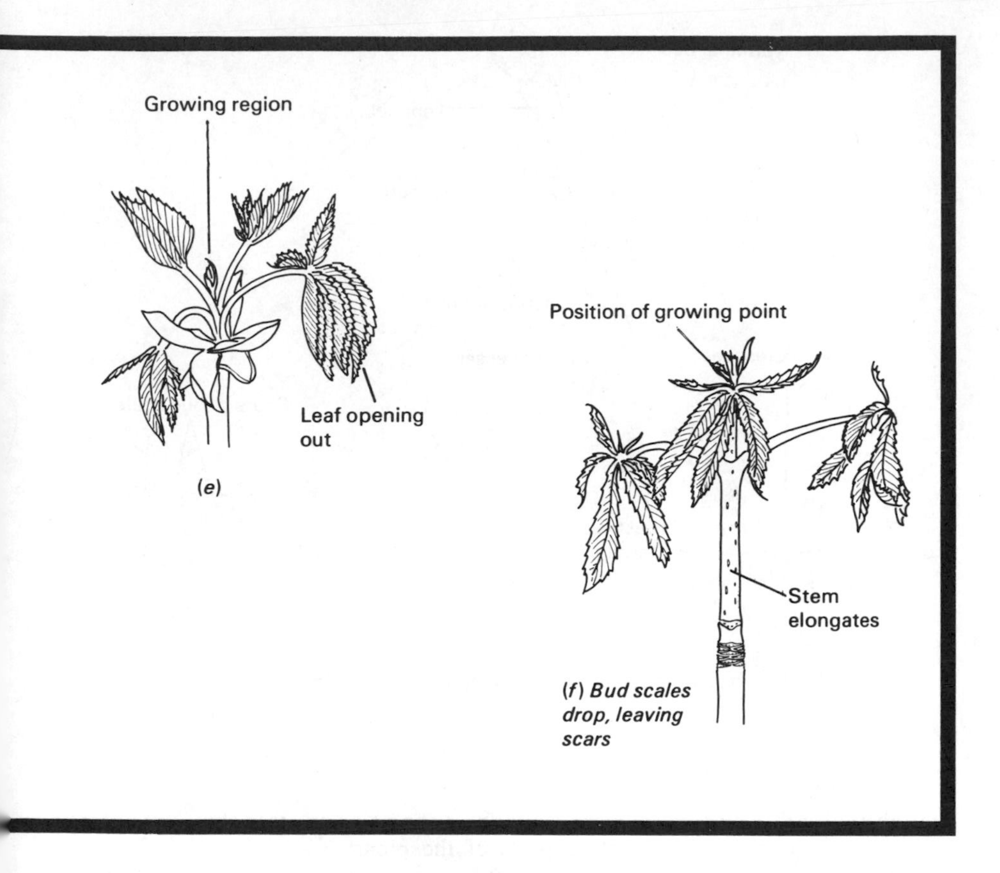

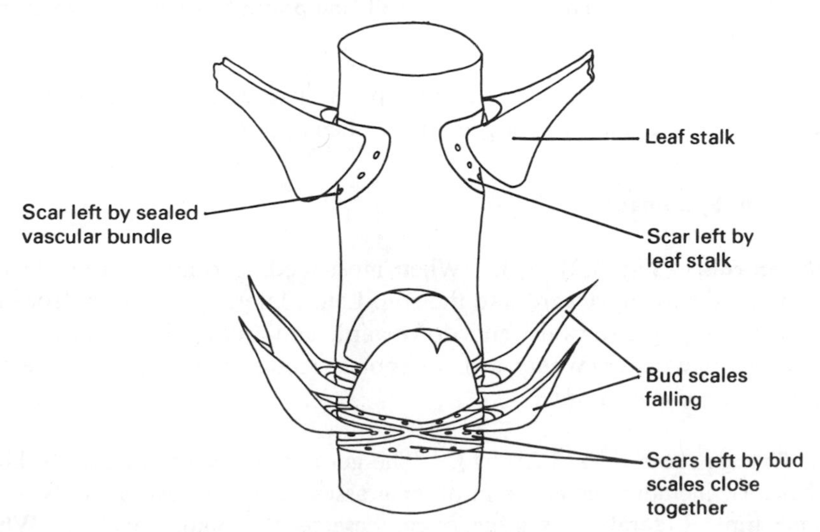

Fig. 5.18 *The formation of "girdle" scars and leaf scars*

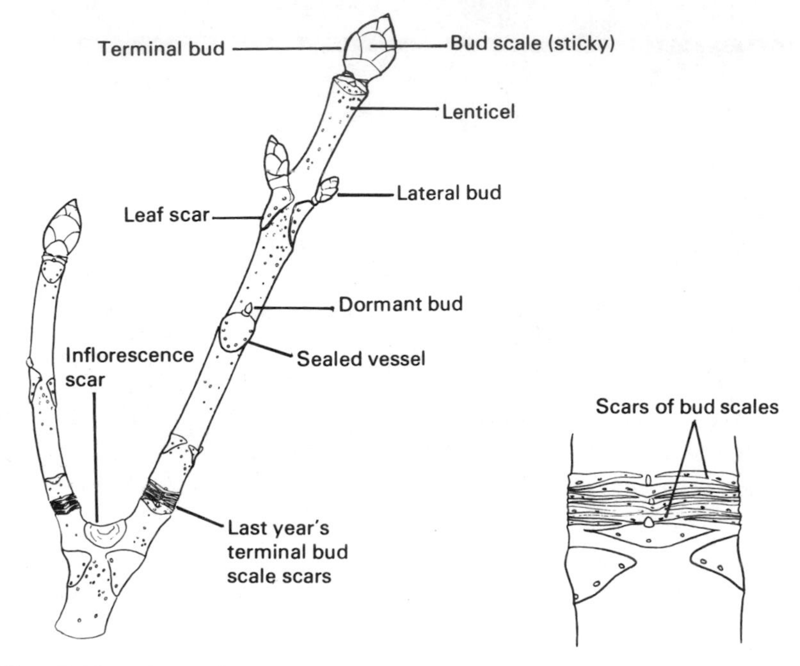

Fig. 5.19 *Horse-chestnut twig in winter*

Fig. 5.20 *Girdle scars of horse-chestnut*

(*i*) they absorb water and mineral salts from the soil and pass them into the stem for transmission to other parts of the plant;
(*ii*) they anchor the plant firmly in the soil, and prevent its being blown over by the wind.

In addition, the roots of certain plants like carrots or turnips act as food stores, and become enlarged and swollen as a result.

(*a*) Root Systems

(*i*) **Tap roots** (Fig. 5.21(*a*)). When most seeds germinate, a single root grows vertically downward into the soil. Later, lateral roots grow from this at an acute angle outward and downward, and from these laterals other branches may arise. Where a main root is recognizable, the arrangement is called a *tap-root system*.

(*ii*) **Fibrous roots** (Fig. 5.21(*b*)). The germinating seeds of certain kinds of plants, including cereals and other grasses, produce several roots at the same time. Lateral roots grow from these as the plant develops. Where

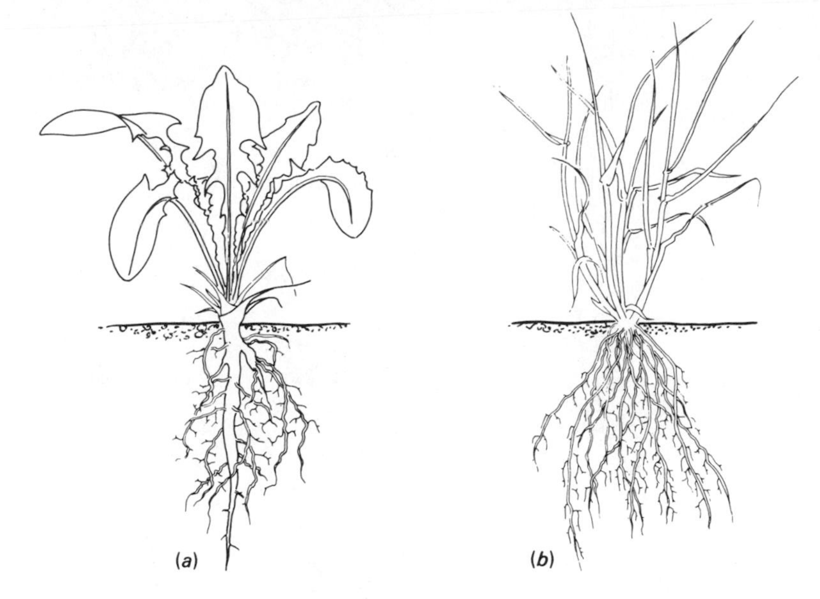

Fig. 5.21 *Types of root system: (a) tap root of dandelion; (b) fibrous root of a grass*

there is no distinguishable main root, the structure is called a *fibrous root system*.

In some plants the roots grow directly from the stem, rather than from a main root; these are called *adventitious roots*. They are to be seen in bulbs like daffodil, in corms like cyclamen, in rhizomes (underground stems) like those of couch grass or iris, and in ivy (see Fig. 6.1, 6.4, 6.5 and 6.6). Systems like these may also be described as fibrous root systems.

(b) **Structure of Roots** (Figs. 5.22 and 5.23)

Epidermis. The epidermis consists of a single outer layer of cells. No cuticle is secreted by the epidermal cells of roots. The younger cells, particularly those with root hairs (see below), are permeable to water and simple salts and permit the uptake of these substances by the plant.

Root hairs. These are the main absorbing region of the root. They are tiny finger-like outgrowths from the younger cells of the epidermis, appearing just above the region of maximum elongation of the root; there are none at the tip of the root nor in its older regions. The root hairs grow out between the soil particles, so that their shape is determined to some extent by the position and shape of the particles between which they grow.

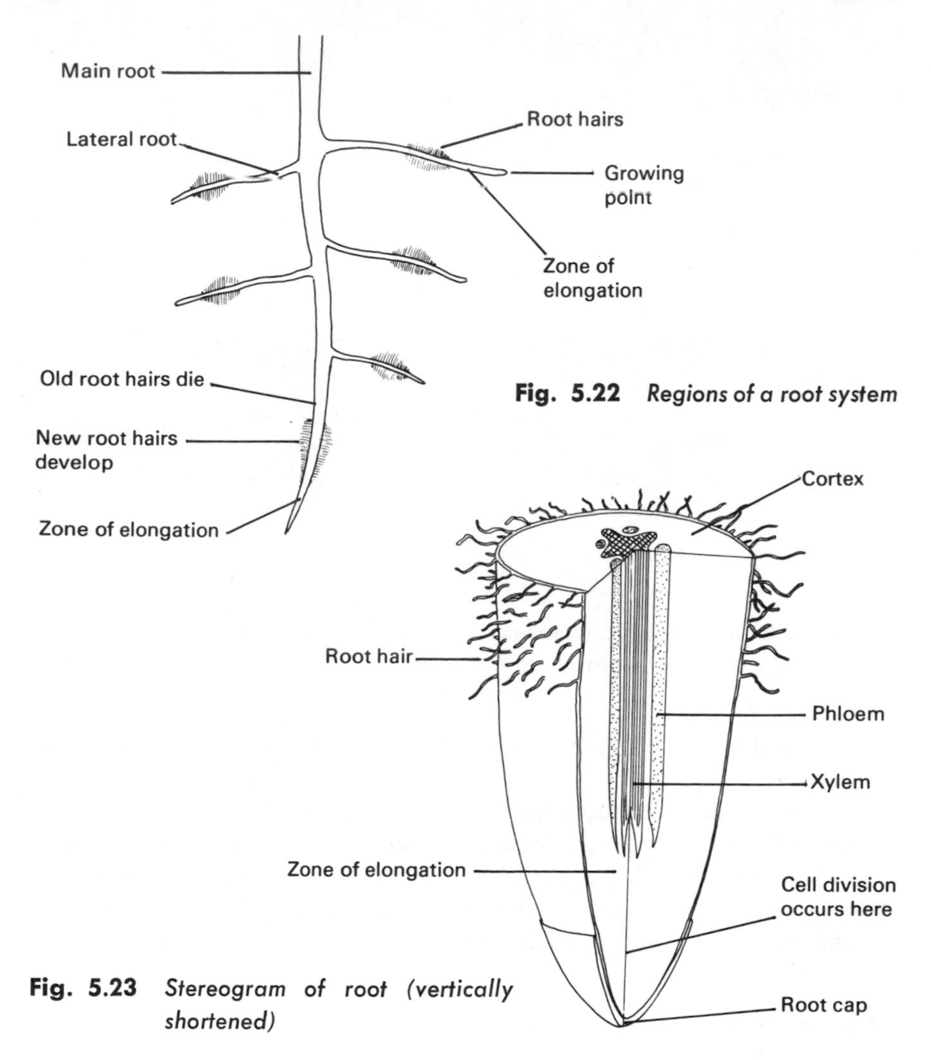

Fig. 5.22 *Regions of a root system*

Fig. 5.23 *Stereogram of root (vertically shortened)*

Their cell walls stick to the soil particles so firmly that they cannot be washed off. This helps to keep the roots firmly anchored in the soil, and incidentally reduces soil erosion by wind and rain. There are millions of root hairs in the root system of even a small plant, and they present an enormous surface area for the absorption of water and salts from the soil (see also Fig. 3.11).

Cortex. The cortex is a tissue consisting of large thin-walled cells with abundant intercellular spaces. These cells store food materials; the innermost layer may be concerned in some way in the regulation of the inward passage of water and dissolved substances.

Vascular tissue. This lies in the center of the root. In young roots the xylem forms a central core of the root, and the phloem lies within the radial arms of this core. The branches that form the lateral roots grow from this region and force their way through the cortex to burst through the outer layer into the soil.

When a plant is blown sideways by the wind, its root experiences strain in a direction along its length. This is in contrast to the sideways strain on the stem in the same conditions. It is interesting to compare the central positioning of the vascular tissue of the root, which is ideally adapted to the kind of strain to which it is subjected, with its cylindrical distribution in the stem.

Growing point (Figs. 5.23 and 5.24). The cells close to the extreme tip of the growing root are dividing rapidly. Behind the root tip, the newly produced cells absorb water and develop vacuoles. This intake of water

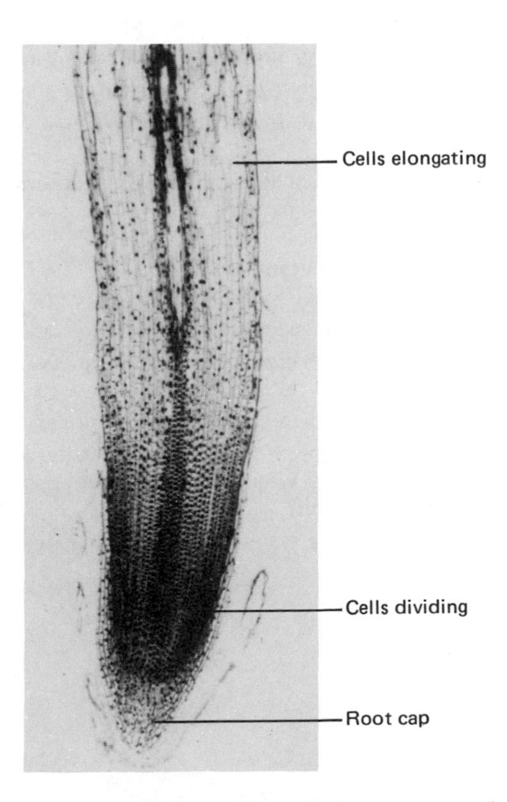

Fig. 5.24 *Longitudinal section of the root tip of an onion*
(GBI Laboratories Ltd.)

causes the cells to elongate, because their walls are still relatively soft and plastic; this part of the root is thus the region of extension. Since the upper part of the root is firmly anchored, the root tip is pushed downward between the soil particles. (The reasons for the characteristic downward growth of roots are discussed in Section 8.5.) The root tip is protected from the abrasion of the soil by the *root cap,* layers of cells which are continually produced in the dividing region and replaced from the inside as fast as the outer ones are worn away by the soil.

5.7 QUESTIONS

1. What are the main functions of (*a*) stems, (*b*) roots, (*c*) leaves?
2. What are the differences in (*a*) structure, (*b*) function, between vessels and sieve tubes?
3. How do roots, stems and leaves obtain supplies of oxygen for respiration?
4. What are the advantages of having a network of veins in a leaf?
5. How is lateral stress resisted by (*a*) a young stem, (*b*) an old stem?
6. In what ways does the broad, thin structure of most leaves adapt them to their functions?
7. From what unfavorable seasonal conditions can autumn leaf-fall protect a deciduous tree? Suggest reasons why most conifers can survive the winter without shedding their leaves.
8. Which plant organs are present in a bud?
9. Bud scales are considered to be modified leaves. In what ways do they differ from the ordinary foliage leaves?
10. When a bud sprouts, what change in form takes place in (*a*) the stem, (*b*) the leaves?
11. What could be the stimuli which cause the buds to open in the spring?
12. The distance between groups of terminal bud scale scars can be used to determine the age of a twig. What is the connection between these "girdle scars" and the seasonal growth?
13. By what means does a rooting system achieve a large absorbing surface?

UNIT 6
STORAGE ORGANS AND VEGETATIVE REPRODUCTION

6.1 INTRODUCTION

In Unit One, we pointed out that the maintenance of the existence of a species through reproduction is an essential characteristic of all living things. Two kinds of reproduction are found in flowering plants, although many species use one method only:

Sexual reproduction involves the flower in its capacity as a reproductive structure; this is discussed in Unit Seven.

Vegetative (or *asexual*) *reproduction* is a term used for all reproductive methods not involving flowers; it includes the artificial methods used by gardeners, like grafting and taking cuttings (see Section 6.8).

Annual plants. Annuals, like poppy or groundsel, have a growth period which lasts less than a year. After the plant has flowered and formed seeds, the shoot and the root die off completely and only the seeds remain alive; they remain inactive until the next growth season, when they begin to develop or *germinate*. In some plants the whole germination–reproduction cycle lasts only a few weeks. Most annual plants germinate and flower in the spring or summer and survive the winter as seeds only, but some *winter annuals* germinate in autumn and flower and die back in the following summer.

Biennial plants. Biennials, like the carrot, do not flower in the first season after germination, but during this time they lay down considerable reserves of food in specially enlarged storage organs in preparation for the following year's production of plants and seeds. After seed formation the plant dies.

Perennial plants. These plants do not die during the winter months but persist from one year to another, often for very many years. They may not flower for several years after germination. *Woody perennial plants* are trees and shrubs in which the trunk and branches remain and grow from year to year. *Herbaceous perennials* do not have woody stems. Some herbaceous perennials, like the iris or certain grasses, persist through the winter with some foliage above the ground but with very reduced growth. In other species, like the tulip or crocus, the parts of the plant growing above the ground wither and die before the winter and the plant spends many months in a dormant underground form such as a bulb or a corm. In either case, the structures which persist below ground usually contain a store of food, so that in the following spring rapid growth and flower formation can proceed without delay. When winter is over, the terminal and lateral buds of herbaceous perennials sprout and produce new shoots. The shoots formed from some of the lateral buds often develop their own adventitious

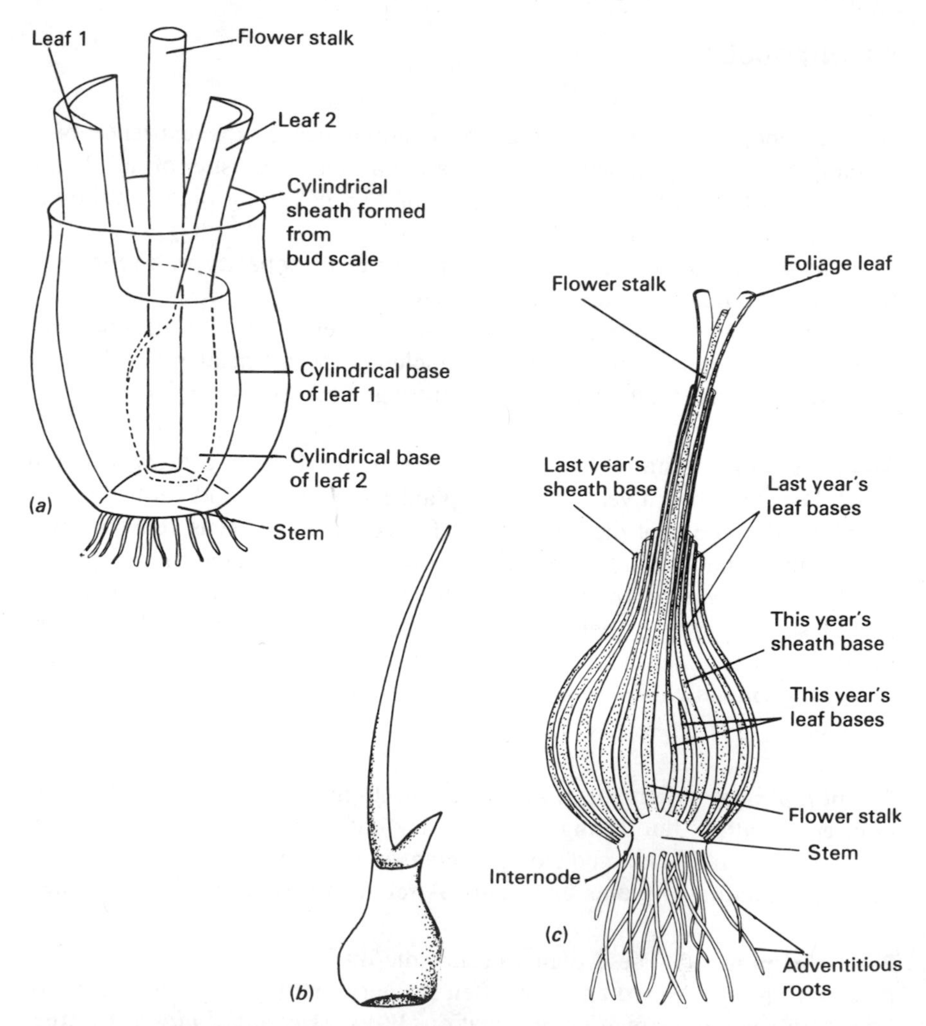

Fig. 6.1 *Snowdrop bulb: (a) stereogram showing structure; (b) detached leaf showing hollow circular base; (c) section through bulb*

roots and by the end of the season have become independent of the parent plant and other lateral shoots. Thus, new plants are produced from buds without pollination or fertilization being necessary: this kind of vegetative reproduction is illustrated by the life cycles described in the next four sections. However, it is important to remember that these plants also produce flowers and seeds, and thus reproduce themselves sexually as well.

6.2 BULBS (Figs. 6.1 and 6.2)

Bulbs are condensed shoots, or buds, with fleshy leaves. The stem is very short and never grows above ground. The internodes are short; the leaves are very close together and they overlap. The outer leaves are scaly and

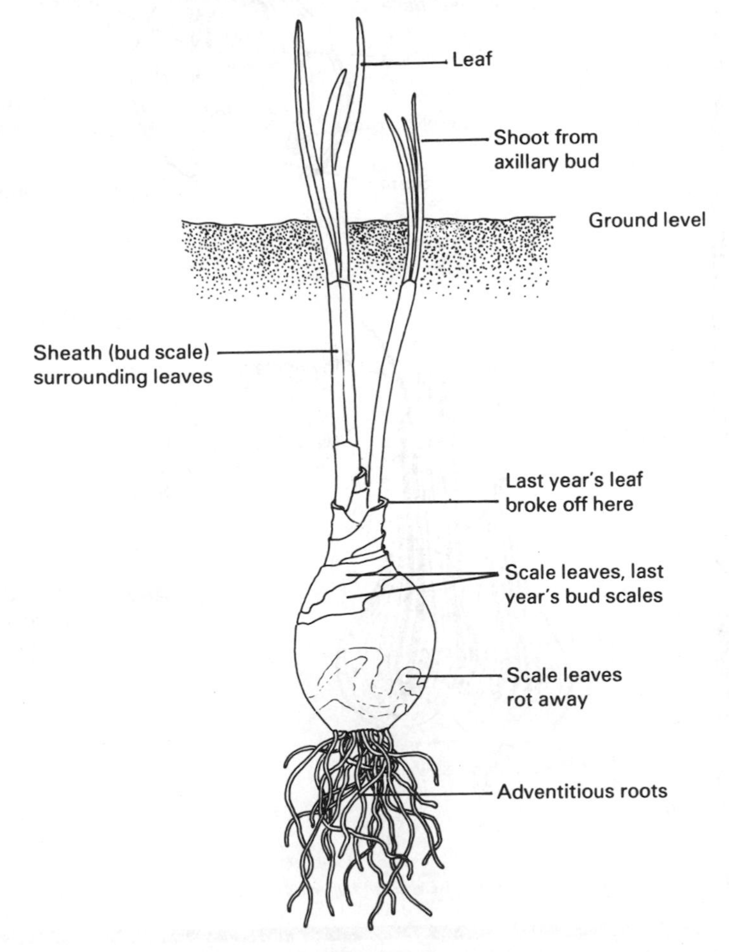

Fig. 6.2 *Daffodil bulb*

dry and protect the inner ones, which are thick and fleshy with stored food. In the snowdrop and daffodil, the bulb is formed by the bases of the leaves which completely encircle the stem, and these cylindrical leaf bases receive food from the rest of the leaf above ground. In the leaf axils of the bulb are lateral buds which can develop into new bulbs and shoots.

Life cycle (Fig. 6.3). In winter, the adventitious roots grow out of the stem and in early spring the terminal bud begins to grow above the ground, making use of the stored food in the fleshy leaf bases which consequently

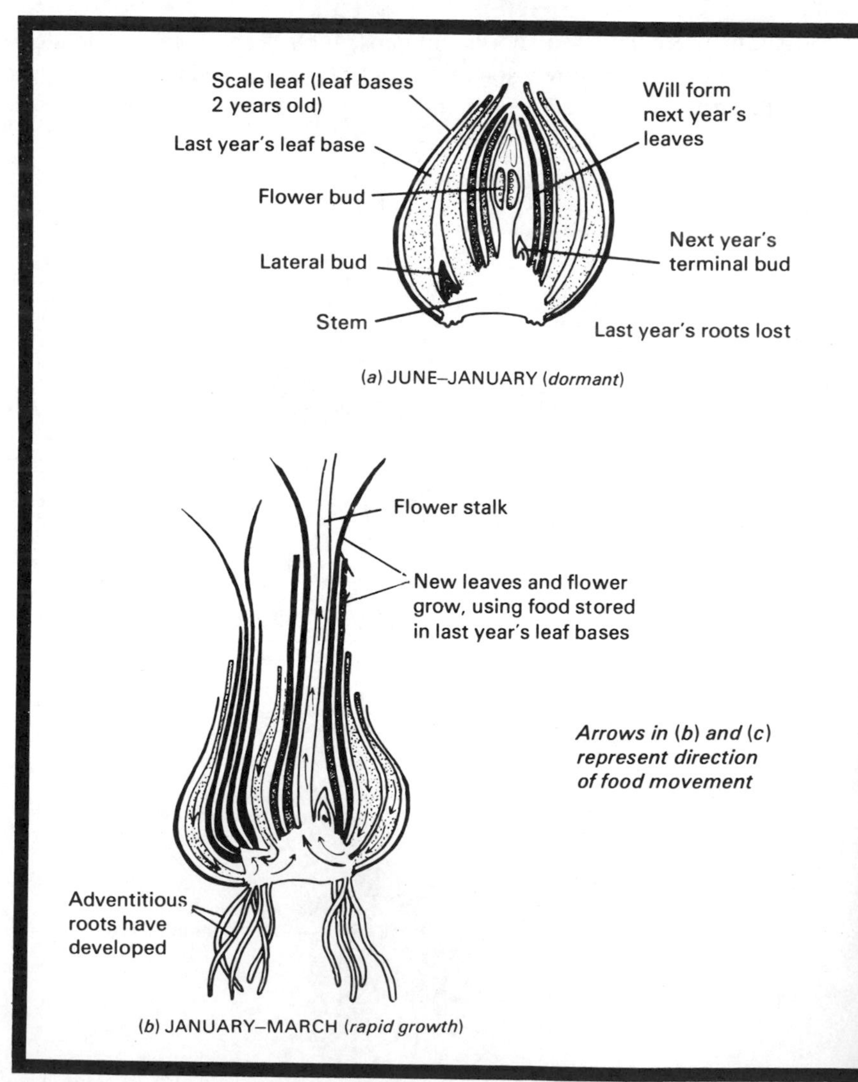

Fig. 6.3 *Annual cycle in a daffodil type of bulb*

shrivel. During the spring some of the food made in the green leaves of the plant is sent to the leaf bases, which swell and form a new bulb inside the old one, and also to one or more lateral buds which then grow to form daughter bulbs. The shriveled storage leaves of the old bulb become the dry scaly leaves around the daughter bulbs. The new bulb and the daughter bulbs produce independent plants next season. Thus bulb formation is a kind of vegetative reproduction.

The daughter bulbs are produced each year on the stem of the old bulb, and therefore are not quite so deeply buried in the soil as the

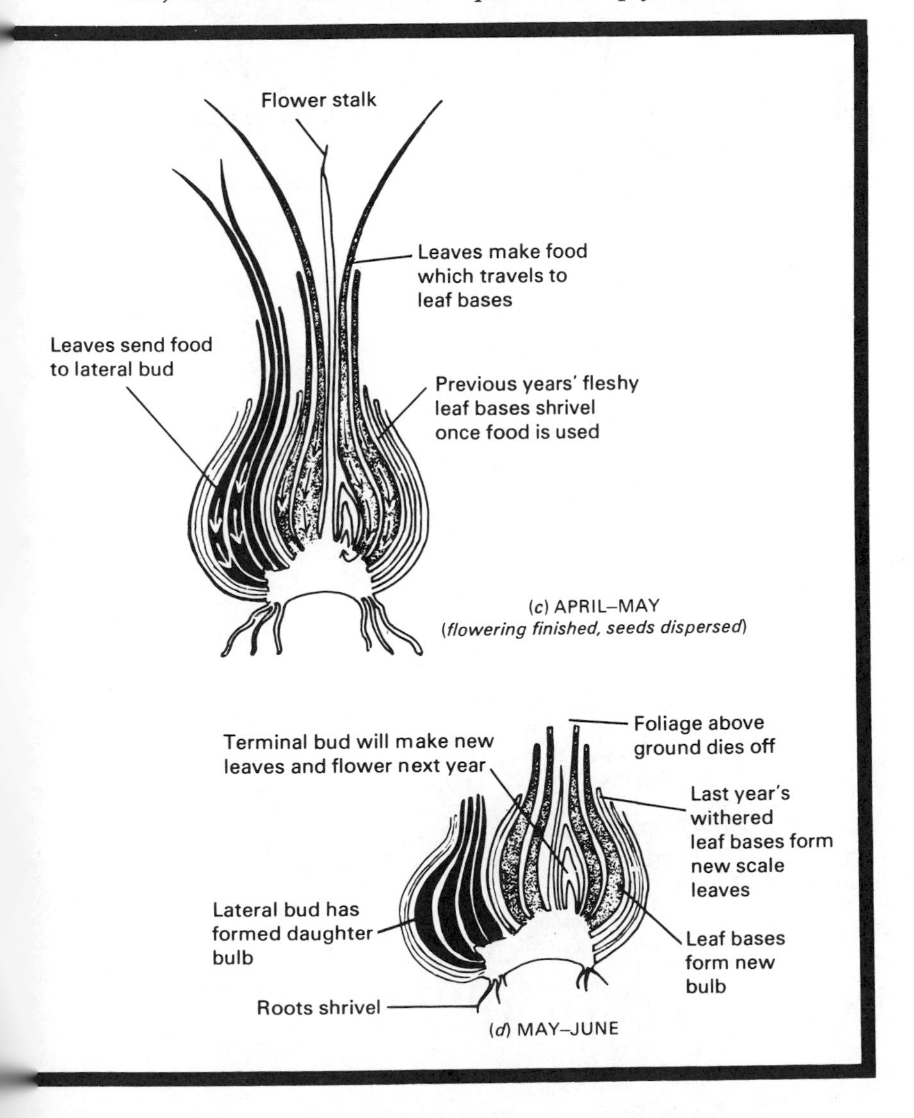

parent bulb. There is thus a tendency for successive generations to lie nearer and nearer to the surface of the soil. However, some of the adventitious roots at the base of the new bulb are specially adapted to counteract this tendency. Once these roots have grown into the soil and become firmly anchored, a region near their junction with the stem contracts so as to pull the bulb downward into the soil. These *contractile roots* can be seen to be wrinkled in the zone where contraction has taken place.

6.3 CORMS

In some plants, food is stored in the stem, rather than in the leaves or leaf bases. When the stem is a short swollen structure, it is called a *corm.* It has a protective scaly covering formed by the leaf bases remaining after the foliage has withered. A familiar corm is that of the crocus (Fig. 6.4). Since

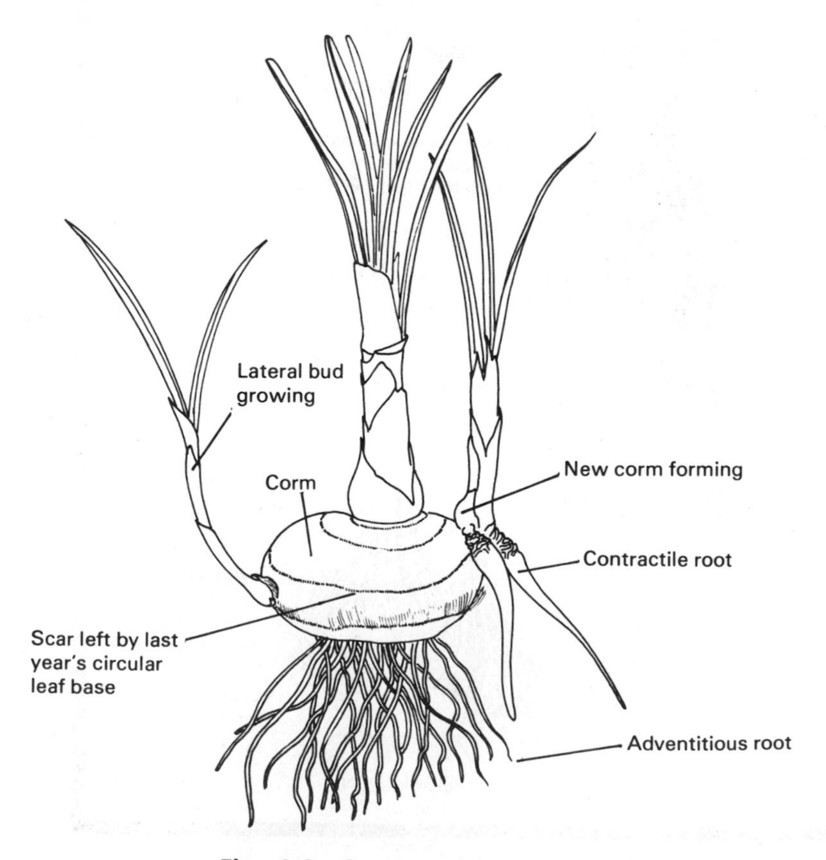

Fig. 6.4 *Crocus corm in spring*

the corm is a stem, it carries lateral buds which can grow into new plants. The stem itself always remains below the ground; only the leaves and flower stalk can be seen above ground level.

Life cycle. In spring, the food stored in the corm enables the terminal bud to grow rapidly and produce leaves and flowers abovc ground. Later in the year, food made by the leaves is sent back, not to the old corm, but to the base of the stem immediately above it. This region swells and forms a new corm on top of the old, now shriveled, corm. The formation of one corm on top of another year by year tends to bring the successive corms nearer and nearer to the soil surface; however, contractile roots like those of bulbs pull the corm downward and keep it at a constant level in the soil.

Some of the food made in the leaves is used by the lateral buds on the old corm; these grow into new corms which product leaves the following spring.

6.4 RHIZOMES (Figs. 6.5 and 6.6)

In certain plants which store their food reserves in underground stems, the old part of the stem does not die away each year as in bulbs and corms, but lasts for several years, continuing its growth horizontally. This kind of stem is called a *rhizome*. In the iris, the terminal bud of the rhizome turns upward and produces leaves and flowers above ground. The old leaf bases form circular scales around the rhizome, which is swollen with food reserves.

Life cycle (Fig. 6.5). The annual cycle of a rhizome is similar to that of a corm. In summer, food from the leaves passes back to the rhizome, and a lateral bud uses it, grows horizontally underground, and so continues the rhizome. Other lateral buds produce new rhizomes which branch from the parent stem. The terminal buds of these branches curve upward and produce new leafy shoots and flowers. Adventitious roots grow from the nodes of the underground stem.

6.5 STEM TUBERS (Fig. 6.7)

In the young potato plant, the whole of the stem grows above the ground. As the plant develops, lateral buds at the base of the stem produce shoots which grow horizontally at first and then turn downward and grow into the soil. These resemble rhizomes, in that they are underground stems with tiny scale leaves and lateral buds; unlike rhizomes, however, they do not swell evenly along their length with stored food.

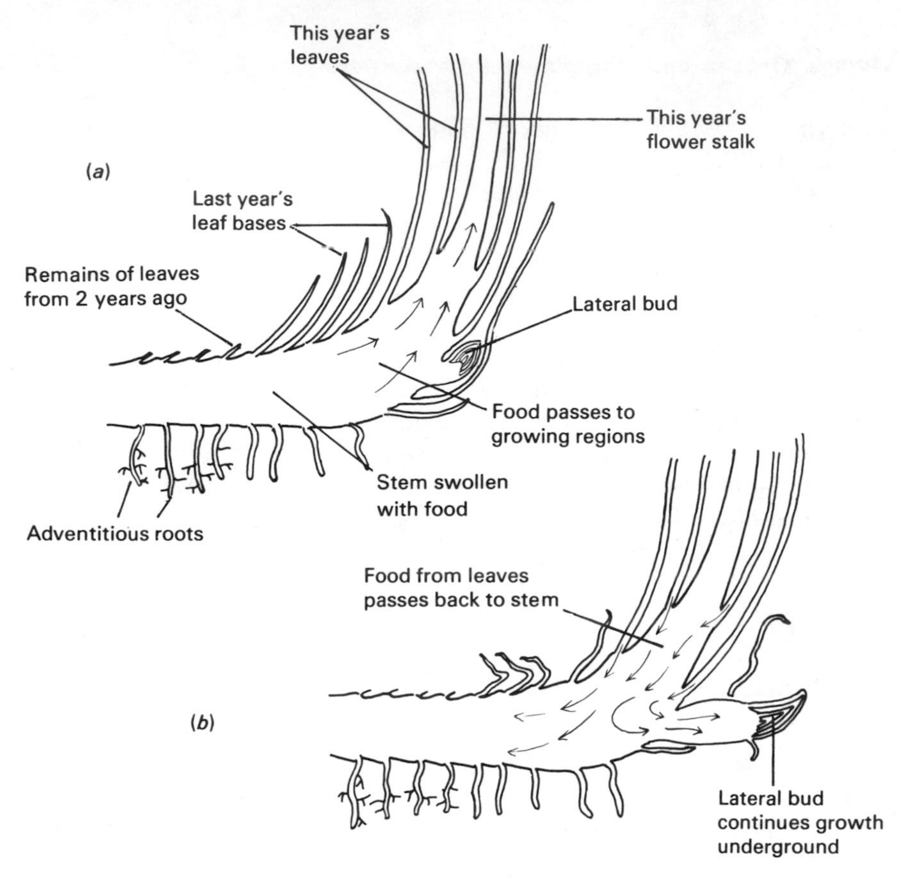

Fig. 6.5 *Diagram showing rhizome growth in iris: (a) the rhizome in spring; (b) the rhizome in summer*

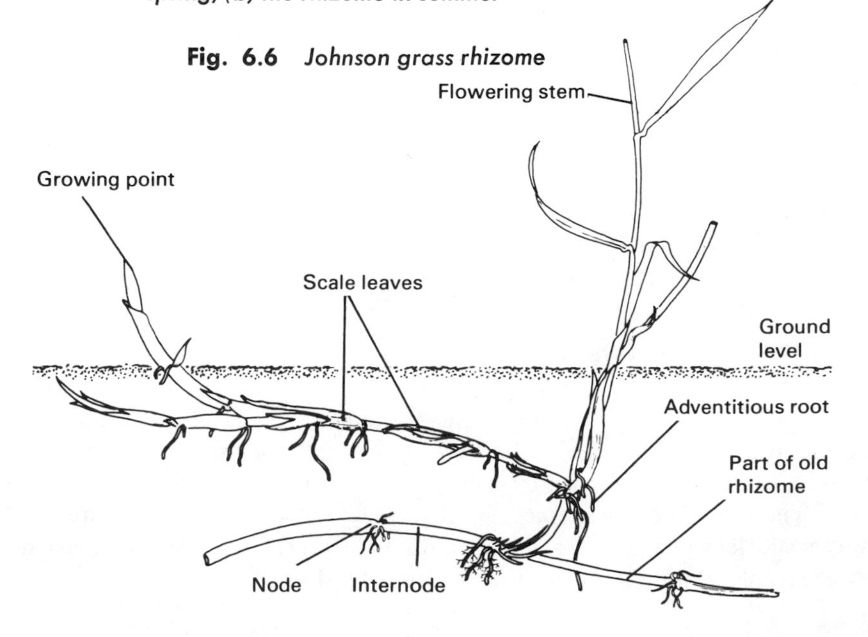

Fig. 6.6 *Johnson grass rhizome*

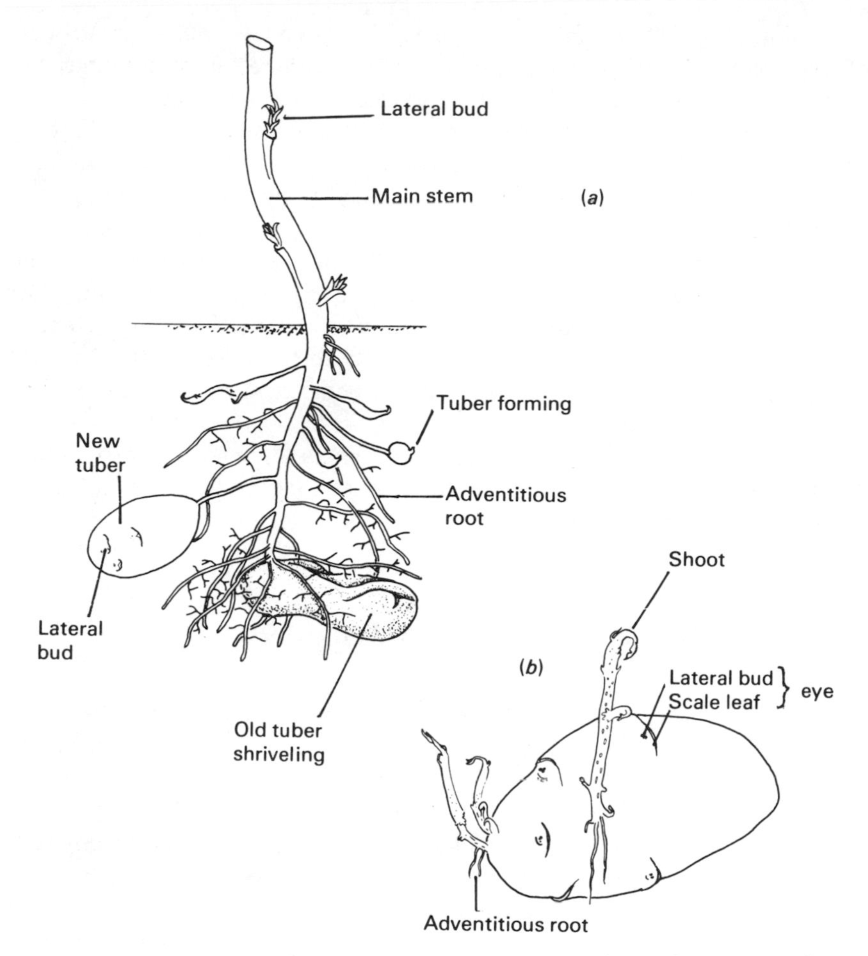

Fig. 6.7 *(a) Stem tubers growing on a potato plant; (b) potato tuber sprouting*

Life cycle. During the summer, food made in the leaves passes to the ends of these underground stems, which swell to form the tubers we call potatoes. Since the potato tuber is a stem, it carries leaves and axillary buds: these are the familiar "eyes." Each of these buds can produce a new shoot in the following spring, using the food supplies stored in the tuber. The old tubers shrivel and rot away at the end of the season.

6.6 ADVANTAGES OF FOOD STORAGE

The ready availability of food in the storage organs means that very rapid growth can be made in the spring: the production of the flower need not wait for a full growth of photosynthesizing leaves to develop. A great many of the familiar flowers of spring and early summer have underground storage organs of some kind: for example, bulbs are produced by the daffodil, snowdrop and bluebell, corms by the crocus and cuckoo pint, rhizomes by the iris and lily of the valley and tubers by the lesser celandine (although these are root tubers, unlike the stem tubers of the potato described in Section 6.5).

This property of early growth means that the plant can flower and produce seeds before other plants begin seriously to compete with it for water, mineral salts and light. This competition is particularly important in woods, where in summer the leaf canopy absorbs much of the sunlight before it can reach the ground and the tree roots tend to drain the soil of moisture over a great area.

Since the beginnings of agriculture, man has exploited many food-storing plants, and bred them so as to produce bigger and more nutritious storage organs for his own consumption.

6.7 ADVANTAGES OF VEGETATIVE REPRODUCTION

Since food stores are available throughout the year and the root system of the parent plant can absorb water from quite a wide area, the hazards of food shortage and drought which beset germinating seeds are substantially reduced for daughter plants produced by vegetative reproduction. These daughter plants grow from buds which are always produced in an environment where the parent plant can flourish; on the other hand, many seeds dispersed from plants die because they never reach a suitable situation for effective germination. Vegetative reproduction does not usually result in rapid and widespread distribution of offspring in the same way as seed dispersal, but tends to produce a dense clump of plants with little room for competitors between them. Such groups of plants are very persistent and, because of their underground food stores and buds, can still grow after their foliage has been destroyed by insects, fire or man's cultivation. Those of them in the "weed" category are difficult to eradicate mechanically, since even a small piece of rhizome bearing a bud can give rise to a new colony. However, selective weedkillers reduce this difficulty.

6.8 ARTIFICIAL PROPAGATION

Gardeners use vegetative propagation when they divide up the rhizomes, tubers or rootstocks at the end of the flowering season. Each part so divided will grow in the following year to make a separate plant. Two other methods of artificial vegetative propagation are also commonly used:

(a) **Cuttings.** It is possible to produce new individuals from certain plants by putting the cut end of a shoot into water or moist earth. Adventitious roots grow from the base of the stem into the soil while the shoot continues to grow and produce leaves.

In practice, the cut end of the stem is usually treated with a rooting hormone (see Section 8.5) to promote root growth; evaporation from the shoot is reduced by covering it with polythene or a glass jar. Carnations, geraniums and chrysanthemums are commonly propagated from cuttings like this.

(b) **Grafting.** A bud or shoot from one plant is inserted under the bark on the stem of another, closely related, variety so that the cambium layers of both are in contact. The rooted portion is called the *stock* and the bud or shoot being grafted is the *scion* (Fig. 6.8).

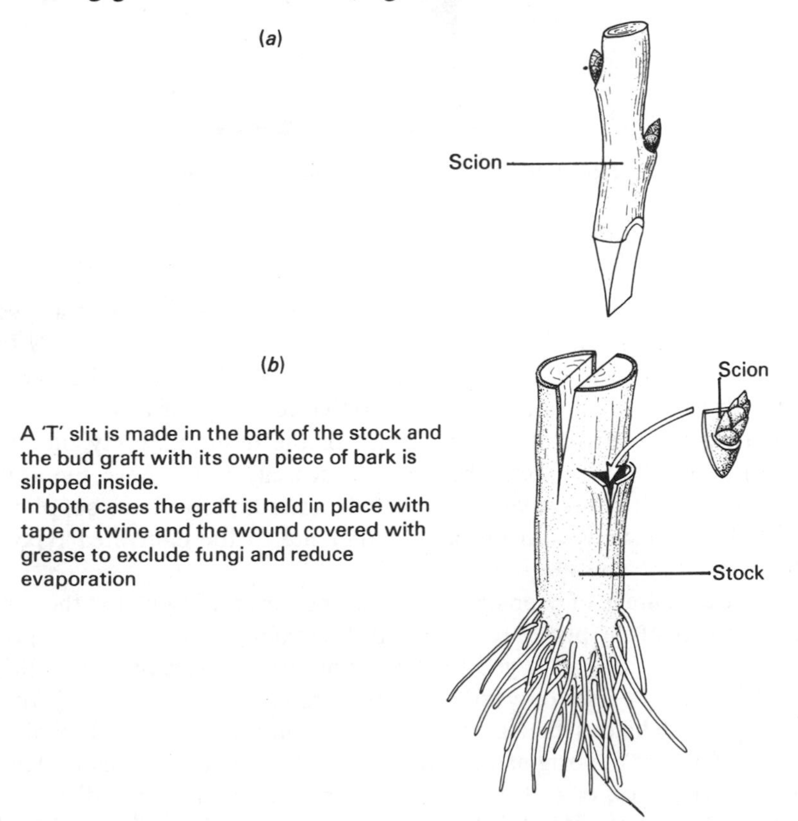

Fig. 6.8 *Two types of grafting: (a) cleft graft; (b) bud graft*

Fig. 6.9 *Cleft graft of pear, six weeks after grafting*
(From Span 15, Shell International)

If a nurseryman wants to produce large numbers of plants of a specific variety of, say, a scented red rose, he cannot do this by sowing seeds from the flowers of this variety: most of the plants so produced would carry only "wild" type roses. Instead, a plant of a cultivated variety is usually obtained by grafting a bud from the parent plant of the desired variety on to the stem of a "wild" rose plant. The stock is then cut away above the grafted bud, and the bud grows, making use of the water and nutrients supplied by the stock. When flowers are produced they are of the cultivated type and color.

The advantage of propagation by cuttings or grafting is that the inbred characteristics of the plant are preserved. For example, an apple tree grown from a seed ("pip") of a good eating apple like a Washington Delicious would probably yield only small sour "crab-apples," but fruit from a Washington Delicious bud grafted on a wild apple tree would retain the size and flavor of the original. However, a plant breeder anxious to develop new characteristics, or to bring together a new combination of them, must propagate many thousands of plants from seed, selecting and breeding from any individual showing the desired feature to some degree—a laborious and uncertain procedure.

6.9 QUESTIONS

1. Which plants organs are modified for storage of food in (*a*) a potato, (*b*) an onion?
2. Plants can often be propagated from stems but rarely from roots. What features of shoots account for this difference?
3. In a bulb-forming plant, what is the principal source of food (*a*) early in the growing season, (*b*) late in the growing season?
4. In bulb-forming plans such as daffodils, you usually see a flower stalk but not a stem (i.e., the leaves seem to emerge directly from the ground). Why is this?
5. Why do bulb-forming plants like daffodils frequently form dense clumps?
6. Annual plants classed as "weeds" (such as groundsel) can be controlled by hoeing them before they flower. Perennial weeds like Johnson grass and Bermuda grass cannot be controlled in this way. What biological principles underlie this difference?

UNIT 7
SEXUAL REPRODUCTION IN FLOWERING PLANTS

7.1 INTRODUCTION

Unit Six discussed the different ways in which plants reproduce themselves by the development of a bud into a daughter plant. But most flowering plants also reproduce by bearing seeds which can grow into new plants; indeed, the primary function of flowers is to act as organs for the production of the reproductive cells or *gametes* which give rise to the seeds.

In both plants and animals, the fusion of two gametes (one male and one female) gives rise to a single new cell or *zygote* from which a new individual can grow; the fusion of two gametes is characteristic of *sexual reproduction* (for a description of sexual reproduction in man see Unit Twenty-five).

A male gamete is usually small, with a nucleus and little cytoplasm; it is normally able to leave the male organ which has produced it and to move about, either by its own power or by the use of external agencies like wind or insects. The male gamete in a flowering plant is a cell in the pollen grain.

The female gamete is generally larger than the male, with a nucleus and more cytoplasm. Sometimes it contains food reserves to supply the new individual during the early stages of its growth. In many species of plants and animals it never leaves the female organ or the body of the female individual where it was formed until after it has fused with the male gamete: even when it does so it is usually immobile. In a flowering plant the female gamete is a cell in the ovule (see Section 7.2), the part of the flower which eventually develops into the seed.

The fusion of the nuclei of the male and female gametes is called *fertilization*. The zygote thus formed undergoes cell division and growth, developing into a new and ultimately independent organism. In plants, fertilization can only take place after pollen has been transferred from the male organ which produces it to the female organ which contains the ovules, a process known as *pollination,* and discussed in Section 7.3.

7.2 FLOWER STRUCTURE

(a) The Parts of a Flower

A flower is a reproductive structure of a plant. Many flowers have both male and female reproductive organs, although some plants carry two distinct kinds of flower, one containing the male organs and one the female. The floral parts are borne in a series of rings (*whorls*) or spirals with short internodes, often at the end of a flower stalk which may be expanded to form a *receptacle*. Sometimes, as in the hyacinth and lilac, the same main stalk carries a group of flowers, called an *inflorescence*.

The calyx. This outermost whorl of the flower consists of *sepals*, which are usually green and small. They enclose and protect the rest of the flower while it is in the bud.

The corolla. Lying within the calyx, the corolla consists of a whorl of *petals*, which are often colored and scented. They attract insects which visit the flowers and collect nectar and pollen, pollinating the flowers as they do so.

The androecium. This male part of the flower, within the corolla, consists of *stamens*. The stalk of the stamen is the *filament*. At the end of the filament is an *anther* which contains pollen grains in four *pollen sacs* (Figs. 7.10 and 7.11). The pollen grains contain the male gametes.

The gynaecium. This female part of the flower consists of *carpels:* some flowers contain only one carpel, some a few and some very many. In some flowers the carpels are separate from each other and in others they are fused together. Each carpel is made up of three parts: an *ovary*, enclosing *ovules* which contain the female gametes, a *style* extending from the ovary, which at its end expands or divides to form a structure called a *stigma*, which receives the pollen when the flower is pollinated. After fertilization the ovules become seeds and the whole ovary the *fruit* of the plant; the ovary wall develops into the *pericarp* or wall of the fruit.

Nectaries. These are glandular swellings, often at the base of the ovary or on the receptacle, which produce a sugary solution called nectar. It is thought that the "honey guides," small grooves or darker lines on the petals of some flowers, serve to direct insects to the nectaries within the flower, and thus encourage pollination.

(b) Number of Parts

In many species of flowering plant, the structures described above occur in definite numbers. For example, if there are five sepals there are likely to be five petals and five or ten stamens.

Whorls may be repeated; for example, there may be two whorls of five petals or two whorls of five stamens. In the flowers of the buttercup and rose families there are numerous stamens and carpels, and the numbers vary from one plant to another. The floral parts usually alternate so that petals do not lie opposite sepals but between them, stamens are borne between petals and so on.

(c) Variations

In a flower like the buttercup the petals are separate from each other, and are all the same size. In many flowers, however, the petals and sepals are joined for part or all of the way along their length, forming tubular structures; the flowers of the foxglove and deadnettle are familiar examples. In other flowers some petals differ in size and shape from the others, as in the violet or the flowers of the pea family like the lupine.

(d) The Half Flower

A drawing of a half flower is a convenient method of representing flower structure. The flower is cut in halves with a razor blade, the outline of the cut surfaces drawn and the structures visible behind these filled in. (A drawing of a longitudinal section of the flower would show only the cut surfaces.)

The following three paragraphs describe three flowers which differ considerably in form. If possible, they should be read with the fresh flowers in hand for comparison.

Buttercup (Figs. 7.1 and 7.2). Five sepals, not joined, curled back in some species but not in others; five petals, with nectaries at their base, not joined up; about 60 stamens, 30 or 40 carpels separate from each other, each with a short style and containing a single ovule.

White deadnettle (Fig. 7.3). Five spiky sepals, joined at the base to make a cup, surrounding a tube made up of five fused petals of which the uppermost is the largest; only four stamens, with their black anthers close together under the top petal and their filaments joined to the petal tube; a forked stigma, with a long style joined to the petals for part of its length and leading to a four-lobed ovary.

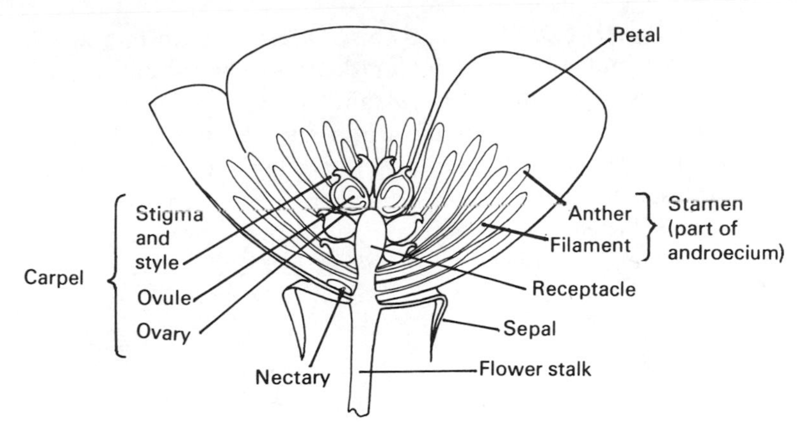

Fig. 7.1 *Half flower of buttercup*

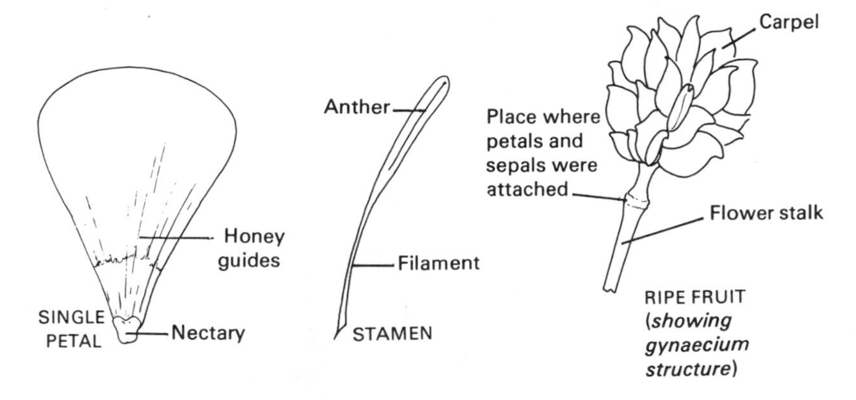

Fig. 7.2 *Floral parts of buttercup*

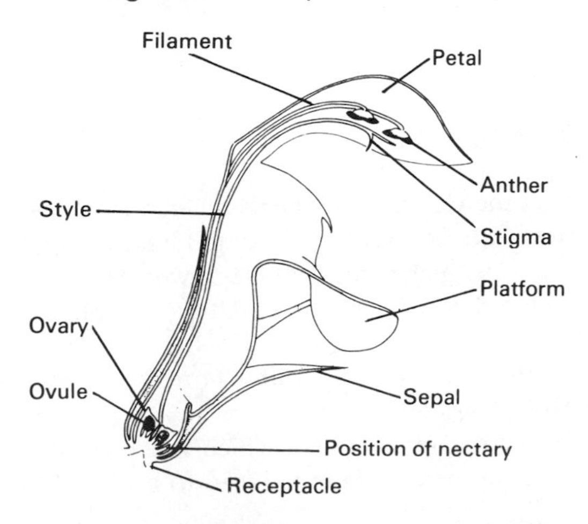

Fig. 7.3 *Half flower of white deadnettle*

Lupine (Figs. 7.4 and 7.5). Five sepals fused together to form a two-lobed calyx; five petals, not all joined but of different shapes and sizes. The uppermost petal is called the *standard,* and the two partly joined petals at the side are the *wings.* Within the wings are two partly joined petals forming a boat-shaped *keel.* Inside the keel are ten stamens, the filaments of which are fused to form a sheath around the ovary. The ovary is long, narrow and pod-shaped, and consists of one carpel containing about ten ovules. The style ends in a stigma, which lies just within the pointed end of the keel.

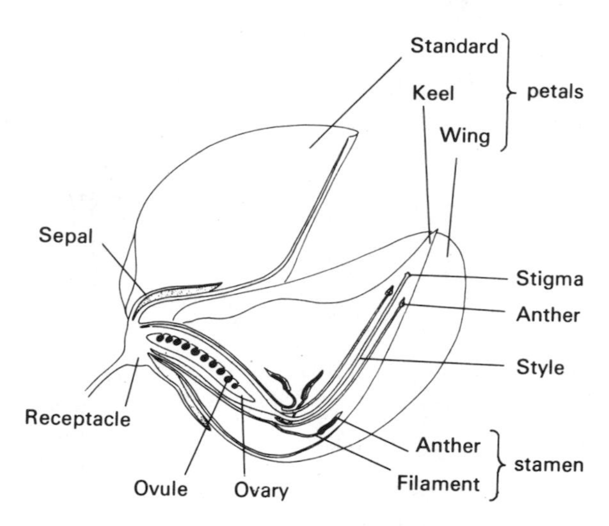

Fig. 7.4 *Half flower of lupine*

(e) Composite Flowers

The flowers of the Compositae family, which includes daisies, dandelions and hawkweeds, are arranged in dense inflorescences (Fig. 7.6) which could be mistaken at first glance for a single flower. However, what appears to be a petal in these flower heads is actually a complete flower, called a *floret* (Fig. 7.7).

In flowers of the daisy type, the outer florets have conspicuous petals but no reproductive organs and are therefore *sterile* (bear no seeds); seeds are only produced by the inner florets which make up the center part of the flower and in which the corolla is reduced to a ring of tiny petals fused together.

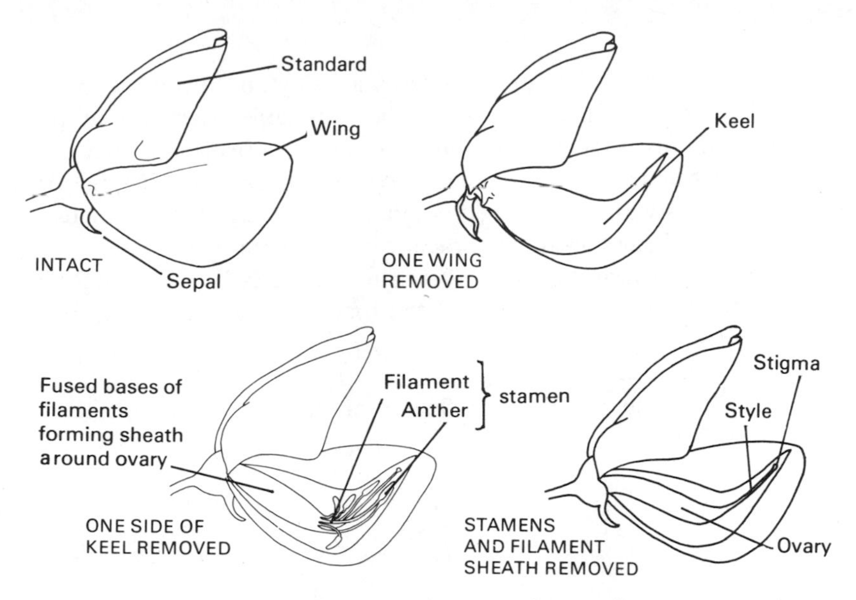

Fig. 7.5 *Lupin flower dissected*

Dandelion. All the florets of the dandelion inflorescence have both male and female reproductive organs. The florets (Fig. 7.7) each have five petals fused together into a broad strap-shaped structure which is joined at the base to form a tube. The sepals are reduced to a whorl of fine hairs called a *pappus,* and the five fused anthers are grouped around the style, which carries a forked stigma. The ovary contains a single ovule.

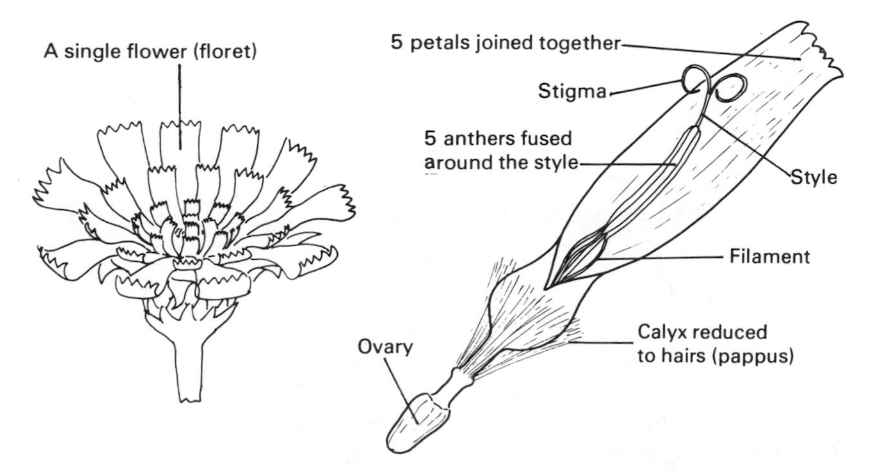

Fig. 7.6 *A hawkweed*

Fig. 7.7 *Single floret of a dandelion*

(f) Grasses

The individual flowers of the grasses are tiny and inconspicuous, but grow in dense inflorescences, which may be quite large. There are no petals or sepals in the usual sense, but the reproductive organs are enclosed in two green, leaf-like structures called *bracts*. The ovary contains one ovule and carries two styles with feathery stigmas. There are three stamens, the anthers of which hang outside the bracts when they are ripe.

Cereals like wheat, oats, barley and maize are grasses especially bred and cultivated by man for the sake of the food stored in the fruits or seeds of their flowers.

Ryegrass. The inflorescence and flower of this grass are illustrated in Figs. 7.8 and 7.9.

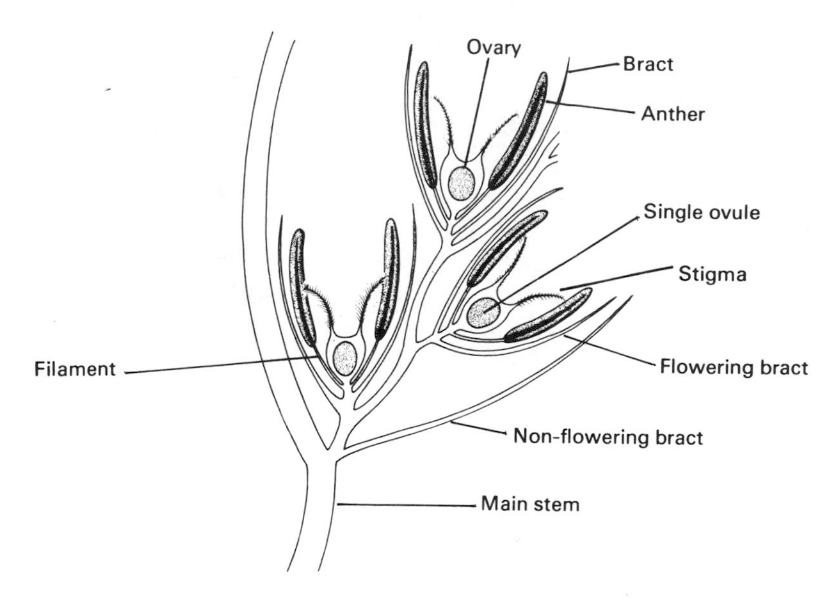

Fig. 7.8 *Expanded diagram of part of a ryegrass spikelet*

7.3 POLLINATION

When the anther is ripe, the pollen sacs split open and expose the pollen grains, which can then be dislodged. The transfer of pollen from the anthers to the stigma is called *pollination*. When the anthers and stigma are parts of the same flower, or of different flowers on the same plant, the process is called *self-pollination; cross-pollination* takes place when pollen from the flowers of one plant is transferred to the stigma of a flower on a different

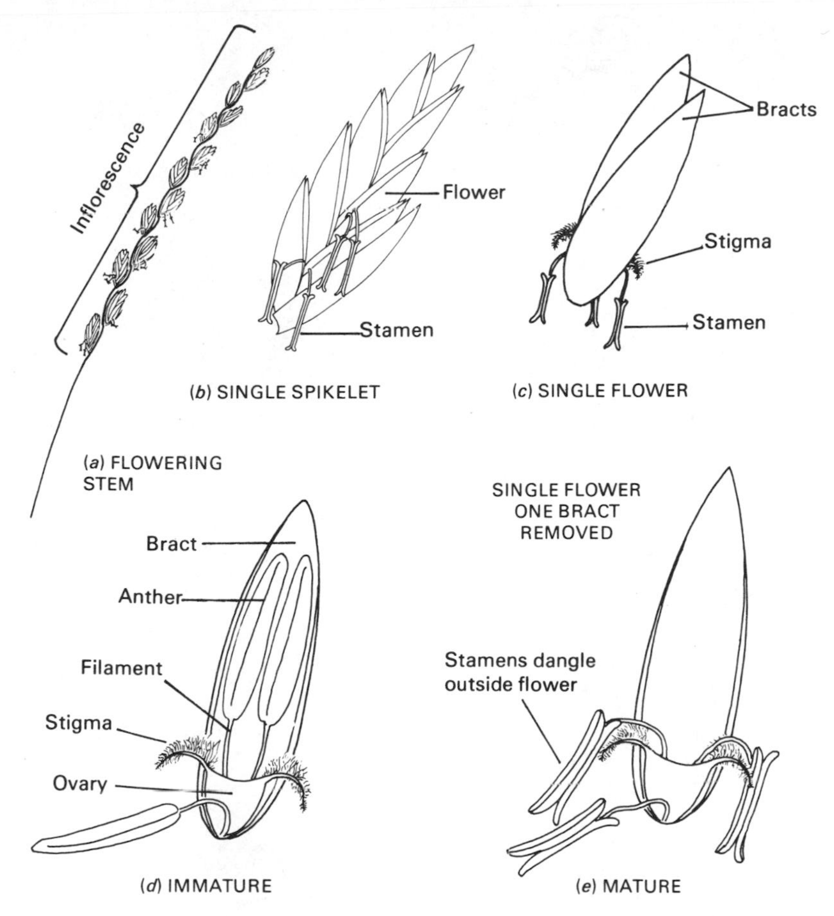

Fig. 7.9 *Ryegrass flower structure*

plant. Cross-pollination increases the possibility of combining the advantageous characteristics of two parents into the daughter plants. The flowers of many species have forms which favor cross-pollination, although in other species self-pollination occurs regularly; in some flowers, self-pollination only takes place when there has been no cross-pollination.

Pollen is usually carried to the stigma of a cross-pollinated flower either on the bodies of insects entering the flower in search of pollen or nectar, or by chance air currents. In either case, pollen from a certain species may reach the stigma of a flower of a different species, but the chemicals present in the stigma cells usually prevent any further development of the "foreign" pollen grains.

Flowers are often specially adapted either to insect or to wind pollination, and Table 7.1 lists the main differences between these two kinds of flower.

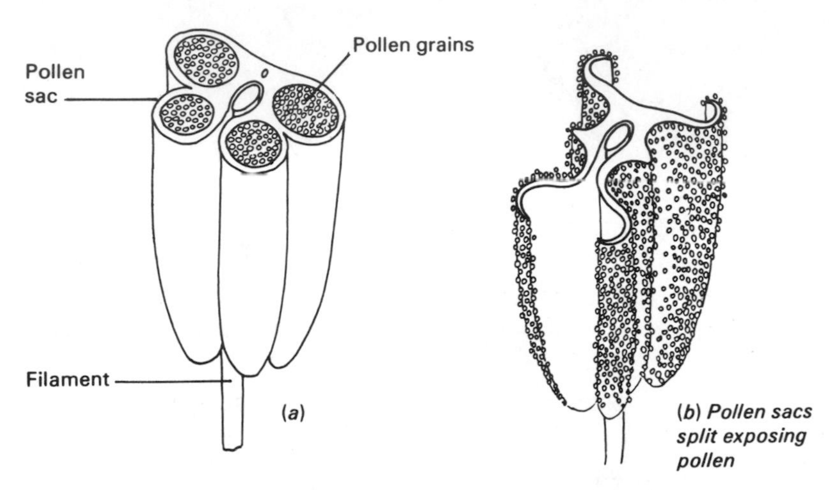

Fig. 7.10 *Structure of anther (top cut off)*

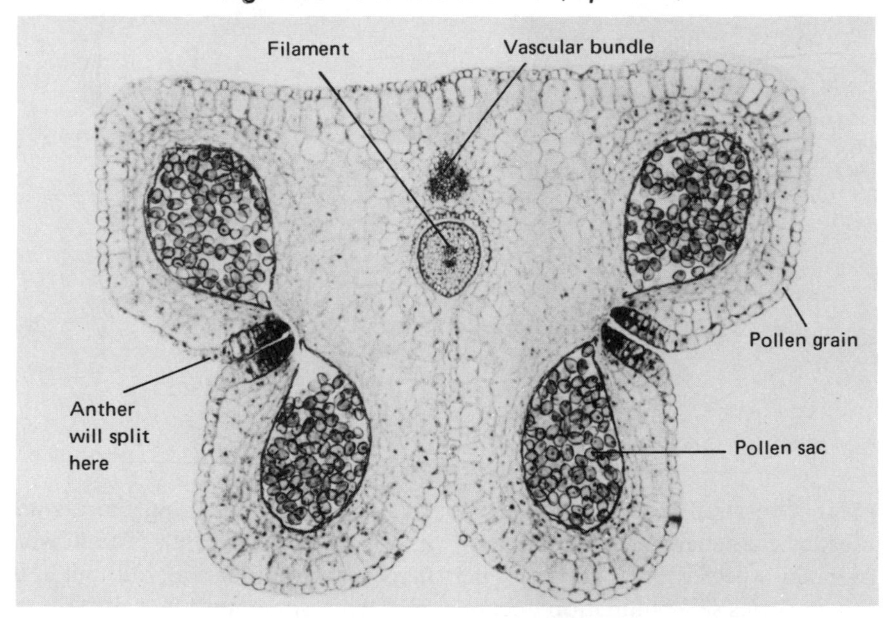

Fig. 7.11 *Transverse section through an anther*
(Gene Cox)

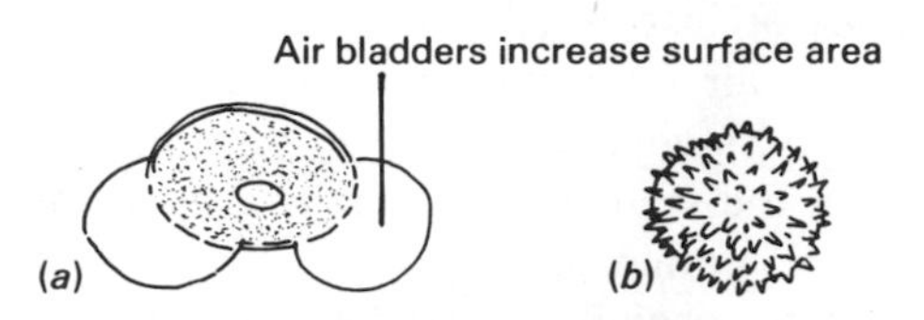

Fig. 7.12 *(a) Pollen of pine (wind-pollinated); (b) pollen of hollyhock (insect-pollinated)*

Table 7.1 *Comparison of pollination methods*

Feature of Plant	*Wind Pollination*	*Insect Pollination*
Flowers	Small, inconspicuous flowers; petals often green; no scent or nectar. There is no need for any mechanism to attract insects.	Relatively large and conspicuous flowers or inflorescences; petals brightly colored and scented; mostly with nectaries, which lie more deeply in the flower than the reproductive organs. Insects are attracted to the flower by the stimuli of color and scent; they enter the flower to collect or eat the nectar or pollen.
Anthers	Anthers are large and loosely attached to the filament so that the slightest air movement shakes them and dislodges the pollen; the stamens hang out of the flower exposed to the wind (Fig. 7.9(*c*) and (*e*)) and the whole inflorescence often dangles loosely (Fig. 7.15).	Anthers are not so large, and are firmly attached to the filament; they are usually carried within the petals of the flower where visiting insects are likely to brush against them.
Amount of pollen	Large quantities of pollen produced; only a very small proportion of wind-borne pollen is likely to reach a ripe stigma of the appropriate species.	Small quantities of pollen produced; the proportion of wasted pollen is much less.
Type of pollen	Smooth, light pollen grains (Fig. 7.12(*a*)) which will not stick together and so are readily carried in air currents.	Sticky or rough-surfaced pollen grains (Fig. 7.12(*b*)), which adhere well to the body of an insect.
Stigmas	Large feathery stigmas hanging outside the flower (Fig. 7.9(*c*) and (*d*)) forming a net to trap wind-borne pollen.	Flat, or lobed, sticky stigmas inside the flower.

(a) Mechanisms of Insect Pollination

These mechanisms always involve an insect such as a bee visiting a flower, becoming dusted with pollen from the ripe stamens and then visiting another flower, or a different part of the same flower, where some of the pollen on the insect's body adheres to the stigma.

The pollen grain can then only develop if the stigma is ripe and fully formed, so that the chemicals needed for its growth are all present. In many species the anthers and stigma of the flower do not ripen at the same time, so that self-pollination is unlikely and cross-pollination is favored.

(i) **Buttercup** (Fig. 7.1). Various insects, such as bees, beetles and ants, enter the flower and wander around it in search of the nectar in the nectaries at the base of the petals. Their movements bring their bodies into contact with the stamens, so that they carry the pollen away from the anthers. In due course they may pass close to a stigma of the same or of another flower, so that either self- or cross-pollination can take place.

(ii) **White deadnettle** (Fig. 7.3). These flowers are so adapted that the nectaries can only be reached by long-tongued insects like bees. When a bee alights on the lower petals and pushes its head into the corolla tube, its back comes into contact with the anthers or stigma lying just under the top petal (Fig. 7.13). In some flowers of the same family, the anthers ripen

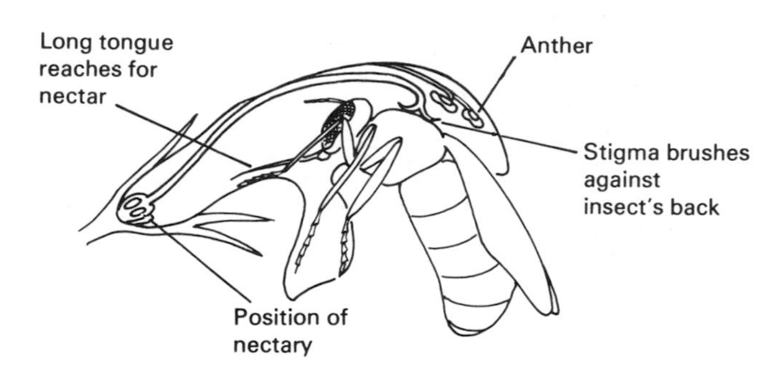

Fig. 7.13 *Pollination of white deadnettle*

before the stigma; the anthers hang down below the unripe stigma, the forks of which remain closed. When the anthers have shed their pollen, the stigma forks open and bend downward so that they, rather than the anthers, touch the back of the insect; most of the flowers are therefore cross-pollinated.

(iii) **Lupin** (Figs. 7.4 and 7.5). The lupin has no nectar, unlike related plants such as clovers; bees visit lupine flowers only to collect pollen.

The bee alights on the "wing" petals of the flower, which are depressed by its weight (Fig. 7.14). Since the wings are linked to the keel

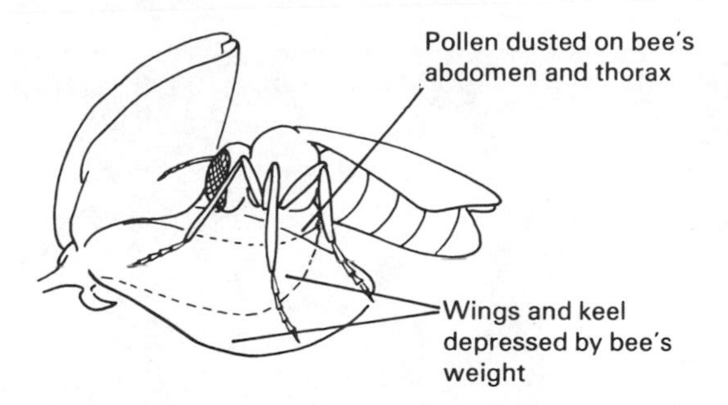

Fig. 7.14 *Pollination of lupin or sweet pea*

petals near their base, these two are forced downward; the stamens and stigma then protrude through the opening between the partly joined keel petals, and touch the underside of the insect. In a young lupin flower, pollen collects in the pointed tip of the keel and is pushed out rather like toothpaste from a tube, so that much of it can adhere to the insect's body. When a bee alights on an older flower which has shed all its pollen, the style and stigma alone protrude from the keel, and thus collect the pollen already carried by the insect.

(b) Wind Pollination

The feathery stigmas of grasses protrude from the flower in such a way as to trap pollen grains floating in the air. Later, the anthers ripen and hang outside the flower, the pollen sacs split and the wind can easily blow the pollen away. The exact sequence varies in different species.

Shaking a hazel branch bearing ripe male catkins, or a flower-head of the ornamental pampas grass, produces a shower of pollen which can easily be seen (Fig. 7.15).

(c) Pollination and Agriculture

After the ovules of the flower have been fertilized, the ovary becomes a fruit: without fertilization, which is always preceded by pollination, no fruit can form. A good yield of fruit—whether from a cherry tree or a field of wheat—can therefore only be obtained if most of the flowers are pollinated.

Most cereals are either self-pollinated or wind-pollinated; planting them very closely improves the effectiveness of the latter. On the other hand, a field of clover or an orchard of apples—both depending on insect pollination—needs plenty of insects, particularly bees, to pollinate the flowers.

Fig. 7.15 *Hazel catkins shedding pollen*
(Stephen Dalton, NHPA)

7.4 FERTILIZATION

The interval in time between pollination and fertilization—the actual fusion of the gametes—varies in different species from as little as sixteen hours or so to twelve months. The pollen grain absorbs nutriment secreted by the stigma, and the cytoplasm of the grain grows out as a *pollen tube*. This tube grows downward between the cells of the stigma and style toward the ovary (Fig. 7.16), obtaining the material needed for its growth from the tissues through which it passes. It grows toward one of the ovules in the ovary and enters it through a minute hole in its outer coat, the *micropyle* (Fig. 7.17). The tip of the pollen tube breaks open inside the ovule, and the male nucleus, which has been moving down the tube, enters the ovule and fuses with the female nucleus in its egg-cell. The egg-cells in an ovary are each fertilized by a male nucleus from a separate pollen grain.

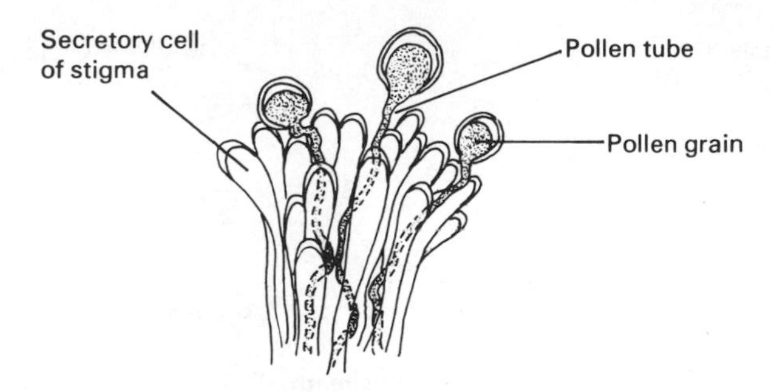

Fig. 7.16 *Pollen grains growing on stigma of crocus*

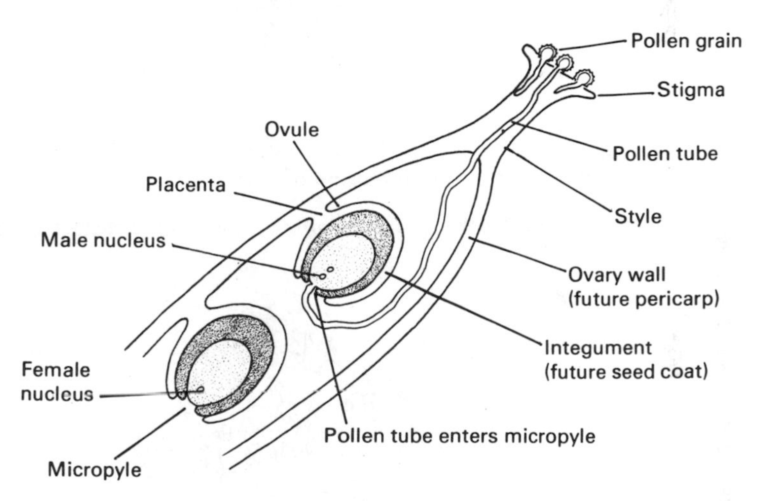

Fig. 7.17 *Diagram of fertilization*

7.5 FRUIT FORMATION

After fertilization the petals, stamens, style and stigma wither and usually fall away. The sepals may persist in a dried and shriveled form (Fig. 7.18). The fertilized ovules and the ovary grow rapidly, using food made by the leaves. Inside the ovule, cell division and growth produce a seed containing the immature plant or *embryo,* consisting of a miniature root and shoot, together with one or two leaf-like structures called *cotyledons,* which often contain food reserves. The outer coat or *integument* of the ovule thickens and becomes tough and hard, to form the protective coat or *testa* of the seed (for a more detailed description see Section 8.1). As the seed ripens, water is withdrawn from it so that it becomes dry and hard; in this condition

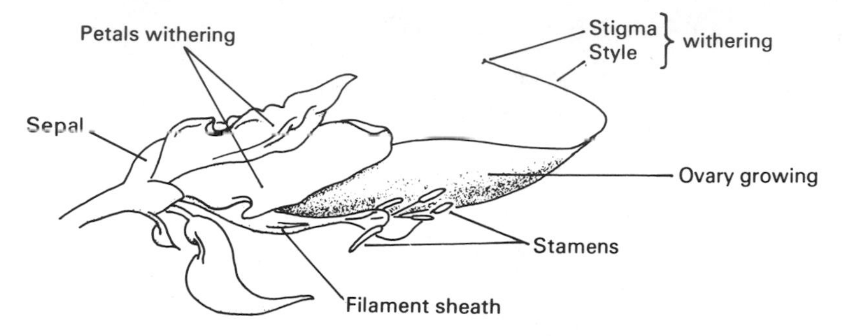

Fig. 7.18 *Lupin flower after fertilization*

Fig. 7.19 *Fruit formation of raspberry*

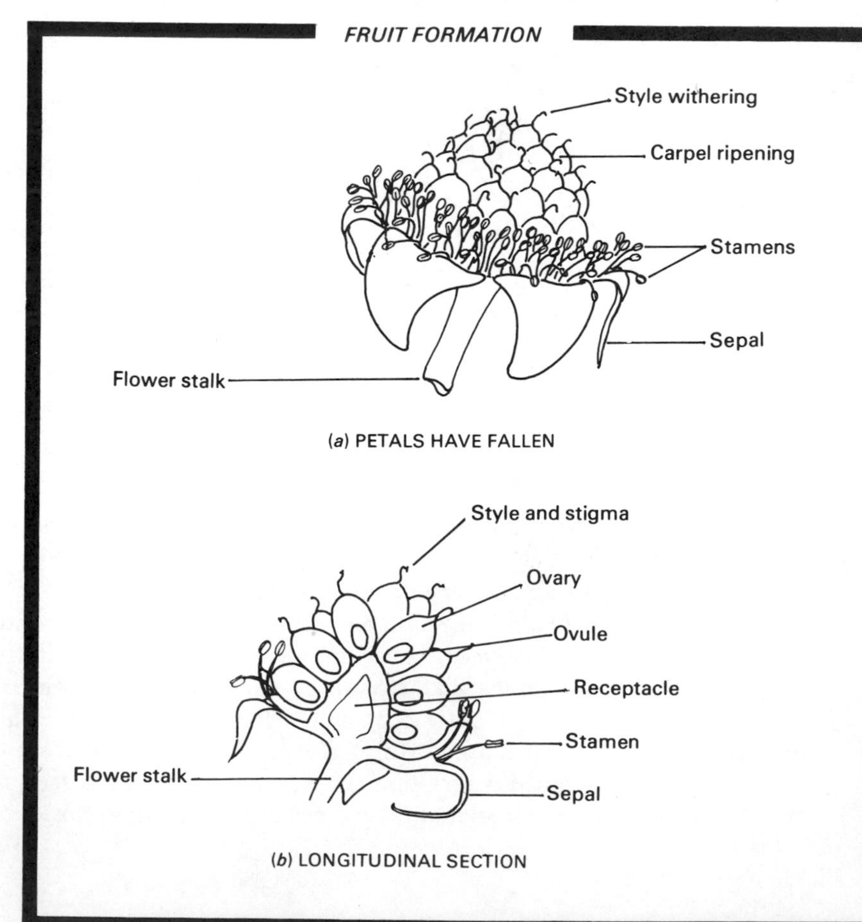

it can withstand extremes of temperature or drought until the next growing season.

The ovary wall becomes the *pericarp* which encloses the seeds. The pericarp may be dry and hard, forming a capsule or pod as in the poppy and lupin, or it may become succulent and fleshy as in the tomato, raspberry and plum.

In the strawberry (Fig. 7.21) the receptacle of the flower swells and becomes fleshy, and the pips of the strawberry, which are the true fruits of the plant, are carried on its surface. The receptacle of the rose encloses the ovaries altogether (Fig. 7.23) and that of the apple becomes juicy and succulent, and is fused to the outside of the ovary wall (Fig. 7.22). Some of these plants have been bred and selected for centuries for their large

Fig. 7.20 *Fruit formation of tomato*

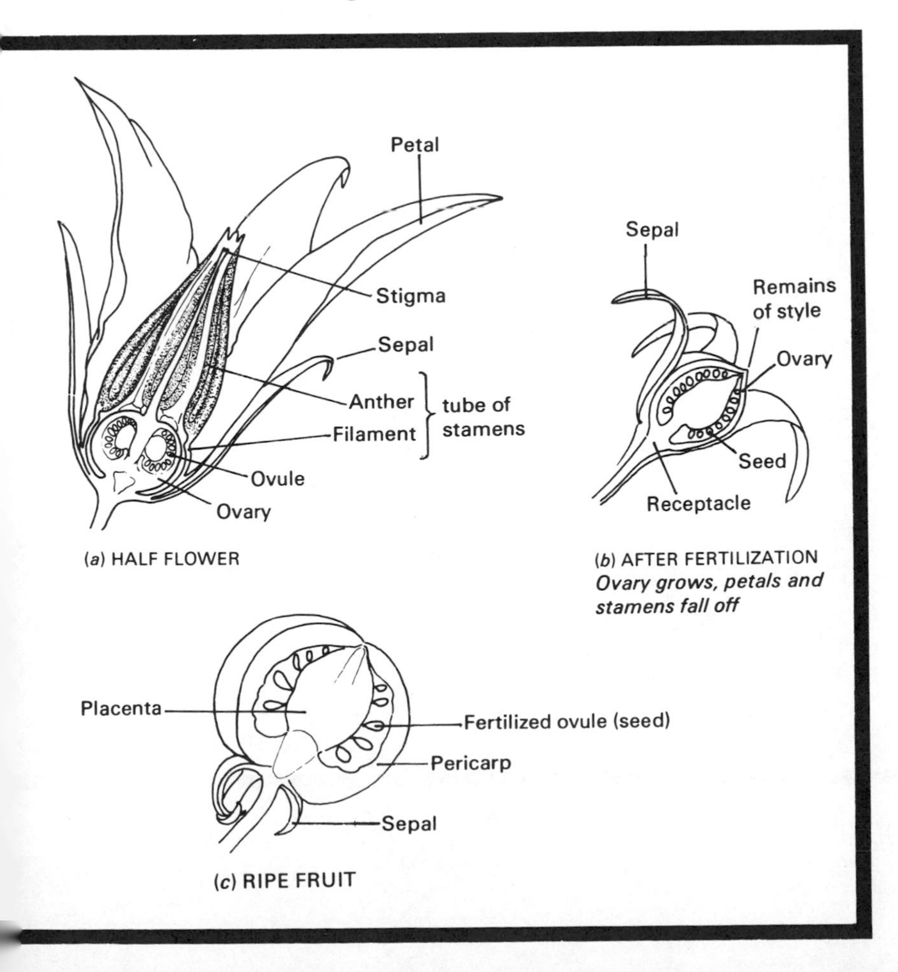

FRUIT FORMATION (cont'd)

Fig. 7.21 *Fruit formation of strawberry*

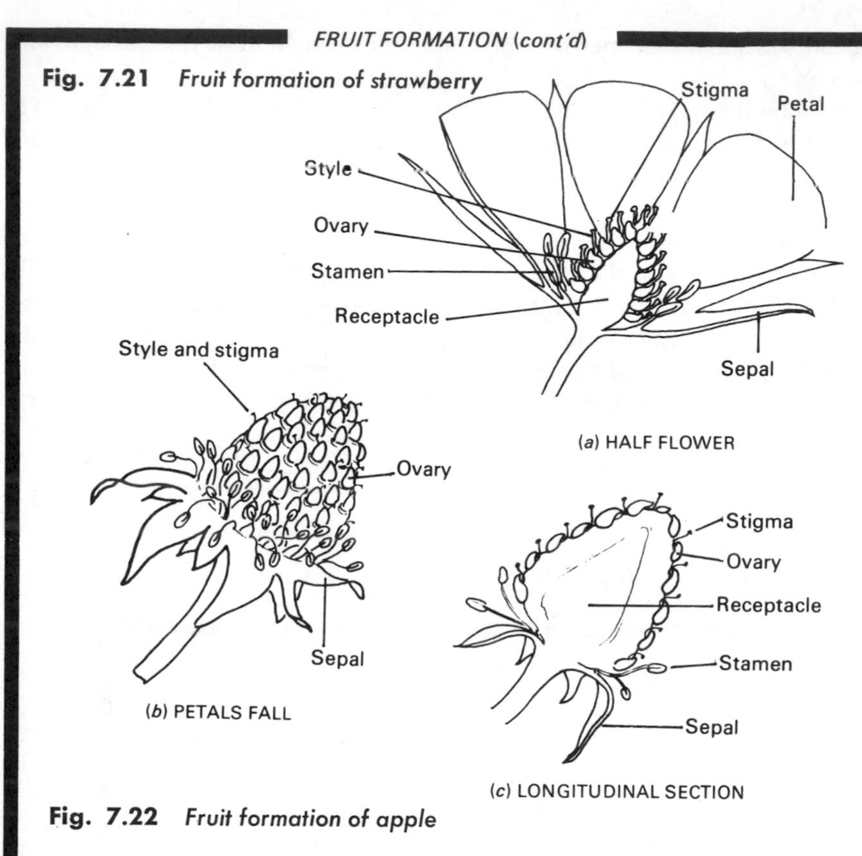

Fig. 7.22 *Fruit formation of apple*

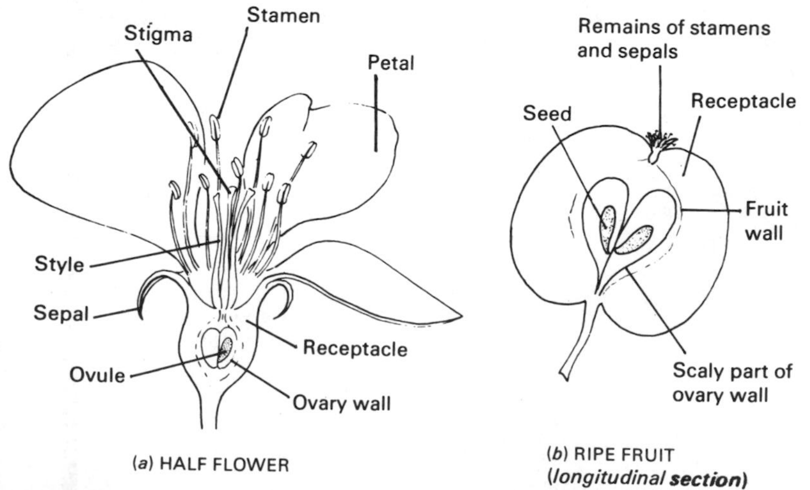

Fig. 7.23 *Fruit formation of rose*

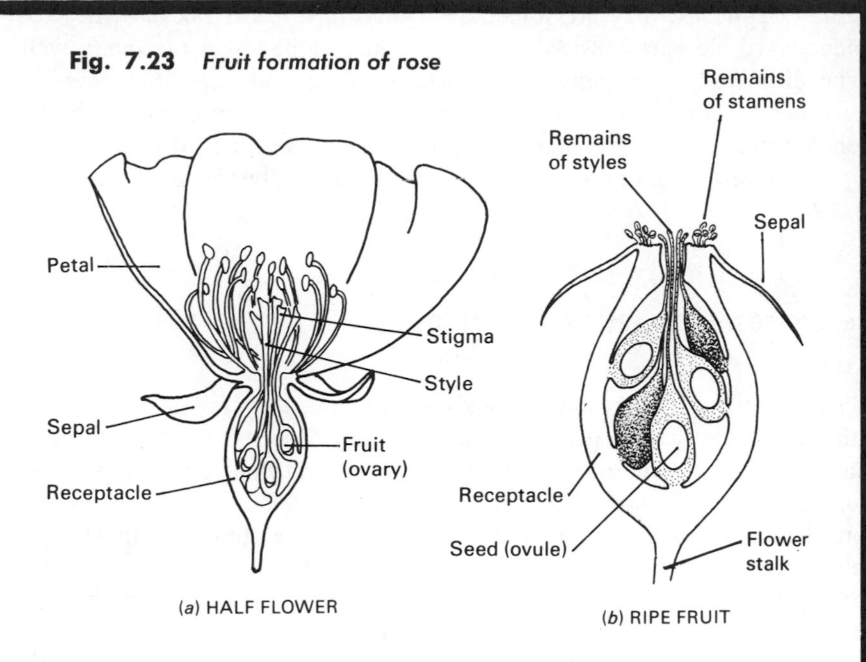

(*a*) HALF FLOWER

(*b*) RIPE FRUIT

Fig. 7.24 *Fruit formation of poppy*

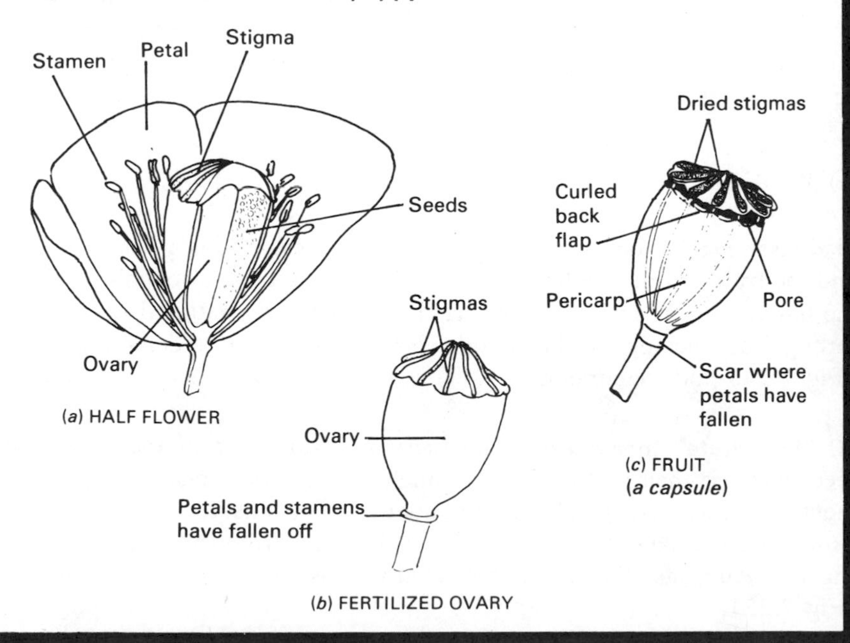

(*a*) HALF FLOWER

(*b*) FERTILIZED OVARY

(*c*) FRUIT
(*a capsule*)

edible receptacles; they are sometimes called *false fruits* because the conspicuous, edible part consists of the receptacle rather than the ovary wall. (The distinction commonly drawn between fruit and vegetable does not correspond to the biologist's definition of fruit; for example, runner beans, french beans, cucumbers, marrows and tomatoes are all fruits.)

The formation of fruit from certain flowers is illustrated in Figs. 7.19 to 7.24.

7.6 DISPERSAL OF FRUITS AND SEEDS

When flowering is over and the seeds are mature, they fall to the ground, either individually or as part of the whole ovary. There they can *germinate,* that is, begin to grow into a new plant, when conditions are suitable. The fruits and seeds of many plants are adapted so that they can be carried for considerable distances from the parent plant before they germinate. This helps to reduce overcrowding among members of the same species, with the consequent competition for light, water and mineral salts from the soil, and results in the colonization of new areas.

The commonest adaptations favor dispersal by wind and by animals, while some plants have "explosive" pods and capsules which actually project the seeds away from the plant, and others have seeds which will float on water and can be carried over long distances by rivers or streams, or even by sea.

(a) Wind Dispersal

(i) **Censer mechanism.** The flower stalk in these plants is usually long, and the fruit is a dry, hollow capsule with one or several openings. When the flower stalk is shaken by the wind the seeds are scattered on all sides through the openings in the capsule. Examples are white campion, poppy (Fig. 7.24) and snapdragon.

(ii) **"Parachute" mechanism.** Feathery projections from these fruits or seeds increase their surface area so much that air resistance to their movements becomes very great. In consequence, they sink to the ground very slowly and are likely to be carried great distances from the parent plant by slight air currents. Examples are clematis, thistles, willow and dandelion (Fig. 7.25).

(iii) **Winged fruit .** These fruits carry wing-like structures, which may be extensions from the ovary wall or leaf-like bracts on the flower stalk; their

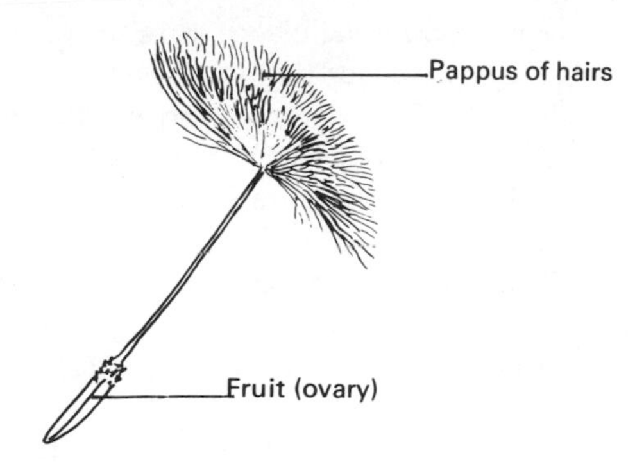

Fig. 7.25 *Dandelion fruit*

presence causes the fruits to spin as they fall from the tree. Their fall is thus prolonged and they have a much better chance of being carried away in air currents. Examples are maple (Fig. 7.26) ash and elm.

(b) Animal Dispersal

(i) **Hooked fruit.** In herb bennet, agrimony (Fig. 7.27), burdock and goose-grass, hooks develop from the style, the receptacle or the bracts around the inflorescence, or on the ovary wall. These hooks catch in the fur of passing mammals or in the clothing of people, and later, at some distance from the parent plant, they fall off or are brushed or scratched off.

(ii) **Succulent fruit.** Fruits like the blackberry and elderberry are eaten by birds and animals. The hard pips containing the seeds cannot be digested and so will pass out of the creature in its feces, away from the parent plant.

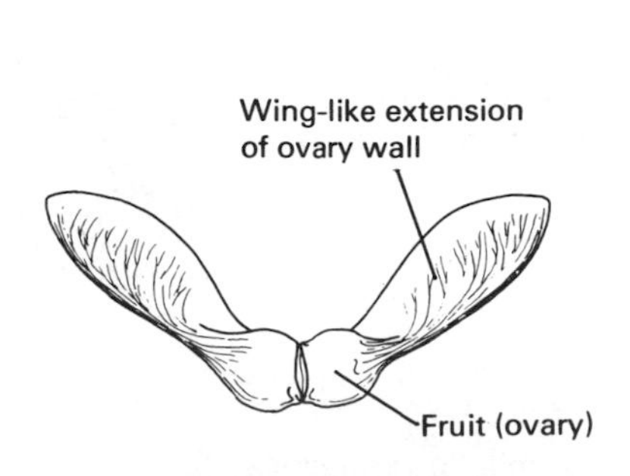

Fig. 7.26 *Maple fruit*

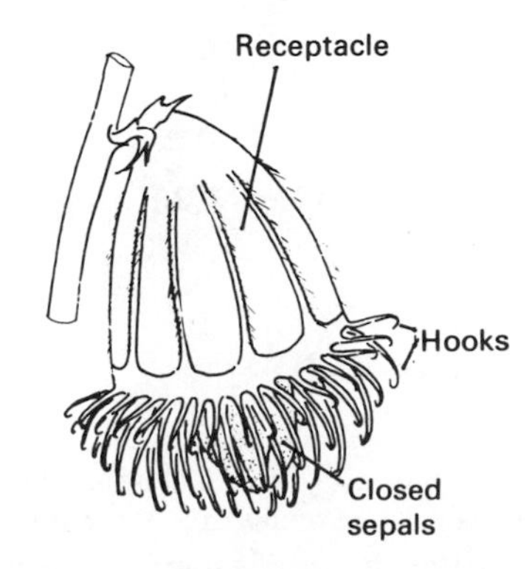

Fig. 7.27 *Agrimony fruit*

Even when the seeds are not swallowed, as in the case of the rose hip, the fruit may be carried long distances before the seeds are dropped. Other succulent fruits attractive to birds are particularly sticky, and the seed of a berry such as yew may adhere to the beak of the bird for some time before being discarded. The flesh of the mistletoe berry is so sticky that the bird can often only get rid of the seed by wiping its beak on a branch, where the seed is ideally placed for its subsequent growth as a parasite on the tree.

The succulent texture and conspicuous color of these fruits may be regarded as their adaptation to this method of dispersal.

(iii) The mud adhering to the hooves or feet of animals, including man, may carry seeds within it, and all kinds of seeds can be dispersed in this way.

(c) Explosive Fruit

The pods of flowers in the pea family, which includes gorse, broom, lupine and vetches, dry in the sun and shrivel. The tough, diagonal fibers in the pericarp shrink and set up a tension. When the fruit splits in half down two lines of weakness, the two halves curl back suddenly and flick out the seeds (Fig. 7.28).

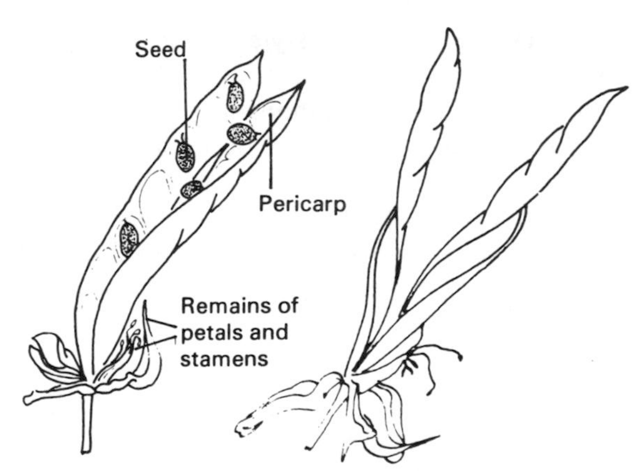

Fig. 7.28 *Lupin: the ripe fruit splits open and the walls curl back, ejecting the seeds*

7.7 QUESTIONS

1. What do you understand by the term "gamete"? What are the male and female gametes in a flowering plant?
2. What is "fertilization" and where does it occur in a flowering plant?

3. Pollination may occur without fertilization taking place but fertilization will not occur without pollination. Explain why this is so.
4. Why should a disease which causes the blossom to fall from apple trees in spring affect the yield of fruit in the autumn?
5. What part in reproduction is played by (*a*) petals, (*b*) stamens, (*c*) carpels?
6. How is self-pollination prevented in the white deadnettle?
7. Only large insects such as bees are likely to effect pollination in the lupine and deadnettle. Why are smaller insects unlikely to do so?
8. Which part of the fruit is considered edible in (*a*) runner beans, (*b*) tomato, (*c*) strawberry, (*d*) apple, (*e*) green peas?
9. Cucumber and marrow are "fruits" to the biologist, while rhubarb is not. Explain the basis for this distinction.
10. Most flowering plants produce many more seeds than are ever likely to grow to maturity. (*a*) What kind of adverse circumstances are likely to prevent successful germination and growth? (*b*) How does seed dispersal contribute to the survival of the species despite these hazards?
11. What kind of competition is likely to take place between seedlings growing closely together?
12. Structures which help to disperse fruits and seeds are modifications of structures present in all flowering plants. Name the structures from which the dispersive features are derived in the case of (*a*) lupin, (*b*) maple, (*c*) dandelion.
13. Distinguish between wind pollination and wind dispersal.

UNIT 8 SEEDS, GERMINATION, AND TROPISMS

8.1 STRUCTURE OF SEEDS

A fertilized ovule develops into a seed, consisting of an immature plant or *embryo* enclosed by a tough coat or *testa.* If conditions are favorable the embryo can begin to grow or *germinate,* eventually becoming a fully independent plant bearing flowers and seeds during its life cycle. The embryo of the seed contains all the potentialities of development and growth to a mature plant resembling other members of its species in almost every detail of leaf shape, cell and tissue distribution and flower color and structure.

Testa. The testa develops from the integument surrounding the ovule (see Fig. 7.17). It is a tough hard coat which protects the seed from injury by fungi, bacteria or insects. It has to be split open by the growing embryo when germination begins. The *hilum* is a scar on the testa left by the stalk which attached the ovule to the ovary wall. The *micropyle* is the minute hole in the integument through which the pollen tube entered at fertilization. It persists as a tiny pore in the testa lying opposite the tip of the embryo root and provides a means for water to reach the embryo before germination begins.

Radicle. This is the embryonic root which grows and develops into the root system of the plant.

Plumule. This leafy part of the embryonic shoot develops into the young plant's first foliage leaves, the shapes of which become apparent as the plumule leaves open and grow. The part of the embryonic stem to which the plumule is attached is called the *epicotyl;* the part below the point of attachment of the seed leaves or cotyledons is called the *hypocotyl.*

Cotyledons. These modified leaves are attached to the embryonic stem by short stalks. They bear no resemblance to the ordinary foliage leaves, and

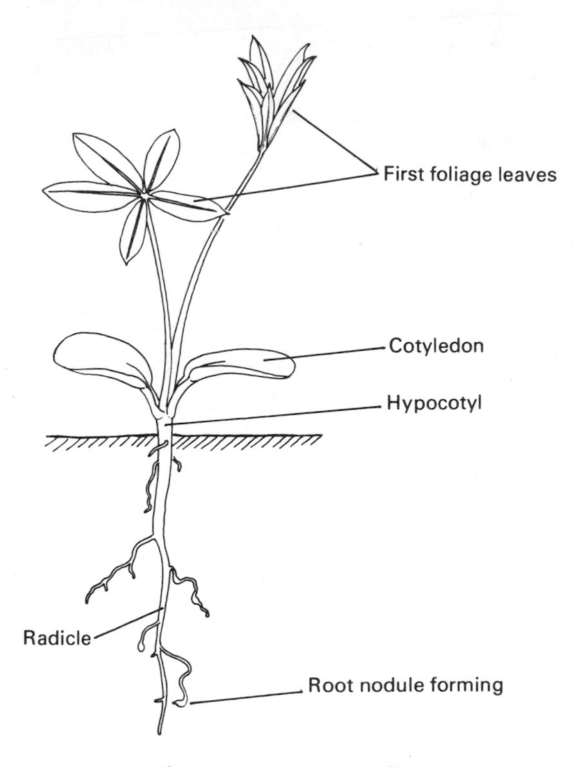

Fig. 8.1 *Lupin seedling*

are often swollen with food reserves which are used up in the early stages of germination. When the first foliage leaves are formed, the cotyledons wither away and die. The seeds of the grasses, including the cereals, and narrow-leaved plants like the iris and bluebell contain one cotyledon only, and these plants are called *monocotyledonous plants* or simply *monocotyledons*. All other flowering plants have two cotyledons in their seeds and are called *dicotyledons*.

Types of Seeds

Figs. 8.2 and 8.3 shows the structure of a dicotyledonous seed such as a french bean, and Fig. 8.4 that of a maize grain which is a fruit containing a single monocotyledonous seed. The maize grain differs from the bean not only in the number of its cotyledons, but also because its food reserves are contained in a separate storage structure called the *endosperm,* whereas the bean's food reserves are contained in its cotyledons. These food reserves

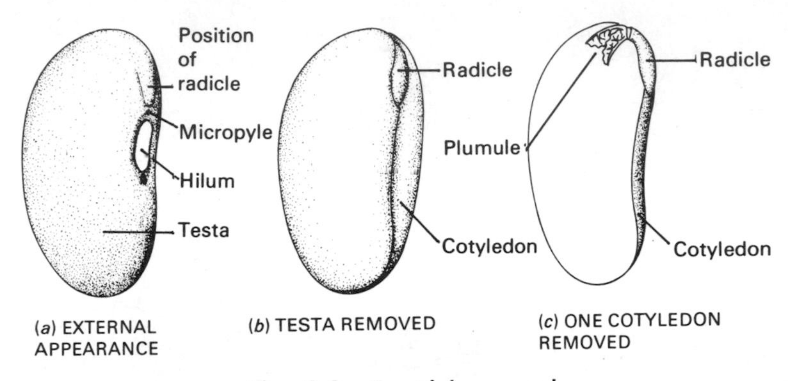

Fig. 8.2 *French bean seed*

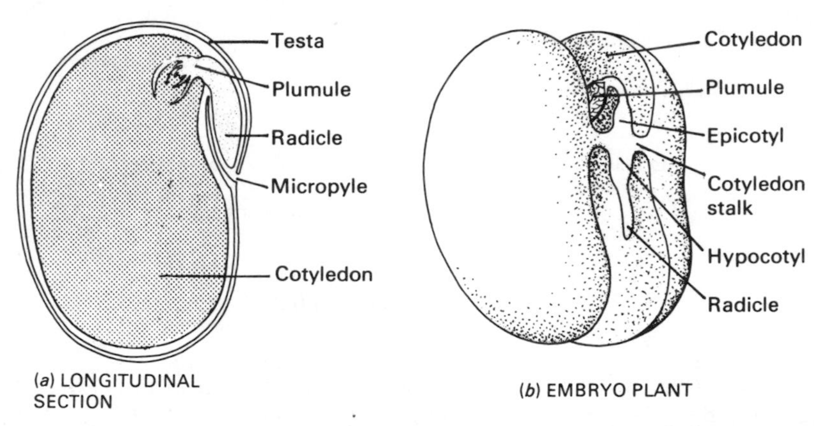

Fig. 8.3 *Diagram to show seed structure in a dicotyledon*

are the reason why the plants are grown as agricultural crops; their constituents can be identified using the food tests described in Section 19.7.

8.2 GERMINATION

Dormancy

Most seeds when shed from the parent plant are very dry; only about ten per cent of their weight is water. In this condition all the chemical processes of living are very slow and little food is used. A dry seed may remain alive but inactive or *dormant* for long periods, without germinating

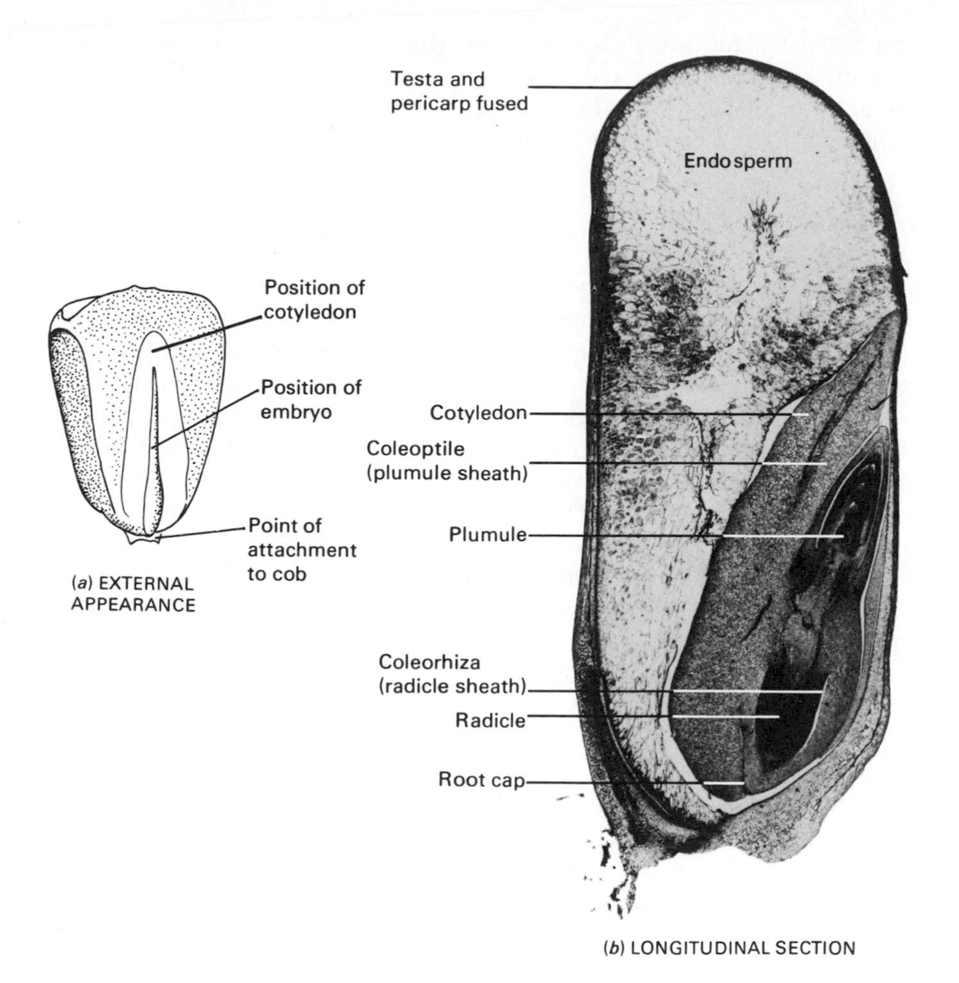

Fig. 8.4 *Maize fruit (corn kernel)*
(Brian Bracegirdle)

but still retaining the power to do so. With a large enough number of seeds it can be shown that they are consuming oxygen and releasing carbon dioxide very slowly while they are dormant. If properly stored, wheat can still be germinated after about fifteen years. On the other hand, the seeds of the Para-rubber plant can germinate only within a few days of being shed, after which the power is lost. The percentage of any quantity of seeds which will germinate decreases with the length of time they are kept dormant.

When conditions for germination are suitable (see Section 8.3) growth can begin. The following paragraphs describe the different ways of germination of the french bean and the maize grain, two seeds which are large enough for their structure to be easily examined and the course of their germination to be conveniently followed in some detail.

(a) **French Bean** (Fig. 8.5)

The seed absorbs water and swells; the radicle is the first part to begin to grow. After three days or so (the period depends on the temperature) the radicle bursts through the testa and grows down between the soil particles, its tip protected by a root cap (see Fig. 5.23). Root hairs grow out of the region of the radicle which has ceased to elongate; they absorb water and salts from the soil, which are passed on to the rest of the seedling plant. Later, lateral roots develop from the radicle, and these anchor the young plant in the ground. The hypocotyl then starts to elongate, and its growth pulls the cotyledons upward; the cotyledons remain pressed together at this stage, so that the plumule lying between them is protected from damage during its passage through the soil. Usually the testa is left behind in the earth, but sometimes it too is drawn upward and is pushed off the cotyledons as they open later on. Once the cotyledons are above the ground level the hypocotyl straightens out; the cotyledons separate and expose the

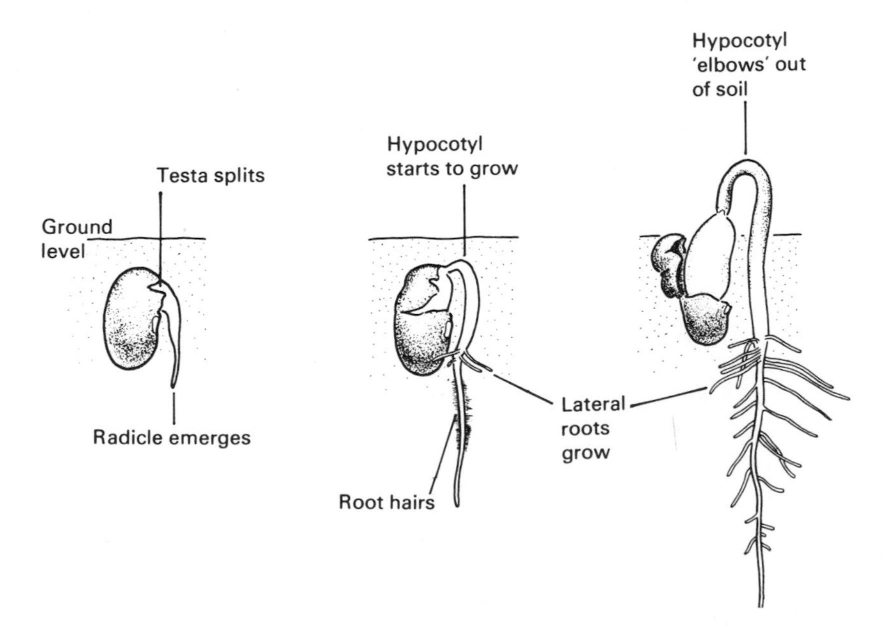

Fig. 8.5 *Germination of french bean (an epigeal dicotyledon)*

plumule, become green and photosynthesize food for the seedling for a day or two. Meanwhile the epicotyl extends and the plumule leaves expand, turn green as their chlorophyll develops, and also start to photosynthesize. The seedling is now an independent young plant.

This kind of germination, in which the cotyledons are brought above the ground, is called *epigeal* germination.

The food reserves in the cotyledons are important to the seedling during its rapid early growth and development. In the french bean these reserves consist mainly of starch and proteins (see Section 19.3) which are insoluble in water; however, when acted on by chemically active substances called *enzymes* (see Section 20.1) produced by the plant, they are converted into soluble products which pass to the growing regions. Here they are used in the build-up of new tissues and also as fuel to provide the

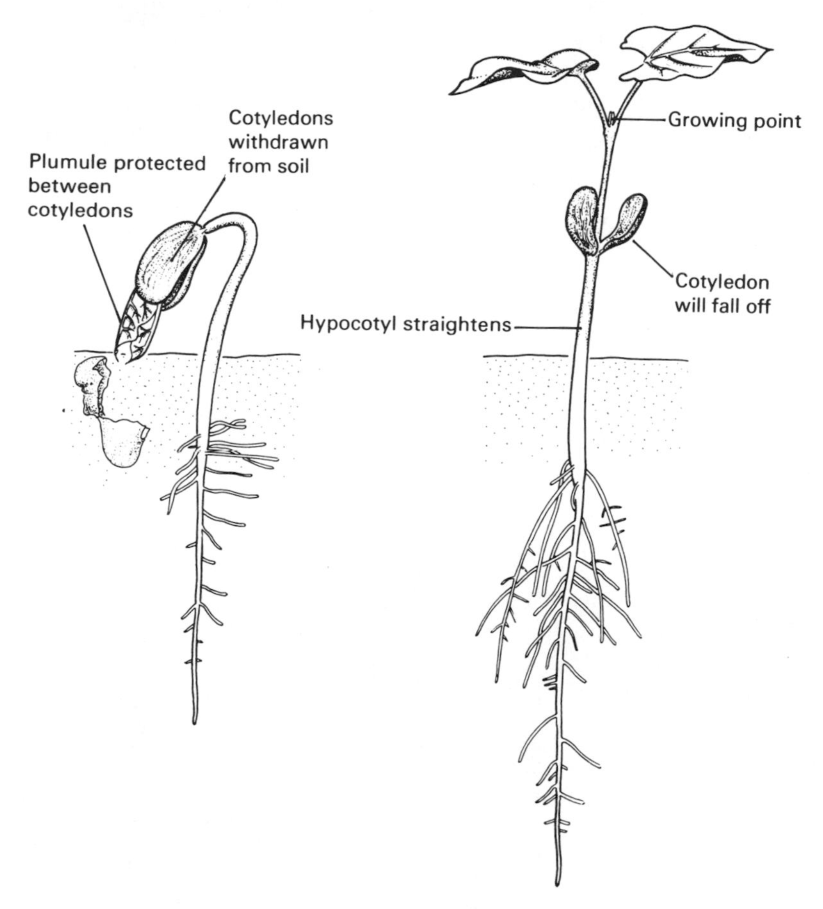

Fig. 8.6 *Germination of maize (a hypogeal monocotyledon)*

energy needed for the building processes. For example, the stored insoluble starch is converted into the soluble sugar glucose: some of the glucose is converted into cellulose and incorporated into new cell walls, some is used in the build-up of new protoplasm, and some is oxidized by respiration, releasing the energy required by the plant. The production of glucose also assists the seedlings to take in water by the process of osmosis (this is discussed in Unit Three): considerable amounts of water are needed as the newly formed cells expand and develop. When the food reserves are exhausted the shriveled cotyledons fall away from the plant.

(b) **Maize** (Fig. 8.6)

Although the maize (corn) grain is actually a one-seeded fruit, the thin ovary wall is fused with the testa, effectively forming a single seed coat which does not interfere with germination. The fruit absorbs water and swells, and the radicle, which in cereals and grasses is protected by a sheath, the *coleorhiza* (Fig. 8.4), grows and bursts through the fruit wall. Root hairs, and later lateral roots, grow out from the radicle. Meanwhile the

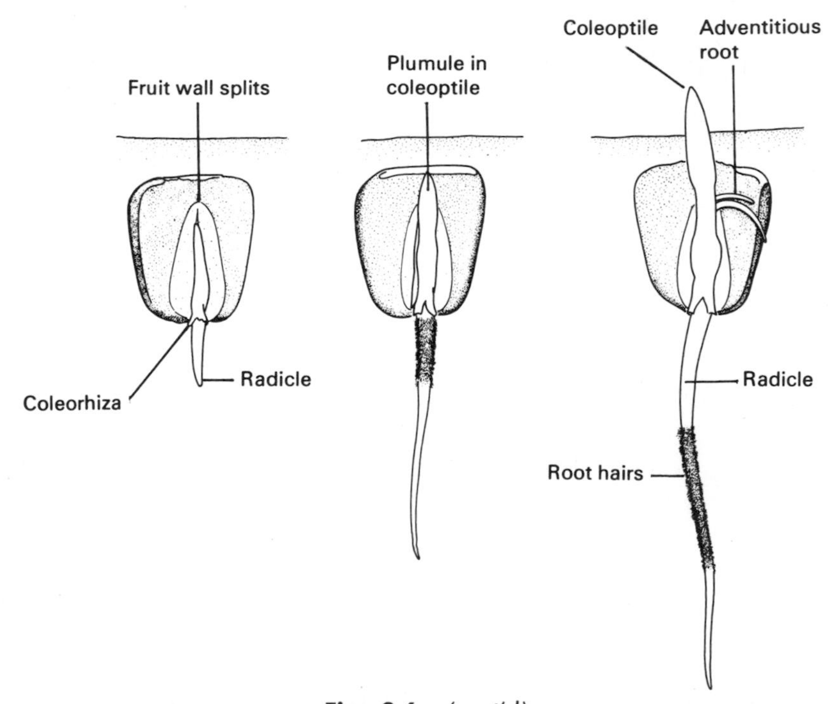

Fig. 8.6 *(cont'd)*

plumule grows straight upward through the fruit wall, its growing point and first leaves protected by a sheath, the *coleoptile,* which has a hard pointed tip. Adventitious roots grow out from the base of the epicotyl. Once above the soil, the first foliage leaves burst out of the coleoptile, which remains as a sheath around the leaf bases. The cotyledon remains below the soil (this is characteristic of *hypogeal* germination), absorbing food from the reserves in the endosperm and transmitting it to the growing root and shoot. Eventually both the cotyledon and the exhausted endosperm rot away.

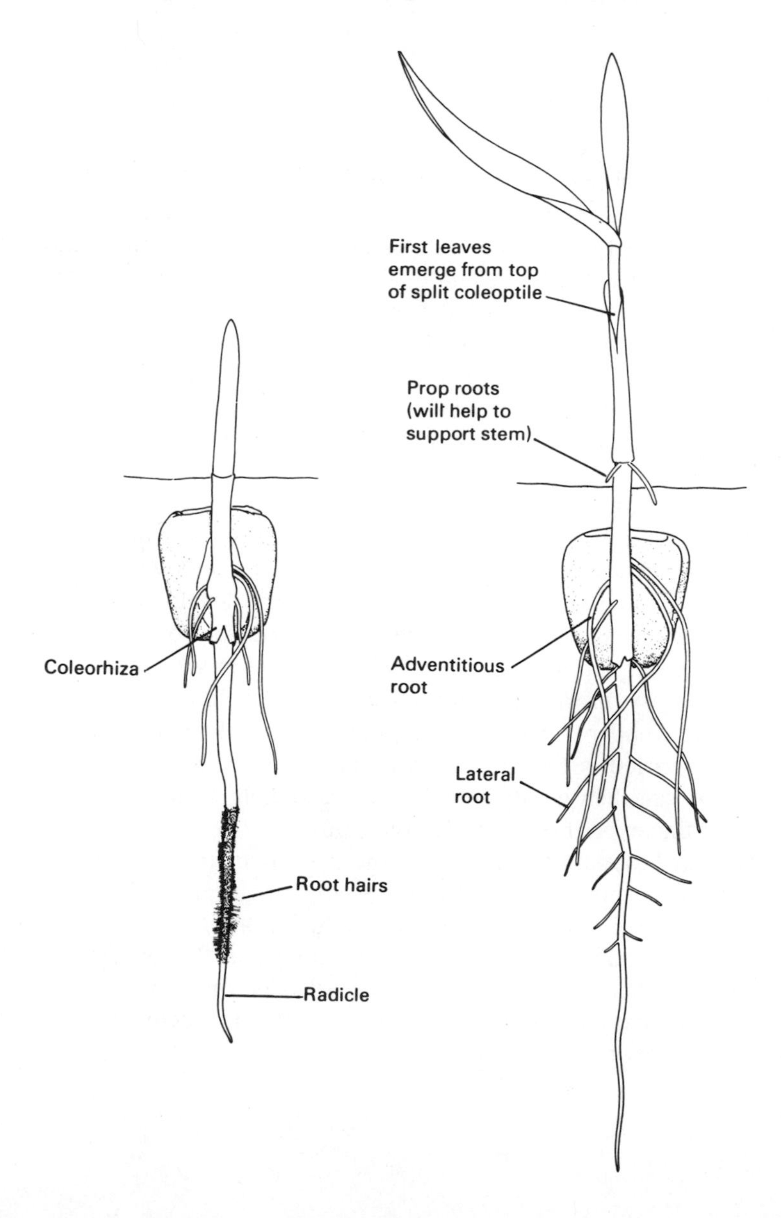

8.3 EXPERIMENTS ON GERMINATION

It is interesting to try to establish what are the conditions which encourage germination. We might guess, for example, that the presence of water might be necessary, or perhaps that there might be an optimum temperature at which germination is most rapid. Experiments to test the validity of guesses like these are fairly easy to carry out, but they must be carefully designed: if they are to be really useful they must be properly *controlled*.

Controlled Experiments

If a wild flower sheds 50 seeds, and it is found that only 20 of these germinate and that only half of these young plants eventually produce flowers and fruit, it is possible to guess at the factors that might have been responsible for the failures. Some of the seeds may have been imperfectly formed, and some may have been damaged by attack by fungi or bacteria; alternatively, the conditions around them may have been unfavorable in one or more respects. Variations in the natural environment like these cannot be influenced by the observer who is studying the plant, nor does he know how conditions may have been changing during the plant's earlier development.

In contrast, a *controlled experiment* is one in which the experimenter controls the conditions. In this way he can find out exactly how these conditions influence the plant or animal. The simplest way to study the importance of one particular factor in the normal development and life of a plant or animal is to keep all other environmental conditions constant, and to exclude or vary the factor of interest. Two experiments are therefore usually necessary, in which conditions are identical in every respect except one, the factor under study: this is kept constant in the first experiment and excluded or varied in the second. Any difference in the results of the two experiments can then be attributed with reasonable certainty to the varied factor. The first experiment is called the *control*.

The need to maintain control of the environment often involves keeping the plant or animal in rather unusual or unnatural surroundings—in a box or a cage, for example—and it might be argued that the peculiar experimental conditions could be responsible for the observed results; however, a further advantage of the control experiment is that it enables one to be sure that the effect of varying one particular factor is independent of the other experimental conditions.

Experiment 8.1 ***To find whether oxygen is necessary for germination*** (Fig. 8.7)

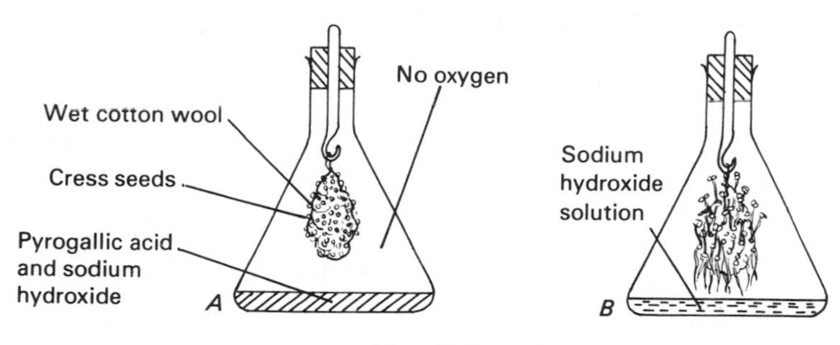

Fig. 8.7

In this experiment, seeds are deprived of oxygen and their power of germination is compared with that of other seeds having a normal supply of oxygen.

A piece of wet cotton is rolled on some cress seeds, which will stick to it. The cotton is suspended by a thread inside a tightly corked flask, *A*, which contains a freshly prepared solution of pyrogallic acid and sodium hydroxide, prepared by dissolving 1 g pyrogallic acid in 10 cm^3 10 per cent sodium hydroxide solution. This solution has the property of absorbing oxygen from the air. (*Caution: the solution is very caustic and attacks skin, clothing and wooden bench tops. If spilt it should be neutralized at once with very dilute hydrochloric acid and washed away with plenty of cold water.*) The cotton must not touch the chemicals. It could be objected that seeds sown in such abnormal surroundings, could hardly be expected to germinate anyway. To check on this a second apparatus is set up using the same size flask, *B*, seeds from the same source, but with only sodium hydroxide solution in the flask. This is the control, in which the experimental situation is the same except that the seeds in it are not deprived of oxygen. Both flasks are placed in the same conditions of light and temperature.

Sodium hydroxide solution absorbs carbon dioxide from the air so that the seeds in flask *A* lack both oxygen and carbon dioxide. In the control flask, only carbon dioxide is lacking.

Result After a few days most of the seeds in flask *B* will have germinated. The seeds in flask *A* will probably not have germinated at all, and any which may have begun to grow will not be as advanced as those in the other flask. It could be argued that they have been damaged or killed by the chemicals in *A*, but this can be disproved by putting the cotton into flask *B;* after a day or two the seeds will germinate normally.

Interpretation. Germination cannot take place satisfactorily unless oxygen is present.

Experiment 8.2 *To find whether water is necessary for germination*

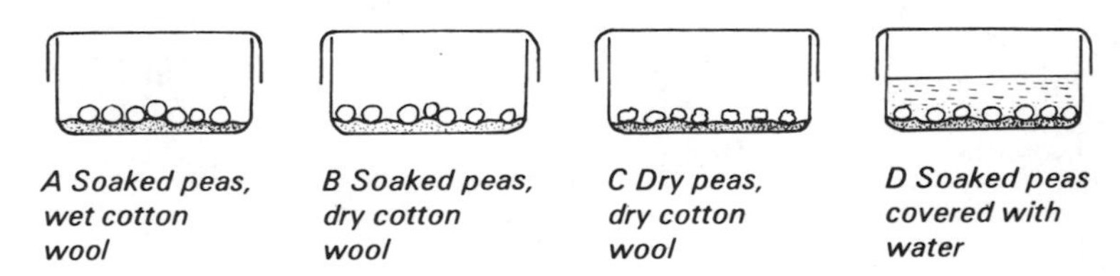

Fig. 8.8

This experiment needs four dishes that can be covered to prevent drying out (clean margarine tubs are satisfactory); they are labeled *A, B, C* and *D,* and blotting paper is placed in the bottom of each. (Cotton can be substituted for blotting paper.) The blotting paper in *A* is moistened and on it are placed seeds which have been soaked overnight—peas should give satisfactory results. Soaked seeds are also placed in *B* but the blotting paper is not moistened. In *C,* unsoaked seeds are placed on dry blotting paper. Soaked seeds are placed in *D,* and these are completely covered with water. There should be equal numbers of seeds in each container, and all the seeds should come from the same source. The four containers are then covered and left in the same conditions of light and temperature for a few days, water being added to *A* and *D* if necessary.

Result. The seeds in *A* should germinate well. Those in *B* may germinate and begin to grow, but will soon shrivel and die. The seeds in *C* will not germinate at all. Some of the seeds in *D* may germinate but if they do so they will soon die and after a few days they will probably begin to rot.

Interpretation. Adequate water must be present for germination to start and continue; excess water prevents germination, probably because it prevents the seeds from obtaining the oxygen they need.

Experiment 8.3 *To investigate the effect of temperature on germination*

Equal numbers of soaked seeds (again, peas should give satisfactory results) are placed on moist cotton in each of three labeled dishes and covered to prevent drying out. The dishes are then placed in situations which differ only in temperature, such as a refrigerator (4 °C), a laboratory cupboard (20 °C), an incubator or an airing cupboard (say 30 °C). The dishes are then left for a week or so, during which the temperatures should be noted daily. The extent of germination of the three groups of seeds is then compared, for example by measuring the lengths of their plumules or radicles.

Result. It will be found that extremes of temperature do not favor germination. Low temperatures prevent it altogether; it is easy to show that the refrigerated seeds have not been killed by the cold by subsequently allowing them to germinate in a warm place. Higher temperatures accelerate germination; but if the surroundings are too warm either the protoplasm of the seeds may be killed or they may be affected by fungal growth, with the consequent eventual death of the seedling.

Interpretation. Each species of seed probably has an optimum temperature for germination, but in this experiment the intervals between temperatures are too widely spaced to determine this optimum.

8.4 EXPERIMENTS ON THE SENSITIVITY OF PLANTS

The *sensitivity* of plants (see Section 1.2(g)) is easily studied by observing the growth of young seedlings because their growing roots and shoots respond readily to the stimuli of light and gravity. Growth movements of this kind, in which the direction of growth is related to the direction of the stimulus, are called *tropisms.* Tropisms are described as *positive* or *negative* according to whether the growth response is *toward* or *away from* the direction of the stimulus.

Experiment 8.4 ***To investigate the effect of one-sided lighting on growing shoots*** (Fig. 8.9)

In this experiment the growth of two seedlings is compared after a period during which one receives light from one side only, while the other is lit equally from all sides. Sunflower seedlings a few inches high are suitable, and two potted, well-watered seedlings at about the same stage of growth are chosen. The first seedling is placed in a cardboard box with a slit cut in one side. The second, the control plant, is placed in an identical box, similarly lit, but standing on a slowly rotating turntable, powered by an electric or clockwork motor. This device is called a *clinostat;* it usually

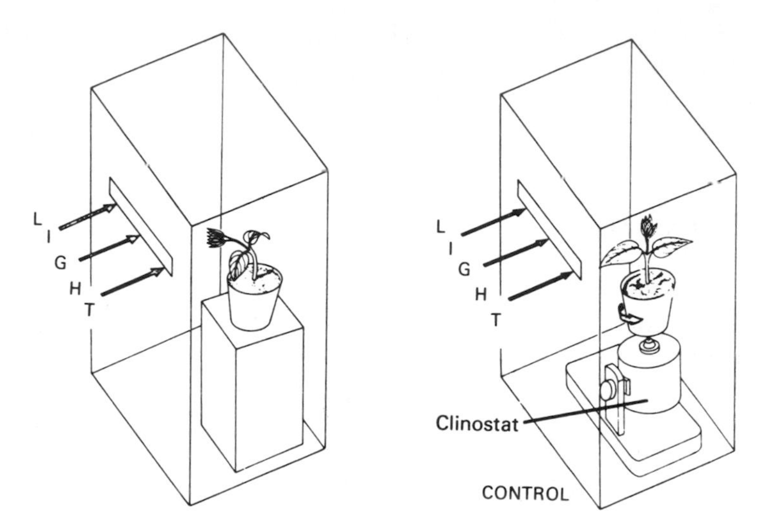

Fig. 8.9

rotates about four times an hour. In this way all sides of the control plant are exposed equally to the light entering the box through the slit. Care should be taken to see that both plants are placed at the same height in the box relative to the slit.

Result. After a few days, the two plants are removed from the boxes and compared, when it will be found that the stem of the plant with one-sided illumination has changed its direction of growth and is growing toward the light.

Interpretation. The results suggest that the young shoot has responded to one-sided lighting by growing toward it. The tendency to grow in response to the direction of light is called *phototropism* and the shoot is *positively phototropic* because it grows toward the direction of the stimulus.

However, the results of one experiment using one species of plant cannot justify drawing conclusions applying to all kinds of green plants. This experiment is a demonstration of the response, rather than a critical study. A thorough investigation of phototropisms would require a large number of experiments on plants of a wide variety of species.

Experiment 8.5 ***To investigate the effect of gravity on shoots*** (Fig. 8.10)

Two equivalent potted seedlings are selected as for Experiment 8.4. One is placed on its side so that the shoot is horizontal, while the second, the control plant, is placed similarly in a clinostat so that although the shoot

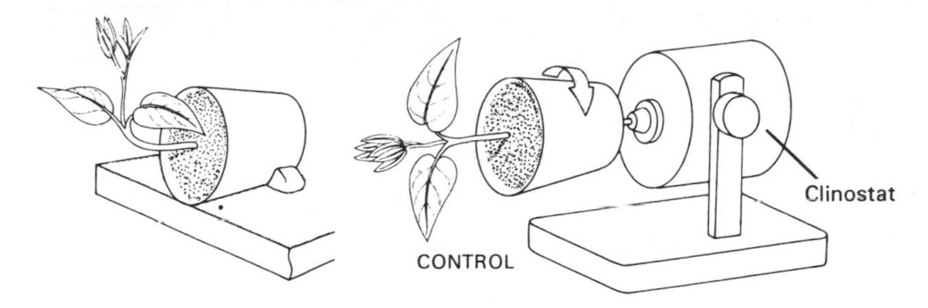

Fig. 8.10

remains horizontal, all sides of it are exposed equally to the pull of gravity. Both shoots should be similarly lit, or alternatively both experiments should be kept in a dark cupboard or covered with cardboard boxes.

Result. After about 24 hours the shoot of the stationary plant will be found to have changed its direction of growth, and its tip will be turned vertically upward. The shoot of the control plant in the clinostat will still be growing horizontally.

Interpretation. The experiment illustrates how growing shoots tend to grow away from the direction of the gravitational pull. This kind of response to gravity is called a *geotropism;* since the shoots tend to grow away from the direction of the stimulus they are said to be *negatively geotropic.* However, as with Experiment 8.4, this limited study does not permit generalizations concerning all kinds of plants.

Experiment 8.6 *To investigate the effect of gravity on roots*

This experiment, and those in Section 8.6, require a supply of germinating seeds with straight radicles. To obtain these, soaked seeds are rolled in blotting paper or newspaper as shown in Fig. 8.11, placed in a

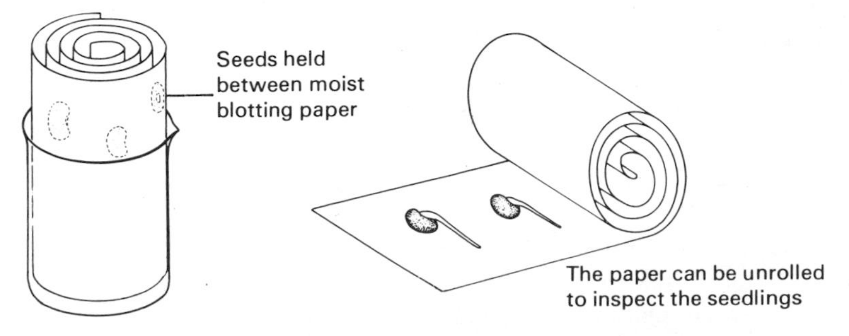

Fig. 8.11 *To obtain seedlings with straight radicles*

beaker and kept moist. If several sets are started at daily intervals there will be plenty of suitable seedlings available for the series of experiments on tropisms.

Two cylindrical boxes are required for the experiment, each with a large well-fitting cork: the boxes are lined with moist blotting paper to saturate the air inside them with water vapor. Bean and pea seedlings with straight radicles are pinned to both corks as shown in Fig. 8.12, and the corks are fitted into the boxes. One box is allowed to rest on its side for 48 hours or so, while the second, the control, is placed on its side in a clinostat as shown in the figure. Both experiments are left in a dark cupboard or covered with a cardboard box to eliminate possible complications by any phototropic response.

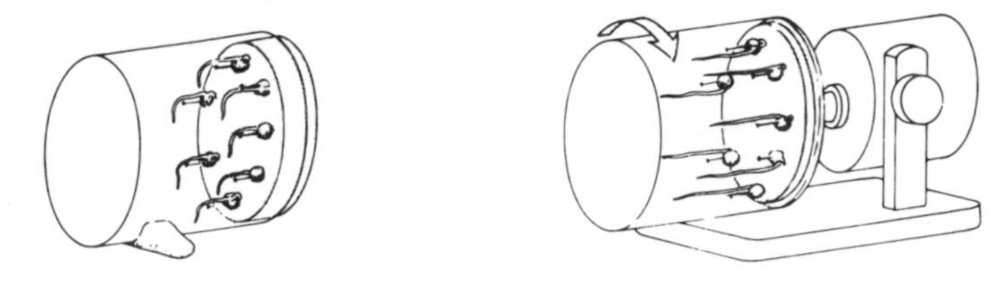

Fig. 8.12 *The effect of gravity on growing roots*

The radicles in the first box, being horizontal, experience a gravitational force in a direction perpendicular to their length, while gravity acts on the radicles of the seedlings in the control box equally in all directions in turn.

Result. The radicles of the seedlings on the clinostat continue to grow horizontally while the direction of growth of those in the stationary box changes, and the tips turn and grow downward.

Interpretation. Since the humidity, lighting and temperature conditions are the same in both boxes, the radicles showing a growth curvature must have responded to the one-sided pull of gravity. This growth response is *geotropism* and the roots are *positively geotropic*. But it is important to remember that lateral roots, which grow more or less horozontally, do not show this positive geotropic response.

Experiment 8.7 *To investigate the effect of light on plants*

The experiment needs two equivalent groups of seedlings. One group is kept in total darkness, and the other group, the control, is allowed to grow in normal lighting conditions; the temperature at which the two groups are kept should be the same, and both groups must be kept moist. After several days the two groups are compared.

Result. Although the seedlings of both groups have continued to grow, the differences between them are very marked. The seedlings grown in darkness are taller than the control plants, and their stems are thinner, with long internodes (the lengths of stem between the leaves). These seedlings carry fewer leaves than the controls do, and the leaves themselves are small and yellowish in color. Such pale, lanky plants are said to be *etiolated* (Fig. 8.13). The control group of seedlings grown in the light have shorter, stouter stems, with internodes of normal length between well-formed green leaves.

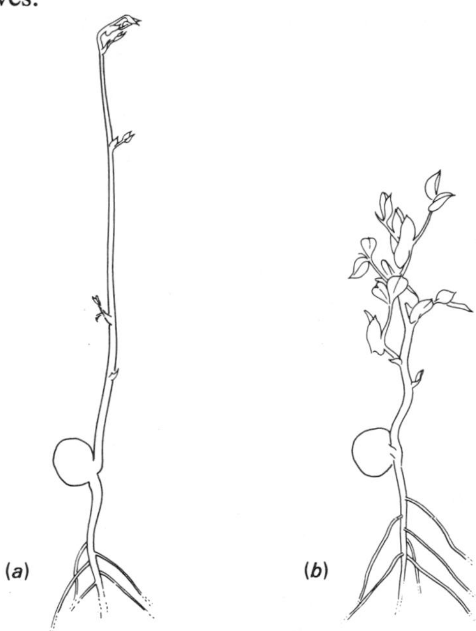

Fig. 8.13 *(a) Etiolated pea seedling; (b) normal seedling for comparison*

Interpretation. The effect of light on plants seems to be
(*a*) to reduce the rate at which stems grow, and
(*b*) to promote chlorophyll production and expansion of leaves.

The phototropic response of shoots studied in Experiment 8.4 is an outcome of (*a*), the effect of light on the growth rate: the side of the stem receiving most light grows more slowly than the other, so that the stem develops a curvature toward the light.

These responses to light have a positive advantage in the case of young seedlings growing in shady or overcrowded conditions. The shoot of a seedling which has germinated in a poorly lit situation or partially obscured by other plants will grow upward rapidly until it reaches the light, whereupon the leaves will expand and develop, chlorophyll will be produced and photosynthesis can begin.

Hydrotropism. This name is given to the growth response of plants to water. Experiments are sometimes described which purport to show that radicles respond positively to the presence of water nearby by growing toward it. In practice it is difficult to design an apparatus which will at the same time keep seedlings in conditions moist enough for growth and also maintain them in a stable moisture gradient pronounced enough to produce a tropic response. If a plant is grown in a pot of soil which is drier on one side than the other, its root system could well develop a lop-sided structure because its growth might be inhibited by the dry soil and encouraged by the moist; such an effect, however, could not be attributed to a directional growth response by the growing root tips.

8.5 THE AUXIN THEORY OF TROPISTIC RESPONSE

The mechanism of the tropistic responses of plants is of interest. There is a good deal of evidence to suggest that the cells near the tip of a growing shoot produce a chemical—called an *auxin* or sometimes a *plant hormone* —which in certain concentrations has the property of accelerating growth in length of the stem. Such auxins probably achieve their effect by delaying the loss of plasticity of the cell walls in the region of elongation, so that at a time when the cells are taking in water by osmosis (see Section 3.4), the increased pressure in the vacuole forces the cell walls to extend.

Several such growth-promoting substances have been identified. One such compound which can be isolated from growing plants is called *indolylacetic acid* (sometimes *indoleacetic acid*), often abbreviated to IAA for convenience. The pure compound is available commercially, and its effect on the growth rate of young shoots is quite easy to demonstrate.

Experiment 8.8 ***To investigate the effect of indolylacetic acid on wheat coleoptiles.*** (Fig. 8.14)

Ten soaked wheat grains are placed on moistened cotton in each of four shallow dishes, which are labeled *A, B, C* and *D*. The fruits are allowed to germinate in darkness for several days, until the coleoptiles are

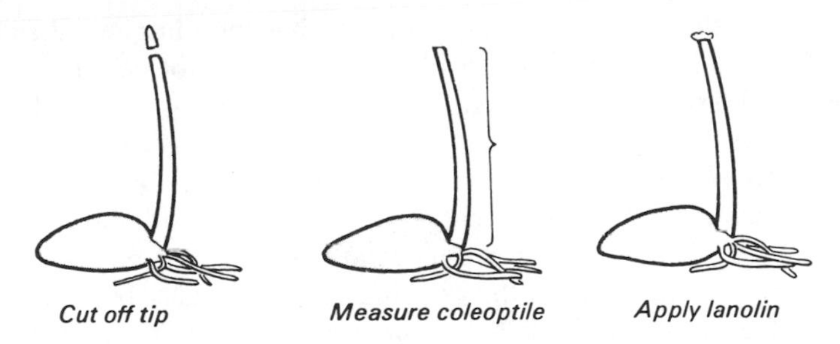

Fig. 8.14 *Studying the effect of IAA on wheat coleoptiles*

about 20 mm long. The control group of seedlings in dish *D* is left to grow undisturbed, but about 2 mm is cut from the tips of the coleoptiles of all the seedlings in the other three dishes. The lengths of all the coleoptiles are then measured, and the average coleoptile length of each group of seedlings is calculated. The cut tips of the coleoptiles in *A* are then treated with a little lanolin containing 0.1 per cent of IAA. Plain lanolin is applied to the cut tips of the coleoptiles in *B*, while those in *C* and *D* are left untreated. The seedlings are then left to continue growing for a further two days or so, and the coleoptiles are measured again. The average increase in length of the coleoptiles in each group can then be calculated.

Result. Usually little difference is found between the growth of the untreated coleoptiles in *C* and that of the lanolin-treated shoots in *B*, while neither of these groups grow as much as the control seedlings in *D* or as the IAA-treated seedlings in *A*—indeed, the seedlings in *A* may well have grown longer than any of the others.

Interpretation. Removal of the coleoptile tip seems to retard the growth of the seedlings in *C* as compared with those in *D*. This might be attributed to the damage or destruction of the growing point, except that the seedlings in *A*, decapitated but supplied with IAA, grow normally. It is therefore possible that the removal of the tip of the coleoptile deprives it of a growth-promoting substance, which could well be IAA itself, or a substance having a similar action. It cannot be the lanolin in which the IAA is dissolved which promotes growth since the coleoptiles in dish *B*, treated with lanolin alone, grow no longer than the untreated ones in *C*.

Experiment 8.8 is one of a wide range of experiments designed to test the auxin theory. Some of these are summarized in Fig. 8.17 (pages 126 and 127).

Assuming that the results with coleoptiles are applicable to other plants (and experiments can be carried out to show that this assumption is justified), it looks as if one-sided lighting alters the production or the distribution of the auxin within the shoot so that the illuminated side contains less of the hormone than the darker side. It is not certain, however, whether the hormone is destroyed by light or whether light causes it to move sideways through the shoot.

The same kind of evidence for the auxin theory can be obtained in the case of the geotropism of roots and shoots. The lower side of a horizontal shoot could, it is suggested, receive more auxin than the upper side, resulting in an upward curvature in the growing region. It must be assumed that the effect of the auxin on roots is to *retard,* not to accelerate, their growth, so that the higher concentration of auxin accumulating on the lower side of a horizontal root produces a *downward* curvature.

Many plant growth substances or auxins are known (although the term *auxin* is occasionally applied specifically to IAA). Other plant hormones influence not only growth but, for example, flowering, rooting, bud sprouting, leaf shedding and seed dormancy, and have wide commercial applications.

IAA promotes the growth of shoots in concentrations of about 10 parts per million (ppm), while roots respond to lower concentrations than this. Higher concentrations may be toxic, but as different plant species vary in their response to specific concentrations of auxin, certain synthetic plant hormones can be used as selective weedkillers. For example, 2,4-D (2,4-dichlorophenoxyacetic acid) can be applied to a lawn in such a concentration that the grasses in it are unaffected while the broad-leaved weeds like dandelions and daisies are killed.

8.6 ADDITIONAL EXPERIMENTS

Experiment 8.9 *To find the region of most rapid growth in radicles*

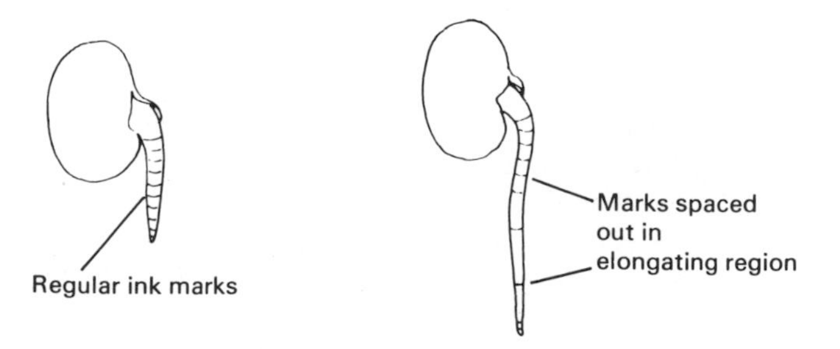

Fig. 8.15 *To find the region of elongation in radicles*

A few pea or bean seeds are germinated between blotting paper as described in Experiment 8.6, so that they develop straight radicles; they are allowed to grow until the radicles are 10 or 20 mm long. The radicles are marked with india ink lines 2 mm apart, and then arranged between two strips of moist cotton in a Petri dish; the lid of the dish is replaced, fixed in position with a rubber band, and placed on its edge with the radicles of the seeds pointing downward. The dish is then left in a dark cupboard or box. After a few days growth the region of most rapid elongation of the radicle will be apparent from the spacing of the ink lines (Fig. 8.15).

Experiment 8.10 *To find the region of geotropic response in radicles*

Seedlings with straight radicles are again set up in the apparatus used in Experiment 8.9, with their radicles marked with india ink in the same way, but arranged horozontally. Some of the radicles have 1 mm cut from their tips before the lid of the Petri dish is closed.

Result. After two or three days the direction of growth of the intact radicles has changed, and their tips are pointing vertically downward. It is clear from the marking on the radicles that the change of growth direction takes place in the region in which the growth is most rapid (Fig. 8.16). The radicles without tips, however, will probably continue to grow horizontally.

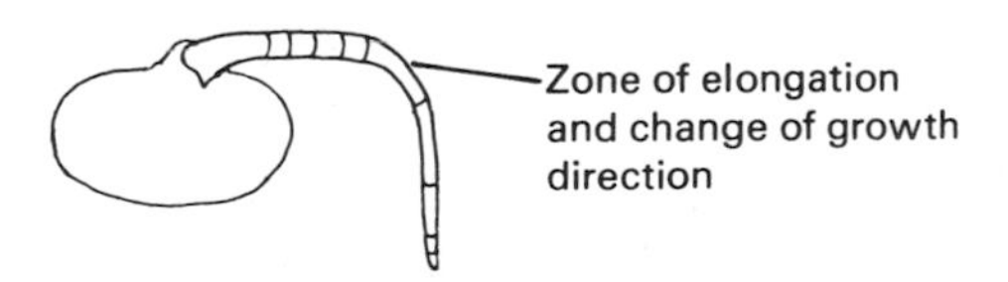

Fig. 8.16 *Some classical experiments to test the auxin theory*

Interpretation. The radicle tip is sensitive to the pull of gravity, but the growth response occurs in the region of maximum elongation.

Note. More experiments relevant to this unit are to be found in the laboratory manual *Germination and Tropisms;* see *Further Reading* at the end of this book.

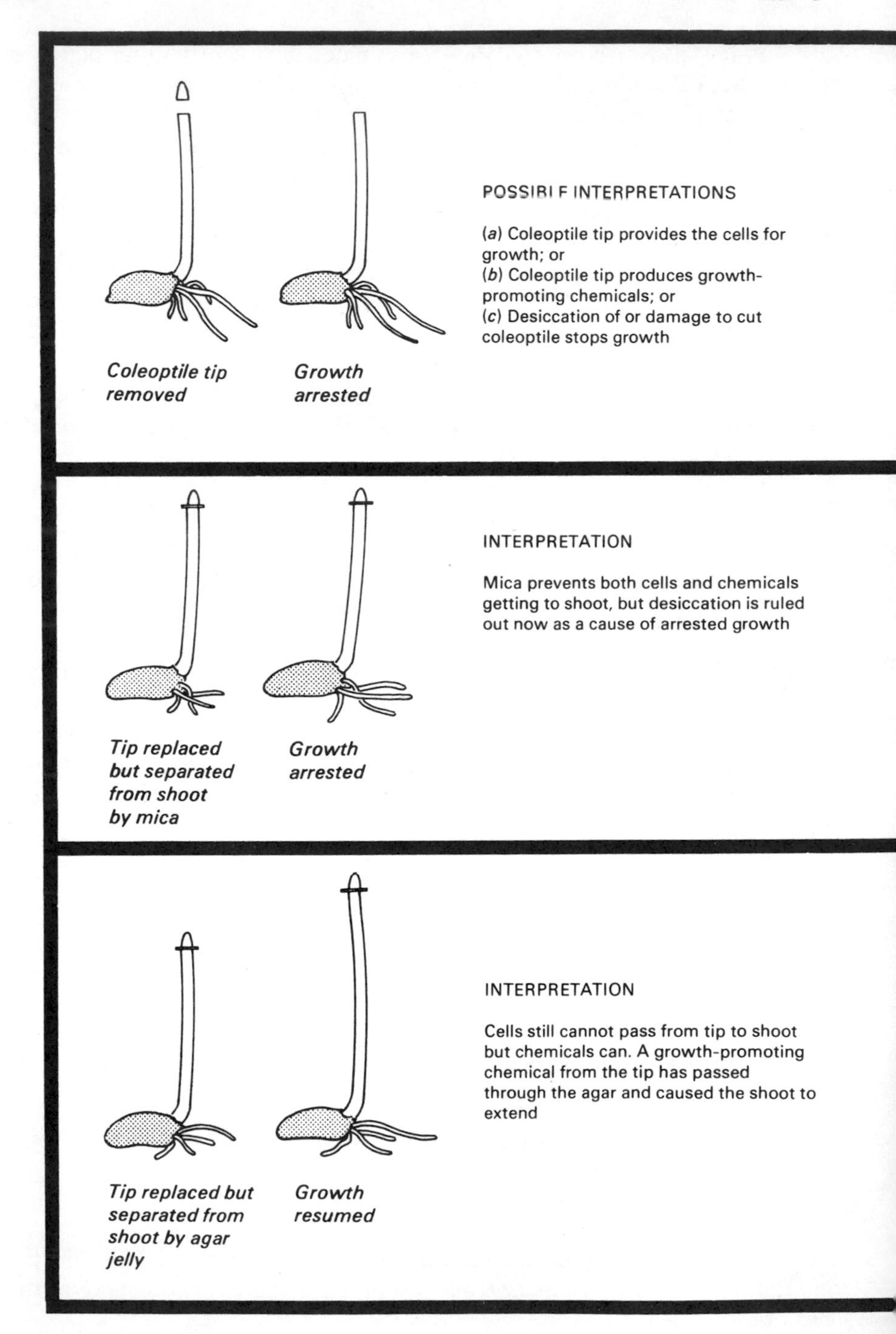

Fig. 8.17 *Response to one-sided gravity*

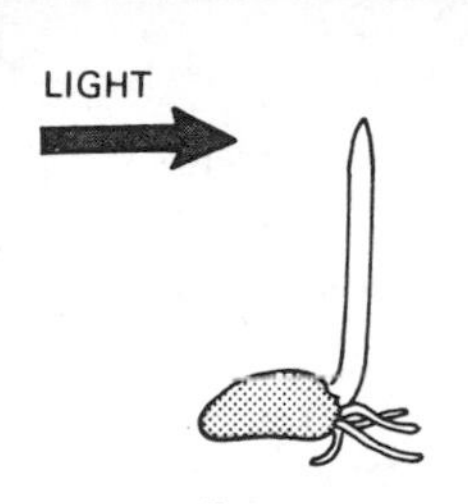

Coleoptile illuminated from one side

Grows toward the light

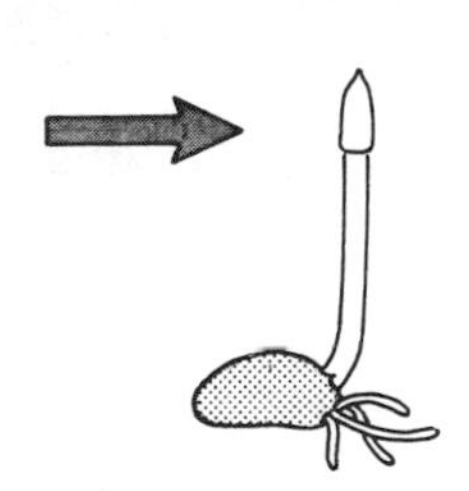

Tip covered with foil cap during illumination

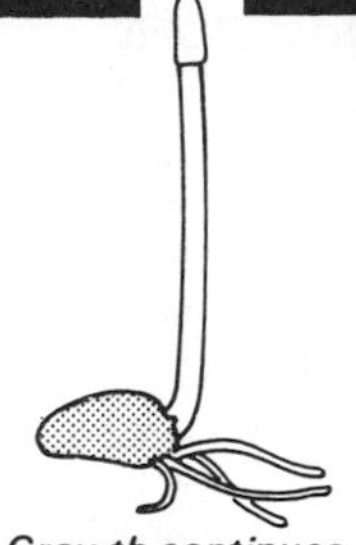

Growth continues but without curvature

INTERPRETATION

The coleoptile tip alone is sensitive to one-sided light but the response occurs below the tip. There must be some form of communication between the tip and the rest of the shoot

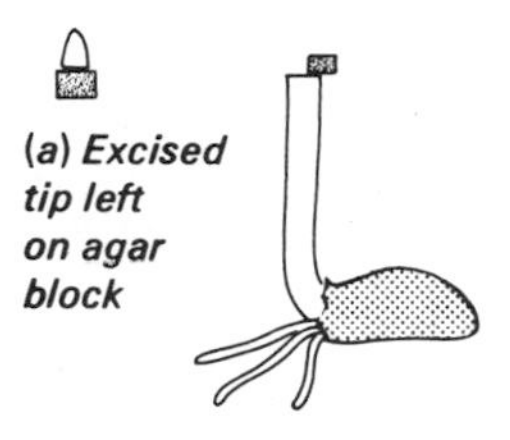

(*a*) *Excised tip left on agar block*

(*b*) *Agar block placed asymmetrically on decapitated coleoptile*

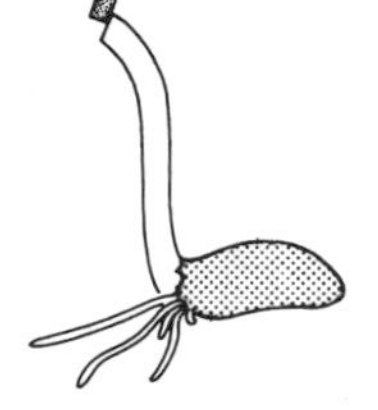

(*c*) *Coleoptile grows and bends as shown*

INTERPRETATION

A growth-promoting chemical has diffused from the coleoptile tip into the agar. In (*b*) the right side of the coleoptile receives more of the chemical than the left. The extra extension on the right side produced the curvature

(*After Went and Thimann,* Phytohormones, *Macmillan, 1937*)

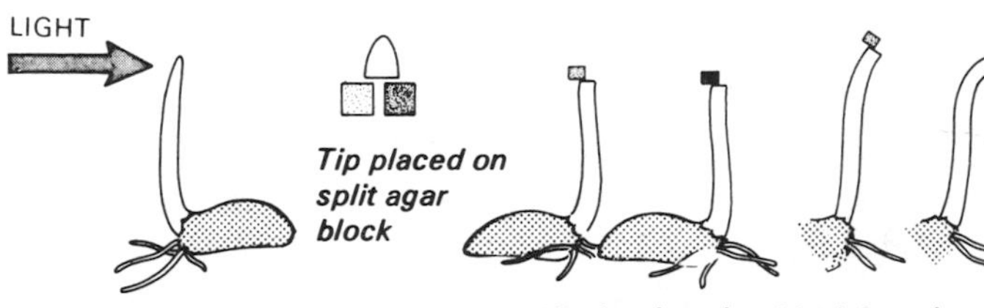

Coleoptile tip illuminated as shown

Tip placed on split agar block

Agar blocks placed on decapitated coleoptiles grown in darkness

Block from beneath illuminated side produces less curvature than that from the dark side

INTERPRETATION

Light reduces the amount of hormone reaching shoot from coleoptile tip. Therefore illuminated side grows less than dark side, so producing curvature

8.7 QUESTIONS

1. Flowering plants are made up principally of root, stem and leaf. In what form are these structures represented in a dicotyledonous seed?
2. How do the functions of cotyledons differ in a dicotyledon such as the french bean and a monocotyledon such as maize?
3. (*a*) How is the food stored in the cotyledons of a bean seed made available to the growing region? (*b*) How is the food used by the seedling? (*c*) At what stage of development does the seedling become independent of this stored food?
4. In terms of the auxin hypothesis explain why (*a*) shoots deprived of light grow very tall and (*b*) shoots illuminated from one side grow toward the light source.

Unit 9 Translocation And Transpiration

9.1 INTRODUCTION

In Unit Three we considered two processes, diffusion and osmosis, which are important in the movement of water and water-soluble compounds (like salts and sugars) into, out of and through cells. This unit is concerned with the way in which these substances move in plants. The transport of dissolved material through the plant is called *translocation,* and the loss of water by evaporation through the leaves is called *transpiration*.

9.2 TRANSLOCATION

In very general terms, water and dissolved salts from the soil travel upward through the xylem vessels while food made in the leaves passes downward or upward in the sieve tubes of the phloem (see Figs. 2.9(*c*) and 5.4).

(*a*) Translocation of Food Substances in the Phloem

The sugars produced in the leaves by photosynthesis are carried in the phloem from the leaves to the stem. They may then travel to any part of the plant that needs food: either up the stem to actively growing regions or to maturing fruits or seeds, or downward to the roots or underground storage organs. It is quite possible for substances to be travelling both upward and downward at the same time in the phloem.

The actual mechanism by which the sugars move is unknown: but it does depend on the fact that the sieve-tube cells in the phloem are *alive*. Anything which kills the phloem interrupts the movement of food. This can be demonstrated by *isotopic labeling*.

All chemical elements consist of atoms, and the differences between the chemical and physical properties of the various elements arise from the differences between their atoms. Most elements, however, are made up of mixtures of different kinds of atoms, even when they are perfectly chemically pure: these differing atoms are called *isotopes*. The isotopes of an element have identical chemical properties but they differ in the weight of their atoms. Naturally occurring magnesium, for example, is a mixture of three magnesium isotopes with atomic weights of 24, 25 and 26; they form 79, 10 and 11 per cent of natural magnesium respectively.

Some isotopes of certain elements are unstable and break down spontaneously: these are called *radioactive isotopes* or *radioisotopes*. When they disintegrate radiation is emitted, together with various kinds of subatomic particles, so that their presence is easily detected using either electronic equipment like the well-known Geiger counter, or photographic plates, which are blackened by some kinds of radiation.

Some radioactive isotopes occur naturally, like those of heavy elements such as uranium, but some of those most useful to biologists are made artificially. For example, natural carbon consists mostly of a single isotope with an atomic weight of 12, which is not radioactive. When it is bombarded with neutrons in an atomic pile part of it is converted into radioactive carbon with an atomic weight of 14 (written carbon-14 or ^{14}C).

By supplying radioactive isotopes to organisms, biologists are able to trace many of the chemical steps in metabolism. For example, if rats are fed with glucose containing radioactive carbon, ^{14}C, it is found that the carbon dioxide they exhale is radioactive.

$$^{14}C_6H_{12}O_6 + 6O_2 \rightarrow 6\,^{14}CO_2 + 6H_2O$$

This is direct evidence that the carbon in glucose is eventually converted to carbon dioxide in respiration.

Although radioisotopes are perhaps the easiest isotopes to detect and estimate, other isotopes can be used if suitable apparatus is available. For example, stable isotopes like oxygen-18 and nitrogen-15 can be determined by an instrument called a mass spectrometer.

If a single leaf of a plant is exposed to carbon dioxide labeled with radioactive carbon-14, it produces carbon-14-labeled sugars which can be tracked with a Geiger counter as they move from the leaf to other parts of the plant (Fig. 9.1). If the phloem in a short length of stem just below the leaf is killed by a jet of steam, the sugars are found only to move upward from the leaf. On the other hand, if the phloem just above the leaf is killed, the radioactive substances only move downward. If the phloem is killed both above and below the leaf, no radioactivity can be detected anywhere in the stem.

A similar experiment can be carried out in which the phloem is killed

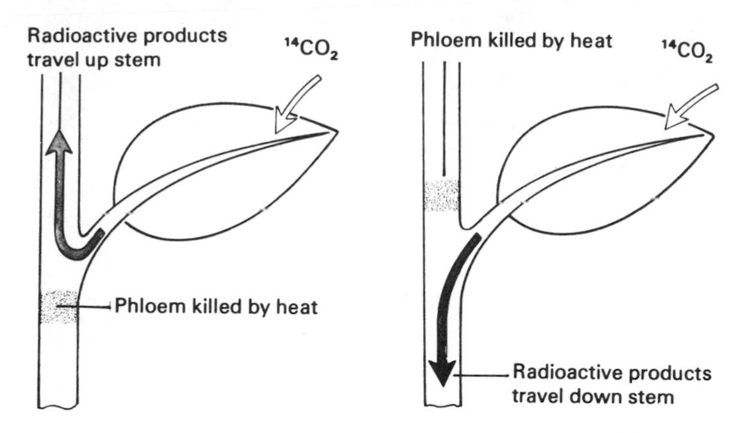

Fig. 9.1 *Experimental evidence for transport in phloem*
(After Rabideau & Burr, *Amer. J. Bot.*, 32, 1945)

by depriving it of oxygen. The results again indicate that translocation can only take place in a living stem.

These experiments confirm the importance of the phloem cells in translocation. It could be argued that the xylem vessels might also have a part in the process: but these vessels consist only of dead cells which could not be affected by heat or oxygen shortage. If translocation of sugars took place in the xylem it is difficult to see why it should stop completely after heat treatment.

Fig. 9.2 illustrates one theory, known as the *mass flow hypothesis,* which attempts to account for the movement of soluble substances in the phloem. It postulates that the driving force of the process arises from the turgor pressures in leaf cells, which are high compared with those in the roots (or other parts of the plant), where sugars are being used up either in respiration or by conversion into insoluble starch. The leaf cell's turgor pressure is high because of its production of sugar by photosynthesis, with the consequent intake of water from neighboring cells or veins (1). In the root cell, on the other hand, the removal of sugar leads to a rise in the osmotic potential of the cell sap and thus to a reduced turgor pressure (2). The mass flow hypothesis suggests that the pressure difference forces liquid carrying food substances in solution through the sieve tubes connecting the leaf with the root (3).

This hypothesis provides at best only a partial explanation of translocation in the phloem. It does not explain how food substances move upward and downward at the same time, nor why translocation stops when phloem cells are killed.

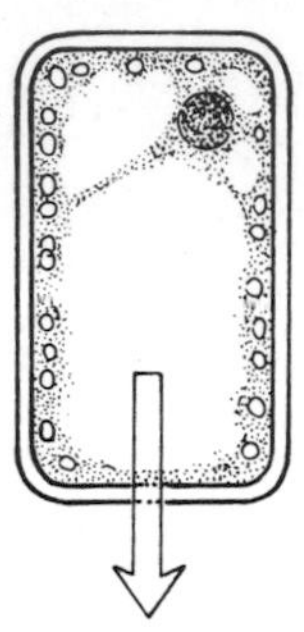

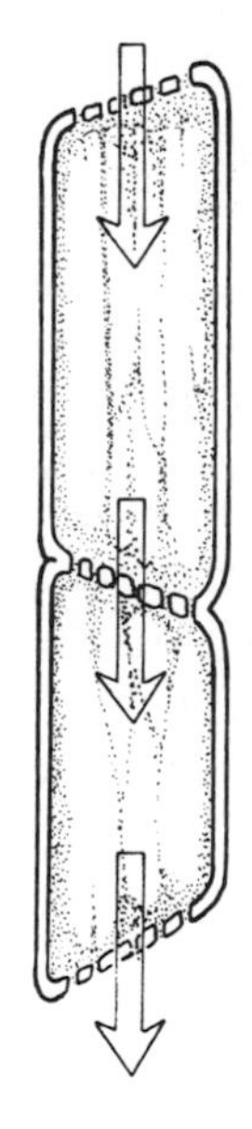

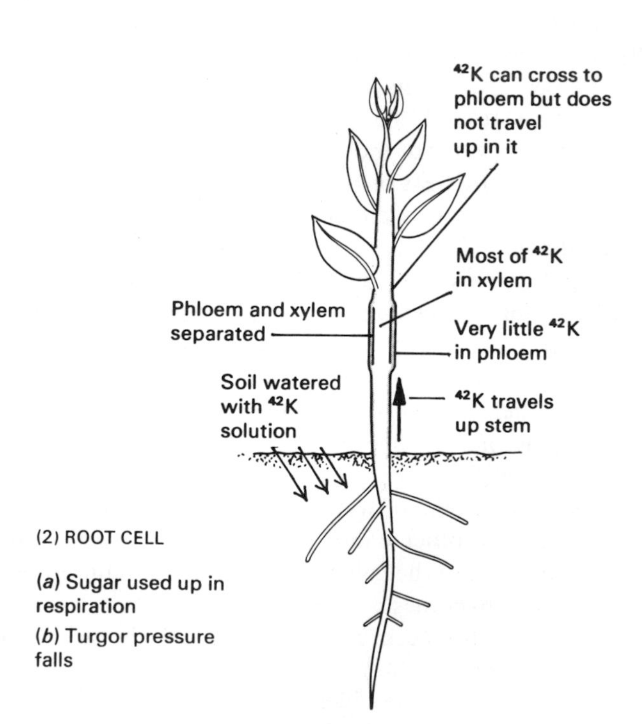

Fig. 9.2 *Translocation in the phloem*

Fig. 9.3 *Evidence for transport of salts in the xylem*

(b) Translocation of Salts in the Xylem

If a ring of bark is removed from a stem, together with its underlying phloem, the movement of salts in the plant is not much affected. But if a core of xylem is removed the upward movement of salts is arrested. Again, this movement is demonstrated elegantly using isotopic labeling.

The bark and phloem of a willow twig are gently raised from the xylem, and a layer of waxed paper is inserted underneath; then the bark and phloem are carefully replaced (Fig. 9.3). (Controls are set up to demonstrate that this treatment does no harm to the shoot.) The twig is watered with a solution of a potassium salt containing radioactive potassium-42 and after a period the potassium contents of the xylem and phloem of the twig are examined.

Although only the roots of the plant receive potassium-42, the radioisotope is found to have moved a considerable distance from the roots up the stem. In the region of separation by waxed paper, only a trace of radioactive potassium appears in the phloem: clearly, therefore, salts are not carried up the stem in the phloem. Considerable amounts of potassium-42 are, however, found in the xylem in this region, confirming its major role in the transport of potassium salts.

Above the region of separation, however, the radioisotope is found almost equally distributed between the xylem and the phloem. This suggests that salts can pass *across* the stem from the xylem to the phloem, and hence to neighboring cells.

The forces moving water and salts through the xylem are described in Section 9.3.

(c) Uptake of Salts by the Roots

There is as yet no wholly convincing explanation of the uptake of mineral salts from the soil by roots. Certainly salts will diffuse from a relatively high concentration in the soil to a lower concentration in the root cells; but there is experimental evidence that

(*i*) plants can absorb salts even when their concentration in the soil is less than that in the roots, and
(*ii*) anything which interferes with the plant's respiration also impairs the uptake of salts.

Uptake of salts, therefore, may well involve some process of "active transport" which can operate against a concentration gradient and which requires energy from respiration. It has been suggested that enzyme-like substances called *carriers* might combine with the salts, carry them across the cytoplasm and release them into the vacuole.

9.3 TRANSPIRATION

Transpiration is the process by which plants lose water by evaporation into the surrounding air. Evaporation takes place all over the plant's surface above the soil, but especially through its leaves.

Water is forced through the permeable walls of the mesophyll cells (see Section 5.3(*c*)) by their turgor pressure, and evaporates from the outer surface of the walls into the intercellular spaces. Thence it diffuses through the stomata into the atmosphere. Closure of the stomata thus greatly reduces evaporation from the leaf, although not entirely preventing it.

(*a*) Significance of Transpiration

Transpiration is probably an inevitable consequence of the adaptations of the leaf to its function of photosynthesis. The large surface area of most leaves is an advantage in the absorption of both sunlight and carbon dioxide, while clearly a leaf can only take up carbon dioxide if it is permeable to the gas. These characteristics of large surface area and ready permeability also encourage the loss of water vapor and hence promote transpiration.

Although excessive evaporation is obviously harmful to the plant, normal transpiration also has beneficial effects.

(*i*) **Transpiration stream.** Evaporation from the leaf cells reduces their turgor and increases the concentration of their cell sap, thus reducing their osmotic potential (see Section 3.4(*b*)). They are therefore able to take in water by osmosis from the neighboring cells and eventually from the xylem vessels. This withdrawal of water produces a tension—a pulling force—in the columns of water in the vessels, so that the pressure in them is actually below that of the atmosphere, and water is pulled up the vessels of the stem from the roots. This flow of water is called the *transpiration stream:* its rate of flow depends on the rate of evaporation from the leaves, and to some extent also on the root pressure.

It may seem surprising that a column of water can be subjected to tension in this way without breaking up, especially a column of great length such as must exist in trees perhaps 100 meters high. These columns, how-

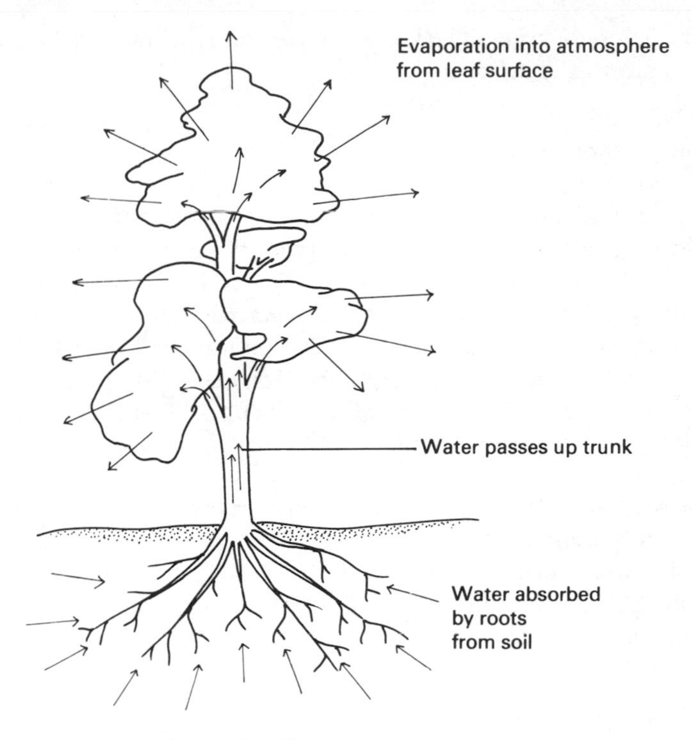

Fig. 9.4 *The transpiration stream*

ever, are contained in exceedingly narrow vessels, and it is suggested that the cohesive forces between molecules in such very thin columns are less easily overcome than those in, say, drinking straws or domestic water pipes. Given this assumption, this *cohesion theory* provides an adequate explanation of water movement through plants and trees.

A large tree on a hot day may evaporate hundreds of liters of water from its leaves; this enormous quantity is replaced by absorption through its roots. Only a tiny fraction of all the water passing through a plant is retained for use in photosynthesis and for maintaining the turgor of the cells.

(ii) ***Transport of salts.*** The transpiration stream is the means by which salts are carried from the roots, where they are absorbed, to the rest of the plant. However, the rate at which salts are absorbed from the soil is not directly dependent on the rate of transpiration.

(iii) **Cooling.** When water changes from a liquid to a vapor it absorbs considerable amounts of heat (*latent heat*) from its surroundings. The cooling effect of rapid evaporation from the surface of the leaves almost

certainly helps to keep the plant's temperature below a damaging level when it is growing in direct sunlight.

(b) **Transpiration Rate**

The conditions which affect the rate of transpiration are mostly those which have some effect on evaporation from the plant's surface.

(i) **Humidity.** When the air is nearly saturated with water vapor, it can absorb little more from the plants and transpiration will be reduced. In dry air the water diffusion gradient across the stomata is "steeper," diffusion of water out of the leaf is faster (see Experiment 3.1) and evaporation and hence transpiration are also rapid.

(ii) **Temperature.** Warm air is able to hold more water vapor than cold, therefore evaporation is faster when the air temperature is high. When the leaf itself is warm evaporation is similarly speeded up. Even on a cold day direct sunlight has this effect, since the leaf absorbs radiant energy and is warmed up in this way.

(iii) **Wind.** On a still day, the air around a transpiring leaf gradually becomes saturated with water vapor until no more can be taken up from the leaf; in consequence transpiration is much reduced. In moving air, the water vapor is swept away from the leaf as fast as it diffuses out, so that the diffusion gradient is maintained and transpiration continues rapidly.

(c) **Control of Transpiration**

(i) **Stomata.** Evaporation from a plant takes place mostly through its stomata, so that when these close transpiration slows down considerably. There is little evidence, however, to suggest that a high rate of evaporation results in the stomata closing, although they do close under extreme conditions: when loss of water greatly exceeds uptake the plant wilts, the leaf cells, including the guard cells, become flabby, and the stomata close. The movements of the stomata normally depend on light intensity, so that they are generally open during the day and closed at night. Water loss is therefore at a minimum during darkness.

(ii) **Leaf-fall.** Deciduous trees shed their leaves in winter (Section 5.3(*d*)). They are thus protected from excessive water loss at a time when its replacement could be difficult because of frozen soil, or because the water in the vessels of the stem or trunk is itself frozen.

(iii) **Leaf characteristics.** Most evergreen plants—that is, plants which retain their leaves throughout the winter—have leaves which are adapted in ways which reduce transpiration to a minimum. These adaptations include thick waxy cuticles which are relatively impermeable to water, stomata sunk below the level of the surrounding epidermis, and reduced surface area, as in the "needles" of pines and other conifers. Characteristics like these are especially useful to the plant in the difficult conditions of winter and are also often seen in plants which grow in dry conditions.

9.4 EXPERIMENTS ON TRANSPIRATION

Experiment 9.1 *To show that water is given off by a plant during transpiration*

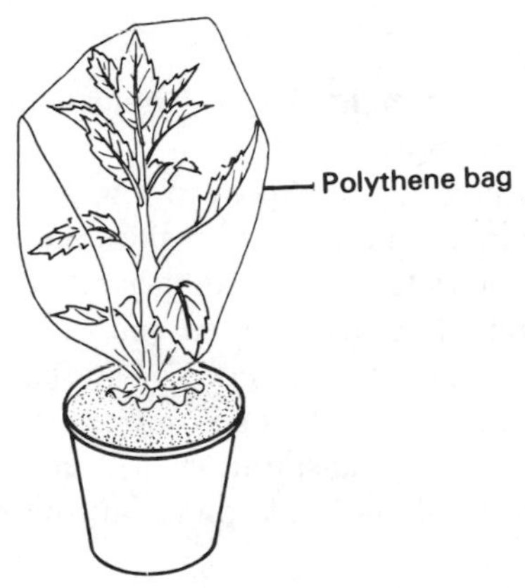

Fig. 9.5

The shoot of a recently watered potted plant is completely enclosed in a polyethylene bag, which is tied around the base of the stem (Fig. 9.5) and the plant is left for an hour or two in bright sunlight. The air in the bag soon becomes saturated, and droplets can be seen to condense on the inner surface of the polyethylene. The condensate can be collected by removing the bag and shaking all the liquid into one corner, and its identity can be confirmed by testing it with anhydrous copper sulphate, a white powder which turns blue in the presence of water.

Experiment 9.2 *To measure transpiration rate by loss of weight*

(a) **Cut shoot.** The only direct method of measuring transpiration is to determine the weight of water lost by a plant in a given time. It can be

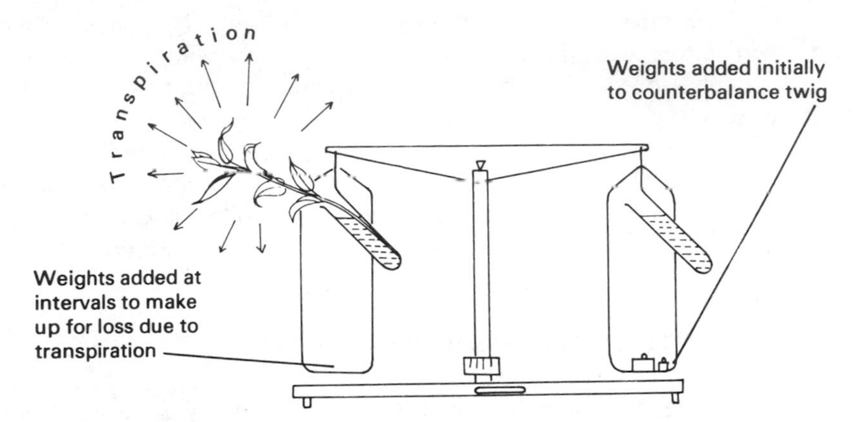

Fig. 9.6 *To measure the rate of transpiration in a cut shoot*

assumed that changes in weight due to photosynthesis or respiration are small compared with transpiration losses.

The experiment is set up as shown in Fig. 9.6, and weights are placed on the right-hand scale-pan until the two sides are just balanced. The purpose of the test-tube of water on the right-hand side is to compensate for water loss by evaporation from the open end of the test-tube on the left; this effect will be the same on both sides of the balance. Thus any loss of weight on the left-hand side will be due only to transpiration by the shoot. The weight of water lost by transpiration in a measured time is found by adding weights to the left-hand scale-pan until the pans are just balanced again.

By setting up the apparatus in various atmospheric conditions the effect of these on the transpiration rate can be estimated, but comparisons are more easily made using a potometer (see Experiments 9.3 and 9.4).

(b) **Potted plant.** The pot of a well-watered potted plant is wrapped in polythene sheet or other impermeable material, which is tied firmly around the base of the stem to prevent direct evaporation from the soil or pot. The whole plant is then weighed at intervals to find the loss in weight due to transpiration.

Experiment 9.3 *To demonstrate water uptake by a cut shoot*

This experiment uses a *potometer* (Fig. 9.7). The advantage of the type of potometer illustrated here is that it uses quite small amounts of water, so that volume changes produced by temperature variations are unlikely to be large enough to interfere with the results.

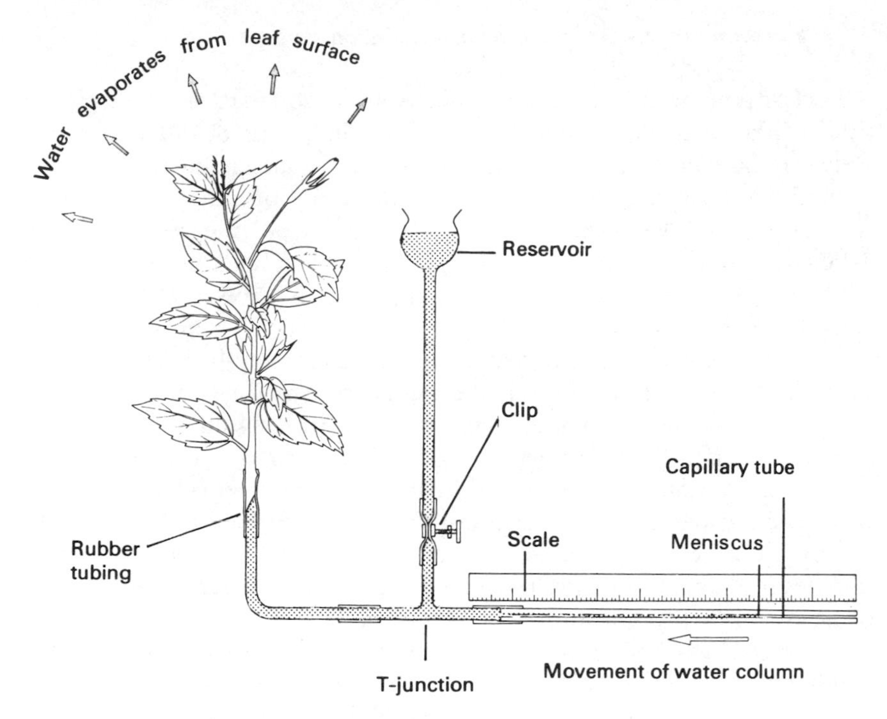

Fig. 9.7 *The potometer*

A leafy shoot is cut from a plant and at once placed in water to prevent air entering the xylem vessels in the stem. The potometer is filled with water, and the cut end of the shoot fitted into the rubber tubing, care being taken to avoid the inclusion of any air bubbles.

The water lost by evaporation from the leaves is replaced by water from the stem, which in turn draws water from the potometer tube. The clip below the funnel is closed, so that all of this water is withdrawn from the capillary tube. Here the meniscus at the end of water column can be seen to move quite rapidly as water is taken up by the shoot and air is drawn in behind the retreating column. The rate of water uptake can be determined by timing the movement of the meniscus over a fixed distance on the scale.

It must be emphasized that the potometer measures water uptake rather than water loss (compare Experiment 9.2). Not quite all the water taken up escapes from the leaves into the air: a small proportion reacts with carbon dioxide to produce sugars by photosynthesis. However, if the light intensity is kept constant this proportion is fixed and can therefore be disregarded when the potometer is used to compare transpiration rates.

Experiment 9.4 *To compare transpiration rates*

The potometer can be used effectively to compare the transpiration rates of the same shoot under different conditions, or of different shoots. Between experiments, the meniscus is sent back along the capillary tube toward the beginning of the scale by momentarily opening the clip below the funnel, thus letting in a little more water from the reservoir. (The T-junction connection to the funnel has the further function of preventing air from reaching the cut stem; if the apparatus is allowed to run for too long the excess air drawn into it will collect in the funnel stem.)

The statements made in Section 9.3(*b*) can easily be tested using the potometer. The rate of water uptake can be measured when the apparatus is standing in a warm room, in a cold room, in front of an electric fan and so on. As long as the light intensity is kept more or less constant, the readings on the potometer scale made after the same fixed period of time are in proportion to the differences in transpiration rate. Before timing is started in any given set of conditions, the apparatus should be left for several minutes in the new situation to allow the new rate of transpiration to become established and steady.

As well as studying the transpiration rate of one shoot under different conditions, it is interesting to compare the transpiration rates of different species, say of shoots of laurel (an evergreen) and beech (deciduous) which have roughly the same leaf area. The dependence of transpiration rate on leaf area can be studied by observing how the rate varies in a shoot while the leaves are picked off one by one over a period of time, or by coating the leaves one at a time with an impermeable layer of petroleum jelly.

Experiment 9.5 *To find which surface of a leaf loses more water vapor*

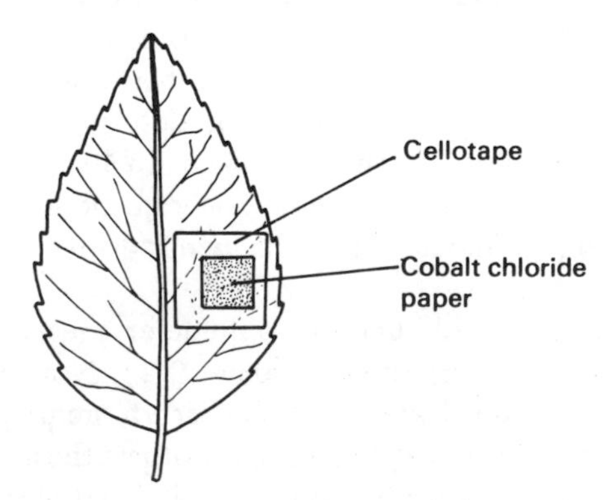

Fig. 9.8 *To find which surface of a leaf loses more water vapor*

Cobalt chloride paper is used in this experiment to detect the presence of water: when quite dry it is blue in color, but in damp conditions it turns pale pink.

Small squares of cobalt chloride paper taken straight from a desiccator are stuck as quickly as possible to the upper and lower surfaces of a leaf blade with "Cellotape" (Fig. 9.8); prominent veins should be avoided to ensure an air-tight seal. The difference in the rates of water loss from the two surfaces can be estimated by comparing the times taken for the two squares of paper to change color.

Experiment 9.6 *To investigate the distribution of stomata in a leaf*

A small leaf is held in forceps and plunged into a beaker of very hot water (about 80° C). The rise in temperature expands the air in the intercellular spaces of the leaf, and it escapes as tiny bubbles through the stomata. The number of bubbles appearing on the upper and lower surfaces of the leaf can be counted and compared; generally they are considerably more numerous on the lower surface. (It could be argued that the air bubbles might not come from inside the leaf but from air dissolved in the water collecting on the leaf's surface. How would you set up a control experiment to eliminate this possibility?)

The result of this experiment can be checked by stripping pieces of epidermis from the two surfaces of the leaf and examining them under the microscope: in this way the number of stomata in, say, 1 cm^2 of epidermis can be determined directly.

If the same species of leaf is used in both Experiments 9.5 and 9.6, it is interesting to compare the ratio of the numbers of stomata on the two surfaces with the ratio of the times taken for the color change of the two squares of cobalt chloride paper.

Experiment 9.7 *To demonstrate water transport in the vascular bundles*

In this experiment the water is made more easily visible by coloring it with a dye such as neutral red or methylene blue. Several cut shoots are placed in a solution of the dye. One is removed after five minutes, a second after ten minutes and so on. The stems are then sliced across with a razor blade: the stained vessels can be clearly seen. The height to which the dye solution has risen in the stem can be determined by cutting sections higher and higher up the stem.

9.5 QUESTIONS

1. Why should cutting a deep ring of bark from a tree cause its death in view of the facts that (*a*) water and salts can travel up to the leaves in the xylem and (*b*) the leaves can still manufacture food by photosynthesis?
2. Explain how it could come about that a plant could (*a*) take in more water than it is losing by transpiration, or (*b*) lose more water by transpiration than it is taking in from the roots, at least for a time.
3. Outline the path taken by water from the soil through the roots, stem and leaves of a plant and into the atmosphere as vapor. Explain briefly the forces causing its movement at each stage.

Unit 10
The Nutrition Of Green Plants

10.1 TYPES OF NUTRITION

Section 1.2 discussed briefly how animals and plants differ in the way in which they feed. In general, animals feed by taking in complex food materials from the tissues of plants or other animals: these substances are broken down by digestion into simple compounds which are then absorbed and either incorporated into the cells of the animal's body or oxidized to obtain energy. These processes are described as *holozoic nutrition.*

Most green plants, on the other hand, feed by taking in very simple substances—carbon dioxide, water, inorganic salts—and building them up into complex compounds which are either synthesized into protoplasm and cell walls or oxidized to release energy for the plant's needs. This method of feeding is called *holophytic nutrition.*

Certain plants, however, can feed by taking in fairly complex compounds from the dead tissues of animals or plants: some food substances are absorbed directly, while others are broken down outside the plant by enzymes produced by it. These plants are called *saprophytes* and their method of feeding *saprophytic nutrition. Parasites* derive their food from other individuals or *hosts* without necessarily killing them, by absorbing blood, sap, digested food or the actual tissues of plants or animals. For example, fleas suck the blood of mammals or birds, aphids and mistletoe both feed on the sap of green plants, and tapeworms live in the alimentary canals of their hosts and absorb the digested food which is there. Some parasites are so well adapted to their hosts that they do not produce any obvious symptoms of disease.

The rest of this unit is concerned with the nutrition of green plants, or holophytic nutrition.

10.2 PHOTOSYNTHESIS

Plants contain many different kinds of compounds, like carbohydrates, fats and proteins, the nature of which is discussed more fully in Section 19.3.

The synthesis of carbohydrates from carbon dioxide and water is however of especial interest, since from these are made all the other substances needed for the build-up of protoplasm and cell walls.

Carbon dioxide and water will only combine to form carbohydrates if energy is available to drive the reaction. In green plants the energy for the first stages of this synthesis is obtained from sunlight, which is absorbed by their chlorophyll. The reaction, or rather series of reactions, is therefore called *photosynthesis* (from the Greek words meaning *light* and *build*); it proceeds in all the green parts of the plants, but principally in the leaves. The plant obtains the necessary carbon dioxide through its stomata, and the water (in land plants) from the soil through its root system.

The process may be summarized as follows:

$$6CO_2 + 6H_2O + \text{energy} \rightarrow C_6H_{12}O_6 + 6O_2$$

though it must be realized that this shows only the beginning and end of a very complicated chain of chemical reactions involving many intermediate compounds, each stage controlled by its specific enzyme.

The leaves of most green plants are well adapted to their function of photosynthesis (see Fig. 5.12).

(*a*) Their broad, flat shape offers a large surface area for absorption of sunlight and carbon dioxide.
(*b*) Most leaves being thin, the distances across which carbon dioxide has to diffuse to reach the mesophyll cells from the stomata are very short.
(*c*) The large intercellular spaces in the mesophyll provide an easy passage through which carbon dioxide can diffuse.
(*d*) The many stomata on one or both surfaces allow the exchange of carbon dioxide and oxygen with the atmosphere.
(*e*) In the palisade cells the chloroplasts are more numerous than in the spongy mesophyll cells. The palisade cells being on the upper surface receive most sunlight and this is available to the chloroplasts without being absorbed by too many intervening cell walls. The elongated shape of many palisade cells may confer the same advantage.
(*f*) The branching network of veins provides a ready water supply to the photosynthesizing cells.

(*a*) Photosynthesis in a Palisade Cell (Fig. 10.1)

The photosynthesizing cell obtains its water from the nearest vein by *osmosis* (see Section 3.3). Carbon dioxide reaches it from the adjacent air spaces by *diffusion* (see Section 3.2) through the cellulose wall and into the cytoplasm. Photosynthesis actually takes place in the chloroplasts, which contain the chlorophyll. Oxygen is released as a by-product; it diffuses out

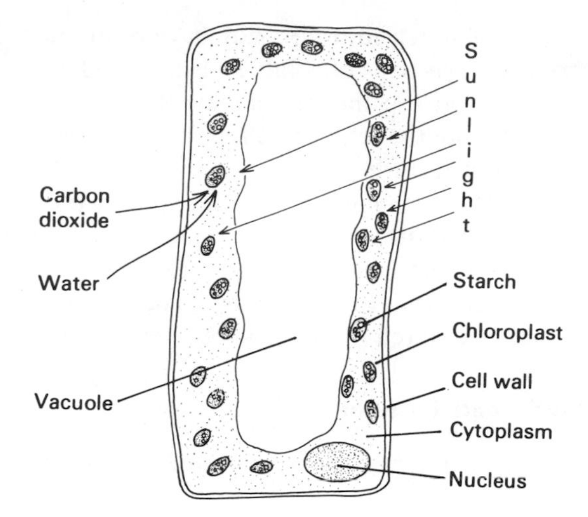

Fig. 10.1 *Photosynthesis in a palisade cell*

of the cytoplasm through the cell wall into the air spaces of the mesophyll, and eventually through the stomata to the atmosphere.

Carbohydrate production by photosynthesis goes on throughout the hours of daylight. Carbohydrates with small molecules, like glucose and other simple sugars, are produced at first but as photosynthesis proceeds the small molecules link together to form much larger molecules like those of starch. Unlike glucose, starch is insoluble in water, so that if a palisade cell is examined under a microscope after several hours of photosynthesis, grains of starch can be seen in the chloroplasts. However, enzymes are acting on the starch all the time, breaking it down to simple, soluble sugars which diffuse in solution out of the cells and are carried away in the sieve tubes of the phloem. The sugars may travel to the storage organs of the plant, where they are converted back into starch, or to the actively growing parts of the plants, where they can be

(*a*) oxidized in respiration to provide energy for chemical reactions in the cell;
(*b*) concentrated in the cell sap; this enables the cell to take in more water by osmosis, increase its turgor and so extend;
(*c*) converted to cellulose to be built into new cell walls or to thicken existing ones;
(*d*) used in the synthesis of other compounds needed by the plant, such as fats, proteins, pigments and so on.

Photosynthesis stops altogether in the absence of light in most plants, but enzyme action continues in the dark, converting starch to sugar and removing it. Little or no starch is left in the leaf after several hours of darkness.

The formation of starch in leaves may have a protective function. In bright light, very rapid synthesis of soluble sugars could produce such high concentrations in the cell sap that the intake of water by osmosis might actually damage or disrupt the cells. Starch, being insoluble, can accumulate without causing osmotic disturbances.

Clearly, if starch is accumulating in a leaf photosynthesis is probably going on. The detection of starch in a leaf, which is quite a simple matter, is therefore reasonably good evidence for photosynthesis. However, the reverse is *not* necessarily true for all plants; some species, including irises and lilies, never produce starch in their leaves, even in bright sunlight.

(b) Photosynthesis and Respiration

The equation summarizing the processes of photosynthesis

$$6CO_2 + 6H_2O + \text{energy} \rightarrow C_6H_{12}O_6 + 6O_2$$

is the reverse of the equation given in Section 10.2 summarizing aerobic respiration:

$$C_6H_{12}O_6 + 6O_2 \rightarrow 6CO_2 + 6H_2O + \text{energy}$$

Although respiration in green plants goes on all the time, photosynthesis can only proceed when the necessary energy is available, that is, when light is reaching the plant. In darkness, therefore, the plant takes in oxygen and produces carbon dioxide as a result of its respiration. On the other hand, in bright sunlight its rapid photosynthesis produces more oxygen than is used up in respiring, and takes up more carbon dioxide than is formed during its respiration: the overall effect is that carbon dioxide is taken in and oxygen given out by the plant. It can be shown experimentally that photosynthesis proceeds more slowly in poor lighting conditions than in bright sunlight. In dim light, therefore, the plant's rate of photosynthesis may just equal its rate of respiration, so that all the carbon dioxide produced by respiration is used up in photosynthesis, while all the oxygen produced by photosynthesis is used up in respiration, and there is no overall exchange of gases with the atmosphere. At this point the rate of carbohydrate formation by photosynthesis just equals the rate of carbohydrate breakdown by respiration, and the plant is said to have reached its *compensation point.*

10.3 EXPERIMENTS ON PHOTOSYNTHESIS

In some of the experiments which follow, the presence of starch is regarded as evidence for photosynthesis. Starch gives a blue coloration with iodine, but this is not easily seen in the presence of chlorophyll, and the following procedure should be used in each of these experiments.

Testing a Leaf for Starch

(*a*) The leaf is detached and dipped in boiling water for half a minute. This kills the protoplasm by destroying the enzymes in it, and so prevents any further chemical changes. It also makes the cells more permeable to iodine solution.

(*b*) The leaf is boiled in methylated spirit until all the chlorophyll is dissolved out. (*Caution: methylated spirit vapor is highly inflammable* and the spirit should *not* be heated over a naked flame: a water bath must be used, as shown in Fig. 10.2.) This leaves the leaf white and makes color changes caused by the interaction of starch and iodine much easier to see.

(*c*) Methylated spirit makes the leaf brittle and hard, but it can be softened by dipping it once more into boiling water. It is then spread out flat on a white surface such as a glazed tile.

(*d*) Iodine solution is placed on the leaf. Any parts which show a blue color have starch in them. If no starch is present the leaf is merely stained brown by iodine.

The following experiments are designed to verify some of the statements made in Section 10.2 concerning the conditions necessary for photosynthesis. Since in each experiment the presence of starch is regarded as evidence for photosynthesis, care must be taken to see that the leaves should not contain any starch at the start of the experiment. If they are

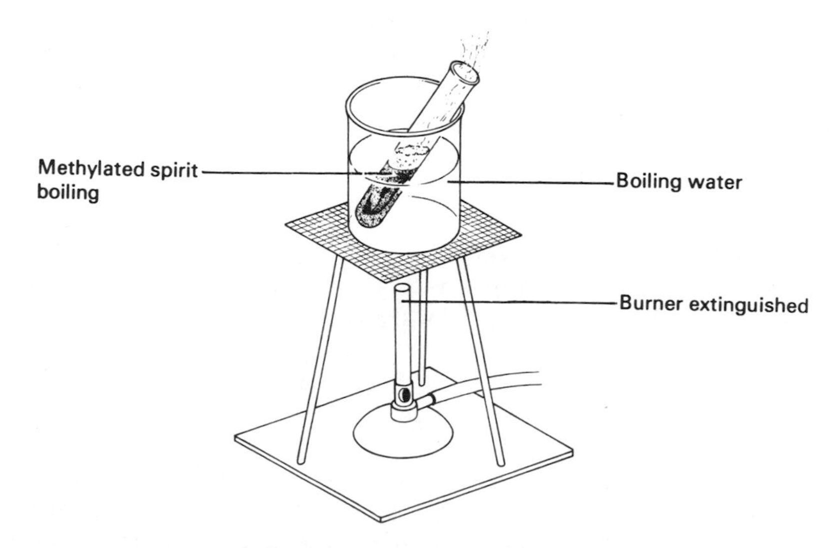

Fig. 10.2 *To remove chlorophyll from a leaf*

potted plants they can easily be *destarched* by leaving them in a dark cupboard for two or three days before the experiment. If plants growing in the open are to be used, the selected leaves should be completely wrapped in aluminum foil for two or three days. In any case one leaf should always be tested before the experiment begins to make sure that destarching is complete: if starch is detected then destarching should be allowed to continue for another day or so.

Experiment 10.1 *To determine whether chlorophyll is necessary for photosynthesis*

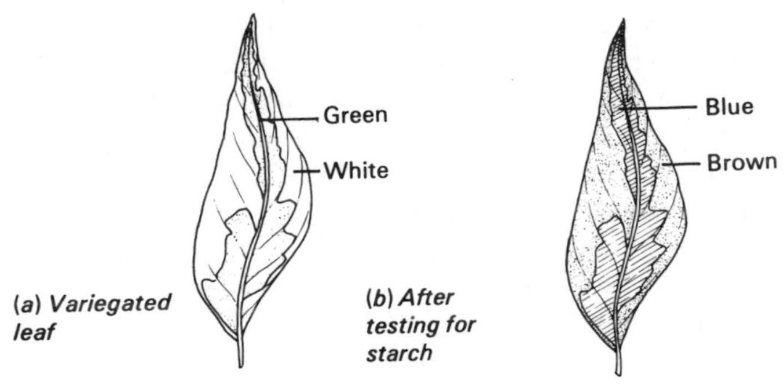

Fig. 10.3

Chlorophyll cannot be removed from a leaf without killing it, and dead leaves cannot photosynthesize. A naturally variegated (green and white) leaf, having chlorophyll only in patches, is therefore used: leaves like this are found in tradescantias, some pelargoniums ("geraniums") and the cornus shown in Fig. 10.3(*a*). After a period of destarching, the leaf on the plant is exposed to daylight for a few hours. It is then detached, drawn carefully to show the distribution of chlorophyll, and tested for starch as described above.

Result. When the leaf is tested with iodine solution, part of it turns brown and part turns blue. Comparison with the drawing shows that the part stained blue coincides with the part of the leaf which was previously green.

Interpretation. Starch is present only in those parts of the leaf which were green, and it seems reasonable to suppose that photosynthesis goes on only where chlorophyll is present. However, there are other possible interpretations of the result: for example, that sugars alone are made in the white parts of the leaf and that chlorophyll is needed only for their conversion into starch. Alternative explanations like these can be tested by further experiments.

Experiment 10.2 *To determine whether light is necessary for photosynthesis*

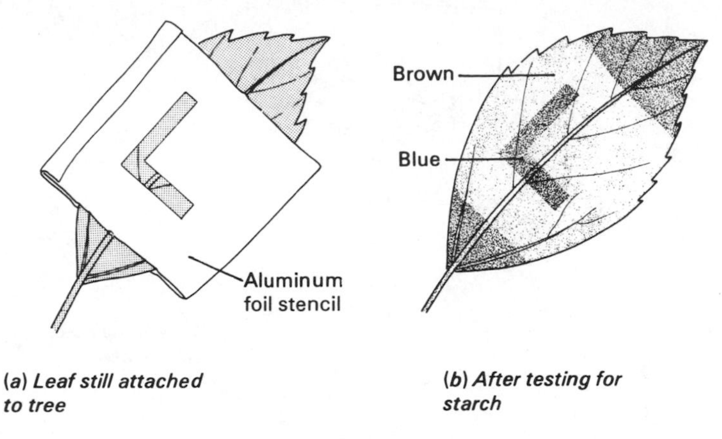

Fig. 10.4

A simple shape is cut out from a piece of aluminum foil, making a stencil which is attached to a previously destarched leaf (Fig. 10.4). The leaf is left on the tree or plant for four to six hours, and then detached and tested with iodine for starch as before.

Result. Starch is detected only in those parts of the leaf that received light.

Interpretation. As no starch is found in the parts of the leaf which were kept in darkness, it can be assumed that light is necessary for starch formation and hence probably for photosynthesis. However, it might perhaps be argued that the foil prevented carbon dioxide from reaching the leaf, and that it was shortage of the gas, rather than the absence of light, which prevented photosynthesis. A control experiment could therefore be carried out using a transparent stencil made of polythene sheeting in place of the aluminum stencil. (In any case, since the leaf is producing carbon dioxide by its own respiration, it could not be completely deprived of the gas by the stencil.)

Experiment 10.3 *To determine whether carbon dioxide is necessary for photosynthesis*

Two destarched potted plants are watered and the shoots enclosed in plastic bags (Fig. 10.5). One bag contains a dish of soda lime, which rapidly absorbs carbon dioxide from the air, and the other contains sodium bicarbonate solution, which tends to lose carbon dioxide to the atmosphere.

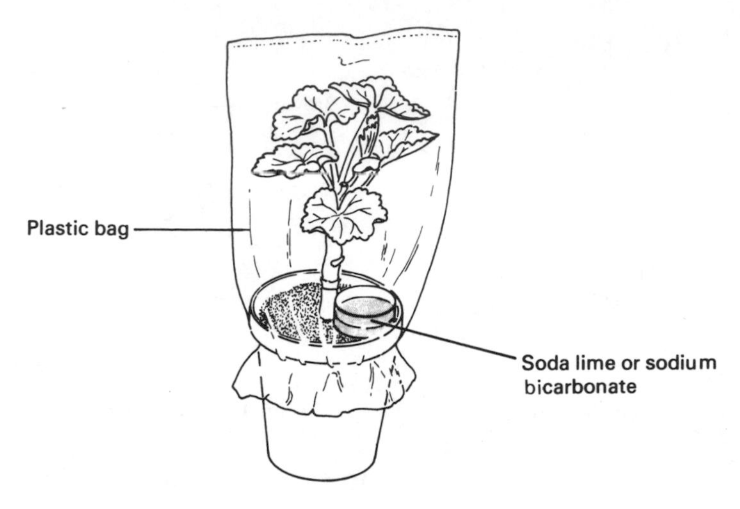

Fig. 10.5

Thus the air around one plant is kept free of carbon dioxide, while a supply of the gas is maintained to the other. Both plants are placed in sunlight or under a fluorescent light for several hours and a leaf from each is then detached and tested for starch.

Result. Starch is detected only in the leaf that was kept supplied with carbon dioxide.

Interpretation. As no starch is found in the leaf which was deprived of carbon dioxide, it is reasonable to assume that carbon dioxide is necessary for photosynthesis. The production of starch by the control plant rules out the possibility that any unusual humidity or temperature conditions in the plastic bag prevent normal photosynthesis.

Experiment 10.4 *To determine whether oxygen is produced during photosynthesis*

A short-stemmed funnel is placed over some Canadian pond weed in a beaker of water, preferably pond water, and a test-tube filled with water is inverted over the funnel-stem (Fig. 10.6). The funnel is raised above the bottom of the beaker to allow free circulation of water. The apparatus is placed in sunlight, and bubbles of gas soon appear from the cut stems, rise and collect in the test-tube. When sufficient gas has collected the test-tube is removed. A wooden splint is lit and the flame blown out so that only a red glow remains, and the glowing splint inserted into the tube of

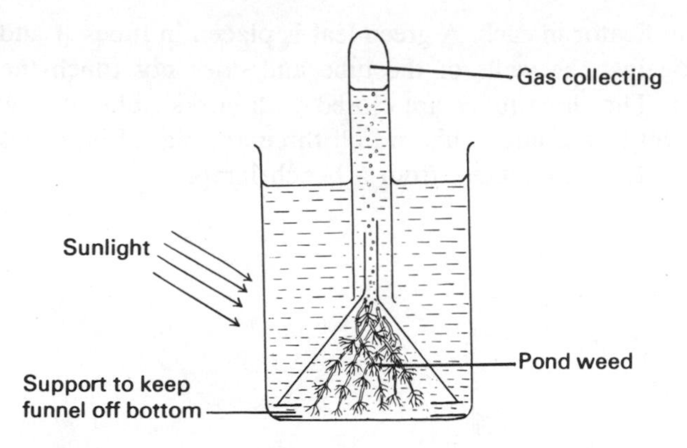

Fig. 10.6 *To demonstrate oxygen production during photosynthesis*

gas. (A control experiment may be set up and placed in a dark cupboard; little change takes place, although a small quantity of gas may possibly collect in the test-tube.)

Result. The glowing splint bursts into flames.

Interpretation. Oxygen will relight a glowing splint. The experiment does not prove that the gas collected is *pure* oxygen but it does show that in the light, and only in the light, this particular plant gives off a gas which is considerably richer in oxygen than is atmospheric air.

Experiment 10.5 *To demonstrate gaseous exchange during photosynthesis*

This experiment is based on the fact that a solution of carbon dioxide in water acts as a weak acid. Changes in acidity can be detected using a solution of two dyes, cresol red and thymol blue, mixed with sodium bicarbonate solution (the mixture is known as *bicarbonate indicator*). The original orange color of this solution represents the acidity produced by normal amounts of carbon dioxide present in air. If the carbon dioxide concentration rises some of the excess dissolves in the indicator solution; making it more acid, and its color changes from orange to yellow. If, on the other hand, the level of atmospheric carbon dioxide falls the solution loses carbon dioxide to the air, becomes less acid, and its color changes to red or purple.

Three test-tubes, *A*, *B* and *C*, are washed out successively with tap water, distilled water and bicarbonate indicator before placing 2 cm^3 bicarbonate indicator in each. A green leaf is placed in tubes *A* and *B* so that it is held against the walls of the tube and does not touch the indicator (Fig. 10.7). The three tubes are closed with corks, tube *A* is wrapped in aluminum foil to exclude light, and all three are placed in a rack in bright sunlight or a few centimeters from a bench lamp.

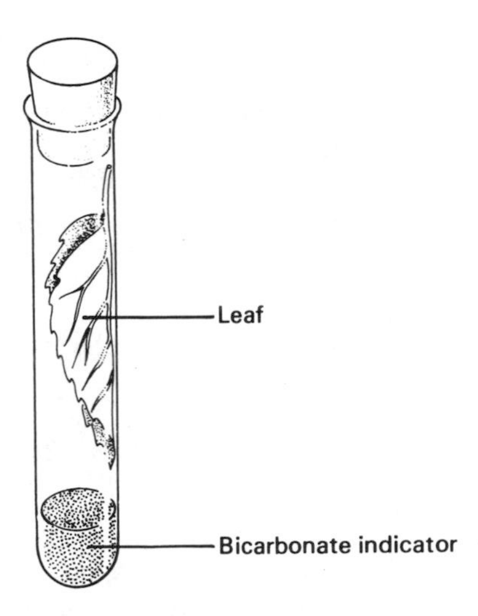

Fig. 10.7

Result. After about 40 minutes the bicarbonate indicator in tube *A* changes color from orange to yellow, and that in *B* to scarlet or purple. The color of the solution in the control tube *C* does not change.

Interpretation. The control experiment proves that it is the presence of the leaf which has produced the color changes in the indicator solution. The color changes in *A* and *B* are evidence for the production of carbon dioxide during respiration and for the overall uptake of carbon dioxide when respiration and photosynthesis proceed together in bright light.

This interpretation of the experiment can be criticized on the grounds that color changes in bicarbonate indicator are produced not only by carbon dioxide but by any acid or alkali. The result in tube *B* would be the same if during photosynthesis the leaf was producing an alkaline gas like ammonia, and that in tube *A* would be the same if the leaf were producing any acid gas during respiration. However, our knowledge of the metabolism of the leaf suggests that changes in carbon dioxide concentration in the tubes are the most probable explanation of the observations.

10.4 DIRECT EVIDENCE FOR PHOTOSYNTHESIS

The simple experiments described in Section 10.3 offer only indirect evidence about the chemistry of photosynthesis. Direct evidence can be gained by experiments based on *isotopic labeling* (see Section 9.2(*a*)).

Fig. 10.8 illustrates an experiment on photosynthesis using oxygen-18. Both tubes contain chloroplasts, isolated from their cells, suspended in water enriched with $H_2{}^{18}O$, and supplied with carbon dioxide. The tubes are illuminated and the oxygen evolved is collected and analyzed in a mass spectrometer: it is found to contain oxygen-18 in almost the same proportion as the original water. However, increasing the $C^{18}O_2$ content of the carbon dioxide has no effect on the isotopic composition of the oxygen evolved.

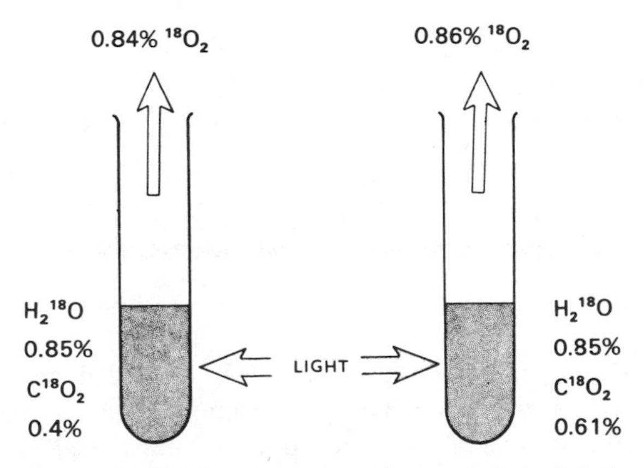

The percentage of radioactive oxygen given off corresponds closely to the percentage of $H_2{}^{18}O$ and does not change significantly when the percentage of $C^{18}O_2$ is altered

Fig. 10.8 *The source of oxygen produced during photosynthesis*

This result provides direct evidence that

(*a*) the chloroplasts and light both play a vital role in photosynthesis;
(*b*) water is essential for photosynthesis; and
(*c*) the oxygen produced during photosynthesis comes from the water and not from the carbon dioxide.

This experiment is concerned with the stage of photosynthesis called the "light reaction," the break-up of water molecules by light energy into oxygen, which is released, and hydrogen, which reacts with carbon dioxide

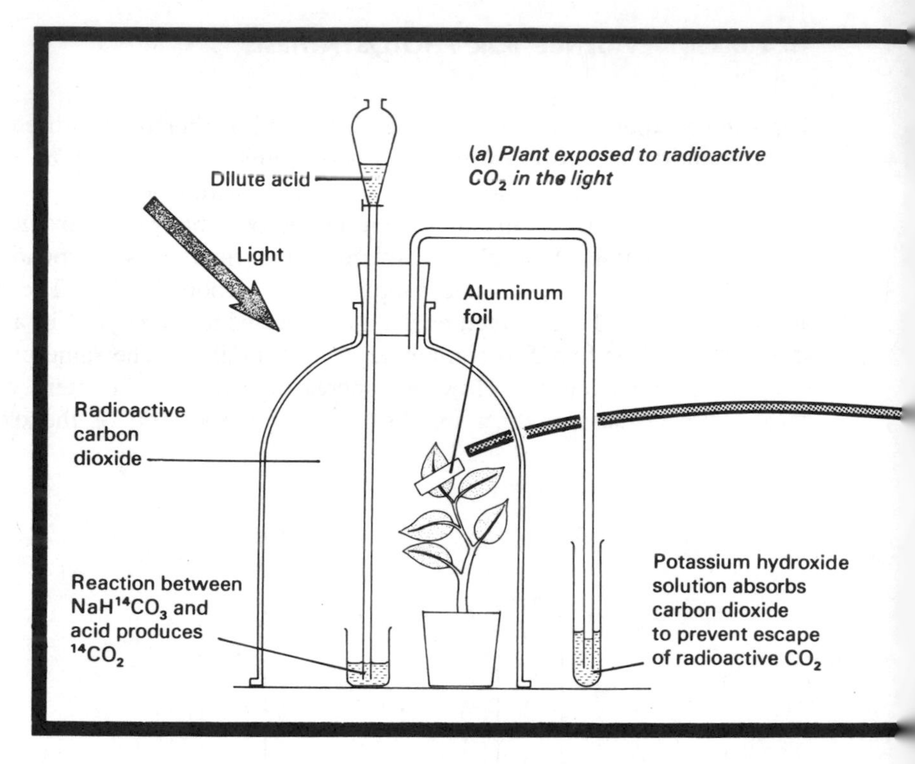

Fig. 10.9 *The use of radioactive carbon to investigate photosynthesis*

in the first stages of carbohydrate synthesis. Provided enough hydrogen atoms are made available by the light reaction, carbohydrate production from carbon dioxide can take place even in darkness (the "dark reaction"). The conversion of carbon dioxide into carbohydrate can be directly demonstrated using radioactive carbon, carbon-14 (Fig. 10.9).

A potted plant is enclosed under a bell-jar, and supplied with carbon dioxide which contains some $^{14}CO_2$, prepared by the action of dilute acid on a solution of carbon-14-labeled sodium bicarbonate. The plant is placed in sunlight, since the "dark reaction" cannot proceed until hydrogen atoms are made available by the "light reaction." After a few hours a leaf is detached and pressed on to a piece of unexposed photographic film. When the film is developed, the shape of the leaf appears as a dark area on the film, an *autoradiograph*. This shows that the radioactive carbon dioxide has been taken up by the leaf.

Part of the leaf can be used as a control by keeping it wrapped in aluminum foil during the experiment. This part does not affect the photographic film, since without light it cannot take up the radioactive carbon dioxide.

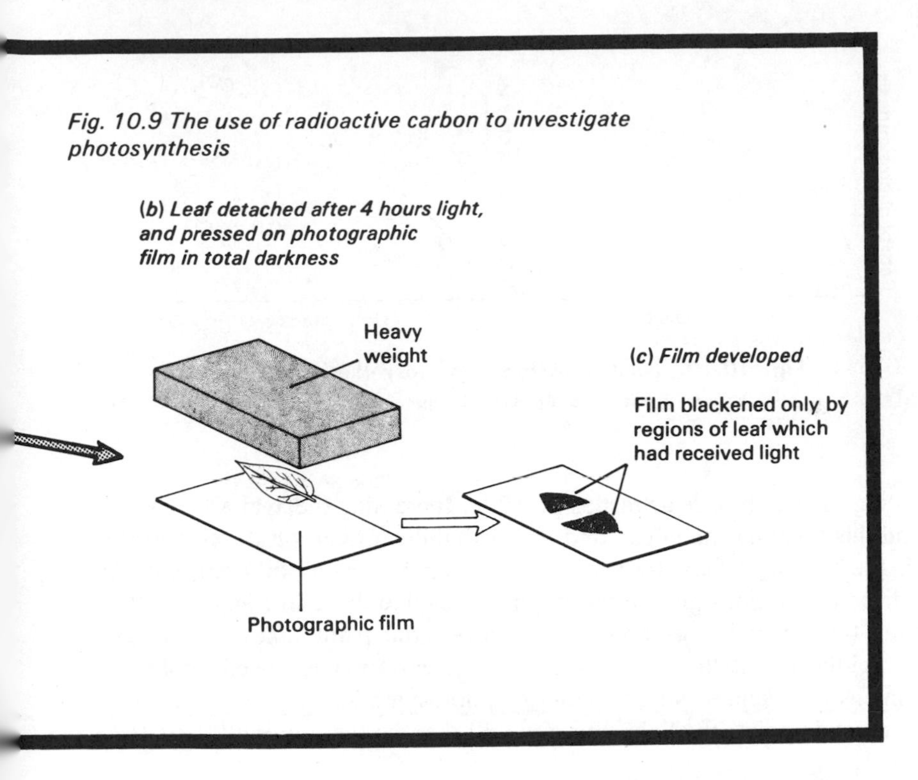

Fig. 10.9 The use of radioactive carbon to investigate photosynthesis

Isotopic labeling can also be used to trace the intermediate steps in photosynthesis. A suspension of single-celled green plants (algae) is supplied with radioactive carbon dioxide and illuminated for varying periods: 5 seconds, 30 seconds, 90 seconds, 5 minutes and so on. The cells in each sample are then analyzed to find which of their constituent compounds contain the carbon-14. The chain of reactions by which sucrose is built up from carbon dioxide can be elucidated in this way.

10.5 LIMITING FACTORS

The rate at which photosynthesis is proceeding within a plant can be measured, for example by observing the rate at which oxygen is being produced. If the rate is measured under different lighting conditions and then plotted against the light intensity, the graph usually appears something like Fig. 10.10(a). Clearly, increasing the light intensity only raises the rate of photosynthesis up to a certain point: this is the point at which the plant is

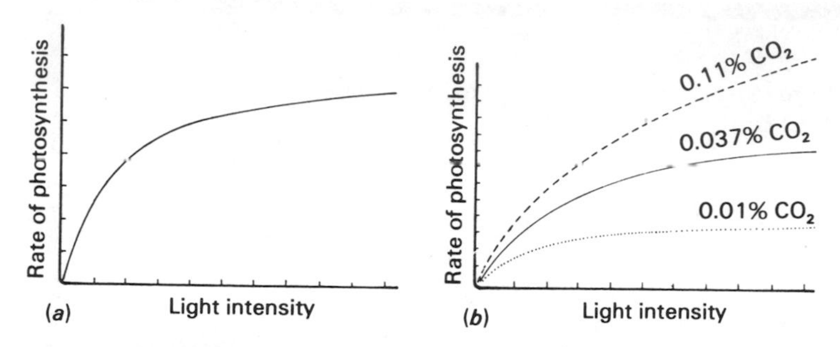

Fig. 10.10 *Limiting factors in photosynthesis*
(After Hoover, Johnston, & Brackett, *Smithsonian Miscellaneous Collection*, 87, 1934)

absorbing carbon dioxide as fast as possible. Increasing the available energy beyond this point has no effect, and the availability of carbon dioxide to the leaf is the *limiting factor*. Repeating the experiment using different carbon dioxide concentrations gives a family of curves like those in Fig. 10.10(*b*), showing that at higher carbon dioxide levels the plant takes up the gas more rapidly. But if the carbon dioxide concentration is raised while the light intensity is kept steady, the rate of photosynthesis still only increases up to a certain point; the light intensity will then be the limiting factor (see Unit 33).

Temperature can be another limiting factor, and the physical properties of the leaf itself also limit the rate of photosynthesis whatever the environmental conditions. For example, there is a limit to the rate at which carbon dioxide can diffuse through the stomata, and into the cells themselves through their walls, and a limit also to the area of chloroplast surface available for the absorption of sunlight.

10.6 THE ELEMENTS ESSENTIAL FOR PLANT GROWTH

Green plants use the sun's energy directly for certain stages in the build-up of sugars. The other constituents of plant tissues, like cellulose, proteins, fats and so on, are built up from these simple carbohydrates. The energy needed for the manufacture of the substances comes from respiration, that is, from the breakdown of carbohydrates, and so this energy too is derived, although indirectly, from the sun.

The plant takes up carbon, hydrogen and oxygen during photosynthesis, but it also requires many other elements and these it obtains from the soil. Nitrogen and sulphur are essential for protein synthesis; the chlorophyll necessary for photosynthesis contains magnesium; phosphorus is a vital constituent of many enzyme systems; calcium is used in the material between cell walls; potassium is concerned in the control of the rates of photosynthesis and respiration. Certain elements, including copper, manganese, boron and many others, are also needed in exceedingly small quantities for healthy growth. These are called *trace elements.*

These elements are absorbed from the soil as soluble salts. The plant's root system collects the salts in very dilute solution and they are concentrated within the plant. Potassium, magnesium and calcium salts are present in most soils. Sulphur is absorbed as sulphates and phosphorus as phosphates. Plants cannot use the abundant free nitrogen in the atmosphere directly, but take up the element from the soil in the form of nitrates.

Water cultures. Since a plant can build up its tissues using nothing but carbon dioxide, water and salts, some plants can be grown quite satisfactorily without soil, in solutions of these salts in water. Solutions like these are called *culture solutions,* and by carefully controlling the salts present their relative importance in the growth of the plant can be studied.

Experiment 10.6 *To demonstrate the importance of certain elements in plant growth*

About a week before the experiment is planned to begin, a number of seeds are germinated as suggested in Experiment 8.6. It is best to choose seeds without large food reserves in their cotyledons or endosperm, so that the reserves are quickly exhausted and the plants soon come to depend on the culture solution for their mineral-element requirements. Wheat grains, for example, are suitable.

A "complete" culture solution is prepared by dissolving 2 g calcium nitrate, 0.5 g each of magnesium sulphate, potassium nitrate and potassium phosphate and a trace of iron(III) chloride in 2000 cm^3 of distilled water. Further solutions are prepared in the same way, but with their composition modified as follows:

(*a*) a sulphur-free solution in which magnesium sulphate is replaced by magnesium chloride;
(*b*) a calcium-free solution from which calcium nitrate is omitted and a total of 2.5 g potassium nitrate is used;
(*c*) a phosphorus-free solution in which potassium phosphate is replaced by potassium sulphate;
(*d*) an iron-free solution from which iron (III) chloride is omitted;
(*e*) a nitrogen-free solution in which calcium nitrate is replaced by calcium chloride and potassium nitrate by potassium chloride;

(*f*) a potassium-free solution in which the potassium salts are replaced by the corresponding calcium salts;

(*g*) a magnesium-free solution in which magnesium sulphate is replaced by potassium sulphate.

A series of test-tubes are filled with culture solutions and labeled: one tube contains the "complete" solution, one distilled water, and the rest contain different solutions each deficient in one particular element. One seedling is placed in each of the culture solutions, supported by cotton (Fig. 10.11). Each tube is wrapped in aluminum foil to exclude light from the solutions and so prevent the growth of algae which might compete with the seedling for the minerals in the solution. The tubes are placed where the shoots receive sunlight equally, and the solutions are kept topped up with distilled water. After a few weeks the seedlings are compared by counting and measuring their leaves and roots as far as possible.

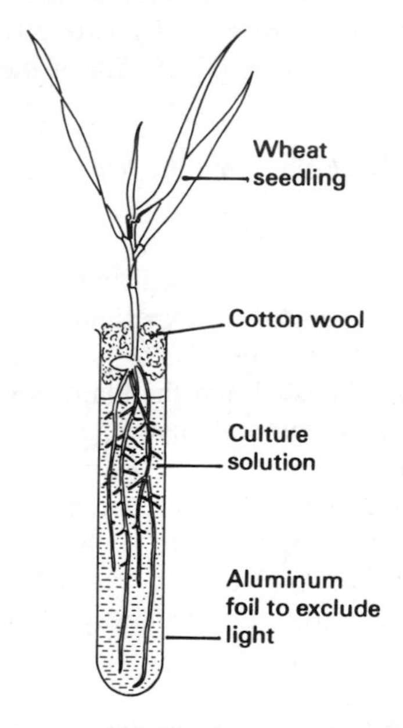

Fig. 10.11 *To set up a water culture*

Result. The seedling in distilled water hardly grows at all, and is stunted and weak even after some weeks. The one in "complete" culture solution is large and healthy with dark green leaves, but the others have probably grown less well, and have leaves which are smaller, fewer in number and

Fig. 10.12 *Bean plants growing in culture solutions*
(Rothamsted Experimental Station)

paler in color. Fig. 10.12 shows the results of one experiment of this kind. However, the results may be less clear-cut than in this photograph, especially if the chemicals used are not very pure. When several experiments have been carried out together, the results can be assessed more precisely by sorting the seedlings into batches—one batch grown in distilled water, one batch grown without potassium, and so on—putting the batches into labeled beakers and leaving them for 12 hours in an oven at 120 °C. They are then cooled and weighed, and the *dry weights* of the batches are compared. It is usually found that the seedlings provided with the full range of mineral salts show the greatest increase in dry weight.

10.7 THE SOURCE OF MINERAL SALTS IN THE SOIL

Soluble mineral salts are continually being lost from the soil: some are taken up by growing plants, and considerable amounts are washed away by rain-water into streams and rivers. Clearly these salts must be replaced or the growth of plants would come to a stop.

The salts are partly replaced by the products of animal excretion and of the decomposition of dead plants and animals. Further replacement comes from the action of rain-water on rocks: rain is always slightly acid because it contains dissolved carbon dioxide. As the rocks are weathered by rain and frost action the salts slowly dissolve and eventually soak into the soil where they become available to the plants.

The cycle of events which constantly adds nitrogen to the soil and then removes it is discussed in Section 18.3.

10.8 QUESTIONS

1. What are the requirements for photosynthesis? How are these requirements met in (*a*) a land plant, (*b*) an aquatic plant?
2. What is meant by "destarching" a plant? In what circumstances can the presence of starch in a leaf be regarded as evidence that photosynthesis has occurred?
3. If a plant was producing carbon dioxide and taking in oxygen why would you *not* be justified in assuming that no photosynthesis was taking place?
4. It can be claimed that the sun's energy is indirectly used to produce a muscle contraction in your arm. Trace the steps in the conversion of energy which would justify this claim.
5. Proteins contain carbon, hydrogen, oxygen, nitrogen and often sulphur. Name the source, for green plants, of each of these elements.

Unit 11 Bacteria And Blue-Green Algae

11.1 CLASSIFICATION

Bacteria and blue-green algae are closely related structurally even though their life-styles are quite different. They are placed presently in a separate kingdom, *Monera,* because they possess an indistinct nucleus and lack many of the complex organelles discussed in Unit Two. This type of cell is termed a *procaryote* (*procaryotic*). Since viruses are not cellular, they are not included in our present system of classification.

11.2 STRUCTURE OF BACTERIA

Bacteria are very small organisms, each consisting of a single cell (Fig. 11.1). They are rarely more than 0.01 mm in length, and so are visible only under the higher powers of the microscope. Bacteria are usually classified as plants rather than as animals, but they differ from the plants discussed in previous units in several respects. A bacterium has a cell wall similar to that of a plant cell, but it consists of fat and protein rather than cellulose. Bacteria have no chloroplasts, and their carbohydrate food reserves consist not of starch but of granules of glycogen, a compound characteristic of many animal cells, together with droplets of oil. Unlike both plant and animal cells, the nuclear materials are not enclosed in a nuclear membrane.

Bacteria may be of various forms—spherical, rod-shaped or spiral (Fig. 11.2); some have filaments of cytoplasm or *flagella* (singular, *flagellum*) protruding from them, the lashing action of which propels the bacterium about. Some species of bacteria require oxygen to respire, while others can respire anaerobically, that is, obtain their energy by breaking down food substances without using oxygen (see Section 4.3).

Bacteria reproduce themselves by simply dividing into two: under

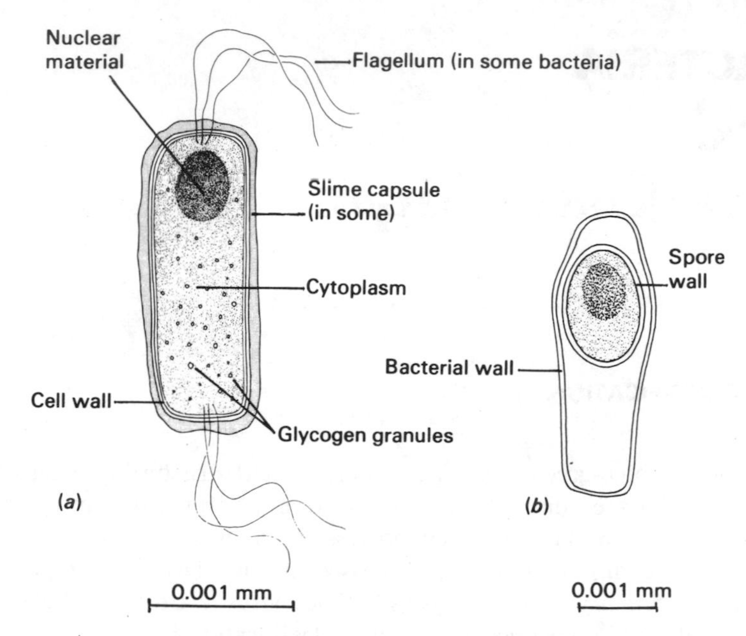

Fig. 11.1 *Generalized diagram of a bacterium: (a) normal form; (b) spore*

favorable conditions this can happen as often as once every 20 minutes, so that during the course of one day a single bacterium can give rise to many millions of daughter cells, which may form chains, clumps or films of individuals.

Spore Formation

When conditions are unfavorable, some kinds of bacteria can form *spores*. A spore is distinct from the usual form of the organism in that all its protoplasm is concentrated in one part of the cell, and it is surrounded by a thick wall (Fig. 11.1(*b*)). Bacterial spores are very resistant to drought and to extremes of temperature; some can even withstand the temperature of boiling water for long periods, although normal bacteria are killed by temperatures above about 50 °C. Spores are easily carried in air currents and are therefore very widespread, occurring on most sufaces, in the soil and dust and in the air. When conditions are suitable, the spores break open, the bacteria take on their normal form again and begin to grow and reproduce in the usual way.

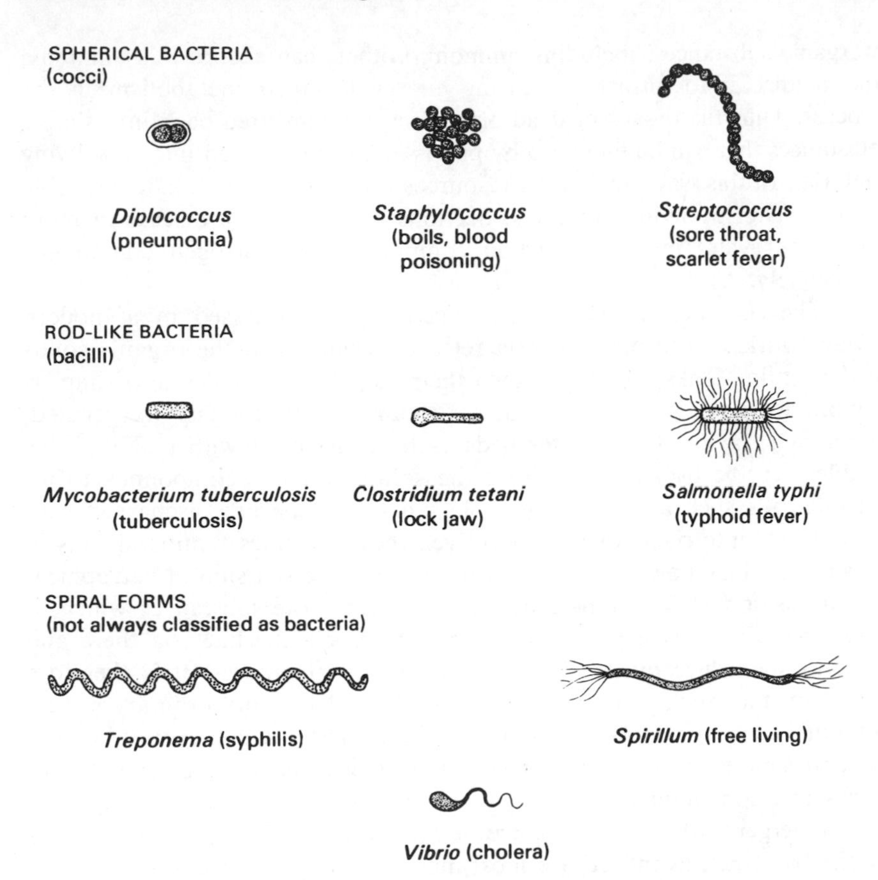

Fig. 11.2 *Bacterial forms*

11.3 BACTERIA IN NATURE

Most kinds of bacteria live freely in water or the soil, or wherever food is available in the form of dead organic matter. These *saprophytic* bacteria (see Section 10.1) are responsible for the decomposition and decay of the dead remains of plants and animals. Enzymes produced in the bacterial cells are released into the food, and break it down into soluble organic compounds such as sugars or amino acids (components of proteins), which the bacteria can absorb through the cell wall. Some kinds of bacteria obtain the energy they need by converting these soluble compounds into simple inorganic substances, including ammonia; others can absorb this ammonia and oxidize it to nitrates, releasing energy for their metabolism in the process. Thus the tissues of dead organisms are converted back into simple substances that can be taken up by plants and resynthesized into new living material. In this way the Earth's resources are recycled and made available for use again by living creatures. Sections 18.2 and 18.3 discuss in more detail the cycles involving carbon compounds and nitrogen compounds respectively.

The activities of these *decay bacteria* are harnessed in a modern sewage works. Anaerobic bacteria release enzymes into the organic solids in the settling tanks, breaking down their complex molecules into simpler, soluble compounds. The solution so obtained is thoroughly oxygenated, either by spraying it over filter beds or by churning it with paddles: this enables aerobic bacteria to oxidize the soluble organic compounds to inorganic salts such as nitrates and phosphates, which are discharged with the effluents into rivers or the sea. (Even these "harmless" mineral salts in the effluents can, however, create hazards: see the discussion of eutrophication in Section 18.5.) Some industries produce waste substances which, if they reach the sewage plant, can inactivate or kill the bacteria there and so seriously reduce the efficiency of the treatment process. It is therefore important that such products are not discharged into the ordinary sewers without some prior purification. Similarly, substances which cannot be used as food sources by the bacteria will pass unchanged through the sewage works and contaminate the effluent. Hence the importance of using only those detergents which are *bio-degradable* (that is, which can be metabolized by the bacteria), as indeed are most modern detergent products.

Some kinds of bacteria live in the *alimentary canals* of animals, obtaining their food from the products of digestion which surround them. Although a few of these can cause diseases such as cholera or typhoid, most are harmless and some are even beneficial. For example, bacteria living in the intestine of man appear to play a part in the synthesis of vitamin K and the vitamins of the B_2 complex (see Section 19.3(*e*)). Herbivorous animals rely on the dense populations of bacteria in their alimentary canals for the proper digestion of their plant food. The most nourishing part of plant tissues is enclosed within cellulose cell walls, but few animals produce an enzyme which can digest cellulose. The cell walls are broken down partly by efficient chewing and partly by the action of cellulose-digesting enzymes produced by the bacteria and other single-celled organisms living in the gut: in the caecum and the appendix of the rabbit, for example, and in the rumen (or paunch) of cows and sheep.

Certain bacteria are able to produce their food by *photosynthesis* (although the chlorophyll pigments they contain are different from the chlorophylls of the higher plants); some of these can use hydrogen sulphide (H_2S) instead of water (H_2O) as a source of hydrogen atoms.

Parasitic bacteria live in or on living animals or plants, and are of especial importance because they may be harmful or even fatal to the host organism.

11.4 HARMFUL BACTERIA

The harm done by parasitic bacteria arises either from damage to the tissues of the host, or, in certain cases (diphtheria and tetanus, for example), from the effect of the highly poisonous proteins or *toxins* formed as waste products of their metabolism. As little as 0.0005 mg dried tetanus toxin will kill a mouse, and 0.23 mg is fatal to a man.

Many serious diseases in both animals and man are caused by bacteria: they include typhoid, cholera, bubonic plague and tuberculosis. These diseases develop when the harmful bacteria or their spores enter the body and multiply rapidly; they may be "caught"

(*a*) by eating food containing living bacteria;
(*b*) by "droplet infection" (inhaling bacteria in droplets of moisture in the air breathed, coughed or sneezed out by infected persons), particularly in a crowded, humid place such as a theater;
(*c*) by touching the skin of an infected person;
(*d*) by bacterial invasion of wounds.

Resistance to Disease

Bacteria are almost ubiquitous in our environment, yet most individuals remain healthy and free from disease for the greater part of their lives. This is because the body maintains a resistance to bacterial invasion in several ways.

(*a*) **The skin** forms a defense against the entry of bacteria into the blood and body cavities, partly by the physical barrier of its dead, cornified outer layer (see Section 24.2(*a*)) and partly by producing chemicals which can destroy bacteria. The eyes, for example, produce a protective enzyme in normal tears. The lining of the alimentary canal and the respiratory passages also resist the entry of bacteria.

(*b*) **The blood** of the higher animals contains large numbers of white blood cells or *leucocytes* which can engulf and destroy bacteria (see Section

21.1(*b*)). In addition, the blood can produce chemicals called *antitoxins* which can effectively neutralize the poisonous proteins given out by the bacteria. Once an animal has recovered from an attack of a bacterial disease, its blood is much better able to combat the particular bacteria producing that disease, and to neutralize their toxins: the animal is said to have acquired *immunity* to the disease (for a fuller discussion see Section 21.2(*d*)).

11.5 PREVENTION OF INFECTION

(*a*) Food

Most uncooked food contains bacteria, but usually in insufficient numbers to cause disease. In man, the digestive juices in the stomach contain substantial amounts of hydrochloric acid, which destroys many bacteria, while the lining of the gut usually resists entry by those which remain.

The growth of dangerously large numbers of bacteria in food can be prevented by *preservation techniques,* which destroy both the decay bacteria producing obvious food spoilage, and also disease bacteria. The high temperatures used in *cooking* kill most bacteria and their spores if continued long enough. *Bottling* and *canning* also kill bacteria by the use of high temperatures, and then prevent their access to the food by sealing it in a vacuum. Poisonous substances like sulphur dioxide are sometimes added to canned or bottled fruit in a concentration high enough to kill bacteria but too low to harm the consumer.

Certain preservation methods aim to produce food in a form which is comparatively hostile to bacterial growth. Thus bacteria cannot grow in *dried food* because of the absence of water, or in *pickled food* because of the high concentration of acid, while the low osmotic potential of foods preserved in *salt* or *sugar* would plasmolyze (see Section 3.4(a)) any bacteria which entered them.

Refrigeration does not kill bacteria, but slows down their growth; at the temperatures used in *deep-freezing,* bacterial growth and multiplication are at a standstill.

Experiments are being conducted on preserving food by exposing it to *radiation* from radioactive cobalt. This treatment destroys bacteria effectively but the quality of some foods is affected by it. The process is not yet in general use.

Food should always be eaten as soon as possible after it has been cooked or removed from a container, so giving invading bacteria, or any surviving the cooking process, little opportunity to multiply. People who handle food, particularly when it is to be eaten without subsequent cooking, should take great care over their personal cleanliness, and especially that their hands do not carry harmful bacteria. Flies are notorious carriers of bacteria and should never be allowed to settle on food.

(b) Water Supplies

Water for human consumption is filtered through beds of sand containing films of protozoa (simple animals) which trap and eat bacteria. Any bacteria which pass through the filters can be destroyed by chlorinating the water.

(c) Sewage Disposal

Feces and urine from people suffering from disease may be heavily infected by bacteria and spores, and also act as breeding grounds for the small number of harmful bacteria which are often present in the excreta of even healthy people, or which may reach them later. Sewage must therefore be disposed of in such a way that the bacteria present cannot be transmitted by flies or other means to food or people. Modern sewage-disposal methods do this effectively. Where these are not available, strong disinfectants or fairly deep burial should be used.

(d) Antiseptics

Healthy skin is a barrier against bacterial invasion, but when it is broken by a cut, graze or burn, precautions should be taken to remove as many bacteria as possible from the injury by washing. The remainder can be inactivated by applying a mild antiseptic such as proflavine. Dressings and bandages can be used to prevent the entry of further bacteria.

(e) Antibiotics

Antibiotics such as streptomycin are chemicals produced by some bacteria (or fungi) which destroy other species of bacteria (see Section 12.3). They can be extracted from the bacteria and used against infections without damaging the cells of the body, either externally in creams or ointments or, more often, internally in tablets or injections. Terramycin, for example, can be used to control an infection of the middle ear, either by giving the antibiotic to the patient to eat or by injecting it into his blood.

(f) Cleanliness

Bacteria and bacterial spores are normally present all over the surface of the skin. Regular washing and bathing remove many of these and so reduce the chances of infection should the skin be damaged. The hands, particularly, should be kept clean since they handle food and manipulate articles that will be used or have been used by other people. Doctors and nurses who move from sick to healthy patients have to be scrupulously clean, and their instruments repeatedly sterilized, to avoid transmitting infectious bacteria.

(g) Sterilization

Sterilization is the removal of bacteria from an object by killing them. Very high temperatures can be employed, for example in an *autoclave,* which uses steam under pressure. *Chemicals* like strong acids can also be used, but when bacteria are to be destroyed in foods or in wounds, the chemicals used must be of such a nature and at such a concentration that they are lethal to harmful bacteria but harmless to man. No such "perfect" chemical has yet been found, but many are suitable for use against certain bacteria under certain conditions. The use of *radiation* from radioactive elements is now in regular use for the sterilization of surgical dressings and instruments.

(h) Isolation

People carrying a disease which can be transmitted by touch or droplet infection should not mix with others at all if the disease is serious, and as little as possible if the disease is mild. In particular they should not mingle with crowds where they could pass on the bacteria to a great many people. The international quarantine regulations are designed to prevent the spread of diseases by infected persons travelling from one country to another.

Wearing gauze masks across the nose and mouth, or merely covering them with a handkerchief when coughing or sneezing, helps to reduce the chances of inhaling or exhaling bacteria.

(i) Immunization

When an individual is suffering from disease, the blood produces substances called *antibodies* (see also Section 21.2(*d*)) which fight the infection; even after a recovery the ability to produce antibodies remains at a raised level. By deliberately infecting a person with a mild form of a potentially dangerous disease, antibodies are produced in his blood which will protect him against subsequent infections. The serious disease of diphtheria has been almost eliminated by the widespread use of immunization.

11.6 CULTURING BACTERIA

Bacteria are identified and studied by *culturing* them, that is, by keeping them in conditions in which they can grow and reproduce rapidly. This is done by supplying them with a mixture of food substances, usually in an agar jelly base, and keeping them in a warm place for a time; the bacteria multiply and form visible groups or *colonies,* each containing thousands of cells. The species making up a colony can be identified from its color and chemical reactions. The optimum conditions for the growth and reproduction of a specific bacterium can be found by varying the temperature or the composition of the culture medium, and the effects of various concentrations of chemicals such as antiseptics or antibiotics on a particular species can be investigated.

11.7 STRUCTURE OF BLUE-GREEN ALGAE

Blue-green algae are procaryotic plants that grow as single cells or in slimy filaments. Their cell walls differ from most bacteria in that they are composed usually of cellulose. They possess blue and green photosynthetic pigments, but these are not enclosed in chloroplasts as in the nucleated (*eucaryotic*) plants. Blue-green algae often contain various other pigments, which give them different colors and hues such as orange, yellow, red, black, and other colors.

Blue-green algae reproduce by simple cell division as in bacteria, but commonly parts of filaments break away from attached filaments to form new colonies.

11.8 BLUE-GREEN ALGAE IN NATURE

Blue-green algae occur in a wide variety of conditions. They are abundant in fresh and salt water environments attached to rocks and other objects. Wet soils and temporary pools often support large numbers of blue-green algae, giving the surface a dark greenish hue. They are also common in and around hot springs and geysers, as in the Mammoth Hot Springs of Yellowstone National Park in Wyoming. Certain species of blue-green algae also live in the snow. In western mountains it is not unusual to find "watermelon" or "red" snow over many square meters due to great numbers of these algae, which possess a red pigment. The Red Sea in Israel obtains its color from another red pigment-possessing blue-green algae. Many blue-green algae live associated with other organisms, such as fungi. Many *lichens* are composed of a fungus and a blue-green algae living for each other's mutual benefit.

Of great importance wherever they occur is the nitrogen-fixing ability of many blue-green algae. By fixing atmospheric nitrogen gas into organic form, the blue-green algae nourishes itself and enriches the water or soil in which it resides. Many aquatic blue-green algae also deposit carbonate, and therefore have played a part in the formation of limestone. When blue-green algae occur in large numbers in reservoirs, substances given off often impart a bad taste and odor to the water and may be toxic to livestock. Copper sulfate is often added to water supplies to prevent this from occurring.

11.9 VIRUSES

Viruses are best mentioned here because they are simple forms, far smaller than the smallest bacteria. The fact that they are not cellular organisms makes them difficult to classify, and in fact they are beyond our present system of classification. Yet they possess the ability to reproduce by using their genetic material to assemble more virus particles from their host cells. Apart from living host cells, a virus cannot reproduce or "act" alive. In fact, viruses can be taken from their host plants or animals, crystallized like so much salt or sugar, and stored on a shelf indefinitely. But once introduced into another living host, the "dormant" virus can "come alive" and reproduce in minutes and cause the host to have the symptoms of a virus disease, such as polio, the flu, smallpox, common colds, and many plant diseases.

11.10 QUESTIONS

1. How does a knowledge of the methods of transmission of bacteria lead to the variety of aseptic precautions taken in a modern operating room?

2. Substances such as wood, paper and cloth will rot away in time. Materials like polystyrene and polythene will not. Why should there be this difference and why is it a cause for concern in industrialized countries?
3. Exposure to mild infectious diseases such as German measles is considered to be a normal or even desirable hazard of communal living. Cases of serious infectious diseases such as smallpox are kept in isolation so that they cannot infect other people; however, one may deliberately be infected with a form of smallpox when being vaccinated.

    Discuss the biological principles underlying these social attitudes and activities.
4. Articles which would be decomposed by the temperatures in an autoclave can be sterilized by heating them to 100 °C for one hour on three successive days, incubating them at 25–30 °C after each steaming.

    What is the biological basis for this method of sterilization?
5. What kind of health checks are desirable on people who are preparing and serving food to a large number of others, or who are working in food-processing factories?
6. Why are blue-green algae classified with bacteria instead of with the other algae?
7. What roles do the blue-green algae play in nature that are of importance to other organisms?
8. Are viruses alive or non-living? Defend your answer.

UNIT 12
FUNGI

12.1 THE CHARACTERISTICS OF FUNGI

(a) Structure

The fungi are included in the plant kingdom in this book, but in many ways they are quite different from green plants. The basic unit of a fungus is not a cell but a threadlike tubular structure called a *hypha* (Fig. 12.1); although in some fungi the hyphae are divided by cross-walls which give them a "cellular" structure, these divisions are not true cells. In some fungi the walls of the hyphae contain cellulose, but in most species they

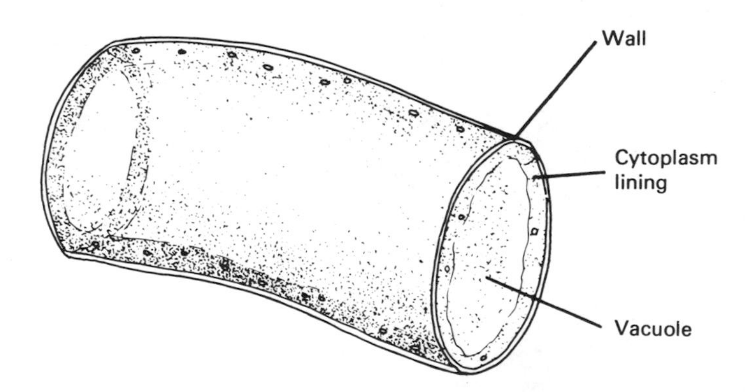

Fig. 12.1 *Part of fungal hypha*

consist mainly of *chitin,* a very resistant nitrogenous substance. The composition may vary with age and environmental conditions. The tips of the growing hyphae are full of cytoplasm, but in other regions there may be a central vacuole. Many very small nuclei are present in the cytoplasm. Fungi contain no chloroplasts and no chlorophyll, and the food particles in the cytoplasm may be oil droplets or glycogen, but not starch.

The hyphal threads spread out and grow over and into their food

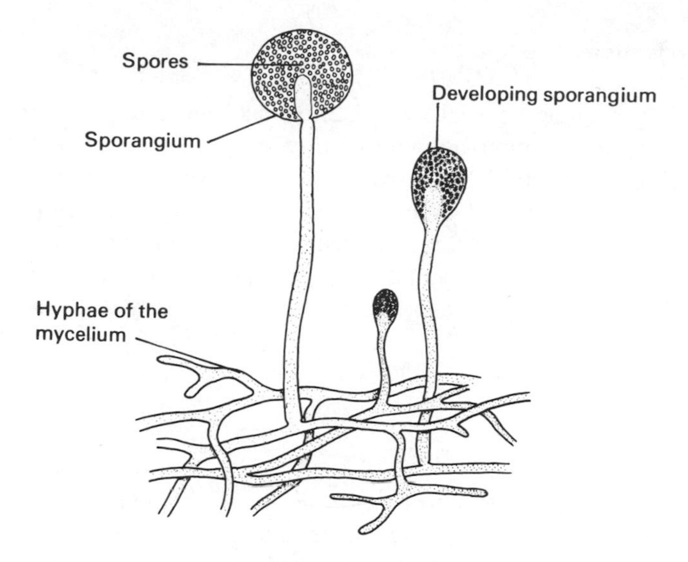

Fig. 12.2 *Asexual reproduction in* Mucor

material, and may make up a visible mesh or *mycelium* (Fig. 12.2). In the familiar structures of puffballs and mushrooms, the hyphae are massed together to form a "fruiting body," but the kind of specialization of cells and tissues seen in the organs of a flowering plant is not so evident in fungi.

(b) Feeding

The absence of chlorophyll means that fungi cannot photosynthesize their food from simple substances like carbon dioxide and water, but must take in food materials that have already been synthesized by other organisms. Fungi do this either parasitically or sapophytically. *Parasitic fungi* live on or in the tissues of another living organism or host, absorbing nourishment from its body. Some of the most devastating diseases of crops are caused by parasitic fungi; potato blight, discussed below in Section 12.4, is one example. The *saprophytic fungi* derive their food from dead and decaying materials. Examples are the molds which develop on stale, damp food and the many fungi which live in the soil and feed on the humus there. Without the decomposing action of soil fungi, vast quantities of dead wood and organic litter would not decay or be returned to the soil.

The growing tips of the hyphae of both parasitic and saprophytic fungi produce enzymes which can break down complex organic materials into simpler soluble substances that can be absorbed through the cell walls, and which also enable parasitic hyphae to break down and penetrate the cell walls of the host.

(c) Reproduction

Both asexual and sexual reproduction take place in fungi. In *asexual reproduction,* an enormous number of tiny spores is produced, discharged into the air and dispersed by air currents. If the spore lands in a suitable situation a new hypha, and eventually a new mycelium, develops from it. *Sexual reproduction,* involving the fusion of two nuclei, takes place in specialized branches of the hyphae; in some species the product is a thick-walled resting spore, rather than a form adapted for easy dispersion.

12.2 *MUCOR*

Mucor is the name given to a group of *mold fungi* which grow on the surface of decaying fruit, bread, horse manure and other organic matter. The hyphae, which have no cross-walls, grow rapidly under favorable conditions and branch repeatedly, within a few days forming a dense white or gray mycelium covering the surface of the food.

Reproduction. Asexual reproduction takes place rapidly after the establishment of a mycelium. Vertical hyphae, a little thicker than the rest, grow up and become swollen at their ends. These swellings contain dense cytoplasm and many nuclei (Fig. 12.2). The swelling becomes the *sporangium* and the protoplasm inside breaks up into elliptical *spores,* each containing several nuclei and surrounded by a wall. In some species of *Mucor* the sporangium wall breaks away and the powdery spore mass is dispersed by air currents. In other *Mucor* species, the sporangium wall liquefies, releasing the spores, but the method of spore dispersal is uncertain; they may be carried by insects.

The spores of *Mucor* are said to be very resistant to adverse conditions such as drought or cold, and can remain dormant for years. When conditions are favorable the wall of the spore breaks open and the protoplasm inside grows out into a new hypha and eventually into a mycelium. Germination is greatly favored by warm, damp conditions.

12.3 ANTIBIOTICS

Certain mold fungi, notably species of *Penicillium,* have become of economic importance owing to the antibacterial chemicals or *antibiotics* they produce (see also Section 11.5 (e)). Antibiotics such as penicillin are manufactured by growing the appropriate species of mold on nutrient broths; the active compound is extracted from the culture fluid and purified.

Streptomycin and many other antibiotics are produced by soil-dwelling microorganisms called *actinomycetes,* which have characteristics of both bacteria and fungi. It may be that by producing antibiotic chemicals in their natural environment these actinomycetes restrict the growth of bacterial colonies in the soil and so reduce the competition for food. There is little direct evidence, however, to support this assumption.

Penicillin seems to attack bacteria by damaging their cell walls, but the mechanism of action of most antibiotics has not been elucidated.

12.4 PARASITIC FUNGI

Parasitic fungi are the principal disease-causing organisms in plants, and fungal attacks can result in devastating agricultural losses. They also produce a few diseases in animals, some of which can be serious. There are countless fungal diseases in humans in the moist tropics.

Potato blight. This disease was responsible for the disastrous Irish potato famines of the mid-nineteenth century and is caused by the parasitic fungus *Phytophthora infestans.* If a spore of the fungus lands on a leaf of a healthy potato plant in warm moist conditions, a new hypha grows out of it and enters the leaf via a stoma (Fig. 12.3). Short branches grow out of the hypha and penetrate the walls of the leaf cells, presumably with the aid of enzymes (Fig. 12.4). The hyphae absorb nutrients from the cell contents

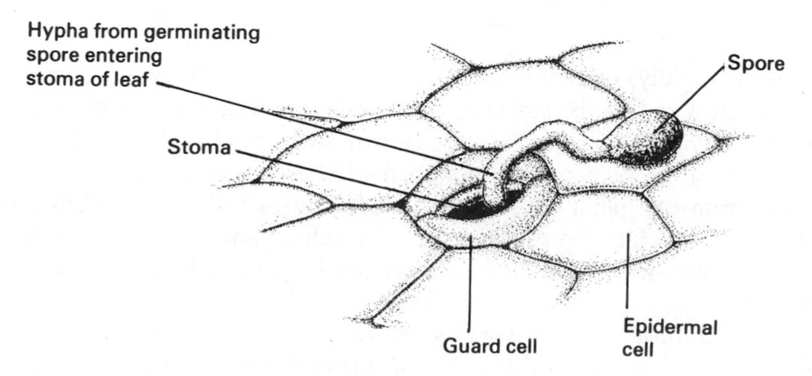

Fig. 12.3 *Infection of leaf by parasitic fungus*

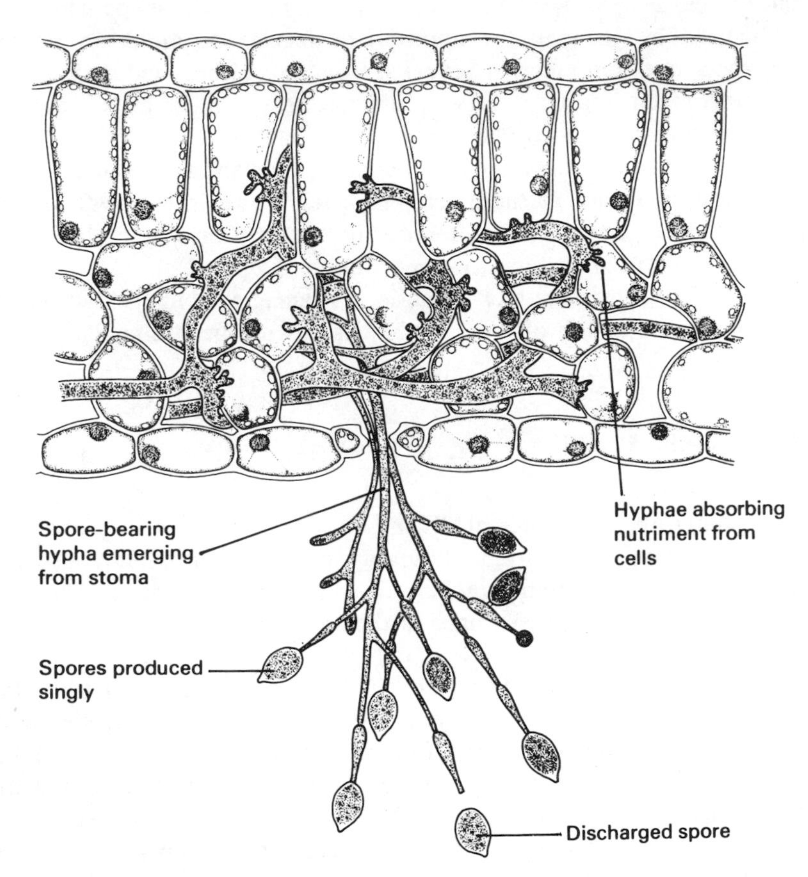

Fig. 12.4 *Diagrammatic section of leaf attacked by parasitic fungus*

and grow rapidly, spreading throughout the tissues of the plant. Branching hyphae grow out of the stomata; constrictions develop near their tips and cut off individual spores which are blown away in air currents. The close proximity of the plants in the potato field allows very rapid spread of the fungus from one plant to the next. When spores fall on the ground they may be washed into the soil by rain, so reaching and directly infecting the potato tubers, which rapidly rot. The whole plant is soon killed by the parasite.

Agricultural research workers are constantly trying to find varieties of food plants which are resistant to the many types of fungal disease, and to develop chemical sprays which will destroy harmful fungi without causing damaging side effects on the crop or on other organisms in the area.

12.5 YEAST

The yeasts are a widely distributed and rather unusual family of fungi. Only a few of the several species can form true hyphae; the majority of them consist of single-celled organisms which can be seen only under the microscope.

Structure. The thin cell wall encloses the cytoplasm, which contains a vacuole and a nucleus. In the cytoplasm are granules of glycogen and other food reserves.

Reproduction. The cells reproduce asexually by *budding,* in which an outgrowth from a cell enlarges and is finally cut off from the parent as an independent cell. When budding occurs rapidly the individuals do not separate at once, and as a result, small groups of attached cells may sometimes be seen (Fig. 12.5).

Yeasts also reproduce sexually: in certain conditions, two cells may *conjugate,* that is, they join together and their cell contents fuse. Later, the cell contents divide into four individuals, each developing a thick wall (Fig. 12.6). These are spores and may constitute a resting stage. When

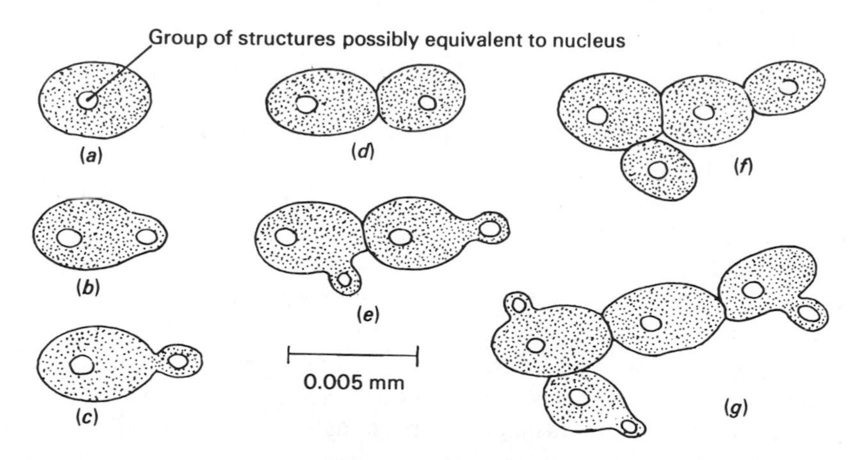

Fig. 12.5 *Yeast cells budding*

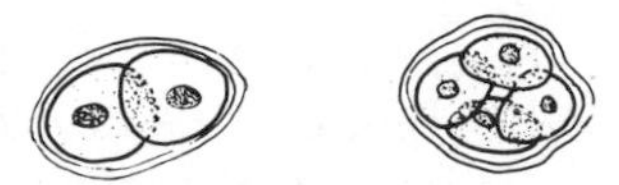

Fig. 12.6 *Yeast: resting stage with division into spores*

the old cell wall enclosing them breaks open, the spores are set free and can germinate to form normal, budding cells. Such spores often arise without any previous conjugation.

Fermentation

Yeasts, in nature, live on the surface of fruits and in other situations where sugars are likely to be available. They have been of economic importance for thousands of years in promoting alcoholic fermentation (see Section 4.4). Yeasts contain a group of enzymes which can break down sugar into carbon dioxide and ethanol ("alcohol"), making energy available for metabolism. However, far less energy is released from a given amount of sugar by fermentation than by aerobic respiration. The later stages of fermentation can be regarded as a kind of anaerobic respiration, but unless fermentable sugars are present the yeast cells require oxygen for the preliminary conversion of other available carbohydrates into compounds that can be used directly as energy sources.

In brewing, barley grains are allowed to germinate and in so doing convert their starch reserves into the sugar maltose. The germinating barley (*malt*) is killed and the sugars extracted with water. Yeast is added to this solution and ferments the maltose to ethanol and carbon dioxide. Hops are added to the mixture to give it the characteristic bitter flavor of *beer,* and the liquid is bottled under pressure so that the carbon dioxide is still very much in evidence. Spirits like *whisky* are made by allowing the fermentation to continue longer, and distilling off the ethanolic product, thus raising the alcohol content.

Wine is made by fermenting fruit juices, many of which will ferment of their own accord if the crushed fruit is left in suitable conditions for the development of the wild yeasts always present on the skins. If too much oxygen is admitted to the mixture, however, or if it is contaminated with the wrong species of yeast or with other fungi or bacteria, the oxidation of the alcohol may continue until acetic (ethanoic) acid is formed: this produces *vinegar.*

In baking, yeast is added to uncooked dough, and the mixture left to ferment in a warm place. The bread "rises" because of the carbon dioxide given off during the fermentation. The dough becomes filled with gas bubbles, which expand rapidly when the mixture is put into a hot oven. The ethanol which is also produced is driven off at the high temperatures used in baking.

Yeast is an excellent source of the B-group vitamins, and is valuable in the treatment of vitamin deficiencies. The development of yeast strains that can be used as a source of low-priced protein is the subject of present research.

12.6 QUESTIONS

1. When something "goes moldy," what is actually happening to it?
2. Suggest why bread, wood and leather are to be seen going moldy while metal, glass and plastic are not.
3. Suggest why mushrooms may be found growing in very dark areas of woodland where green plants cannot flourish.
4. The fungi are not easy to classify as plants but they bear little resemblance to animals. Discuss the ways in which they (*a*) resemble, (*b*) differ from green plants and animals.

Unit 13
Algae

13.1 INTRODUCTION

The higher plants discussed in Units Five through Ten are complex in structure and function. They consist of specialized cells, tissues and organs that perform vital life processes. Absorption of water and nutrients, their transport throughout the plant body, the synthesis of new food stuffs and reproduction are but a few of the processes performed by specialized parts of the plant. But with the algae little specialization is found, and most cells of their simple bodies need to perform a variety of vital tasks.

Algae are generally small aquatic plants with simple bodies, but some attain large proportions, Regardless of size, the algae are characterized by lacking tissues and organs or organ systems, and having few specialized cells. They are the primary food and oxygen producers in both fresh and salt water environments. Over half of the world's oxygen supply comes from these simple plants.

The algae are divided into several groups based primarily on the pigments they contain. They all possess chlorophyll and some have one or more accessory pigments. They have organelles, but these are not seen with normal light magnification.

13.2 GREEN ALGAE

Green algae are common to fresh water mainly, but some occur in coastal waters. Some grow on the moist, shady bark of trees. One group is the main algae component of lichens.

The plant body may be one-celled, colonial or multicellular. The multicellular green algae are either filamentous (with cells attached end-to-end forming a thread) or leaf-like. Many one-celled green algae are beautiful in symmetry and color under the microscope (Fig. 13.1). Movement

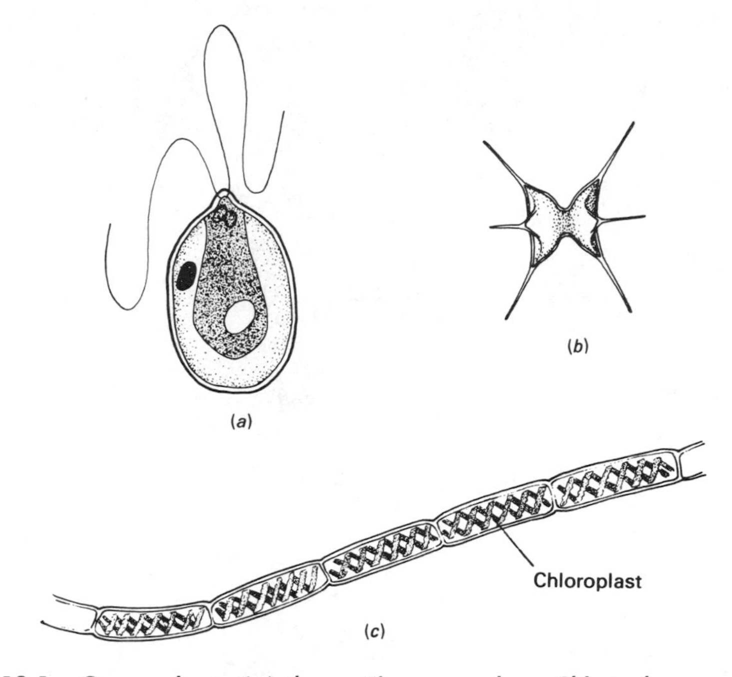

Fig. 13.1 *Green algae: (a) the motile green algae* Chlamydomonas; *(b) a non-motile desmid; (c) the filamentous* Spirogyra

(motility) is accomplished by means of long whip-like appendages called *flagella* (*flagellum* singular).

Spirogyra. This simple green plant is often seen as a filmy, rather slimy, growth in ponds; on closer inspection it shows long slender filaments, visible to the naked eye. Each filament consists of a long row of cylindrical cells joined end-to-end (Fig. 13.1c). All the cells in the filament are identical and none is specialized for any particular function. Thus *Spirogyra,* although not a unicellular organism, is very like one. Indeed, if any one cell becomes detached from the filament, it can lead an independent existence and give rise to a new plant.

The structure of an individual *Spirogyra* cell is shown in Figs. 13.2 and 13.3. The cellulose cell wall is lined with a thin layer of cytoplasm which contains two long ribbon-like chloroplasts, arranged spirally along the cell. Each contains numbers of pyrenoids, protein bodies in which starch is synthesized. The nucleus lies centrally in the cell.

The cells are not mobile at any stage of their development, although pieces of the filament may break off and float away.

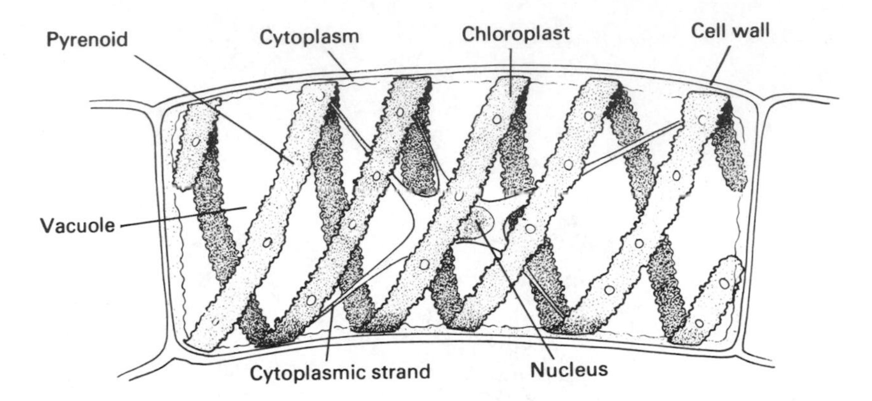

Fig. 13.2 Spirogyra *cell (×320)*

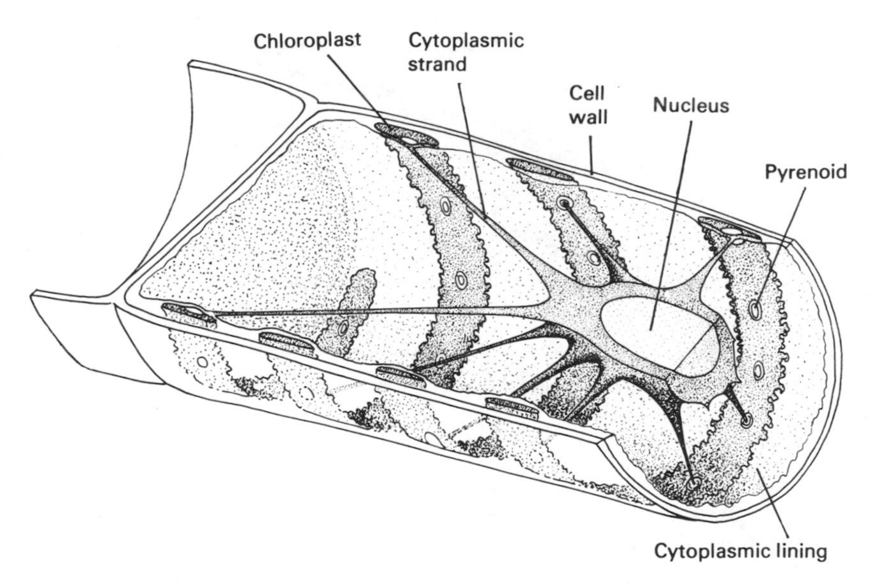

Fig. 13.3 *Stereogram of* Spirogyra *cell in section*

13.3 DIATOMS

The diatoms are a fascinating and important group of algae. They are one-celled, having a cell wall composed of two halves and heavily silicified (impregnated with silica), which is intricately sculptured. Diatoms are common to both fresh and salt water. The silicified parts accumulate in great deposits called *diatomaceous earth.* This material has numerous industrial and commercial uses including insulation and high grade filters.

Diatoms are of great importance both in nature and to the world's food supply since they are the basis of the food chains upon which animals, especially fish, feed.

13.4 BROWN ALGAE

The largest algae are the seaweeds or kelp common to coastal waters. Brown algae may attain lengths in excess of 50 meters, but their body structure is still relatively simple. Most of them must grow attached to rocks or other solid objects. They have complicated life cycles that involve swimming reproductive cells. Commercially some kelps are a source of *algin* used in a great many prepared food products. In nature, kelp is not only used as a food source by many animal forms, but it is an important habitat for coastal fish and other marine life.

13.5 RED ALGAE

Most red algae are small, marine plants that grow attached to rock surfaces along sea coasts. They are often delicate in appearance, being finely branched. Even though smaller than brown algae, many red algae can grow to greater depths, as deep as 175 meters. Red algae are the source of agar, the media universally used in the culture of bacteria and molds. *Carrageenan* is a substance from red algae used as a stabilizer to keep solids suspended in liquid, such as chocolate drinks. Some red algae are grown commercially in the Orient on wooden frames and are harvested and used in a variety of food dishes.

13.6 QUESTIONS

1. In nature what would be the primary benefits of having algae present?
2. What groups of algae have commercial value to man?
3. What is the potential value of algae as a major food source for man in the future?

UNIT 14
THE ANIMAL KINGDOM

14.1 INTRODUCTION

The animal kingdom is a vast, diverse assemblage of organisms that ranges from one-celled protozoans to gigantic whales. The advanced animals generally have more complex internal systems than plants, but in truth there are great numbers of animals that are rather simple in organization. All animals are heterotrophs in that they must depend upon organic materials produced by others as their food source. Most animals move about, although some are sedentary like most plants. But all have sensory organs, and those that move possess structures and systems for locomotion.

Our everyday acquaintance with the animal kingdom is limited usually to encounters with birds, dogs and cats, and other human beings. This gives us a distorted view of the animal kingdom because there are probably over 2 million different kinds of animals on earth. Our view is further distorted by the fact that most animals (95 per cent) do not have a backbone, and yet in our everyday experience we see and think mostly of animals that have backbones—*vertebrates.* This word comes from the bones of the back called vertebrae, which make up the vertebral column or backbone. The vast number of animals forms that lack a backbone are called *invertebrates.* Most animals are best suited to a watery environment rather than being land creatures. Only two major groups of animals—the insects and the vertebrates—have large numbers of their kind on land.

14.2 PROTOZOANS

Protozoans are minute, one-celled creatures that are sometimes placed in the kingdom Protista along with fungi and algae. But here we consider these "first animals" as the simplest animals since they have all of the characteristics of animals. They move about under their own locomotion powers and are classified according to their types of locomotion, namely flowing, cilia and flagella—all live and move in liquid environments. They have no cell walls and no chlorophyll; they feed by taking in and breaking down complex organic materials. Within this group are found some of man's most devastating parasitic diseases.

Ameba

Ameba is a protozoan animal, different species of which are widely distributed in ponds, ditches and other moist places, and in the soil. Some are animal parasites and are the cause of amebic dysentery in man. Its structure is shown in Fig. 14.1. The cytoplasm is bounded by a thin, flexible membrane, and is made up of a fluid, granular substance, the *endoplasm,* surrounded by a clear jelly-like layer, the *ectoplasm*. It contains a single nucleus and several droplets of fluid or *vacuoles*. These are quite different from the vacuoles of plant cells: they are much smaller, and both their size and their position in the cell can change quite quickly.

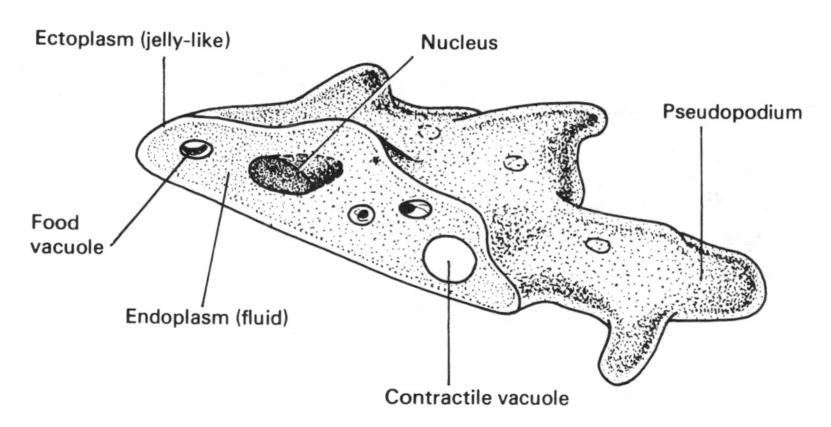

Fig. 14.1 *Structure of* Ameba *(in section) (×200)*

Locomotion. The shape of an *Ameba* is constantly changing. It has no specialized organs of locomotion like flagella (Section 11.2), but it can move across the surface of the mud or soil by the flowing of its cytoplasm (Fig. 14.2). It begins to move by developing a protuberance of *pseudopodium* at one point of its surface, into which the fluid cytoplasm flows. In time the entire cell contents are transferred into the pseudopodium, so that the *Ameba* in effect reaches a new position. Changes of direction are effected when a new pseudopodium begins to form at another part of the *Ameba's* surface. The direction of movement is probably determined by local differences in the composition or temperature of the water. Slight acidity or alkalinity may cause the cytoplasm to start flowing or prevent its

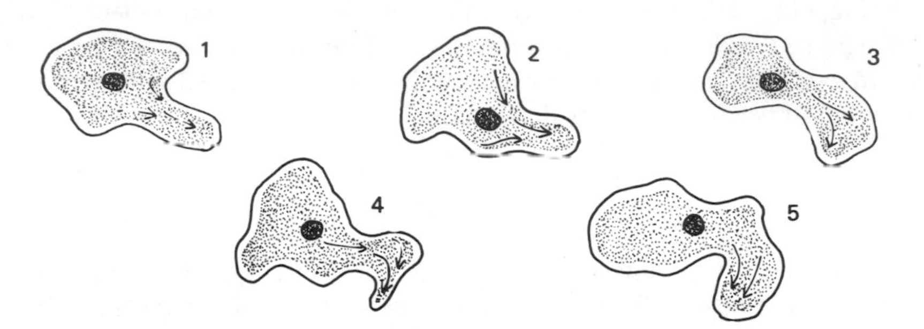

Fig. 14.2 *Diagrams to show flowing movements of* Ameba

doing so altogether. The chemicals diffusing from suitable food material may cause the cytoplasm to flow in that direction.

Feeding. When an *Ameba* encounters a fragment of food such as bacterium or a microscopic plant, pseudopodia flow out rapidly and surround it, so that the particle is ingested into the cytoplasm of the *Ameba,* together with a drop of water, forming a *food vacuole.* Enzymes are secreted into the food vacuole by the surrounding cytoplasm, breaking down the food material into simple, soluble substances which are absorbed into the cytoplasm. Any undigested residue is left behind (*egested*) as the *Ameba* flows on its way. Food vacuoles with material in various stages of digestion can usually be seen in the cytoplasm of an *Ameba.* Ingestion and egestion can take place at any point on the surface of the animal; there is no "mouth" or "anus."

Osmoregulation. The cell membrane surrounding an *Ameba* is semi-permeable, with the result that the low osmotic potential of the solutions in the endoplasm causes water to be taken in. This excess water is collected up in a spherical *contractile vacuole,* which gradually swells, and then seems to contract or burst, expelling the accumulated water from the cell.

Reproduction. The *Ameba* does not reproduce sexually. It reproduces asexually by dividing into two (*binary fission*). The animal stops moving; first its nucleus and then its cytoplasm divide, to make two daughter individuals which move away independently (Fig. 14.3).

In unfavorable conditions some species of *Ameba* form *cysts,* that is, the *Ameba* secretes a thick resistant wall around itself, within which it can survive for a considerable time and in which form it can be carried to other situations. Within the cyst the *Ameba* divides repeatedly, so that when the wall breaks open a fairly large number of daughter *Amebae* are released.

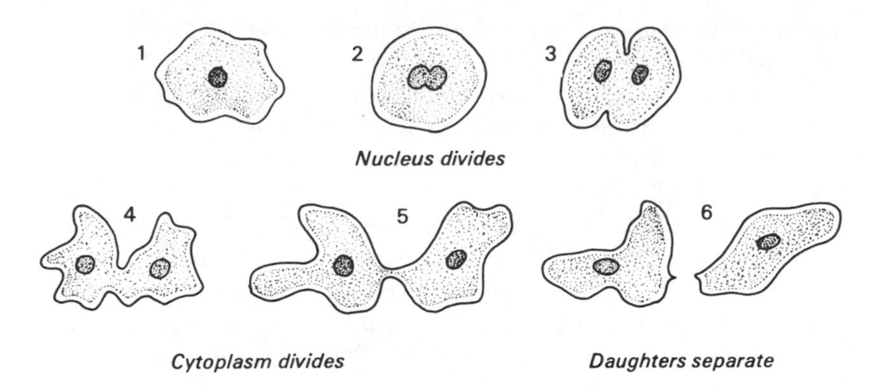

Fig. 14.3 Ameba *reproducing by binary fission*

Ciliates

These are the single-celled animals most frequently encountered and are so called because they possess rows of cytoplasmic filaments, *cilia,* extending from their surface (Fig. 14.4). In *Paramecium,* the cilia are

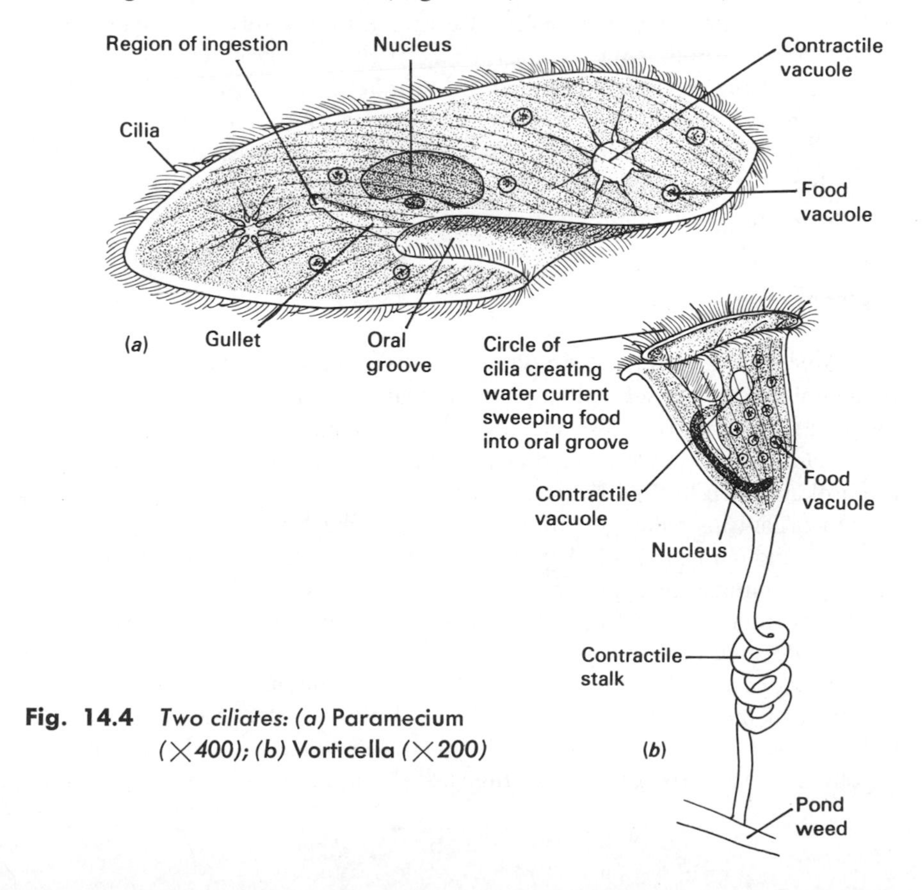

Fig. 14.4 *Two ciliates: (a)* Paramecium *(×400); (b)* Vorticella *(×200)*

arranged in rows along the cell and by rhythmic waves of flicking movements propel the animal smoothly forward or backward. There is an efficient co-ordination of the movement of the cilia brought about by a network of special cytoplasmic strands near the surface, connecting the bases of the cilia. These probably have conducting properties. The motion of a special row of cilia creates a water current which sweeps food particles through a funnel-shaped *oral groove* into a shallow *gullet;* ingestion can only take place at the end of this gullet. The food vacuoles move in a very definite path through the body of the *Paramecium,* and egestion takes place at one point only, near the region of ingestion. Clearly, the organization of the bodies of ciliates is more complicated than that of *Ameba.*

Protozoa Living in Other Animals

There are many protozoa which live only within the bodies of higher animals, frequently in the gut but sometimes elsewhere. They are often harmless or even beneficial, and some play an important part in the digestion of cellulose in the rumen of the cow and the cecum of the rabbit, to mention only two examples (compare Section 11.2), which deals with bacteria that have a similar role). Certain termites rely entirely on the protozoa in their alimentary canals to digest the woody material on which they feed; if the protozoa are killed the termite dies of starvation.

Some parasitic protozoa, however, can seriously damage their host organisms. For example, the malarial parasite, which lives in the red cells of the blood, and sleeping sickness, which is also a blood parasite in warm-blooded animals and man.

Euglena

Spirogyra and related simple green organisms like *Pleurococcus* can be unambiguously described as plants, because of their method of feeding by photosynthesis. It is also clear that *Ameba* and *Paramecium* are correctly said to be animals, since they feed on complex food substances which are broken down in their bodies. However, there is an interesting group of related organisms called *flagellates* which are harder to classify with certainty. Although nearly all of them have characteristics which render them more like animals than plants, many contain chlorophyll and can photosynthesize their food, while others feed *saprophytically* (like fungi) and others *holozoically* (like animals).

Euglena gracilis (Fig. 14.5), which is abundant in polluted water, has most of the structural features of the family. It has no rigid cell wall, unlike *Spirogyra* and other green plants. It can move about by lashing its flagellum to and fro, and can change its shape as it swims along. It has

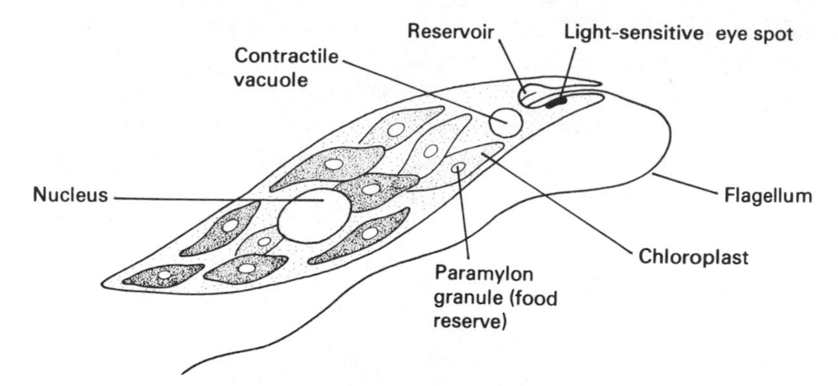

Fig. 14.5 Euglena gracilis (× *800*)
(From Vickerman & Cox, *The Protozoa*, John Murray)

chloroplasts containing chlorophyll and can photosynthesize its food, but if it is kept in the dark for some time it loses both its green color and its photosynthetic ability. However, it will continue to live if suitable organic material is available in the surrounding water. In this respect it is very similar to its relatives in the genus *Astasia,* flagellates which never possess any chlorophyll and which live only in water rich in organic materials such as the fluid in the puddles around cowsheds. They absorb the organic substances from their environment and use them as food, that is, they feed saprophytically. However, they still have the ability to synthesize some complex materials from relatively simple inorganic chemicals, although they cannot build up starch from carbon dioxide and water.

Other colorless flagellates feed by taking in solid particles of food material and digesting them, very much in the same way as *Ameba.* This holozoic feeding is essentially an animal characteristic. The distinction between holozoic and holophytic feeding (see Section 10.1) is the fundamental difference between animals and plants, and this leads some biologists to suggest that the remote ancestors of present-day flagellates such as *Euglena* and *Astasia* could have given rise to the plant and animal kingdoms in the early stages of evolution; for it is only in this group of unicellular creatures that closely related individuals, or even single creatures, exhibit both plant and animal characteristics.

14.3 SPONGES

Sponges are primitive water animals that at one time were thought to be plants. There are about 5,000 species and all are composed of many cells (multicellular), but these are rather loosely arranged and are irregular in form. Their most obvious feature is their holey or perforated body. Some simple sponges encrust objects, whereas others are attached to objects only at their base and grow to a considerable size, up to a meter in height and width. They are filter feeders, circulating water through their porous bodies and then collecting small organisms and organic matter by means of ameba-like cells. They have no movable appendages. Sponges reproduce both sexually by shedding sex cells into the water and asexually by fragmentation. They produce an internal framework or skeleton composed of various substances. Commercial sponges are only the dead skeletal remains of the animal.

14.4 COELENTERATES

Coelenterates are a diverse group of aquatic animals numbering over 10,000 species. Jelly-fish, sea anemones and coral are examples of this group of interesting animals. They exhibit in simple form most of the basic systems found in higher animals. They have a two-stage life cycle in that they reproduce sexually and yet have a plant-like stage, which buds-off new individuals asexually (Fig. 14.6). All coelenterates have a single opening (mouth-anus) into which foods enter and undigested matter leave the digestive cavity. They all have *tentacles* for food gathering, and these are equipped with stinging cells that are used in defense and food capture.

14.5 FLATWORMS

Flatworms (about 10,000 species) are simple worms that are truly flat and thin and have bilateral symmetry, i.e. they have a right and left side, top and bottom, and a head and tail end. About one-third are free-living (i.e. not parasites) worms that live in fresh water or in damp places on land. The well-known *Planaria* belongs to this group, yet the most numerous and important flatworms are parasites to other animals. *Flukes* invade the liver or blood; *tapeworms* reside in the small intestine of their host.

In all cases of human infestations of flatworms, the problem is related to poor sanitary conditions, including contaminated food or water supplies. In China and other Oriental countries where human wastes are used as fertilizer and where fish is eaten raw, the human liver fluke is a major health hazard.

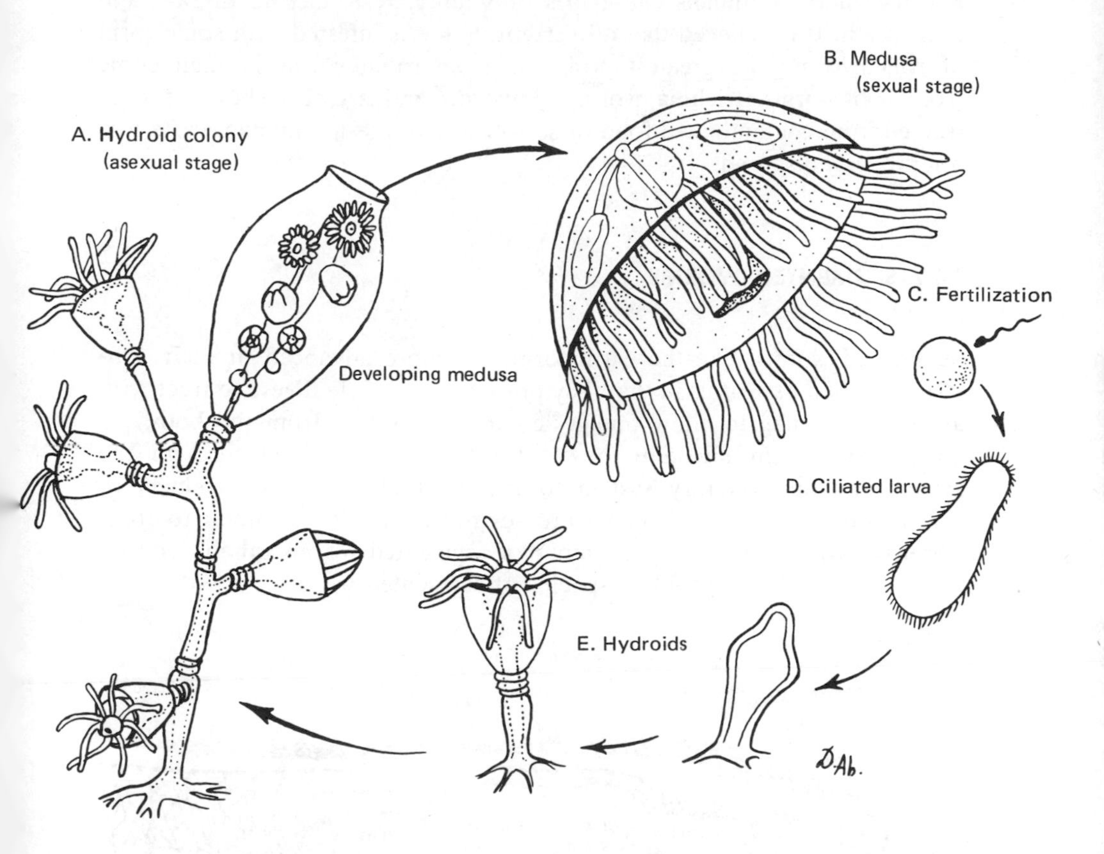

Fig. 14.6 *Obelia* *life cycle. Young medusae are budded off by the asexual hydroid or polyp stage*

14.6 ROUNDWORMS

Roundworms are a large (12,000 species) group of small, cylindrical, unsegmented worms that pervade the soil, water environments and the bodies of most other animals. Careful examination of moist soil reveals large numbers of white, thread-like worms. Some are free living (non-parasitic) and are called *nematodes,* but the great majority of roundworms are parasites. Crop damage from nematode infestation of the roots of plants in sandy soils is common and costly. Roundworm infestations in domestic livestock and in humans cause loss of vitality, resistance to disease and even death. It is believed that all vertebrate life is infested with some form of roundworm. The greatest problems from roundworms in man come from hookworms, trichina worm, pinworms and ascaris. They are contracted from contaminated food or soil in which infested human waste was not properly disposed.

14.7 SEGMENTED WORMS

Segmented worms, like the earthworm, are more advanced in their construction than flatworms in that they possess a complete digestive tract with an anus (opening to allow undigested matter to pass from the body), a circulatory system including hearts that pump the blood through the system and an excretory system to rid the body of toxic products of respiration (Fig. 14.7). Leeches are segmented worms common to fresh water environments. A large number of segmented worms inhabit coastal waters. There are about 9,000 species of segmented worms.

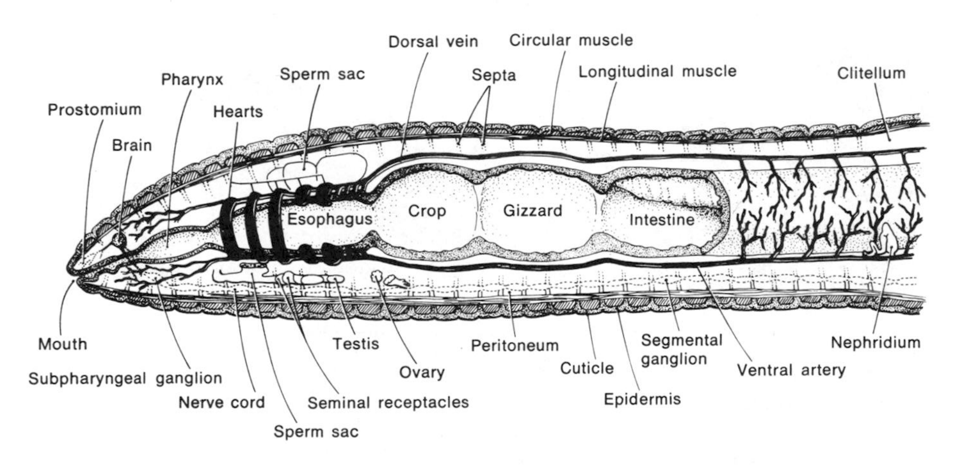

Fig. 14.7 *Side view of the internal anatomy of an earthworm showing major body systems*

14.8 CRUSTACEANS

Crustaceans are a group of jointed-legged arthropods numbering over 25,000. They are mainly aquatic animals and many are of great importance in nature and to man. Common crustaceans are lobster, shrimp and crab.

But there are vast numbers in lakes, streams and oceans that feed upon microscopic plants (*phytoplankton*) and are in turn fed upon by larger animals. Hence these crustaceans provide a vital bridge between the simple plant food supply and the higher animals. Their key anatomical features are their two sets of antennae (used as touch and smell sensory organs) and their numerous legs.

14.9 INSECTS

The number of insects is staggering, over 925,000 species. Perhaps eight out of every ten kinds of animals on earth are insects. This group is so vast and diverse that special attention is given to it in the following chapter.

Insects have six jointed legs, three body regions, a pair of antennae and usually wings. This group dominates the land (terrestrial) habitats and is well represented in fresh water environments, but is hardly present in the oceans of the world. Insects are the dominant invertebrates on land; crustaceans dominate the seas.

The question is often asked why are insects so successful on earth. Surely, many of them are man's greatest competitors for food and fiber. Of the many reasons that account for the success of insects, the following are perhaps the most important:

1. Their small size allows them to hide and escape both their enemies and the elements.
2. They are very responsive to climatic and other environmental factors, thus being active only when food supply and other conditions are most favorable.
3. Their exoskeleton is a coat of armor against the drying effect of air, unfavorable temperatures and physical attack.
4. Their musculature and wings give them power and agility.
5. An elaborate hormone system moves them through developmental stages in synchrony with the seasons.
6. They have exceptional sensory powers that aid them in survival and reproduction.

14.10 ARACHNIDS

Arachnids are creatures with eight legs, two body regions and no antennae. The 60,000 species include spiders, harvestmen, scorpions, ticks and mites. Most of these are terrestrial, but, like the water mites, some are aquatic. Spiders and scorpions are predators on other invertebrates, but some are parasitic, such as ticks.

14.11 MOLLUSKS

Next to insects, mollusks are the largest group of invertebrates, numbering over 110,000 species. They are a soft-bodied but muscular group that is successful both on land and in aquatic environments. Many of them produce shells of limestone, but mollusks, like the squid, octopus and slugs, do not. They all have an outer covering, the *mantle,* which protects them or in turn produces the shell for protection. The formation of pearls in oysters results from secretions from the mantle. The largest invertebrate animal in the world is a mollusk, the giant squid of the South Pacific Ocean. This rare creature may exceed 20 meters in length.

14.12 ECHINODERMS

Echinoderms have hard spiny plates and coats, a hydraulic system that operates numerous suction cups (*tube feet*), and spines, but no head. They are all marine animals that are bottom dwellers and are associated with rocks and tide pools. Representatives are sea stars (starfish), serpent stars, sea urchins, sand dollars and sea cucumbers.

14.13 FISH

Fish are the simplest vertebrates; they have an internal skeleton (*endoskeleton*), bilateral symmetry, and well-developed sensory organs and muscles for locomotion. Primitive fish such as lampreys are jawless and scaleless and have a skeleton made of a flexible material (*cartilage*). Most fish, however, have bony skeletons, jaws, a gill covering (*operculum*), an air bladder for buoyancy and well-developed paired fins. The bony fish number about 20,000 species and are common to both fresh and salt water (see Unit Sixteen).

14.14 AMPHIBIANS

Amphibians are vertebrates that live mostly on land but require moist to wet environments to live and reproduce. Representatives are toads, frogs, salamanders and newts. They lay small gelatinous eggs, which must develop in water. These hatch into larvae (such as *tadpoles*), which are gill-breathers and feed upon plants and animal remains. In some, such as salamanders, they develop air-breathing *lungs* and four legs, but remain basically similar to the larval form. In others, such as frogs, there is an absorption of the tail, the emergence of four limbs and the development of lungs. All amphibians have scaleless skins that are porous, and they lose vital body fluids rapidly if in a dry air environment. Their skin may contain glands that produce bad-tasting or noxious substances, and these are the main means of defense for these rather docile creatures (see Unit Seventeen).

14.15 REPTILES

There are over 6,000 reptiles. They are a mysterious group of animals that holds a certain intrigue for most people. They are truly land animals—they have dry, scaly skin, lungs, powerful muscles for locomotion and food-getting, and they lay large, dry-skinned eggs suitable to development on land. The most common reptiles are lizards, snakes, alligators, crocodiles and turtles. The dinosaurs of past ages were reptiles, and some reached sizes not attained by any land animals today.

14.16 BIRDS

Birds are vertebrates that have lightweight skeletons, feathers and musculature that makes them the champions of flight in the animal kingdom. Birds are also able to maintain a constant, high body temperature, which allows them to remain active even in cold conditions. The 8,600 known species of birds are diverse in form, beauty and behavior, ranging from the delicate hummingbird to the powerful ostrich. Their beaks and feet, and often elaborate behavior, suit them to particular habitats, ways of life and mating requirements. Bird watching is a popular and beneficial pastime.

14.17 MAMMALS

When we think of animals we think of mammals, one of the smallest groups in the animal kingdom. However, they are considered the most highly developed and intelligent forms of life on the Earth. Mammals have large brains, maintain a constant body temperature, most give birth to living young, nourish their young on milk from *mammary glands* and are covered by skin or fur. They are classified according to the way the young develop, tooth types, and feet and toe structure.

14.18 QUESTIONS

1. What are the relative numbers or proportions of vertebrates and invertebrates in the animal kingdom?
2. What does "vertebrate" mean? What does "invertebrate" mean?
3. Give several characteristics of protozoans.
4. How do sponges differ from protozoans? How are they similar?
5. What factors perpetuate the existence of parasitic flatworms in humans?
6. What is the largest group of animals on the Earth? What are some reasons for its success?
7. Make a list of five features of fish that contrast with five features of reptiles.
8. What are the features used in classifying mammals?

Unit 15
Insects

15.1 INTRODUCTION

Insects are part of a large group of invertebrate animals known as *arthropods*. Other members of the group are spiders, millipedes, centipedes and crustacea. All arthropods have a hard exoskeleton and several pairs of jointed legs; their bodies are *segmented,* that is, divided down their length into more or less regular sections or segments (Fig. 15.1).

There are a great many different species of insects. All have three distinct regions in their bodies: the *head,* the *thorax* (the first three segments behind the head) and the *abdomen* (the remainder of the body). Mature insects have one pair of *antennae,* sensory organs attached to the head, and *compound eyes* (see Section 15.6). They have three pairs of jointed legs carried on the thorax, and usually two pairs of wings. In some insects like beetles, grasshoppers and cockroaches, the first pair of wings is modified to form a hard outer covering over the second pair. Some species, including the houseflies, crane-flies and mosquitoes, have lost one pair of wings during the course of evolution, and other species, including parasites like fleas, have lost both pairs of wings.

15.2 LIFE HISTORY

(a) Complete Metamorphosis

Many insects lay eggs which hatch out into *larvae,* immature forms which are usually quite unlike the adult. According to the species of insect they are known as grubs, maggots or caterpillars. Generally the larva eats voraciously and grows rapidly, shedding its cuticle several times (see Section 15.3). When it has reached its full size, its form changes again and it becomes a *pupa.* The pupa does not feed and in most species does not

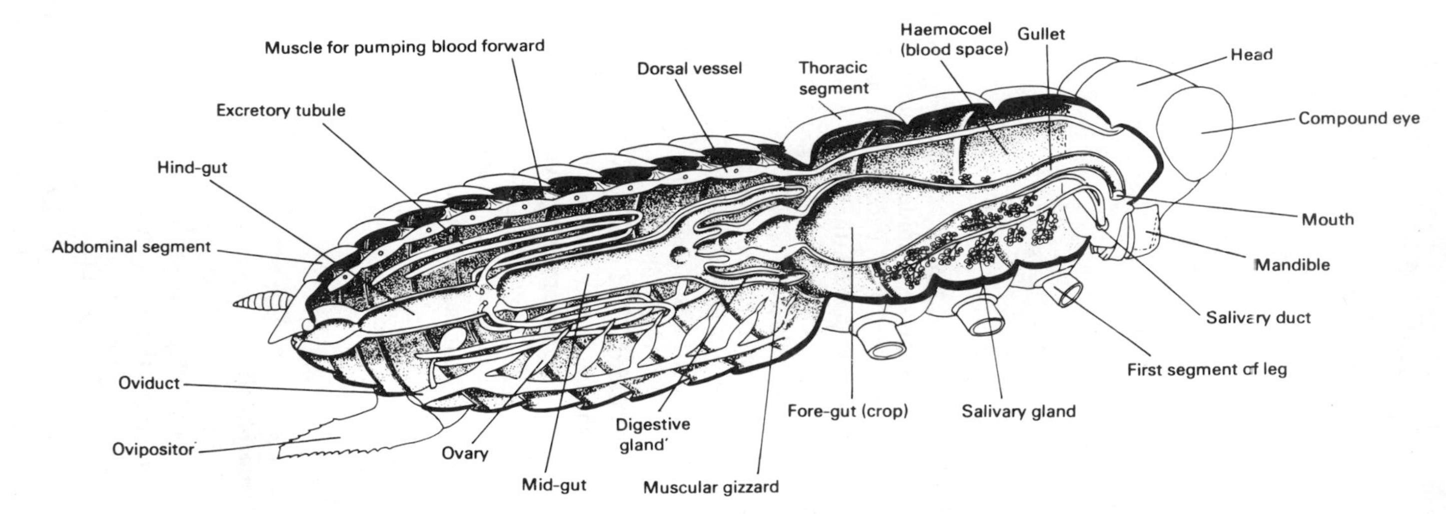

Fig. 15.1 *Anatomy of the adult female insect* (*After Nicholas Jago*)

move, but there is at this stage an extensive breakdown and reorganization of the tissues of the body, giving rise eventually to the adult form or *imago*: the adults then mate and lay eggs. The life history of the large white butterfly is of this kind, and is described in Section 15.10. The change from the immature form to the adult is called *metamorphosis*; the sequence of changes just described is *complete metamorphosis*.

(b) Incomplete Metamorphosis

Not all species pass through the stage of pupation. The eggs of insects like mayflies and dragonflies hatch into *nymphs* rather than larvae; although very different from the adult, the nymph resembles it more closely than does a larva. Like the imago, the nymph has three pairs of jointed legs and compound eyes; it also has rudimentary wings. Each time the cuticle is shed, changes occur which bring it nearer to the adult form, the final molt revealing drastic changes that have taken place during the last weeks of the nymph's development, but there is no prolonged resting stage. These changes are called *incomplete metamorphosis*.

Larvae and nymphs differ from the adults not only in appearance but also in habitat, behavior, locomotion and feeding habits. For example, a mayfly nymph lives and grows in water for a year or so, but the flying adult form lives only for a few hours, long enough to mate and lay eggs. The description of the general characteristics of insects in Sections 15.3 to 15.8 refers to the adults except where immature forms are specifically mentioned.

15.3 CUTICLE AND ECDYSIS

The hard external cuticle of insects reduces the loss of water vapor from the internal tissues of the body, and also protects them from damage and bacterial invasion. It maintains the shape of the body and provides for muscle attachment for rapid locomotion. However, the cuticle imposes certain limitations on the size of insects and other arthropods, since the cuticle of an animal bigger than a large crab would be so heavy that the creature would be unable to move.

The cuticle is flexible between the segments of the body and at the joints of the legs, antennae, mouth parts and wings, but otherwise it is hard and rigid. Any growth of the insect is therefore prevented except during certain periods of development when the cuticle is shed; the insect can then increase its size before the new cuticle has time to harden (Fig. 15.2). The shedding of the cuticle, or molting, is called *ecdysis,* and takes place only in larvae (or nymphs) and pupae of insects, and not in adults; mature insects do not grow.

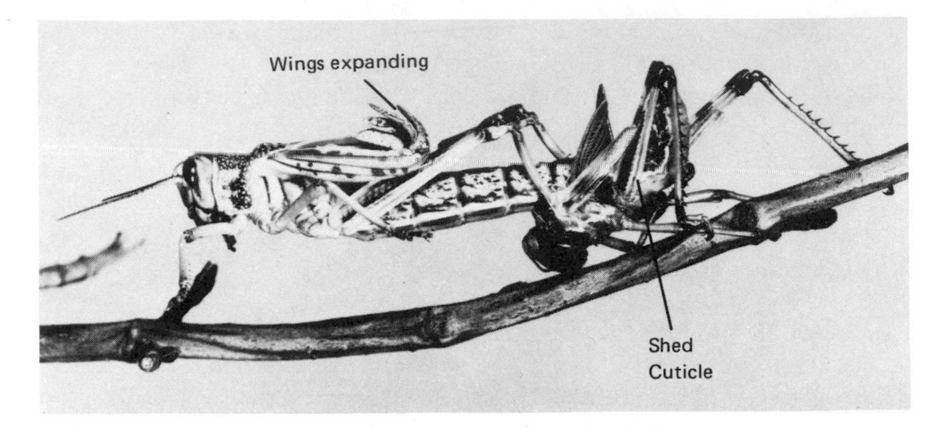

Fig. 15.2 *The final ecdysis of the locust*
(Shell International Petroleum Co. Ltd.)

15.4 BREATHING

The respiratory systems of insects are very different from those of vertebrates, in which oxygen is absorbed in gills or lungs and conveyed in the bloodstream to the tissues. In an insect, oxygen reaches the tissues directly through a system of breathing tubes, and the carbon dioxide produced in respiration is excreted mainly by the same path, although some may leave the body by diffusion from its surface.

The breathing tubes or *tracheae* make up a branching system carrying air from the atmosphere to all parts of the insect's body (Fig. 15.3). They are lined with a thin layer of cuticle, thickened in spiral bands; this thickening keeps them open against the internal pressure of the body fluids (Fig. 15.4). The tracheae branch repeatedly until they terminate in very fine tubes or *tracheoles* which penetrate the tissues of the body. The walls of the tracheae and tracheoles are permeable to gases; oxygen can diffuse through them to reach the cells, and carbon dioxide can diffuse in the reverse direction into the tubes. The network of tracheoles is most dense in the region of very active muscle, for example, in the flight muscles in the thorax.

The tracheae open to the atmosphere by pores called *spiracles* (Figs. 15.3 and 15.27(a)). Often there is one spiracle on each side of every segment of the body, but in some insects there are only one or two spiracles on either side. The entrance to the spiracle is usually supplied with muscles which enable it to open and close. The spiracles close when the insect is not active and therefore needs little oxygen, and this closure helps to reduce the loss of water by evaporation from the internal tissues.

The movement of oxygen from the atmosphere through the breathing

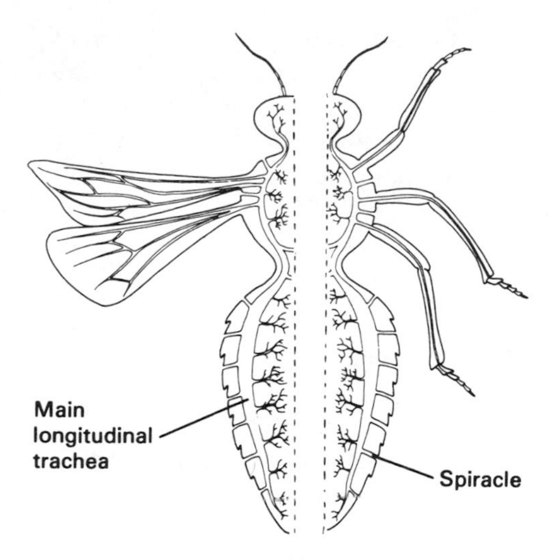

Fig. 15.3 *Tracheal system of insect (wings and legs each shown on one side only)*

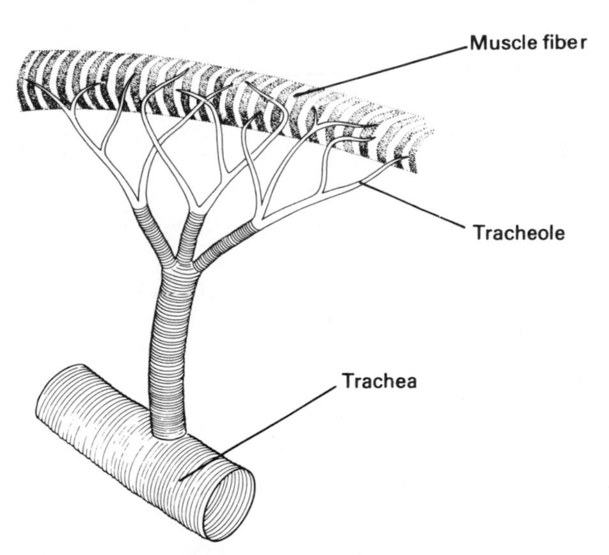

Fig. 15.4 *Diagram to show how the tracheae supply oxygen to the muscles*

(From V. B. Wigglesworth, *Principles of Insect Physiology*, Chapman & Hall)

tubes to the tissues, and of carbon dioxide in the opposite direction, can be accounted for by simple diffusion. In active adult insects, however, there is often a ventilation process which exchanges up to 60 per cent of the air in the tracheal system. Ventilation is brought about by rhythmic contractions of the abdomen, produced by the contraction and relaxation of the internal muscles. When the muscles relax, the abdomen springs back into shape, and the tracheae expand and draw in air. Positive muscular action in insect breathing produces expiration, in contrast to the mechanism of mammalian breathing (see Section 22.4).

15.5 BLOOD SYSTEM

The supply of oxygen to the organs through the tracheae means that the blood system of an insect has a different role from that of a vertebrate. Insect blood has little need to carry oxygen from one part of the body to another, except in regions where the tracheoles terminate at some distance from a cell, and it contains no hemoglobin and no cells corresponding to the red blood cells of mammals. There is no network of blood vessels; the free space or *hemocoel* between the cuticle and the organs in the body cavity is filled with blood. The single dorsal blood vessel propels the blood forward by contractions of its muscular walls, drawing in blood through pores in its sides and releasing it into the hemocoel in the region of the head.

The main functions of the blood are the distribution of the products of digested food, which diffuse into the blood through the walls of the gut, and the collection of the excretory products of the tissues. It has in addition certain hydraulic functions: during ecdysis, the expansion of the thorax by the accumulation of blood initiates the splitting of the cuticle, and, in metamorphosis, the pumping of blood into the crumpled wings of the newly emerged adult insect serves to expand them fully.

15.6 SENSORY SYSTEM

(a) Touch

There are a great many fine bristles on the body surface of insects, especially on the lower segments of the legs, the head, the thorax, the wing margins and the antennae, according to the species. Most of these bristles have a sensory function, responding principally to touch, vibration or the presence of certain chemicals.

(b) Proprioceptors

Certain small oval or circular areas of cuticle are slightly thicker then the surrounding parts, and are supplied with sensory fibers. These organs probably respond to distortions of the cuticle resulting from pressure, and so feed back information to the central nervous system about the position of the limbs. Organs of this kind respond also to deflections of the antennae during flight and are thought to "measure" the air speed and help to adjust the wing movements accordingly. In some insects there are nerve endings associated with muscle fibers which apparently have the same function as the stretch-receptors of vertebrates.

(c) Sound

Low-frequency vibrations are sensed by the touch or tactile bristles on the cuticle and antennae, but in many insects there are additional specialized organs sensitive to high-frequency sounds; some of these organs can distinguish between sounds of different frequencies. Each consists of a thin area of cuticle overlying a distended trachea or air sac, supplied with sensory fibers; they appear on the thorax, on the abdomen or on the legs, according to the species. They can be used to locate the source of a sound, for example, when a male cricket "homes" on the chirping sound made by the female.

(d) Taste

Experiments show that certain insects can differentiate chemicals which we describe as sweet, sour, salt and bitter, and that some insects can distinguish other substances as well. The taste-sensitive organs are most abundant on the mouth parts and in the mouth itself, and on the lower (tarsal) segments of the legs. A house fly walking over its food is actually tasting it with its feet.

(e) Smell

Most of the sensory organs concerned with smell are on the antennae. These *chemoreceptors* are in the form of bristles, pegs or plates, the cuticle of which is very thin and contains minute perforations through which project nerve endings sensitive to chemicals. Sometimes these sense organs are grouped together and sunk into depressions in the cuticle called *olfactory pits*. The sense of smell is very highly developed in certain moths:

an unmated female emperor moth will attract males from distances of up to one or two kilometers away by the scent she exudes, although this may be quite imperceptible to humans. The antennae of a male moth carry thousands of chemoreceptors and are large and beautifully feathered.

(f) Sight

The *compound eyes* of mature insects are made up of thousands of identical units packed closely together (Figs. 15.5 and 15.6). Each unit contains a lens system, formed partly by a thickening of the cuticle, which is transparent in this region, and partly by a special *crystalline cone.* This lens system, unlike that in the human eye, has a fixed focal length which cannot be varied for viewing objects at different distances from the eye.

Each eye unit can record whether any light is emitted or reflected from a small patch in the eye's field of view and, if so, what is its direction and intensity, and sometimes also its color. However, although the number of units in an insect's compound eye is very large—between 2,000 and 10,000 according to the species—it cannot reconstruct a very accurate or detailed impression of the outside world. The impression is produced from a mosaic of tiny images, each representing a part of the field of view, overlapping to some extent in certain species but not in others (Fig. 15.7). Nevertheless a crude impression of the form of well-defined objects can be

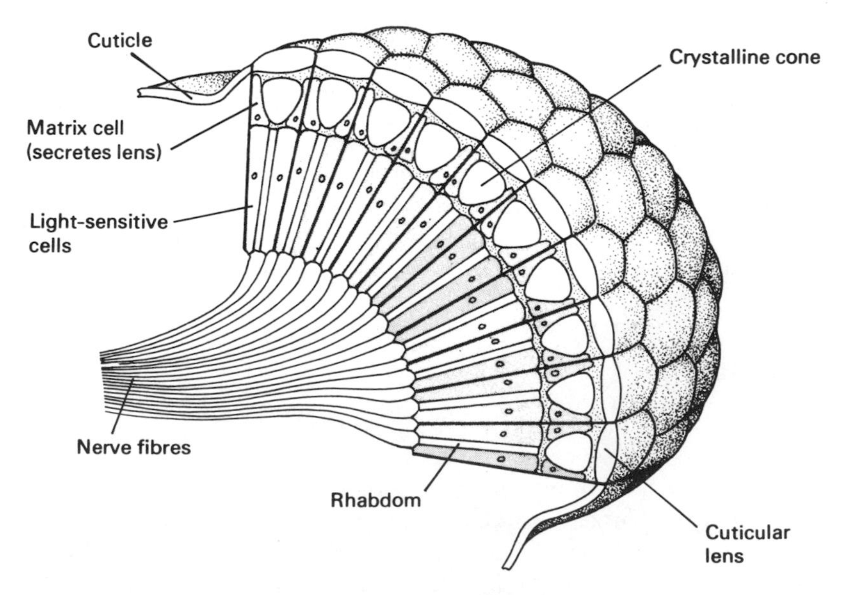

Fig. 15.5 *Compound eye*

Fig. 15.6 *Head of housefly, greatly magnified to show compound lens*
(Shell International Petroleum Co. Ltd.)

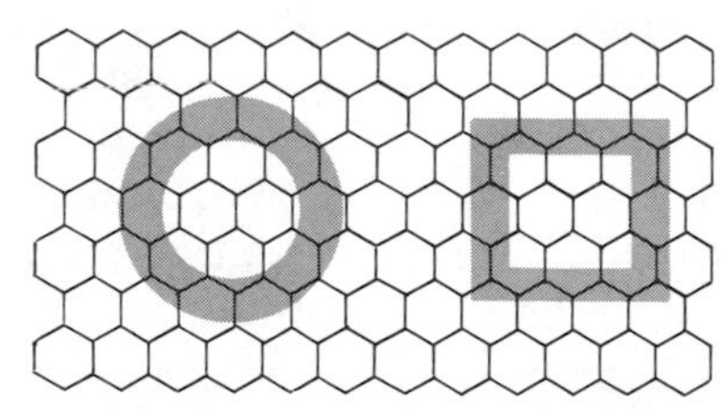

The shapes projected on to equal areas of compound eye will

stimulate the same pattern of eye units and
so produce indistinguishable mosaic images

Fig. 15.7 *Limitation of interpretation by compound eye*

obtained over a wide field, sufficient, for example, to enable bees to seek out flowers, and to use landmarks for finding their way to and from the hive. It is likely that the construction of compound eyes makes them particularly sensitive to moving objects; for example, bees are more readily attracted to flowers when they are being blown by the wind.

Color vision. Flower-visiting insects, at least, can be shown experimentally to be able to distinguish certain colors from shades of gray of equal brightness. Bees are particularly sensitive to ultraviolet light (invisible to man) and to blue and violet, but cannot differentiate red and green from black and gray unless the flower petals are reflecting ultraviolet light as well. Some butterflies can distinguish yellow, green and red.

Caterpillars and most other insect larvae have small simple eyes (Fig. 15.22), consisting of a cuticular lens with a group of light-sensitive cells beneath, rather like a single eye unit. They show some color sensitivity and, when grouped together, some ability to discriminate form.

Many flying insects also have simple eyes or *ocelli* in their heads, in addition to their compound eyes; the ocelli probably only respond to changes in light intensity.

15.7 LOCOMOTION

Movement in insects depends, as it does in vertebrates, on muscles contracting and pulling on jointed limbs or other appendages. Many of the joints in an insect's body are of the *peg-and-socket* type, which permit movement in one plane only, like a hinge joint (Fig. 15.9). However, there are several such joints in an insect limb, each operating in a different direction, so that the limb as a whole has considerable directional freedom of movement (Fig. 15.8). The muscles are attached to ridges and projections on the inside of the cuticle: pairs of antagonistic muscles act across each joint, one bending and one straightening the limb.

(a) Walking

The characteristic walking pattern of an insect is shown in Fig. 15.10. The body is supported on three of its six legs while the other three swing forward to a new position, and then take the weight of the body in their turn.

The tarsi (lower segments) of the legs carry small claws and, in some species, glandular pads which secrete an adhesive substance (Fig. 15.8).

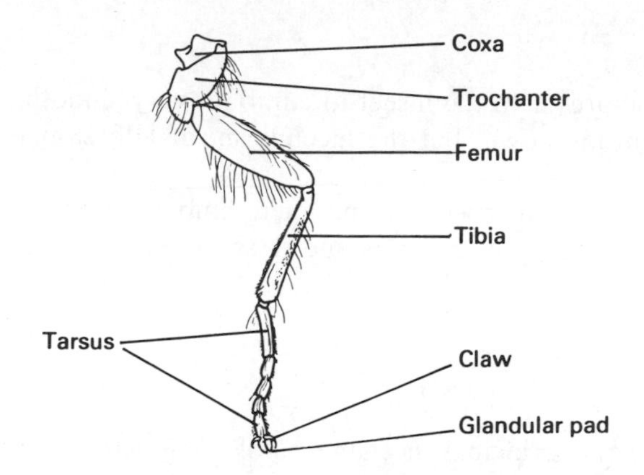

Fig. 15.8 *Leg of housefly*

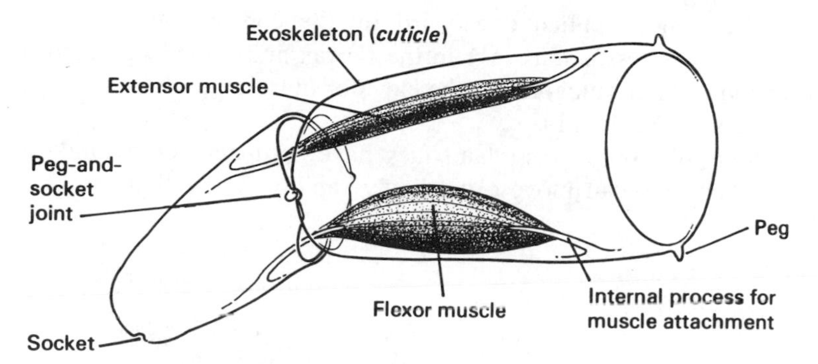

Fig. 15.9 *Muscle attachment in an insect limb*

The shaded legs are moving more or less at the same time, the others are in contact with the ground

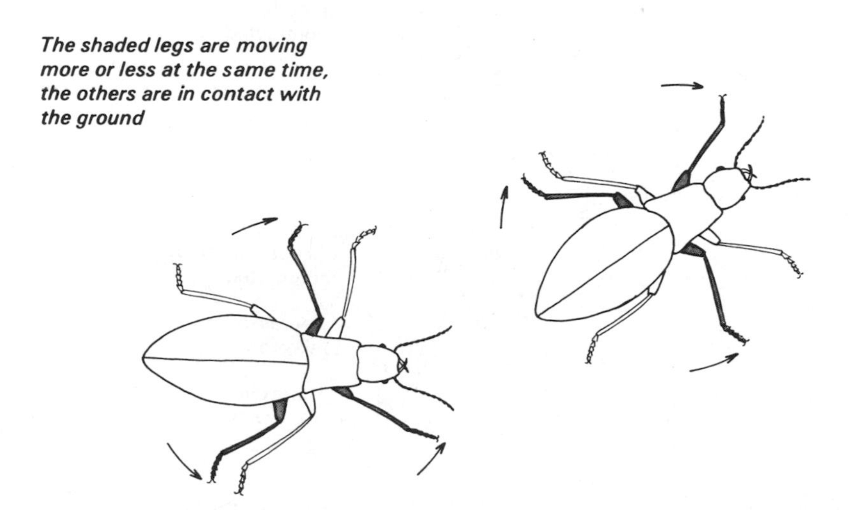

Fig. 15.10 *Walking pattern in a beetle*

Both these structures help the insect to climb on very smooth surfaces and even to hang upside down, but the mechanism of adhesion is not known with certainty.

Some insects have specially modified limb structures and muscles enabling, for example, grasshoppers and fleas to leap and water-beetles to swim.

(b) Flying

There are two principal mechanisms of wing action, depending very much on the relative proportions of the insect's body. Insects like butterflies and dragonflies have fairly light bodies and large wings; in these insects the wing is pulled downward by the contraction of one set of muscles (the depressor muscles) in the thorax and pulled up again by the contraction of their antagonists (the elevator muscles), while the depressor muscles relax (Fig. 15.11).

Insects like bees, wasps and flies have compact bodies and smaller wings, which they can move exceedingly rapidly. Their flight muscles act

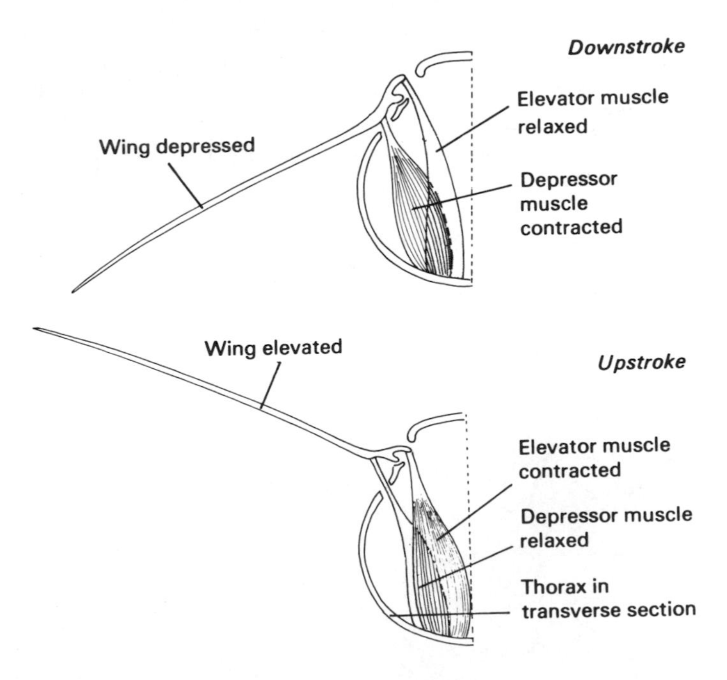

Fig. 15.11 *Action of direct flight muscles (e.g., butterflies)*
(Adapted from Snodgrass, *Principles of Insect Morphology*, 1935, by permission of McGraw-Hill)

indirectly by pulling on the walls of the thorax and changing its shape. Contraction of the elevator muscles pulls the roof of the thorax downward, elevating the wings; when the elevator muscles relax and the depressor muscles (running longitudinally down the thorax) contract, the thorax returns to its original shape and the wings are depressed again (Fig. 15.12).

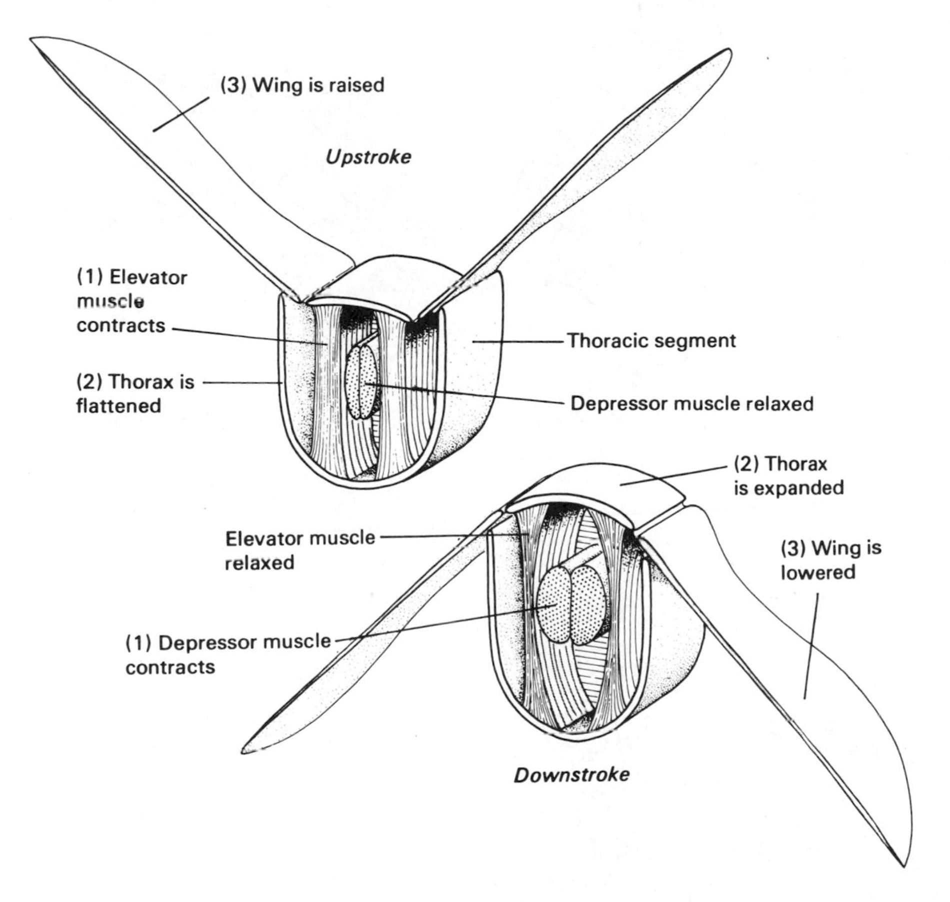

Fig. 15.12 *Action of indirect flight muscles (e.g., bees, wasps, flies)*

Both kinds of insect have muscles which alter the angle of the wing as it passes through the air. During the downstroke the wing is held horizontally, thrusting downward on the air and producing a lifting force. At the bottom of the stroke the wing is rotated into a vertical position and remains vertical during the upstroke, thus offering little resistance as it is drawn through the air.

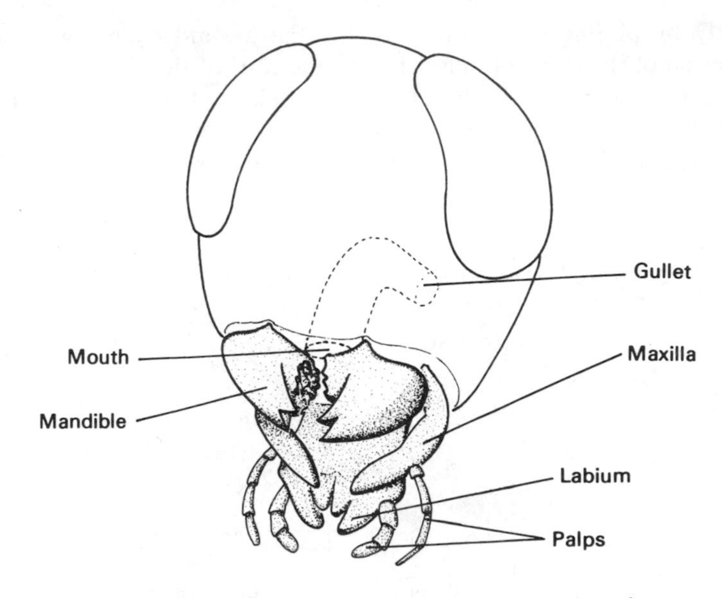

Fig. 15.13 *Mouth parts of insect (e.g., cockroach)*

15.8 FEEDING METHODS

The feeding methods of different insect species vary greatly. However, all insects have three pairs of *mouth parts,* hinged to the head below the mouth, which are used to extract or manipulate food in one way or another. In many species these are extensively modified and adapted to exploit different kinds of food source. The least modified are probably those of grasshoppers, locusts and cockroaches (Fig. 15.13). In these insects the first pair of mouth parts, the *mandibles,* work sideways across the mouth, cutting off pieces of vegetation, which are manipulated into the mouth by a second pair called the *maxillae* and a third pair fused together into a structure called the *labium*. The mouth parts of some insects carry jointed sensory organs or *palps*.

Aphids (Fig. 15.14). These insects feed on plant juices which they suck from leaves and stems. Their mouth parts are greatly elongated to form a piercing and sucking structure called a *proboscis* (Fig. 15.15(*a*)). The maxillae fit together to form a tube which can be pushed into plant tissues to reach the sieve tubes of the phloem and so extract nutrients (Fig. 15.15(*b*)); the labium is modified to form a sheath around the proboscis, enclosing it when it is not being used.

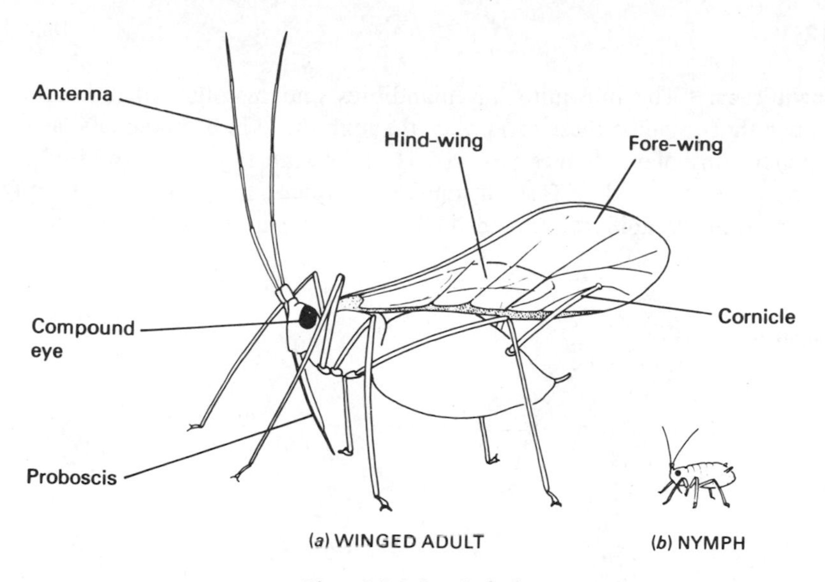

Fig. 15.14 *Aphids*

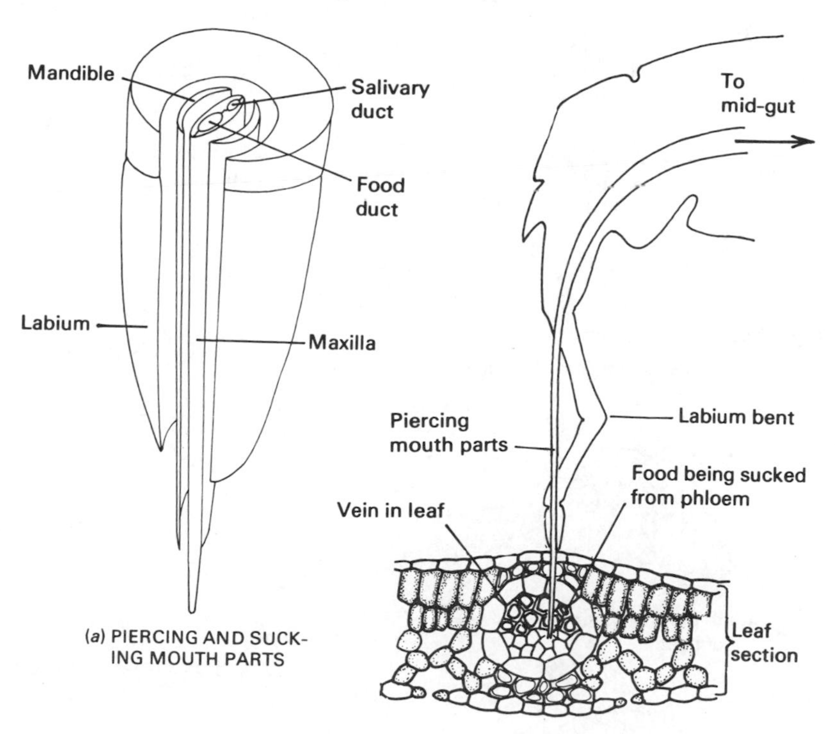

Fig. 15.15 *Feeding methods of aphids*

Mosquitoes. The mosquito has mandibles and maxillae in the form of slender sharp stylets; these can pierce through the skin of mammals, as well as penetrating plants tissues (Fig. 15.16). The *labrum* or "upper lip" of a mosquito is modified to form a long tube which can be inserted through the skin into a capillary, and through which the insect sucks blood (Fig.

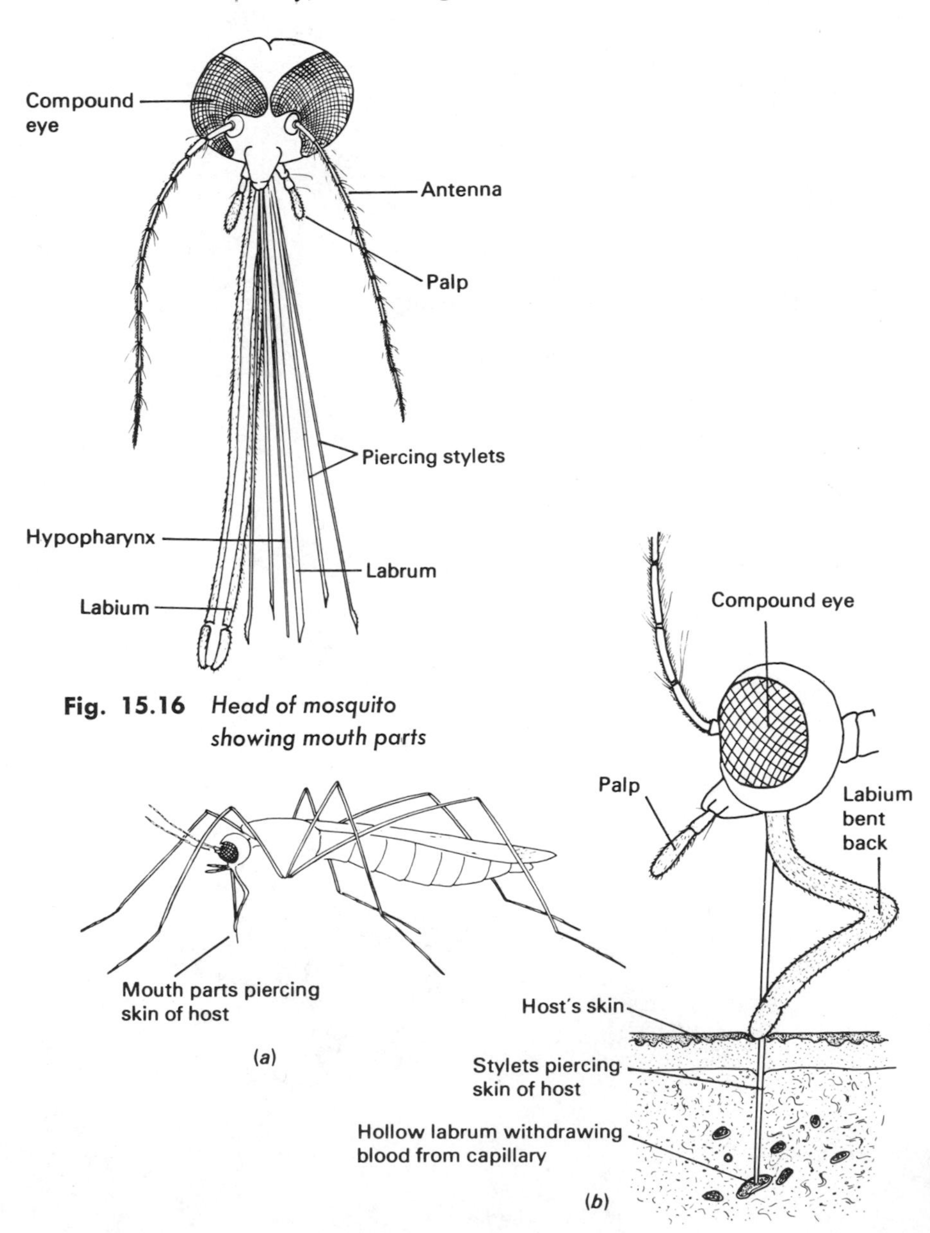

Fig. 15.16 *Head of mosquito showing mouth parts*

Fig. 15.17 *Mosquito feeding*

15.17). Another tubular structure, the *hypopharynx,* is used to inject the wound with a substance which prevents the blood from clotting and so blocking the narrow tube of the labrum. As in aphids, the labium is rolled around the other mouth parts, sheathing and protecting them when they are not in use.

Butterflies. These insects feed through a proboscis, formed from the greatly elongated maxillae which are in the form of half-tubes, like a drinking-straw split down its length. The two halves fit closely together to form a tube through which nectar is sucked from the flowers (Fig. 15.18). When not in use the proboscis is coiled up underneath the butterfly's head.

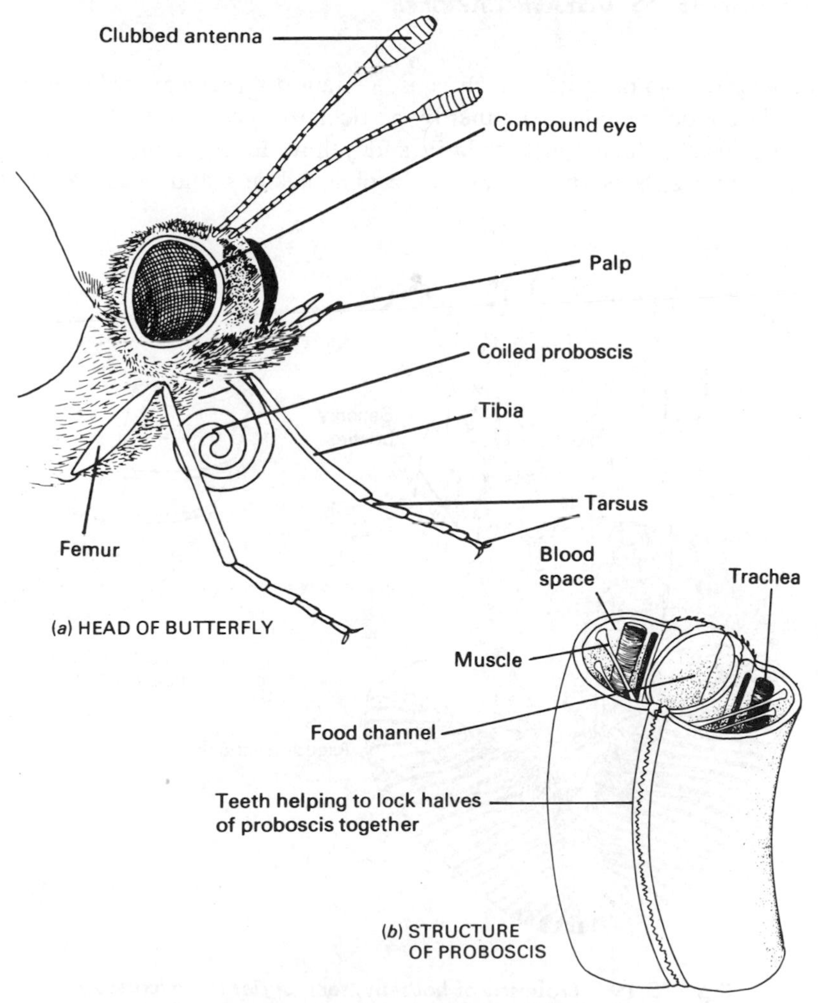

Fig. 15.18 *Feeding apparatus of butterfly*

Houseflies. The housefly also feeds by sucking liquids. The labium is modified to form a relatively short proboscis which cannot penetrate tissue or reach the nectar of flowers; it terminates in two pads whose surface is channeled by grooves called *pseudotracheae* (Fig. 15.19). The fly applies its proboscis to the food (Fig. 15.21) and pumps saliva along the channels and over the food. The saliva dissolves soluble parts of the food and is thought to contain enzymes which digest some of the insoluble matter. The nutrient liquid is then drawn back along the pseudotracheae and pumped into the alimentary canal (Fig. 15.20).

15.9 INSECTS AS DISEASE CARRIERS

The microscopic parasites which cause certain diseases can only be transmitted from one person to another by particular insect species: for example, the organisms which cause malaria and yellow fever in man are carried by mosquitoes, tsetse flies transmit sleeping sickness and fleas harbor the

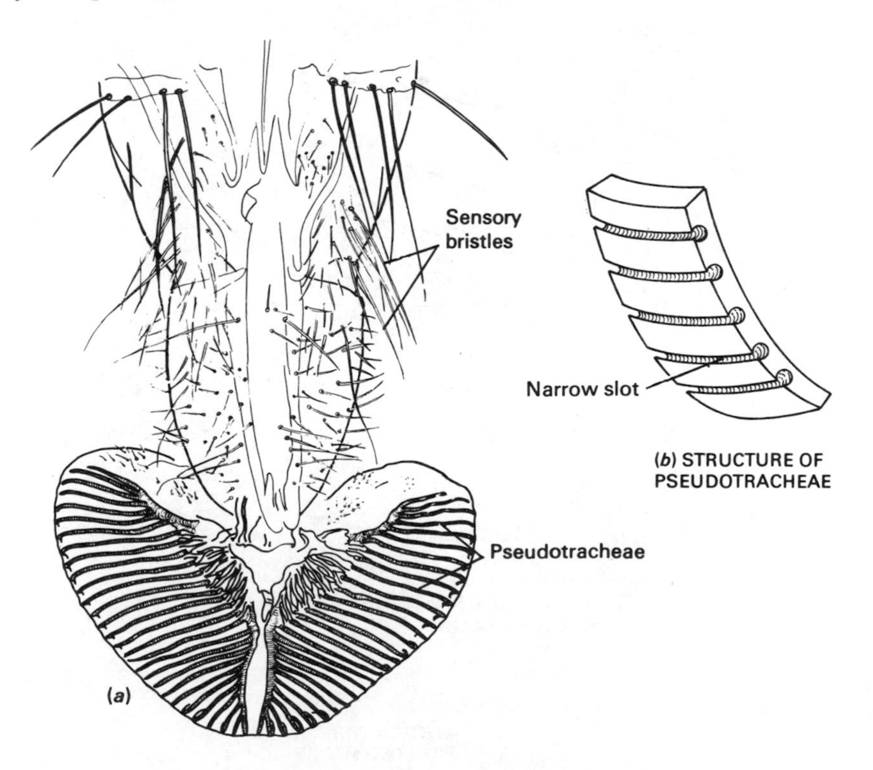

Fig. 15.19 *Proboscis of housefly, seen under the microscope*

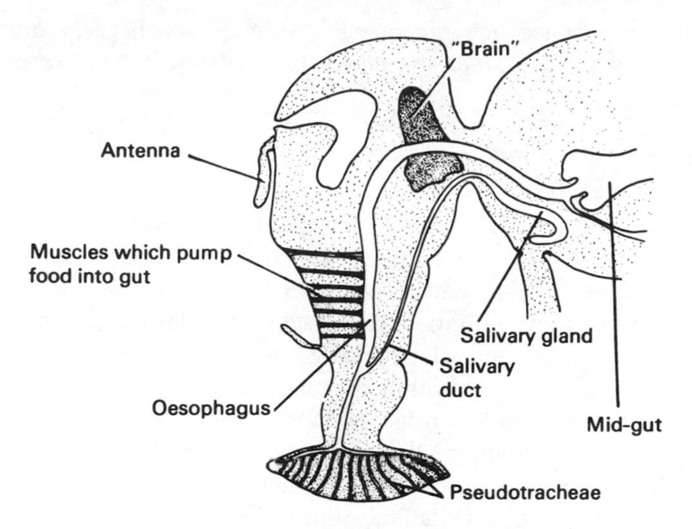

Fig. 15.20 *Section through head of housefly*

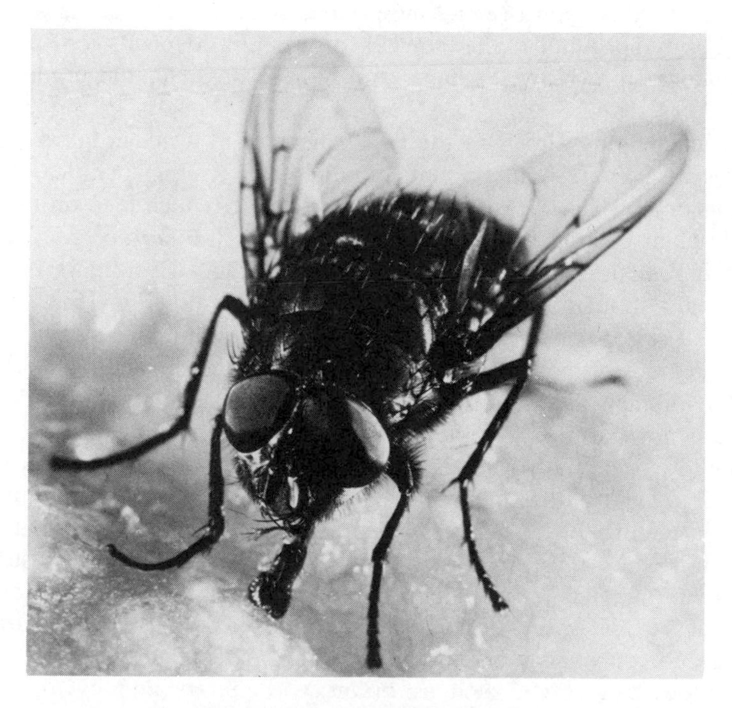

Fig. 15.21 *Blowfly feeding*
(Heather Angel)

bacteria which produce bubonic plague. Animals which carry organisms that can cause disease in other animals or in plants are called *vectors*.

(a) Mosquitoes and Malaria

The serious disease of malaria is caused by a protozoan parasite. The blood of an infected person contains large numbers of these parasites. Female mosquitoes of the genus *Anopheles* feed on human blood, which they suck from the skin capillaries through their tubular mouth parts (Figs. 15.16 and 15.17). Any malarial parasites in the blood so taken begin a new stage in their life cycle within the mosquito, undergoing a series of complicated changes and reproducing themselves in enormous numbers. Eventually they accumulate in the salivary glands of the mosquito. If the insect now bites a healthy person, the parasites are likely to enter his blood via the mouth parts of the infected mosquito. They enter the red blood cells where they feed, grow and reproduce, eventually destroying large numbers of the red cells and producing the characteristic symptoms of malaria.

Malaria can be treated by drugs, but even so the disease is likely to recur again and again over a long period. Since the parasite is only transmitted by mosquitoes, the control of malaria depends on preventing mosquitoes from biting humans, and, where possible, the eradication of mosquitoes.

Female mosquitoes lay their eggs in stagnant water in lakes, ponds or ditches, or even in small amounts of rain-water lying in puddles, drinking-troughs or cans. The eggs soon hatch into larvae, which feed on the microscopic plants in the water. The larva lives at the surface of the water, breathing air through a *tracheal tube* (Fig. 15.22). Eventually it pupates; the pupa, although it does not feed, still breathes air (Fig. 15.23). Finally the pupal skin splits open and the imago emerges and flies away.

Methods of mosquito eradication are directed at each of the stages of its development. The adult insect can be attacked with pesticide sprays that are not harmful to other life forms. The breeding grounds of the mosquito can be destroyed by draining stagnant swamps, turning sluggish rivers into swiftly flowing streams, and by preventing the accumulation of water in puddles or tanks accessible to the mosquito. Treating static water with oil-based sprays produces a film of oil on the water which suffocates the larvae and pupae by blocking their breathing tubes; the addition of pesticides to the sprays increases their effectiveness. Such spraying must include not only lakes and ponds but any accumulation of water which mosquitoes can reach, such as drains and gutters and even the small amounts collected in old tin cans and other rubbish.

As a result of the World Health Organization's program of malaria

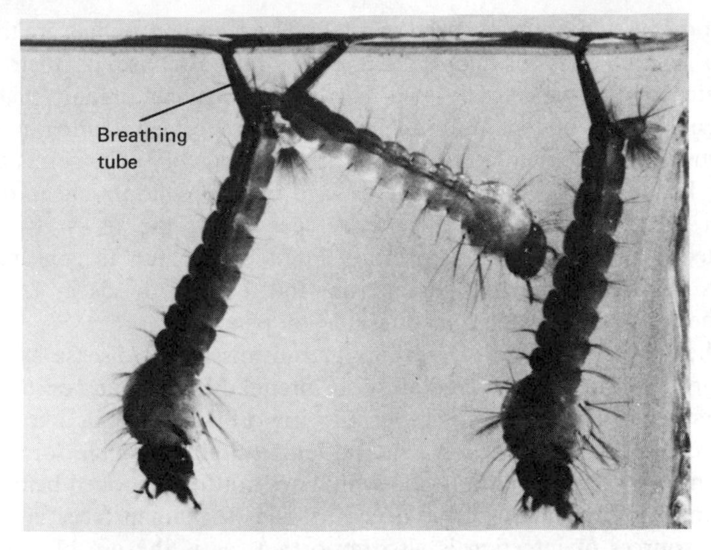

Fig. 15.22 *Larvae of mosquito* (Culex molestus)
(Shell International Petroleum Co. Ltd.)

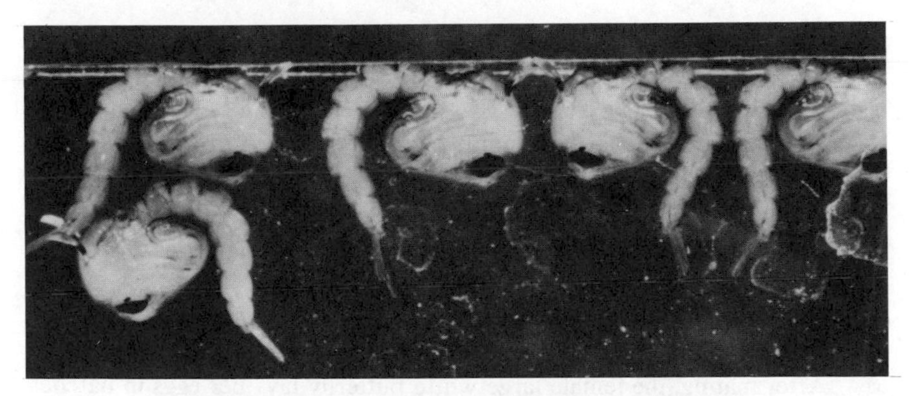

Fig. 15.23 *Pupae of mosquito* (Culex molestus)
(Shell International Petroleum Co. Ltd.)

eradication, the elimination of malaria by these methods is virtually complete in thirty-six countries and well advanced in another forty-six.

(b) Houseflies and Disease

A great many harmful bacteria and viruses are found in or on the bodies of houseflies, but the insects are not essential to the development and transmission of these organisms in the way that the mosquito is essential to the transmission of the malarial parasite. However, houseflies are thought to help spread sixty or more diseases, mainly because of their indiscriminate feeding habits. They will settle on decaying organic matter in which bacteria are abundant, or alight on human feces containing, possibly, the germs of typhoid, cholera, poliomyelitis or dysentery. The germs adhere to their legs or their proboscis; they lodge in the pseudotracheae or pass through the alimentary canal to be released with the feces. If a contaminated fly alights and feeds on food intended for human consumption, it seems very likely that the germs from the feet, proboscis or feces will reach the food and eventually be ingested by people.

The simplest method of preventing transmission of disease by houseflies is probably to keep covered all food intended for human consumption, to prevent their access. This is especially important for foods like cooked meat, cream cakes, or puddings which offer a suitable medium for harmful organisms to grow and multiply and which are not to be cooked before they are eaten. The prevention of the access of flies to human feces and other possible sources of infection is also important, as is the use of pesticides against both adults and larvae to reduce the housefly population.

15.10 THE LARGE WHITE BUTTERFLY

In the preceding sections the general characteristics of insects have been discussed with reference to various species to illustrate certain points of interest. This section describes in more detail the life history and habits of one conspicuous and fairly common insect, the large white butterfly.

(a) **Eggs**

After mating, the female large white butterfly lays her eggs in batches of from 6 to 100, usually on the underside of leaves of plants of the cabbage family, to which they are attached by a sticky secretion. The egg is "teardrop"-shaped, yellow or orange, with a ribbed pattern over its surface (Figure 15.24). It contains yolk, which nourishes the developing embryo. There is a small hole or *micropyle* at the top of the egg, which admitted a sperm when the egg was fertilized and which now allows air to reach the embryo. The eggs hatch about a week after laying if the weather is warm. Small larvae or *caterpillars* (the term used for the larvae of butterflies and moths) eat their way out of the egg shells and then devour the remains of the shells; it seems that one or more of the constituents of the shells is essential for their continued development. They then begin feeding on the cabbage leaf, biting off pieces with their mandibles.

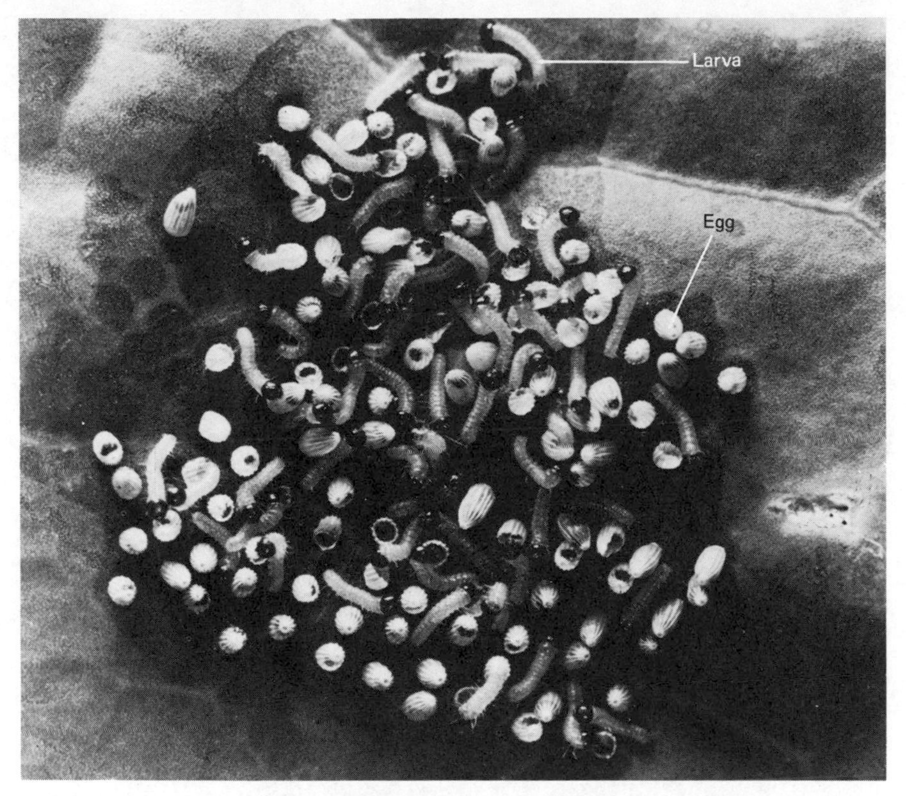

Fig. 15.24 *Large white butterfly; caterpillars hatching*
(Stephen Dalton NHPA)

(b) Caterpillars

The body of the caterpillar of the large white butterfly is cylindrical in shape and yellow in color, with black markings. It consists of a head and thirteen body segments (Fig. 15.27).

The head carries a pair of short antennae and powerful mandibles which move transversely, slicing off small pieces of leaf. There are six simple eyes on each side of the head; each consists of a single lens with light-sensitive cells beneath (Fig. 15.25). Inside the head is a pair of glands which function as silk-producing organs. Their ducts join and project in a tube called the *spinneret* behind the mandibles. Liquid secreted by these glands passes through the spinneret and hardens in the air to form a thread of silk. The caterpillar uses this to attach itself when walking on a slippery surface, secreting a zigzag trail of sticky thread which clings to the surface and which can be gripped by the legs.

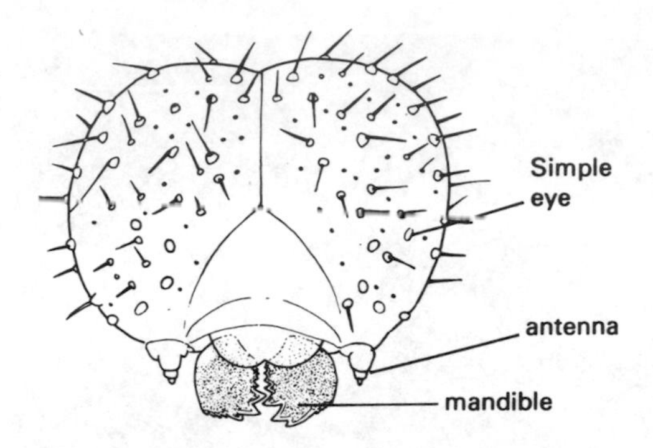

Fig. 15.25 *Caterpillar's head, seen from the front*

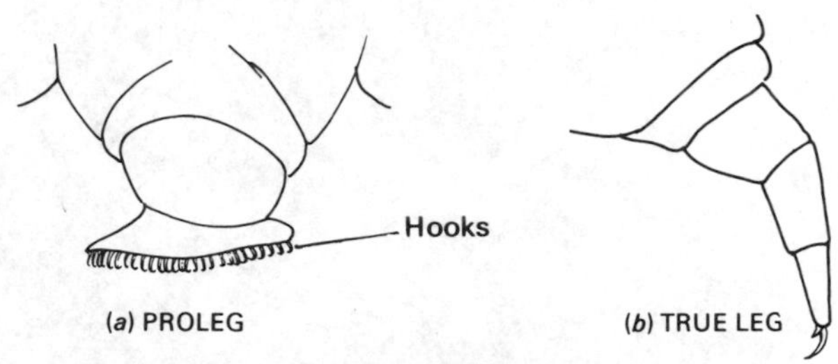

Fig. 15.26 *Caterpillar's legs*

The first three body segments correspond to the thorax of the adult and the remaining ten to its abdomen. There is a spiracle on both sides of the first thoracic segment and of the first eight abdominal segments. The thoracic segments each bear a pair of jointed "true" legs (Figs. 15.26(*b*) and 15.27(*a*)), with claws at the end. These correspond to the legs of the adult. The third, fourth, fifth and sixth abdominal segments each bear a pair of unjointed fleshy projections with rows of tiny hooks at their ends, called *prolegs* (Figs. 15.26(*a*) and 15.27(*a*)); these are not present in the imago. The last segment bears a pair of claspers similar to the prolegs.

The caterpillar is a voracious eater, but because its cuticle is relatively inextensible it can only grow in size at the time of molting (see Section 15.3). During its development it sheds its cuticle four times.

The caterpillars of the large white butterfly are often attacked by a parasitic *ichneumon fly*. The female ichneumon lays eggs inside the body of the caterpillar using her ovipositor to pierce the caterpillar's cuticle. The ichneumon larvae emerge from the eggs inside the caterpillar and feed on the food reserves in its tissues without, however, harming its vital organs; the caterpillar thus continues to live and grow at the same time as the ichneumon larvae are developing inside its body. When the caterpillar is fully grown it dies, and the ichneumon grubs emerge through its skin and spin themselves yellow cocoons upon its surface, within which they pupate.

(c) Pupation

When the caterpillar has reached its full size it leaves the cabbage plant and migrates to a dry, sheltered place such as a wall or a tree. Here it settles vertically, head uppermost, and attaches its thorax to the surface by spinning a girdle of silk (Fig. 15.27(*b*)). The caterpillar's thorax shortens and swells; after about two days its last cuticle splits down the back and is pushed off by rhythmic contractions of the body (Fig. 15.27(c)). On the last segment a group of hooks can now be seen; when the cuticle is shed the hooks anchor into a pad of silk that the caterpillar has previously spun.

The pupa or *chrysalis* which emerges from this last larval ecdysis is pale and soft at first (Fig. 15.27(*d*)) but soon hardens and darkens, often approximately matching the color of its background. It is clear that extensive changes have already taken place for the pupa is very different in form from the caterpillar, the outline of the adult's wings, legs, antennae and proboscis being clearly visible in the cuticle (Fig. 15.27(*e*)). During the next two or three weeks the pupa remains more or less motionless, but inside its body most of the larval organs are being digested away, while certain specialized cells, dormant until now, begin to multiply. These cells give rise to the organs of the new adult, adapted to its different feeding methods, locomotion and habitat.

At the end of the pupation period the cuticle of the pupa splits down the back and the adult butterfly pulls itself out of the skin. Its wings are crumpled, folded and soft at this stage, but they expand as blood is forced into their "veins" (see Section 15.10(*d*)), and finally harden and dry.

(d) Imago

The imago's body consists of a distinct head, thorax and abdomen. There are no legs on the abdomen, but there are three pairs of legs attached to the thorax which also bears two pairs of wings. The head carries a pair of long antennae bearing organs of smell, two large compound eyes, each consisting of about 6,000 lenses with light-sensitive cells beneath them, and a long proboscis. The proboscis consists of the two elongated maxillae, grooved along their inner surfaces so that when they are placed together they form a tube through which nectar can be sucked from flowers (Fig. 15.18).

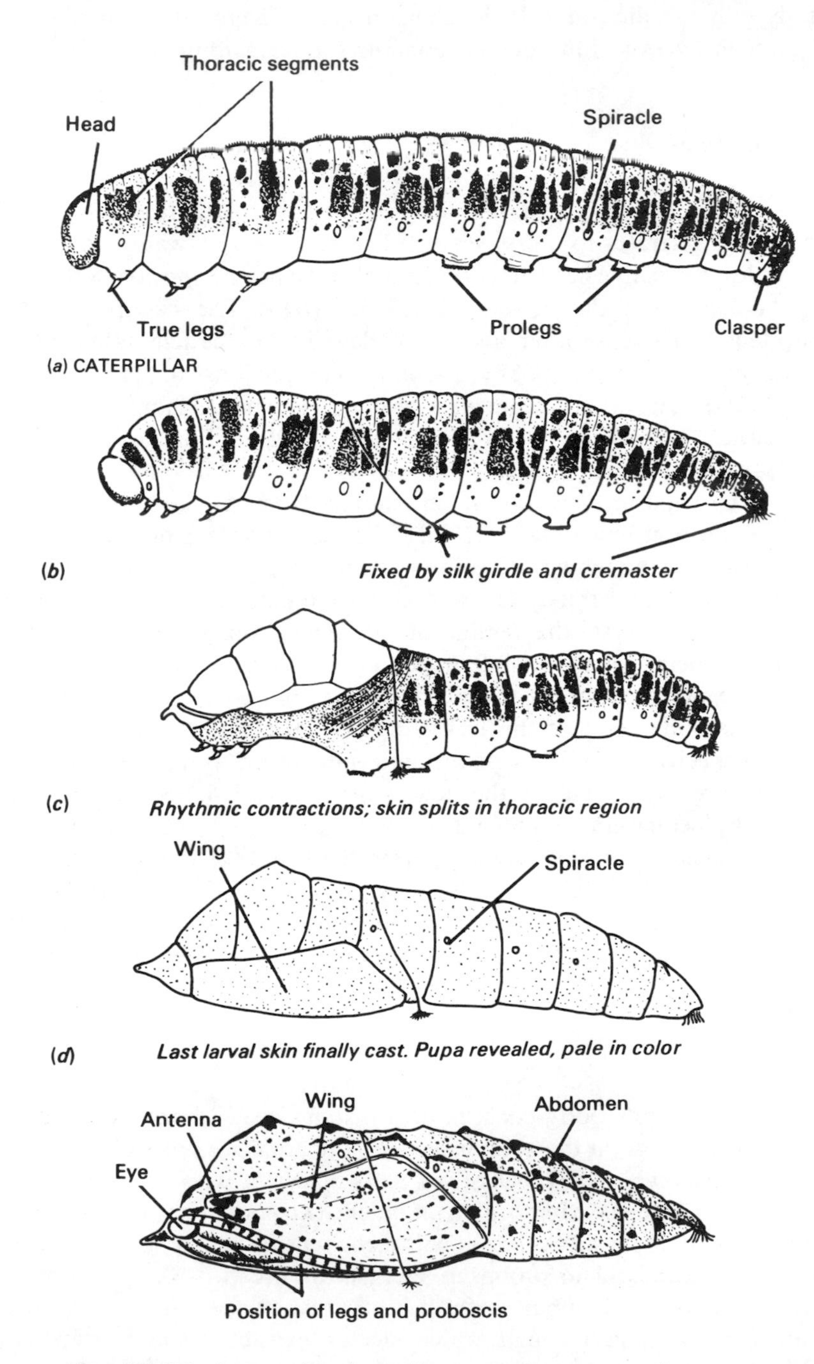

Fig. 15.27 *Large white butterfly: stages in metamorphosis*

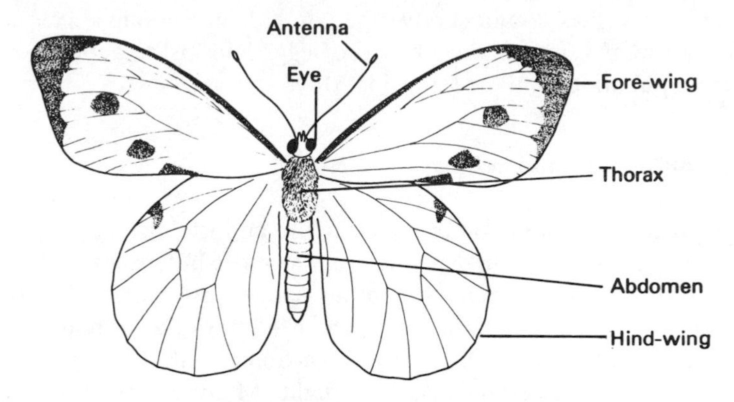

Fig. 15.28 *Large white butterfly (female)*

The broad wings on the thorax are supported by a network of "veins," hollow canals leading from the hemocoel, containing tracheae and nerves. The wings are covered with tiny overlapping scales which give the characteristic wing patterns and colors, partly by the way they reflect and absorb light according to the angle they make with the light, but largely because of the pigment they contain. The fore-wings overlap the hind-wings, so that in flight both pairs move together.

Each of the six legs consists of nine segments joined together. The first two, the *coxa* and the *trochanter,* are very short, the next two, the *femur* and the *tibia,* are long and the foot or *tarsus* is made up of five short segments (Fig. 15.18; compare Fig. 15.8). In some species the tarsi of the last pair of legs bear taste organs, and if these legs are dipped into sugar solution or fruit juice the butterfly will uncoil its proboscis to feed.

The thorax is covered with *setae* or bristles; these are quite unlike the hairs of mammals, and many have touch-sensitive sensory cells at their bases (see Section 15.6(a)).

The adult large white butterfly feeds by sucking nectar from the nectaries of flowers through its long proboscis. The insects are attracted to the flowers by sight and by smell. Butterflies are believed to be able to distinguish color, but not all species are equally sensitive to the same colors. The large white's eye is particularly sensitive to red and purple light.

The male is usually attracted to the female by the scent which she exudes. He flutters around her and stimulates her with his own scent, which is secreted by certain glandular areas beneath the scales on the wings. He grips the female with the claspers at the end of the abdomen and passes sperms into her oviducts. Afterward she will lay fertilized eggs through her ovipositor on one of the food plants suitable to the species, usually a plant of the cabbage family. There may be two or more broods during the season. The butterflies do not survive the winter, but the pupae can do so, in which case adults from the last brood of eggs laid in the autumn do not emerge from pupation until the following May.

(e) Other Butterflies and Moths

The preceding description of the structure and life history of the large white butterfly also applies in general terms to most butterflies and moths. However, some of the differences of other species are worth mentioning.

Frequently, the eggs are not laid in batches but singly, although nearly always on the food plant of the caterpillars. Some butterflies scatter their eggs at random over grassland while in flight. Many caterpillars feed on only one species or family of plants, whereas others can exist on a great variety of food. About half the species pupate in the same way as the large white butterfly, but most of the others pupate hanging head downward, and a few, like the silkworm, spin cocoons of silk around themselves before the last larval ecdysis.

In winter the majority of butterflies hibernate as caterpillars, a number as pupae, a few as eggs, and a few (such as the small tortoise-shell butterfly) as adults. The Monarch butterfly of western North America migrates hundreds of kilometers and hibernates in California and Mexico, then makes a return flight to its birthplace to lay its eggs.

15.11 QUESTIONS

1. What structural features do nearly all insects have in common?
2. The mandibles, maxillae and labium represent the basic external feeding apparatus of most insects. Say how these head appendages are modified to cope with the feeding methods of the butterfly, the aphid, the mosquito and the housefly.
3. Outline the stages in metamorphosis of (*a*) a butterfly, and (*b*) a mosquito.
4. State briefly how the larva and imago of a butterfly are adapted in their feeding and locomotion to exploit differing aspects of their environment.
5. What features of a vertebrate eye enable it to convey a more detailed impression of the environment to the brain than does an insect's compound eye?
6. Mention the sense organs involved and the part they play in the event of a butterfly approaching a flower, alighting and seeking nectar with its proboscis.
7. In what ways is a knowledge of the life history and habits of a harmful insect (for example, a vector of disease organisms) helpful in the design of methods of controlling it?

Unit 16
Fish

16.1 INTRODUCTION

The fishes with a bony skeleton form a large group of vertebrate animals which live in water, breathing by means of special organs called *gills.* Their bodies are streamlined in shape and bear fins, and their skin is covered in scales. Most fishes reproduce by laying eggs.

Fish, like all animals other than mammals and birds, are *poikilothermic,* that is, their body temperature varies with that of their surroundings. The temperature of the body of a poikilothermic animal is usually a few degrees above that of its environment, but a rise or fall in the temperature of the air or water in which it lives will produce a corresponding change in the animal's body temperature. The colloquial term "cold-blooded" is unsatisfactory since it might convey the impression that the animal's body remains at a constant low temperature, whereas the bodies of fish in tropical seas may well be at as high a temperature as those of homoiothermic or "warm-blooded" animals.

Most of the chemical changes that take place in a living organism are accelerated by a rise in temperature; it follows that the rate of activity of a poikilothermic animal depends to a large extent on its temperature and hence on that of its surroundings. Some animals may be reduced to a state of torpor at low temperatures, yet be quickly restored to vigorous activity by a rise in temperature.

16.2 EXTERNAL FEATURES

The bodies of bony fish are covered with *scales,* tough plates produced by cells in the skin. In sharks, rays and dogfish, scales grow out through the skin, but in other fish they are covered by skin. The scales overlap each other and form a tough protective coating to the fish's body. Ring-shaped structures can be seen in fish scales if they are examined under a microscope. The rate of growth of the fish affects the rate at which the rings develop, so that groups of rings lying close together or spaced far apart may appear, representing periods of time when food was scarce or plentiful respectively. The age of the fish can thus be estimated from careful examination of the rings of the scales, although one ring does not correspond to one year.

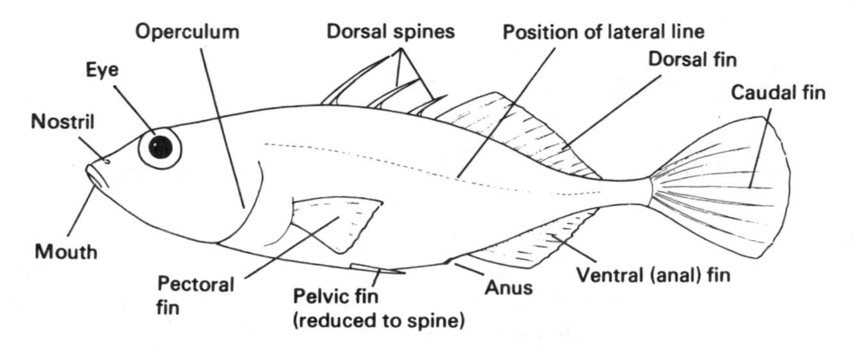

Fig. 16.1 *Three-spined stickleback*

Fish have a number of *fins* on their bodies. The *median fins* lie on the midline of the fish's body, the *dorsal fin* on the back and the *ventral* (or *anal*) *fin* below, and the *caudal fin* at the tail. There are two sets of *paired fins*, the *pectoral fins* lying just behind the head and the smaller *pelvic fins* farther back. The fins give stability to the fish's body and control its direction of movement when it is swimming (see Section 16.3).

The operculum is a bony flap-like structure covering and protecting the gills on each side of the body; the mouth and the opercula act together to maintain the flow of water over the gills (see Section 16.4).

Sensory Organs

Sight. The *eyes* of a fish have large round pupils which do not vary in size.

Hearing. Although fish have no visible external ears, they can hear by transmission of vibrations through the body to sensitive regions of the sacculus or utriculus in the *inner ear*. Sensory receptors in the utriculus and the semicircular canals enable them to maintain their equilibrium. There are no ossicles or cochlea as in mammals.

Smell. The nostrils of fish do not open into the back of the mouth as do those of mammals, and are not used for breathing. They lead into organs of smell which are very sensitive in most species, so that a fish can detect presence of food in the water from considerable distances. The nostrils are double, so allowing water to pass through the organ of smell.

Vibration. The *lateral line* is a fluid-filled tube or canal running along the fish's flank just below the skin (Fig. 16.2). Its function is to detect movements in the water. A disturbance set up in the water will produce vibrations in the fluid in the tube. The lining of the canal contains nerve endings which are sensitive to vibrations and which send impulses to the brain. The fish can thus detect the direction and intensity of water movements, and this sense helps it to navigate around obstacles or to avoid enemies, even if its vision is impaired, for example in darkness or in muddy water.

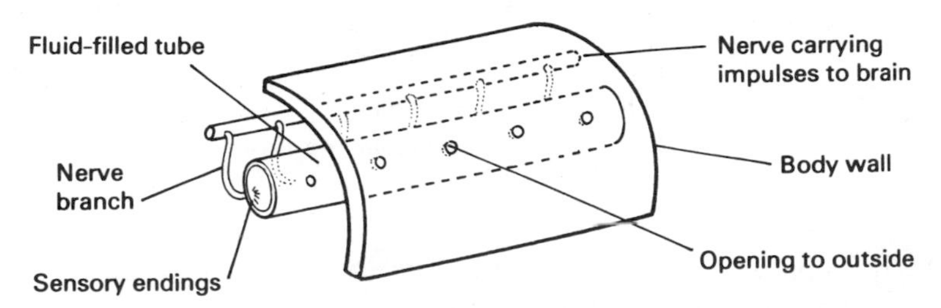

Fig. 16.2 *Diagram of lateral line*

16.3 SWIMMING

The vertebral column of a fish consists of a number of vertebrae held together by ligaments. The ligaments are loose enough to permit slight sideways movement between each pair of vertebrae, so that the spine as a whole is somewhat flexible, especially toward the tail.

The forward movement of the fish through the water is produced by wavelike movements passing down its body, ending in sideways lashing movements of the flexible tail. The movements may be a hardly perceptible rippling motion, as in mackerel, or very pronounced, as in eels, and their frequency varies from about 50/min in dogfish to 170/min in mackerel.

The swimming movements are produced by waves of muscular contraction passing from the head to the tail down each side of the body alternately (Fig. 16.3). A sideways and backward thrust of the tail and body against the water produces a sideways and forward movement of the

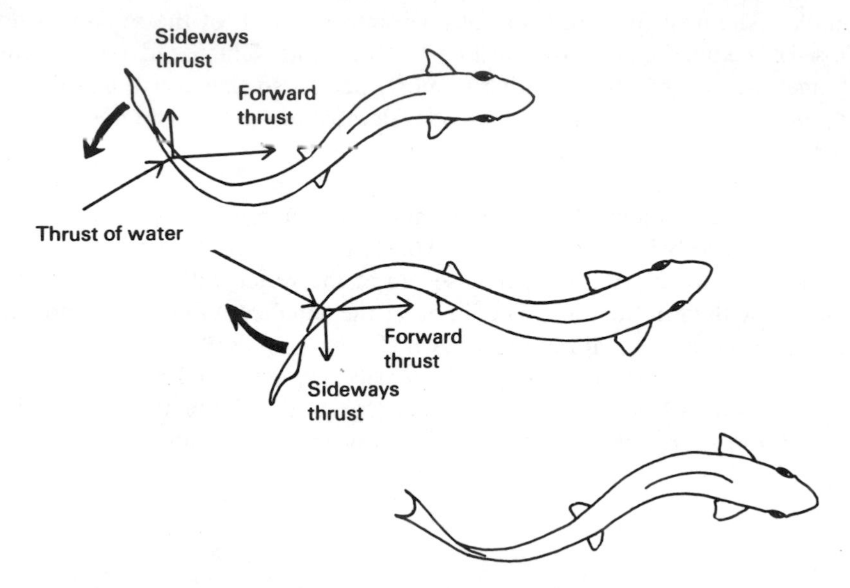

Fig. 16.3 *Diagrams to show how swimming movements produce motion; the bold arrows indicate the direction in which the tail is moving*

fish in the opposite direction (compare the way in which the backward thrust of your foot against the ground produces a forward motion of your body). The final lash of the tail may contribute as much as 40 per cent of the total forward thrust. The sideways movements produced by two successive and similar lashing actions are equal and opposite, and cancel each other out so that the net effect is to move the fish's body straight forward. The fish can turn to the right or left by lashing its body more strongly to one side than to the other.

The swimming speed of fish over long distances is not as fast as one might expect from watching their rapid darting movements in aquaria or ponds. A 10-kg salmon may be able to reach 16 km/h but most small fish probably cannot achieve a speed greater than 4 km/h.

Fins. The force required to propel a fish through the water is produced by movements of its whole body. In most fish the fins do not contribute at all to the propulsive force, but their main function is to control the stability of the fish and the direction of its movements (Fig. 16.4). Rolling and yawing from side to side are controlled by the median fins, which present a large vertical area to the water and resist any tendency to sideways drift. The paired fins act in the same way as the hydroplanes of a submarine, and control the pitch of the fish. The angle at which they are held can be varied by muscular action, causing the fish to swim upward or downward;

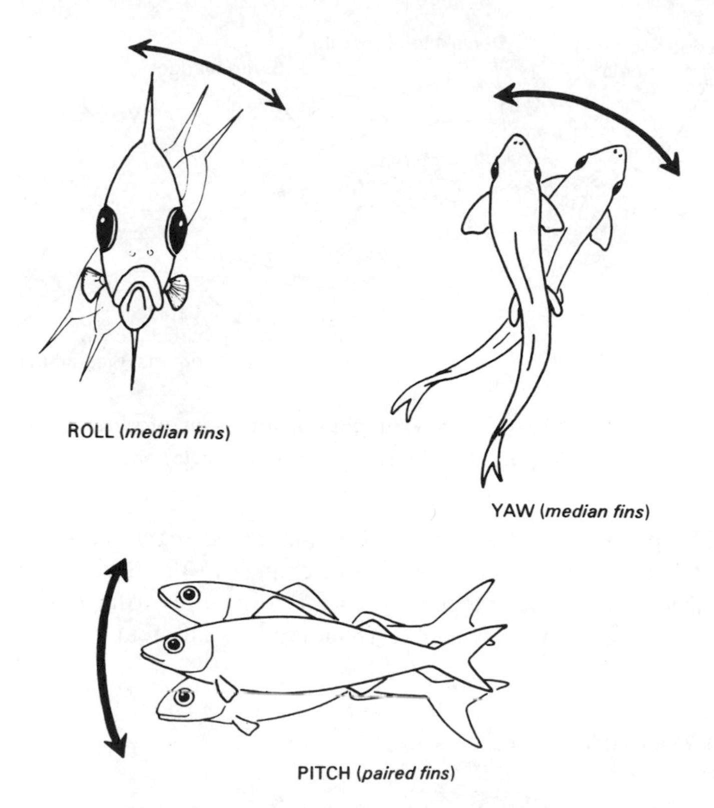

Fig. 16.4 *Movements controlled by fins*

the pectoral fins, which lie in front of the body's center of gravity and are readily rotated, are chiefly responsible for pitch control. When the broad surfaces of the paired fins are turned at right angles to the direction of movement, the fish slows down and stops.

Swim bladder. A long gas-filled bladder lies just below the spinal column in the body cavities of most fish. In some species the bladder opens into the alimentary canal and the pressure in it may be increased or decreased by gulping in or releasing air through the mouth. Some lung-fish which live in swamps where the water is poorly oxygenated obtain air for their respiration in this way.

In other fish the swim bladder has no outlet and the pressure in it is controlled by the secretion or absorption of gases by the blood vessels surrrounding it.

The swim bladder gives buoyancy to the fish's body; fish like sharks, dogfish and related species which have no swim bladder begin to sink as soon as they stop swimming.

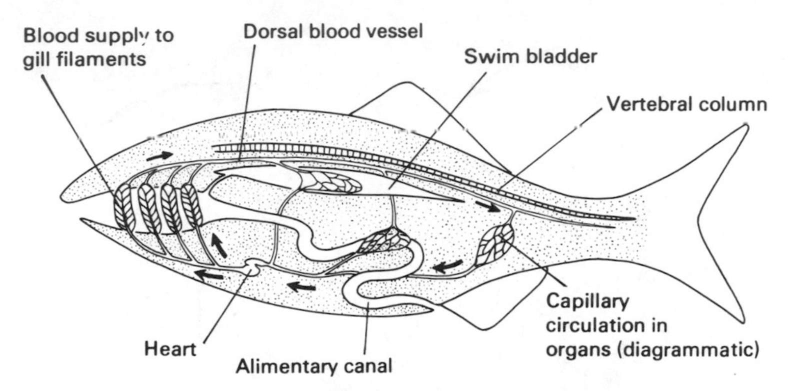

Fig. 16.5 *Diagram showing position of swim bladder (arrows indicate the direction of blood circulation)*

The pressure in the swim bladder has to be regulated by one of the methods mentioned whenever there is a change in the degree of buoyancy of the fish's body. This happens when the fish swims to a different depth, where the pressure of the water surrounding it is different.

16.4 BREATHING

Oxygen is appreciably soluble in water and is absorbed by fish from the water surrounding them through special organs called *gills*. In most fish there are four gills lying below the operculum on each side, each consisting of a curved bony *gill bar* bearing many long filaments in a fringe-like formation (Fig. 16.6); the filaments bear smaller filaments along their length, and these in turn divide into still smaller branches (Fig. 16.7). Blood vessels run through the gill bar, and branches run down from them into the filaments. The great number of filaments have a very large surface, through which oxygen can diffuse rapidly from the surrounding water through their thin walls. The gills can thus be regarded as an orderly system of blood capillaries exposed to the water in such a way as to absorb oxygen rapidly and efficiently. Oxygen is carried in the blood from the gills to the rest of the fish's body in combination with hemoglobin in the red cells, as in mammals.

The movements of the mouth and the opercula are coordinated to produce a stream of water over the gills so that fresh supplies of oxygen are constantly made available. Water enters the body through the mouth and passes over the gills and out of the opercula. The details of the pumping mechanism vary in different species, but the following is a general description (Fig. 16.8).

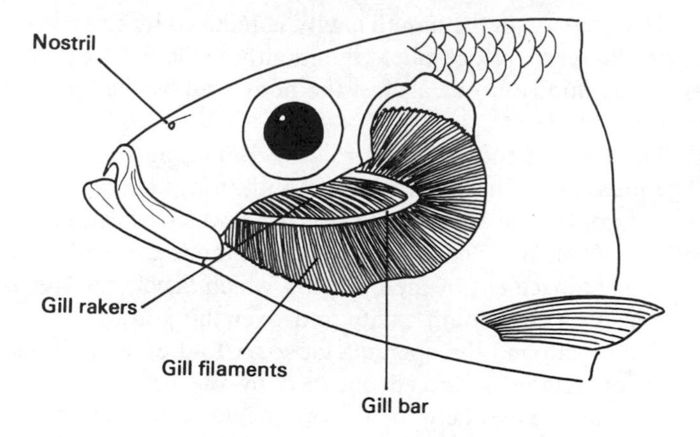

Fig. 16.6 *Herring (operculum cut away to show gills)*

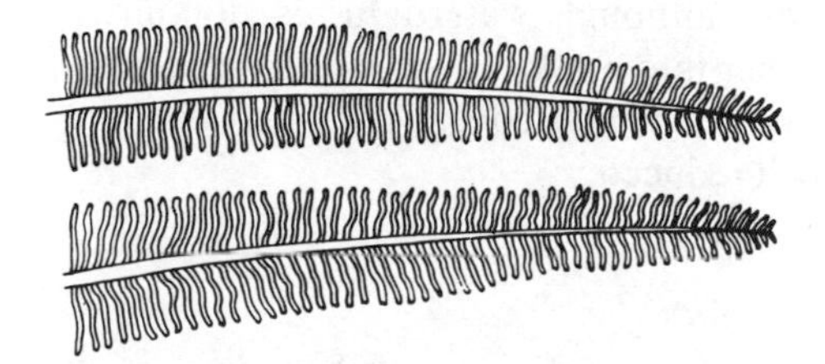

Fig. 16.7 *Tips of gill filaments seen under the microscope*

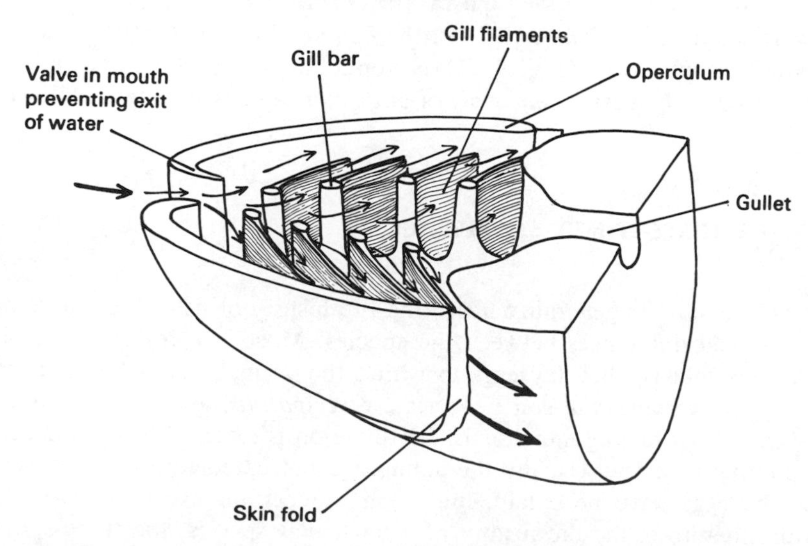

Fig. 16.8 *Diagram to show respiratory currents (gill rakers not shown)*

(*a*) The pressure in the mouth cavity is reduced by lowering the floor of the mouth. Water therefore enters the mouth, while the free edges of the opercula are kept firmly closed against the body wall by the higher pressure of the water outside them.

(*b*) The floor of the mouth is raised, decreasing the volume of the mouth. The pressure of the water in the mouth brings about the closure of two inturned folds of skin along the upper and lower jaws, and forces water across the gill filaments. The movement of water is assisted by a simultaneous outward movement of the opercula which "sucks" water from the front to the back of the mouth cavity and over the gills.

(*c*) The mouth and the opercula close. A fold of skin along the free edge of each operculum is forced outward by the pressure of the water inside it, and water escapes between the operculum and the body wall.

Although there is more oxygen in air than in water, a fish will suffocate in air. This is probably because the muscular system of mouth and opercula which can work in water will not function in air; in other words, the system of valves, although watertight, is not airtight. Also, the gill filaments cling together into a mat when out of water, because of the surface tension of the water film which covers them, so that the total gill surface exposed is very much reduced.

Filter Feeding

The current of water passing into the mouth and out of the opercula also serves as a feeding mechanism in certain fish such as the herring. These *filter feeders* have long gill rakers (Figs. 16.6 and 16.9) projecting forward from the gill bars and when the fish swims through surface waters in which zooplankton (Fig. 33.3) is abundant, the microscopic animals are trapped in the basket-like array of gill rakers and eventually swallowed.

16.5 THE THREE-SPINED STICKLEBACK

It is not possible to generalize about the life history of fish, since there are such marked differences between the species. Most fish, for example, are *oviparous,* that is, they lay eggs in which the young are enclosed in egg cases or membranes, but some species are *viviparous,* and their young are born as free-swimming individuals. Fertilization is external in most species but internal in a few. Usually the young are not cared for by the parents once the eggs have been laid, but there are certain exceptions. In this section, therefore, the life history of a particular species, the three-spined

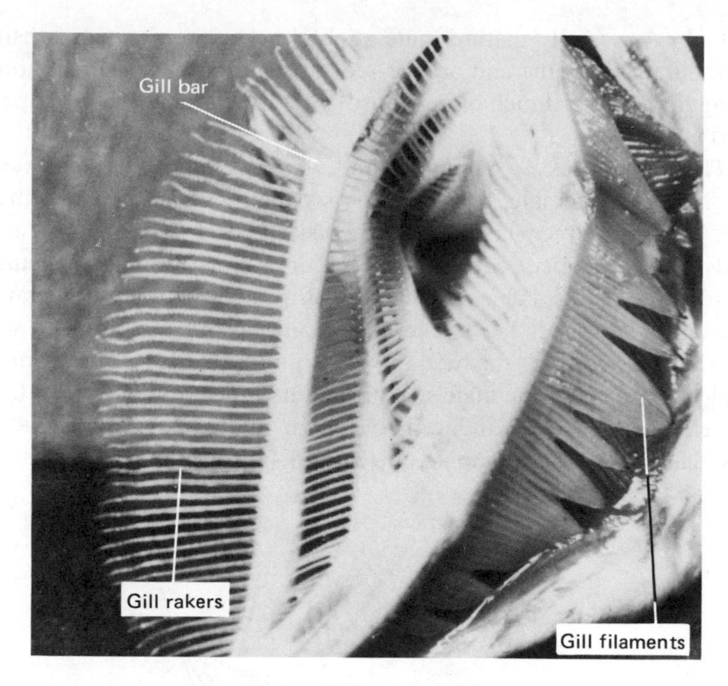

Fig. 16.9 *Gill rakers of herring*

stickleback (Fig. 16.1), is described as an example, but general conclusions concerning other species cannot be drawn from the paragraphs that follow.

Sticklebacks usually live in small shoals in fresh-water streams or ponds, and will be caught most frequently in clumps of weeds. They are carnivorous, eating aquatic insects and their larvae, worms, fish eggs and young fish, and will even attack objects larger than themselves. If kept in an aquarium they will rapidly devour every living animal that is small and mobile. They themselves are eaten by perch, pike and other larger carnivorous fish.

As the days lengthen in spring, the male sticklebacks move into the shallower parts of their stream or pond. Each takes up a small area of territory and begins to build a nest, digging out a small hollow in the bottom of the pond by biting movements, and then roofing it with pieces of vegetation stuck together with a kidney secretion, thus producing a tunnel-like structure.

During this time the males, normally silvery in color, change in appearance and behavior. Their eyes become bright blue and their breasts orange-red. If another male stickleback with this coloration invades the territory, the occupant drives him away by a threatening posture or a fighting attack. A sexually mature male stickleback is particularly sensitive to the combination of blue and orange-red, and will attack a crude model fish carrying these colors much more vigorously than an uncolored but perfectly formed model.

The stickleback is oviparous and its eggs are fertilized externally. When a pregnant female, her abdomen swollen with eggs, enters the territory of a male, he recognizes her by her shape and swims toward her in a special "zigzag" dance, consisting of a series of swoops. The female responds to the male's dance and to his coloring by swimming toward him, and he turns and leads her to the nest, pointing out the entrance with his head. The female enters the nest and is stimulated to lay her eggs or *spawn* by rapid thrustings of the male's snout against her body (Fig. 16.10). Two or three females may be induced to spawn in this way. Immediately the female leaves the nest, the male enters and fertilizes the eggs by shedding sperms on them.

Fig. 16.10 *Male stickleback inducing female to lay eggs in nest* (after Tinbergen)

For the next 7 to 10 days the male keeps the developing eggs supplied with oxygen by maintaining a current of water through the nest with fanning movements of his pectoral fins. The eggs then hatch, and the young fish or *fry* are kept in the nest by the male for about a week. During this time he will protect them by fighting off intruders, but gradually his bright colors fade and eventually he allows the young fish to leave the nest and disperse.

(Questions: see note at end of Unit Eighteen.)

Unit 17
Frogs

17.1 AMPHIBIA

The frogs are some of the few remaining members of the amphibia, a vertebrate group which flourished 250 million years ago. Other present-day members of the group are the toads (Fig. 17.6), newts and salamanders. The amphibia are so adapted that they can, in general, move, feed and breathe equally well on land and in fresh water. In general amphibians are dependent upon moist habitats in both temperate and tropical regions. Exceptions are certain toads associated with deserts, where they are active and reproduce only during wet periods.

17.2 EXTERNAL FEATURES

The frog (Fig. 17.1) is a poikilothermic animal (see Section 16.1). Its *skin* is loose-fitting and moist (see Section 17.6). Its *eyes* are set at the top of its head, and protrude so that they are above water when the rest of the body is immersed (Fig. 17.2). The eyelids are movable, and the whole eye-

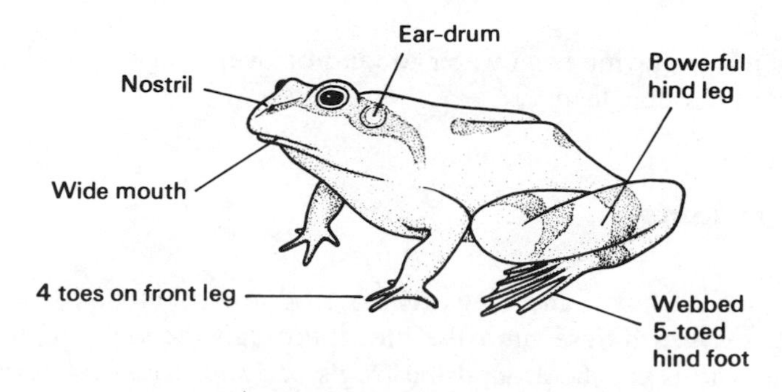

Fig. 17.1 *External appearance of frog*

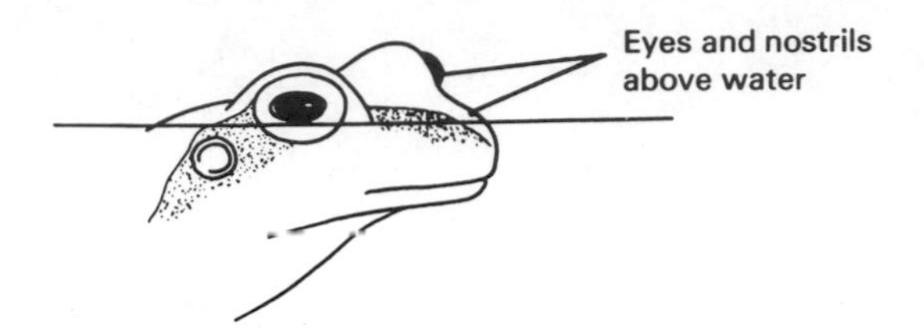

Fig. 17.2 *Head of frog partly submerged*

ball can be drawn into the head to some extent by muscles; this can be seen to happen sometimes when the frog is swallowing. The *nostrils* also lie on the upper surface of the head, and air can be taken in through them when the frog is swimming at the surface; they can be closed when the frog submerges. At the sides of the head behind the eyes are circular *ear-drums,* thin membranes which are set vibrating by sound waves traveling in the surrounding air or water, the vibration being transmitted by a small bone to a sensory organ which sends nervous impulses to the brain.

17.3 LOCOMOTION

The frog usually moves on land by a series of leaps and plops, and in the water by swimming; its powerful hind legs are adapted for both swimming and leaping. When the strong extensor muscles of the thigh contract, the limb extends and the foot exerts a backward thrust against the ground or against the water. The thrust is transmitted to the frog's body through the pelvic girdle and the spine, so that the whole body moves forward (Fig. 17.3; see also Section 26.6). When the frog is swimming, the webs of the hind feet provide a large surface area for pushing backward against the water. The fore-limbs are much smaller than the hind limbs and help to steer when the frog is swimming and to absorb the shock of landing after a jump on land.

When it is moving from water to land or over rough ground, the frog will crawl rather than leap.

17.4 BREATHING

The skin, mouth lining and lungs of the frog are all involved in gaseous exchange. Oxygen diffuses into the blood through the walls of the blood capillaries, and is carried around the frog's body in the circulatory system in combination with *hemoglobin,* the red pigment of the blood. The blood takes up oxygen either through the skin or in the lungs.

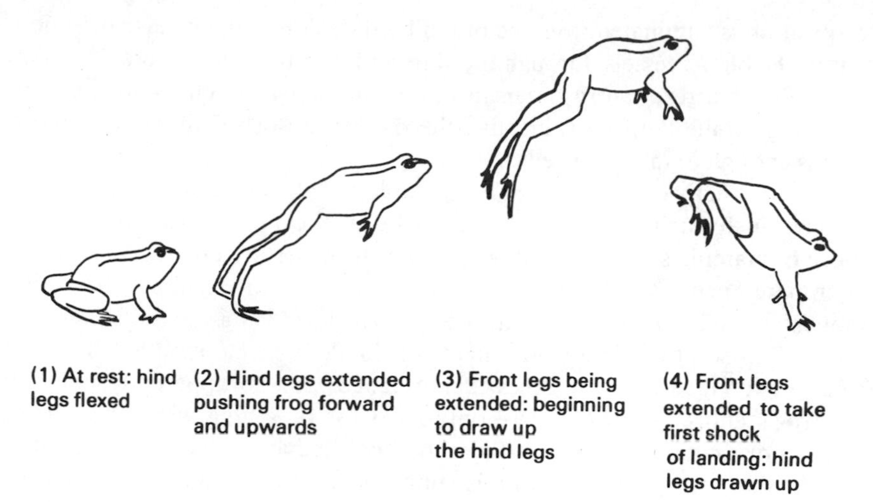

Fig. 17.3 *Frog leaping*
(From James Gray, *How Animals Move*, Cambridge University Press)

The frog's skin is smooth and moist, fairly thin, and well supplied with blood vessels which branch into a network of fine, thin-walled capillaries (Fig. 17.4). Oxygen enters the blood by dissolving in the film of moisture covering the body and diffusing through the skin, through the walls of the capillaries and into the blood. Carbon dioxide produced in

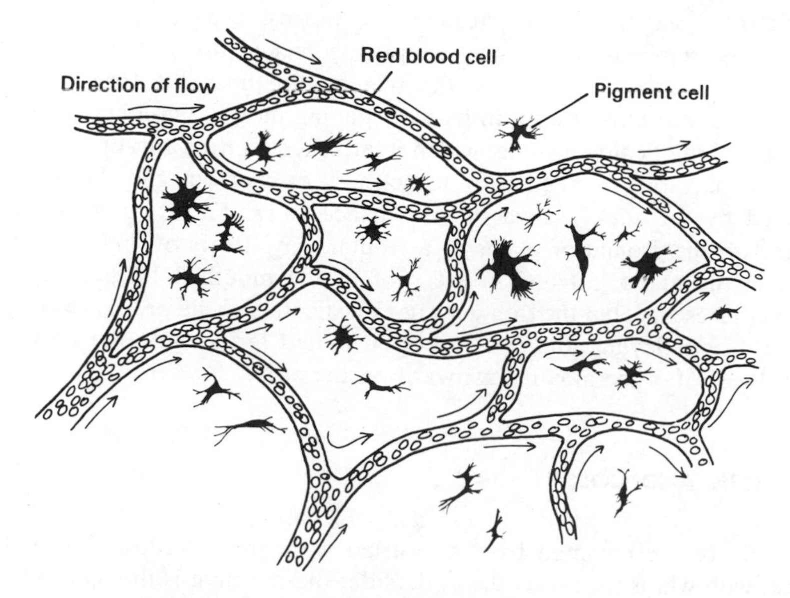

Fig. 17.4 *Capillary circulation in tadpole's tail*

respiration is eliminated from the blood by diffusion in the reverse direction, out of the blood vessels, through the skin and into the atmosphere. The skin is an effective respiratory organ in either air or water and is in use continuously; while the frog is inactive the oxygen absorbed through its moist skin is enough to meet its needs.

When active, however, the frog is able to take in an additional supply of oxygen through its lungs. The lungs lie in the body cavity and, unlike those of mammals, are not separated from the other organs by a diaphragm (compare Section 22.4). Frogs do not make regular, rhythmic breathing movements like mammals, but force air into their lungs spasmodically as the need arises, by gulping movements of the floor of the mouth. The lungs can be expanded to many times their relaxed size, so apparently inflating the frog's entire body. They are only used for breathing when the frog is on land or swimming on the surface; the nostrils have valves which prevent the entry of water when the frog is submerged and which control the flow of air into the lungs.

The moist skin lining the frog's large mouth is also a respiratory surface. Like the skin covering the rest of the body it is in constant use, except when the animal is submerged. The air in the mouth can be exchanged by movements of the mouth floor.

17.5 FEEDING

Adult frogs are carnivorous, feeding on worms, beetles, flies and other insects. Worms and beetles may simply be picked up by the mouth but insects may also be caught on the wing. Sometimes the frog will leap toward the insect and trap it in its wide gaping mouth. On other occasions it uses its long flexible tongue, which is attached to the front of the mouth and can be rapidly extended by muscles. It is shot out and the insect is trapped by the sticky saliva on its surface (Fig. 17.5). Insects can be picked off the ground or plants in a similar way. Rows of tiny closely set teeth in the upper jaw and in the roof of the mouth hold the prey and prevent its escape, but the frog does not masticate and the prey is swallowed whole. In swallowing, the eyes are often pulled farther into the head and press the roof of the mouth downward on the prey.

17.6 SKIN AND COLOR

The skin of the common frog is mottled with green, yellow, brown and black, with whitish areas on the underside; the mottling is thought to be of protective value in camouflaging the animal in its natural surroundings.

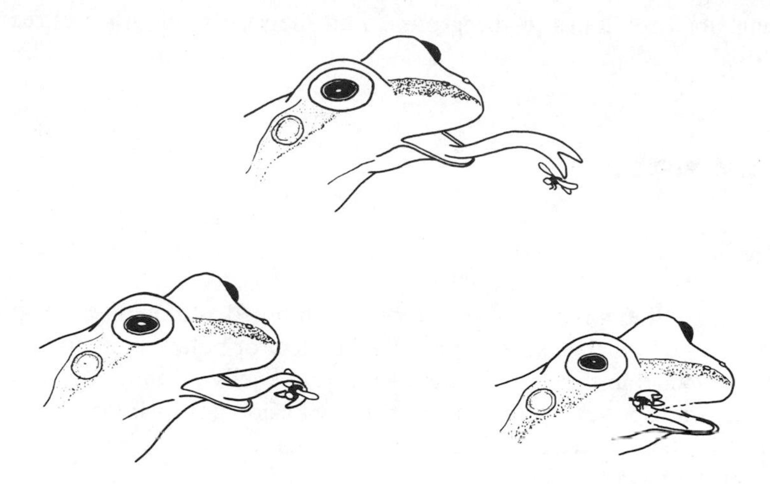

Fig. 17.5 *Diagrams to show a frog catching a fly with its tongue*

The frog's color can change to some extent, pigment cells in the skin (Fig. 17.4) expanding or contracting to make the frog darker or lighter. Such tones may correspond more closely to the frog's background in different circumstances and help to conceal the animal. The color changes are produced when the brain receives impulses from certain sensory organs and stimulates the pituitary gland to manufacture a hormone; this circulates in the bloodstream and affects the pigment cells of the skin.

A frog's body is always covered with a slimy fluid produced by mucous glands in the skin. The sliminess makes the frog difficult to catch and keeps the skin moist. When the frog is on land it relies on the constant film of moisture on its body surface to maintain its oxygen supply; if its skin were dry, oxygen could not go into solution and so reach the blood beneath the skin.

17.7 HABITAT

In early spring, during the breeding season, frogs spend their time in ponds and lakes with a steady flow of water. They are not usually found in swiftly running water. After they have laid their eggs they usually live in damp vegetation rather than in water. They are unlikely to be found in any dry situation where their skins could lose water and dry up and so seriously impair their breathing. In winter they hibernate in the sense that they are dormant and do not feed. They lie up in the mud at the bottom of ponds, in damp moss or holes in the ground and their eyes, mouth and nostrils are closed.

17.8 LIFE HISTORY

(a) Mating

Nearly all amphibians return to water to breed, although some species make a "pond" for themselves in a rolled-up leaf or hollow tree, the "pond" being derived from the liquefaction of a layer of jelly around the eggs. The female of one species of frog places the fertilized eggs in little pouches of skin on her back, where they develop through all the tadpole stages into tiny but fully formed frogs.

Male and female frogs in temperate regions leave their winter quarters in spring, generally during March, and make their way to the nearest pond. The female is usually larger than the male, her body being swollen with mature eggs. The male develops pads of black horny skin on his thumbs, which enable him to grip the female's body just behind her fore-limbs. In this way the male may be carried about on the female's back, mostly in

Fig. 17.6 *Toads mating*
(Heather Angel)

the water, for several days before she lays her eggs. The eggs are laid in water and are fertilized externally (see Section 25.1). As they leave the female's body, the male pours over them a *seminal fluid,* containing sperms; both eggs and seminal fluid leave the frog's body through an opening called the *cloaca,* just above the region where the hind legs join the body. The sperms swim through the jelly that surrounds the eggs, and fertilize them, one sperm fusing with each egg.

The jelly or albumen surrounding the eggs swells on contact with water; fertilization is therefore only possible at the moment when the eggs leave the female's body, and the pairing behavior of the frogs ensures that this can happen.

(b) Development

(*i*) The eggs of the frog are often found in springtime in ponds or streams, as floating masses of *frog-spawn* (Fig. 17.7). The eggs themselves

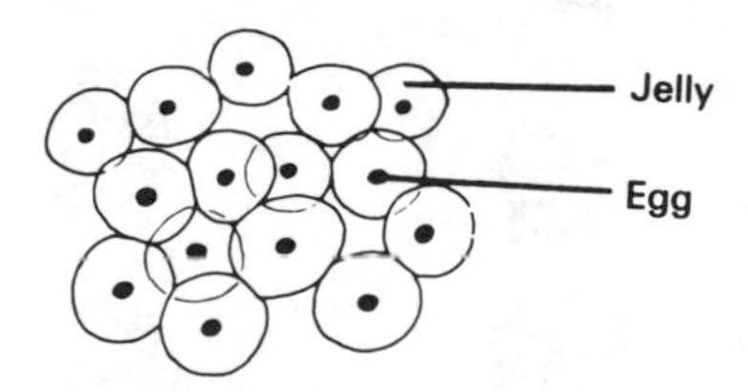

Fig. 17.7 *Frog-spawn*

are small spheres, consisting of semi-liquid cytoplasm which contains a nucleus and which is surrounded by a tough black egg membrane. The lower part of the cytoplasm contains yolky granules which are the food reserves for the first few weeks of the embryo's development (Fig. 17.8(*a*)). Each egg lies inside a ball of clear jelly. This sticks the eggs together, and prevents them from being swept away or eaten. It protects them from mechanical injury and from drying up, and probably also from attacks by fungi and bacteria. The oxygen required by the developing egg for its vital processes reaches it by diffusion through the jelly and the egg membrane.

(*ii*) The egg begins to develop soon after it has been fertilized, the time depending partly on the temperature. The nucleus divides into two smaller nuclei, which separate (Fig. 17.8(*b*)). The cytoplasm then also divides to include each nucleus in a separate unit of cytoplasm so that there now appear two smaller cells, each with a nucleus (Fig. 17.8(*c*)).

(*iii*) The two cells both divide again at right angles to the first division, making four smaller cells roughly equal in size (Fig. 17.8(*d*)).

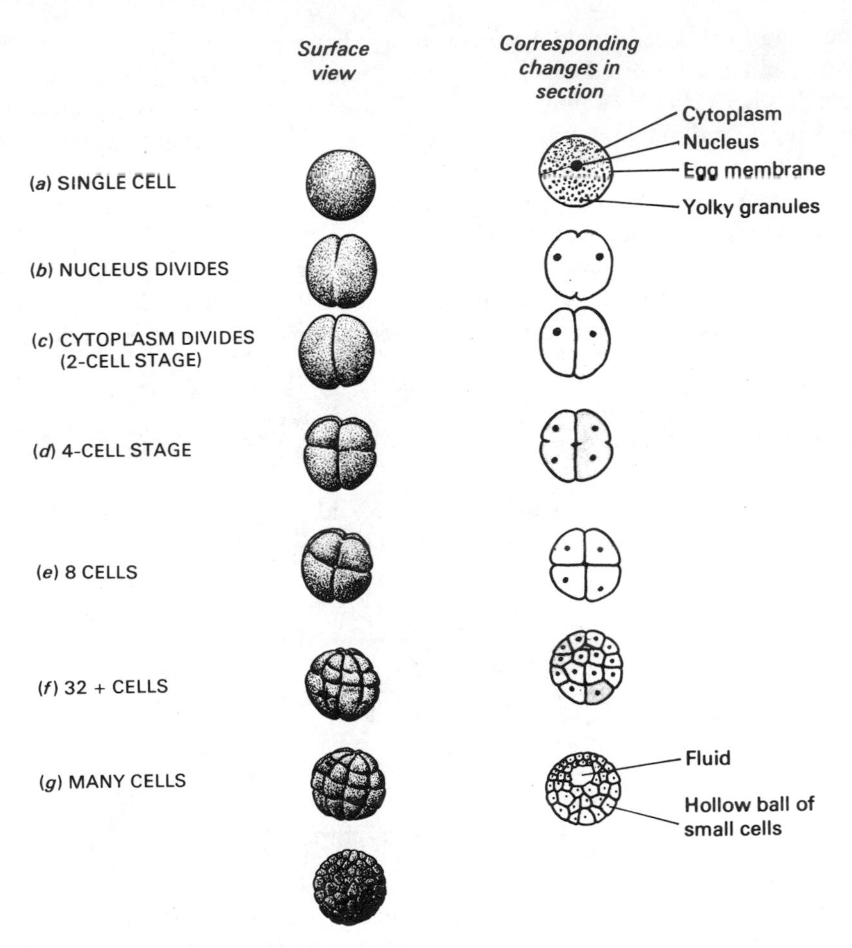

Fig. 17.8 *Changes in the ovum after fertilization*

(*iv*) The four new cells all divide again, this division being at right angles to the other two, forming eight cells of which the lower four are slightly larger than the upper four (Fig. 17.8(*e*)).

(*v*) The cells divide again and again and become very numerous; they are too small to be seen individually at this stage, even with a hand lens. In spite of the great increase in the number of cells and nuclei there is little increase in size during these divisions or *cleavage*. Within a day or two of fertilization the egg has become an *embryo* consisting of a hollow ball of tiny cells (Figs. 17.8(*f*) and 17.8(*g*)), but still appears as a little black featureless sphere to a casual observer, with little evidence of the vigorous activity that has been going on.

(*vi*) The sphere begins to elongate and to develop a distinct head

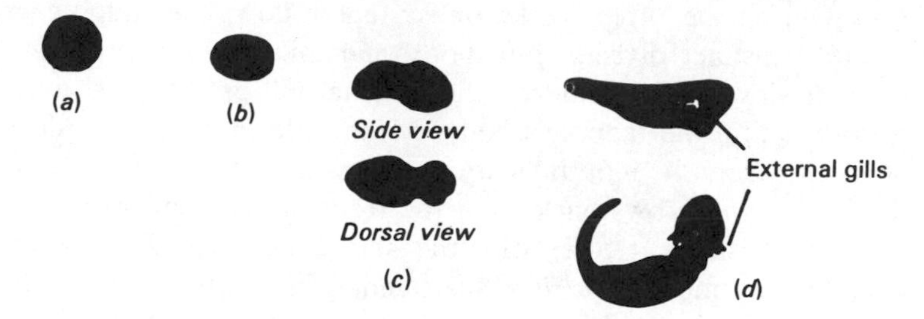

Fig. 17.9 *Changes taking place during the 4 to 5 days after laying (×3)*

(a) Early embryo consists of many tiny cells; (b) elongates; (c) tail and head distinguishable; (d) wriggling in jelly and ready to emerge.

and tail (Figs. 17.9(*b*) and 17.9(*c*)), while the inner cells become organized and adapted to form specialized tissues and organs. The energy and raw materials for these processes come from the yolk.

(*vii*) After about ten days the jelly around the embryo liquefies and the frog larva or *tadpole* can be seen moving about inside (Fig. 17.9(*d*)). Eventually the tadpole wriggles out of the jelly into the water. It is quite black at this stage and breathes through its skin, although it has rudimentary external gills. Its mouth has not yet opened, and it is still obtaining the nutrients it needs by digesting the remains of the yolk in its intestine. It clings to water weed or to the surface of its jelly by a sticky secretion of the *mucus glands* just below the mouth. The tadpole's body wriggles spasmodically, but it does not yet swim freely.

(*viii*) The tadpole's mouth opens two or three days after hatching, and it begins to feed by scraping microscopic plants and other deposits from the surface of pond weeds, using a pair of horny toothed lips. It now has three pairs of external gills, consisting of branched thin-walled gill filaments which present a fairly large surface to the water. The blood can be seen circulating in them under the low power of the microscope (Fig. 17.10). Oxygen diffuses from solution in the water surrounding them through the filament walls into the blood close below the surface.

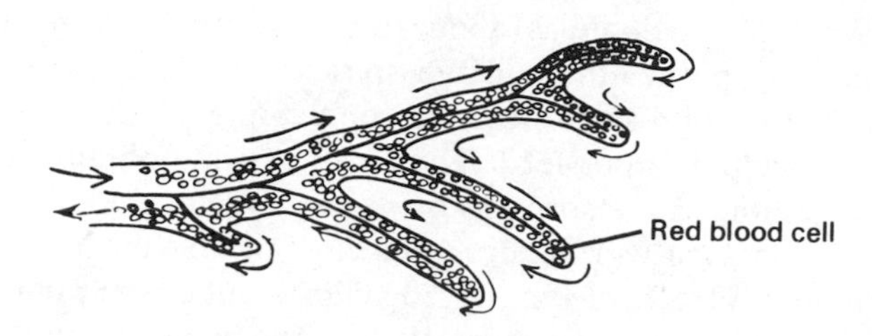

Fig. 17.10 *Diagram of external gill showing blood circulation*

(*ix*) During the three weeks or so that follow, the tadpole grows rapidly and a distinct division into body and tail regions develops. The mucus glands slowly disappear and the external gill gradually shrivel and are reabsorbed into the tadpole's body. Meanwhile internal gills develop, opening by slits from the mouth cavity or *pharynx* to the outside. A fold of skin, the *operculum,* grows back from the front of the head, covering the gill slits on both sides; it fuses with the skin behind the slits to form a continuous gill chamber or *atrium* surrounding the slits, which opens to the outside by a single small hole, the *spiracle,* on the left side of the body.

The tadpole now begins to breathe through its internal gills. Water is taken in through the mouth and passes over the gills, through the gill slits into the gill chamber, and finally out through the spiracle. As the water passes over the gill filaments, dissolved oxygen diffuses through their thin walls into the blood.

(*x*) Over the two or three weeks that follow, the tadpole continues to grow in size. Eyes and nostrils become visible, and the tail elongates and develops a broad transparent web along its upper and lower edges. The tadpole swims by vigorous wriggling movements of its body and tail, in the same manner as a fish but with less speed and precision.

A long coiled intestine develops and becomes visible through the skin of the abdomen during the first few weeks of the tadpole's life. At this stage the intestine is adapted to the digestion of an exclusively vegetable diet.

(*xi*) By the time the tadpole is about two months old, its lungs have begun to function, and the animal comes frequently to the surface to gulp air. Its limbs start to develop; the hind-limb buds near the point where the tail joins the body begin to grow and develop into small perfect legs (Figs. 17.12(*a*) and 17.12(*b*)), but they are not yet used for locomotion and hang limply beside the body, while the tadpole continues to swim by wriggling movements. The front legs also grow but are not easily seen because they are still covered by the operculum, although the bulge they make beneath the skin is visible. Tadpoles in an aquarium begin to nibble at dead animals or raw meat in preference to plants at this stage; the intestine shortens as it becomes adapted to the change of diet and, later, the abdominal region narrows.

(*xii*) When the tadpole is about three months old, its front legs appear; the left leg pushes through the spiracle and is usually seen first, while the right has to break through the operculum (Figs. 17.12(*c*) and 17.12(*d*)). The tadpole stops feeding, and for a while obtains its food by the internal digestion and absorption of its tail, which gradually shortens. The skin is shed, taking with it the larval lips and horny jaws, leaving a much wider mouth. Finally the young frog climbs out of the pond on to the land, still with a tail stump, but using its legs for jumping and swimming. This final sequence of changes or *metamorphosis* is controlled by the hormone *thyroxine* and normally takes about four weeks. The young frogs

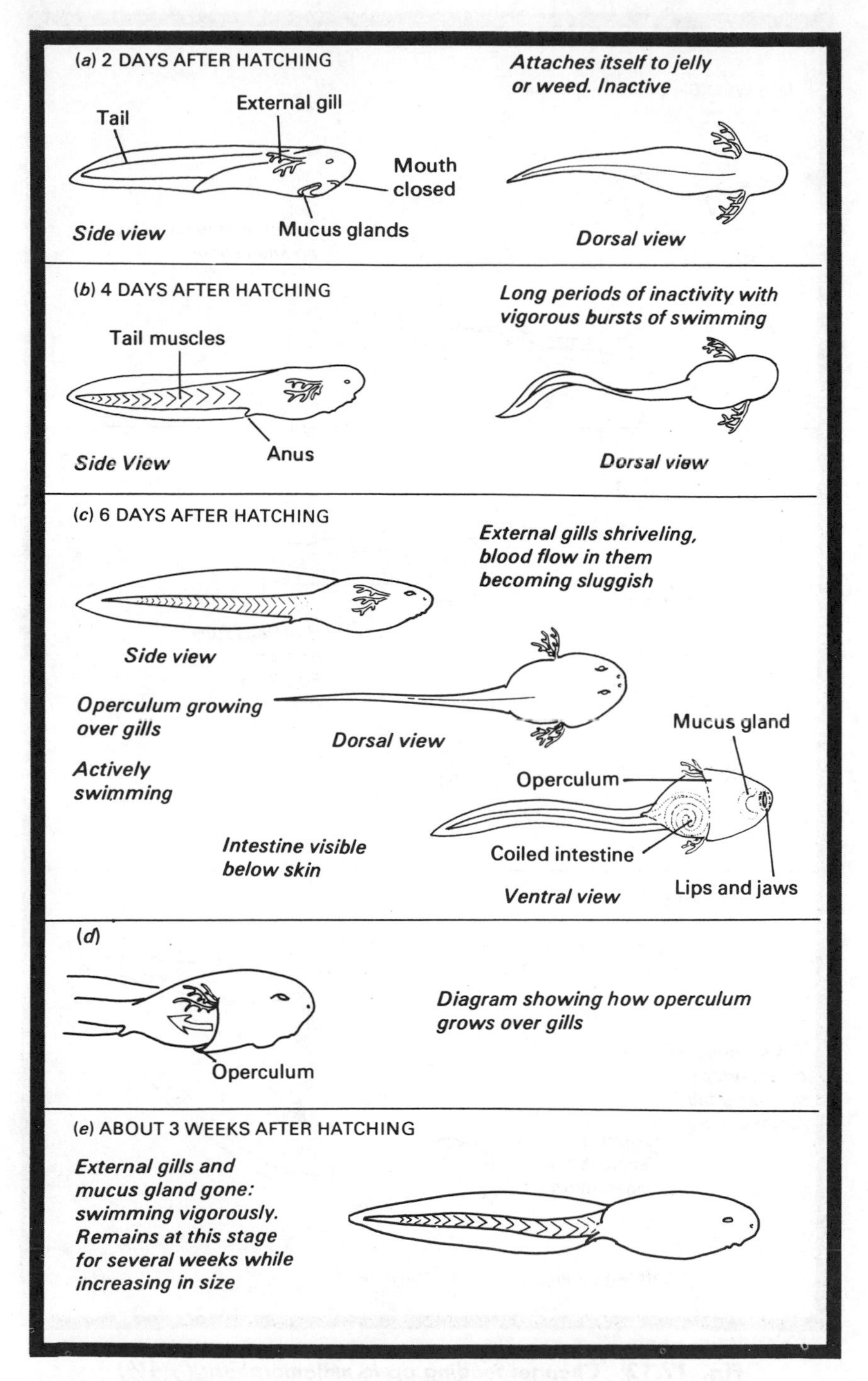

Fig. 17.11 *Changes in the first days after hatching (×3)*

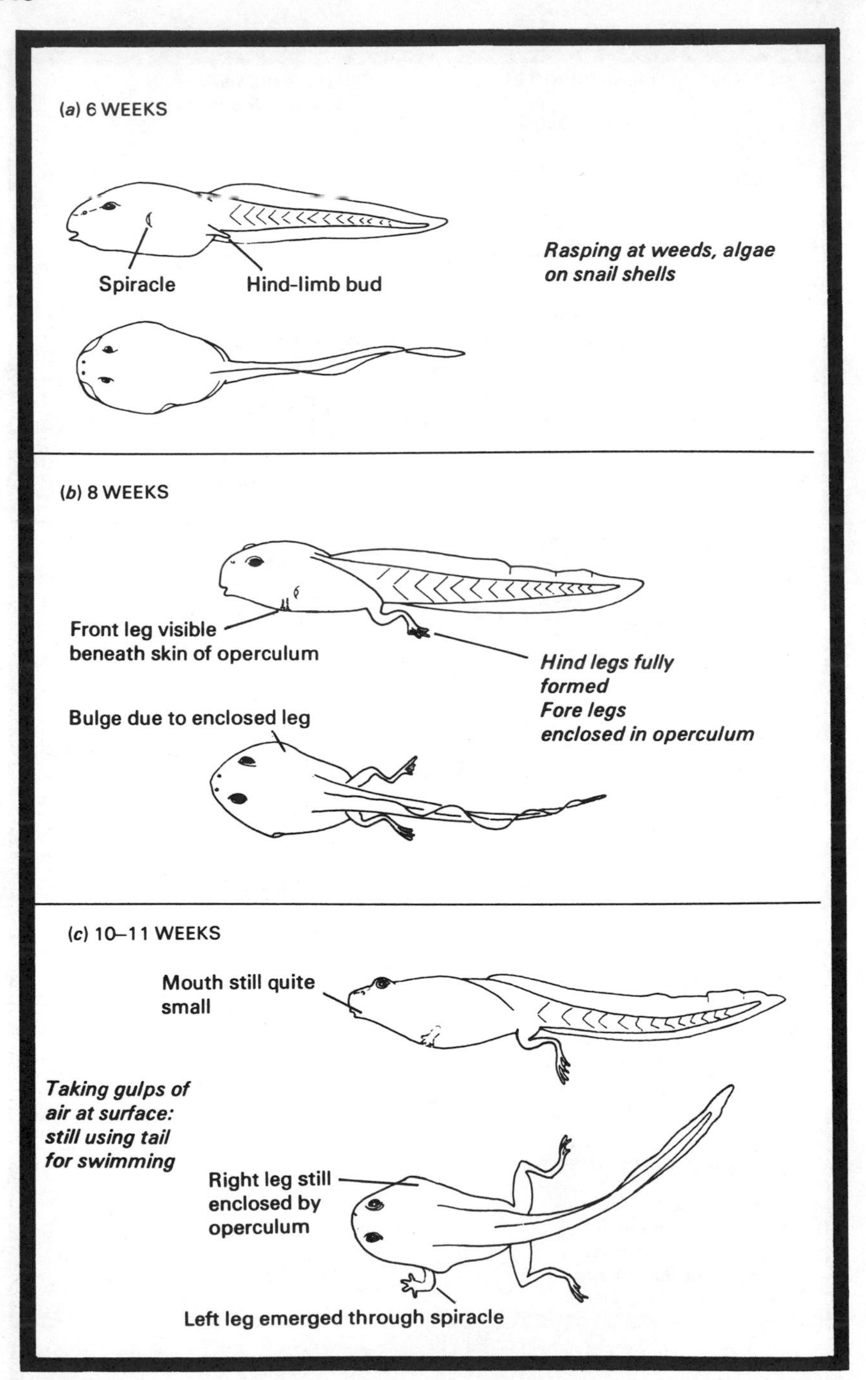

Fig. 17.12 *Changes leading up to metamorphosis (×1½)*

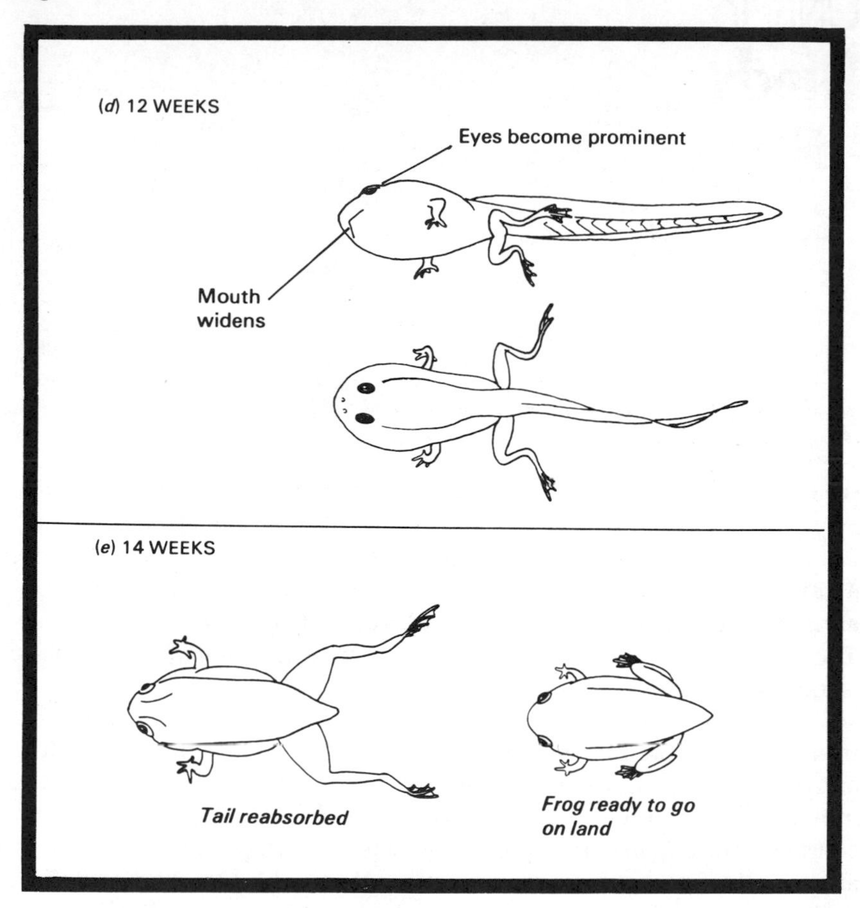

Fig. 17.12 *(cont'd)*

remain in the damp vegetation and long grass in the vicinity of the pond, catching and eating small insects. By the time they are four years old they are fully grown and are old enough to breed.

The times mentioned in the preceding paragraphs for the respective stages of development are only approximate since they are very much affected by the temperature of the pond-water. A fall in temperature can alter the duration of a stage of metamorphosis from days to weeks, while a rise in temperature will speed it up. It is therefore more reliable to indicate a phase of development by mentioning relevant characteristics—for example, the "hind-leg stage" or the "external-gill stage"—rather than by the animal's age in days or weeks.

(Questions: see note at end of Unit Eighteen.)

UNIT 18
Birds

18.1 INTRODUCTION

Birds are *homoiothermic* vertebrates (see Section 24.2), which reproduce by laying eggs. Their skins are covered with feathers, and their forelimbs are modified to wings; most species have the power of flight. Their legs and toes are covered with overlapping scales. A bird's skull and lower jaw are extended forwards into two *mandibles* which together form a *beak*. The eye has three eyelids; the inner third eyelid, the *nictitating membrane*, is a transparent fold of skin, which can move across the eye.

The *feathers* of birds are the external feature which distinguishes them from other vertebrates. Feathers are produced from the skin, which is loose and dry, containing no sweat glands. They repel water, and form an insulating layer around the bird's body, helping to keep its temperature constant. Their quills are supplied with muscles which can alter the angle at which the feathers lie; in cold weather, this enables the bird to fluff out its feathers, trapping more air between them and enhancing their insulating effect. They also have a nerve supply which is stimulated when the feathers are touched, in a similar way to a cat's whiskers.

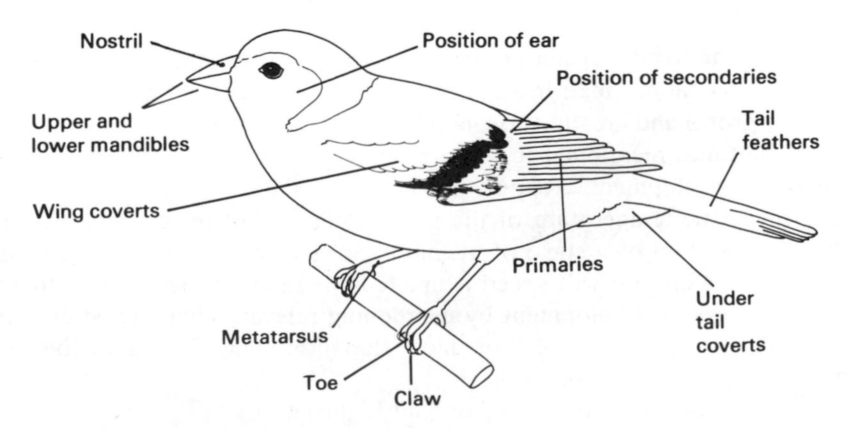

Fig. 18.1 *External features of a chaffinch*

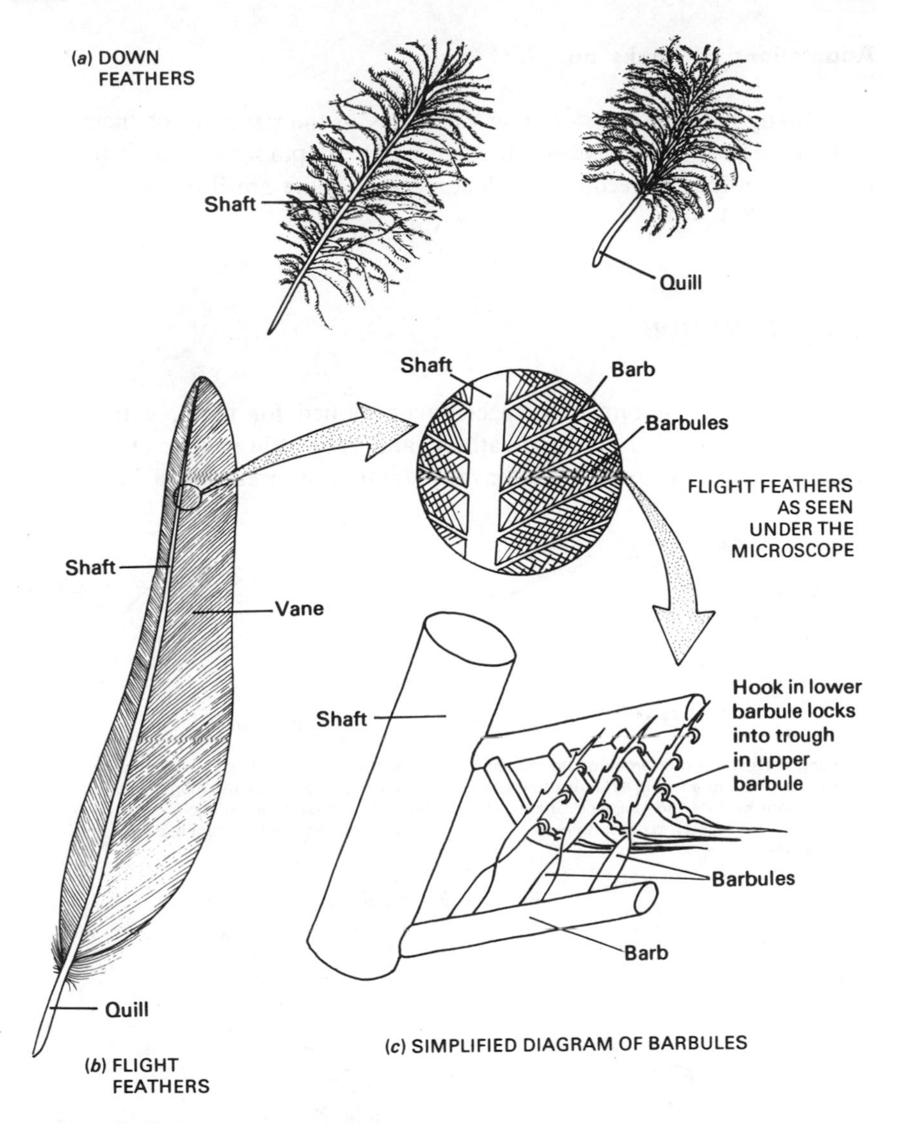

Fig. 18.2 *Feather structure*

Greatly magnified to show interlocking mechanism. Many barbules and hooks have been omitted. (Drawn from a model in the Natural History Museum, South Kensington, England.)

The *flight feathers* and *coverts* are broad and flat. The *barbules* of these feathers interlock in such a way that should a feather be disarranged in flight, for example, preening with the beak will re-form it perfectly (Fig. 18.2). The *down feathers* found between the quills of the flight feathers are fluffy and are particularly effective in holding the insulating layer of air close to the body.

Adaptations of Beaks and Feet

Many birds show interesting variations in the structure of their beaks and feet. These differences are thought to be adaptations to their mode of life and methods of feeding and locomotion. Some are illustrated in Figs. 18.3 and 18.4.

18.2 LOCOMOTION

The wings of most birds are specially developed for flight, with a large surface area provided by their feathers, and very little weight (Fig. 18.5). Generally, the fast-flying birds have a small wing area and a large span,

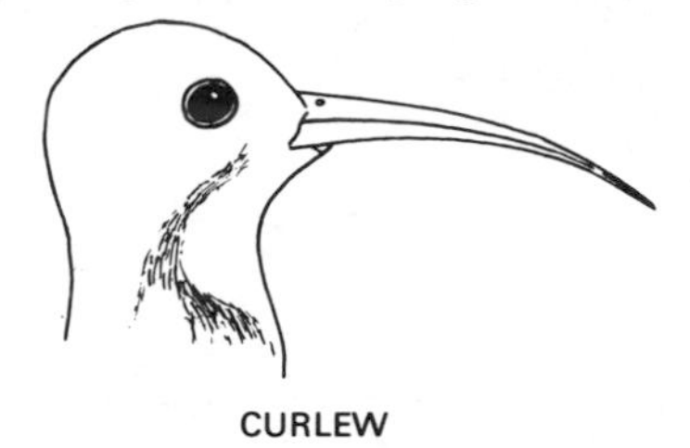

CURLEW

Long, narrow beak for probing into mud and sand on the shore and in estuaries to reach burrowing worms and molluscs. Characteristic of most waders, e.g. sandpipers, redshanks

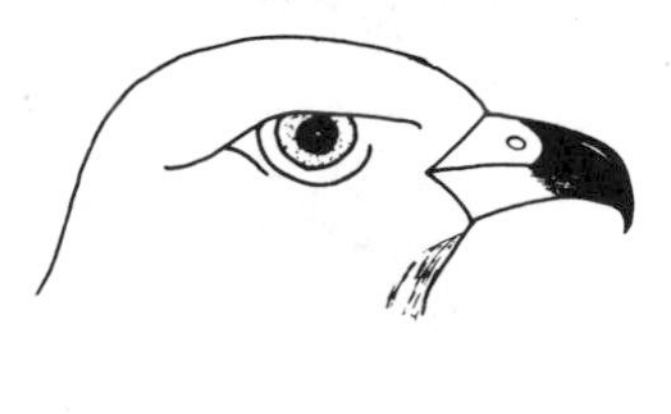

BUZZARD

Powerful, sharp, hooked beak, for tearing the flesh from small birds and mammals. This type of beak is characteristic of most birds of prey, including hawks, falcons, eagles and owls

Fig. 18.3 *Beaks*

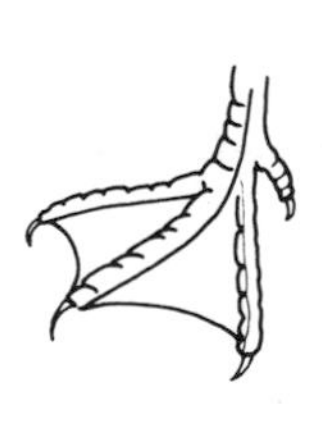

HERRING GULL

Hind toe very small. The web between the three front toes is for swimming, and for walking on soft surfaces. Characteristic of other gulls, sea birds and ducks

LITTLE OWL

Three toes directed forward, and one back, but they can be bent to meet. They are powerful, with sharp, curved talons for catching and killing prey. Characteristic of many other predatory birds such as falcons and hawks

Fig. 18.4 *Feet*

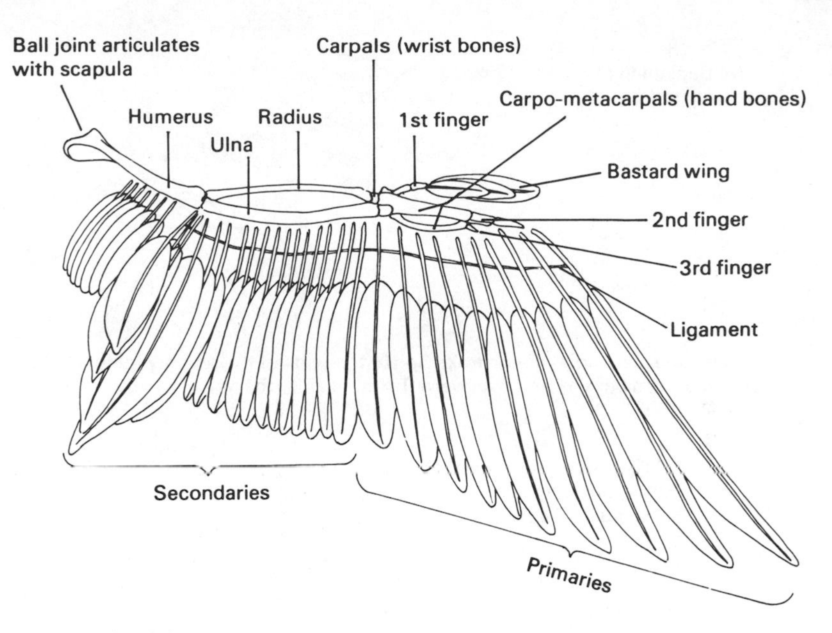

Fig. 18.5 *Skeleton of wing and arrangement of flight feathers (mallard)*

with specially well-developed primary feathers, while the slower birds have shorter, wider wings with well-developed secondaries.

Flight is of two kinds, flapping flight and gliding or soaring flight; different species use the two types to different extents. In *flapping flight* the lifting force is provided by the downstroke of the wing, produced by contraction of the powerful *pectoralis major* muscle. The resistance of the air to the downward movement of the wing produces an upward force

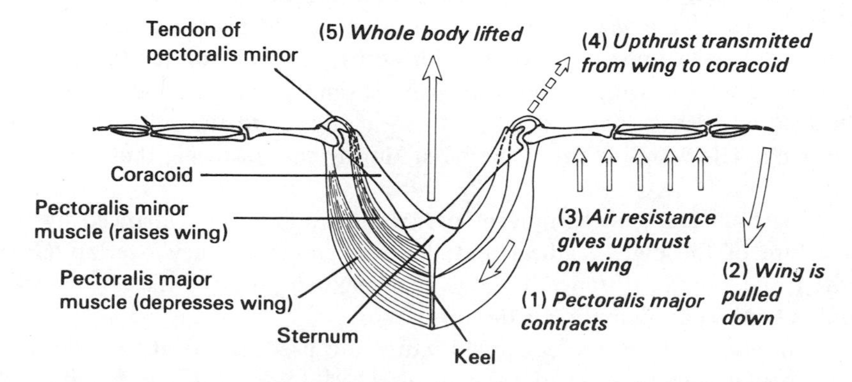

Fig. 18.6 *Front view of skeleton concerned with flight, showing how muscles and bones work together*

(*b*) and (*c*) Near end of downstroke. In (*b*) the ends of the primary feathers are curled up and back, giving a forward thrust against the air.

(*d*) Upstroke. The 'wrists' are leading, flexing the wing and reducing its resistance.

Fig. 18.7 *Flapping flight*

acting against the wing; this force is transmitted through the *coracoid* bones to the *sternum,* and, acting through the center of gravity of the bird, lifts its whole body. In addition to this lifting action, the slicing movement of the wing, especially of the wing-tip, provides forward momentum. During the downstroke, the leading edge of the wing is below the trailing edge, so that the air is thrust backward and the bird moves forward. Roughly speaking, the secondary feathers provide most of the lifting force, and the primary feathers most of the forward component. In flight it is thought to be positioned in such a way that a smooth flow of air is maintained over the wing surface.

The upstroke of the wing is much more rapid than the downstroke. The tendon of the *pectoralis minor* muscle passes over a groove in the coracoid to the upper side of the humerus, so that contraction of this muscle raises the wing. Often the arm is simply rotated slightly so that the leading edge is higher than the trailing edge, and the rush of air lifts the wing. The wing is bent at the wrist during the upstroke, thus reducing the resistance.

The efficiency of the movements of flapping flight is enhanced by the structure of the flight feathers and the way in which they overlap (Fig. 18.8); the wing's resistance to air is at a maximum during the downstroke and is kept to a minimum on the upstroke.

In *gliding flight,* the outspread wings are used as airfoils; as the bird loses height it gains forward momentum (Fig. 18.9). Birds like gulls and vultures can use rising thermal air-currents or intermittent gusts of wind to soar and gain height without moving their wings.

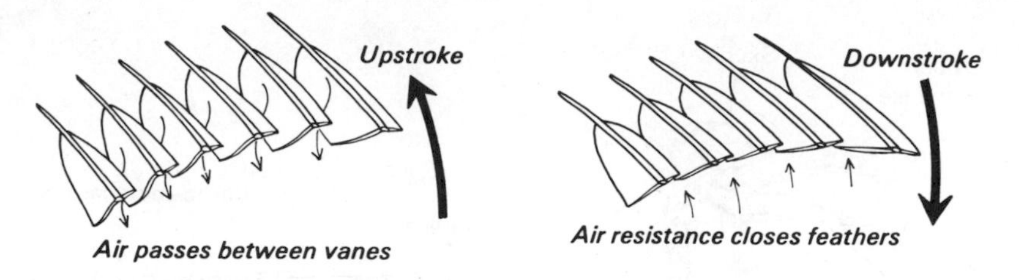

Fig. 18.8 *Diagram showing action of wing feathers during flight*

Fig. 18.9 *Gliding flight: wings and tail outspread; uplift obtained from air currents or by losing height*

In both flapping and gliding flight the tail feathers help to stabilize the bird; they are particularly important in braking and landing. There are varying estimates of the speeds that birds can reach in flight; swifts are said to achieve 160 km/h and racing pigeons 60 km/h.

In *walking,* the posture of the bird brings its center of gravity below the joint of the femur and the pelvis.

Features Adapting the Bird for Efficient Flying

(*a*) The shape of the bird's body is streamlined when in flight.

(*b*) The wings have a large surface area and little weight.

(*c*) The pectoral muscles which move the wings up and down are large and powerful; they account for as much as one-fifth of the body weight of some birds.

(*d*) A bird has a rigid skeleton which provides a firm framework for the attachment of the flight muscles. Many of the bones which can move in mammals are fused together in birds; for example, the vertebrae of the spinal column in the body region.

(*e*) The sternum (breast bone) carries a deep keel-like extension for the attachment of the pectoral muscles.

(*f*) The coracoid bones, which transmit the lift of the wings to the body, are particularly well developed and strong.

(*g*) Many of the bones of the skeleton are hollow, reducing the bird's weight.

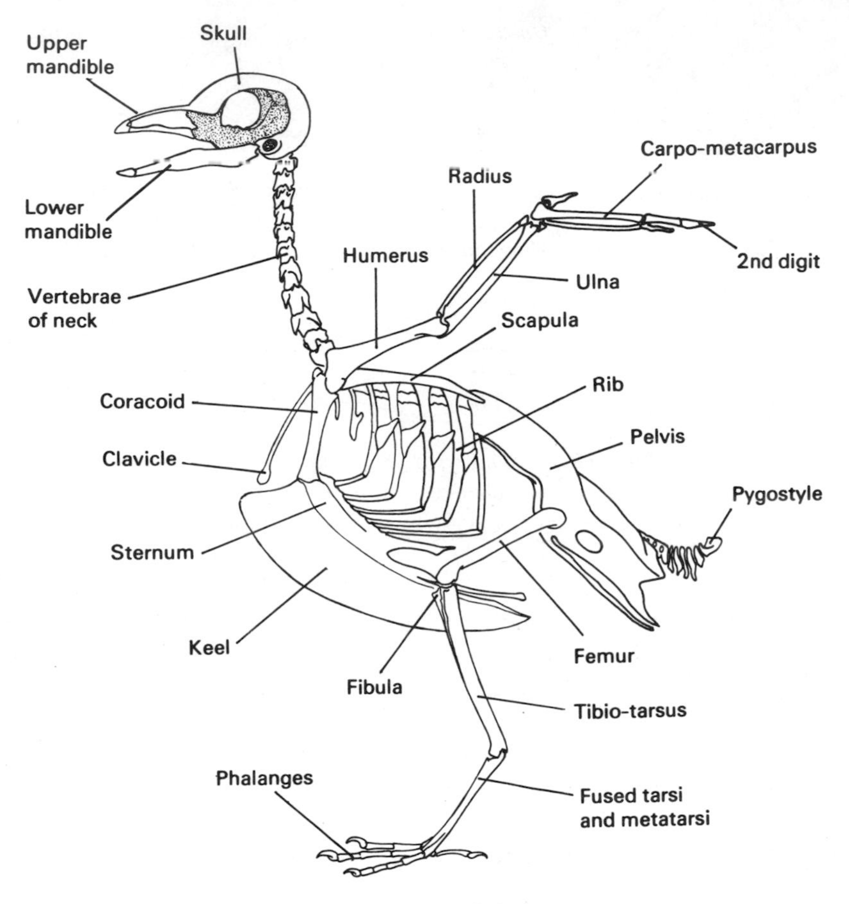

Fig. 18.10 *Bird skeleton*

18.3 REPRODUCTION

The detailed pattern of reproduction and parental care varies widely in different species but, in general, it follows the course outlined below.

(a) **Pairing.** A sequence of behavioral activities or *courtship display* leads to pair formation, a male and female bird pairing at least for the duration of the breeding season.

(b) **Nest building.** One or both of the pair of birds constructs a nest. Nest structures vary greatly from one species to another: for example, some birds

weave elaborate nests from materials like grass, leaves or feathers, while others merely scrape a hollow in the ground.

(c) **Mating.** Further display leads to mating. The male mounts the female, applies his reproductive openings to hers and passes sperm into her oviduct; the eggs are thus fertilized internally.

(d) **Egg laying.** As the fertilized egg passes down the oviduct it becomes enclosed in a layer of *albumen* and then in a hard shell. Finally it is laid in the nest. Usually, one egg is laid each day and incubation does not begin until the full clutch has been laid.

(e) **Incubation.** The female bird is usually responsible for incubation. She keeps the eggs at a temperature close to her own by covering them with her body, pressing them against her *brooding patches,* featherless areas which allow direct contact between the skin and the egg shell. Incubation also reduces loss of water from the egg by evaporation through the shell. Normally the eggs are incubated for from two to four weeks, depending on the species.

(f) **Development.** During the period of incubation the tissues and organs of the young bird develop. The yolk of the egg provides most of the food supply necessary for the embryo's growth and the albumen ("egg-white") is a source of both protein and water (Fig. 18.11). The developing embryo floats in a fluid-filled sac or *amnion*; the fluid protects it from injury or distortion during its growth by supporting the tissues equally on all sides. The embryo is connected by the umbilical stalk to a network of capillary blood vessels spread out over the yolk and over a special sac, the *allantois,* which becomes attached to the air space at one end of the egg (Fig. 18.12).

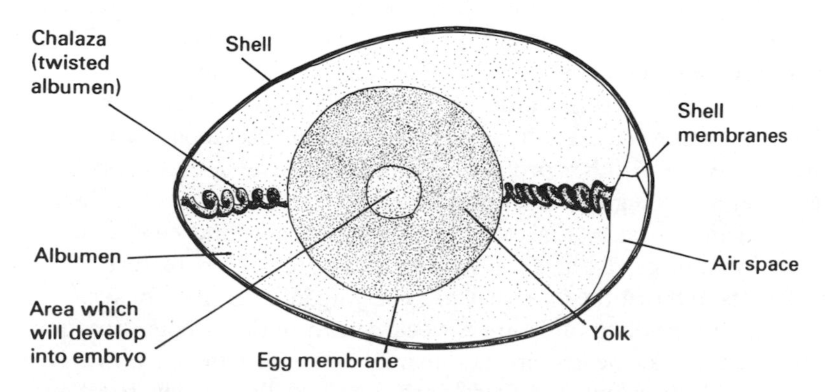

Fig. 18.11 *Bird's egg with top half of shell removed*

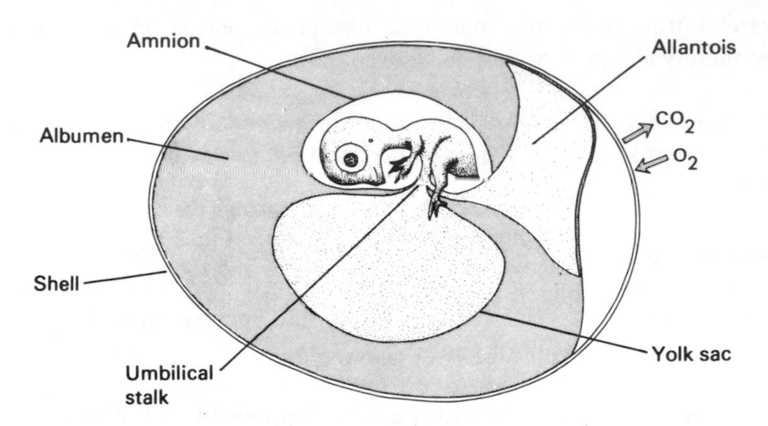

Fig. 18.12 *Chick development*

Since the egg shell and the membranes just beneath it are permeable to gases, oxygen can diffuse from the atmosphere into the air space and hence into the blood in the capillaries; the blood carries the oxygen to the embryo. Carbon dioxide produced in the tissues of the embryo is eliminated from the egg by the reverse process. Eventually the yolk sac is absorbed into the body of the young bird; the amnion bursts and the chick uses its beak to break out of the shell. The allantois remains adhering to the shell.

(g) **Parental care.** The chicks of ground-nesting birds like pheasants and domestic fowl are covered with downy feathers when they hatch. Very soon they can run about and peck at objects on the ground, quickly learning to identify materials suitable for food. They stay close to the hen, responding to her calls by taking cover or seeking her out, according to the circumstances.

The young birds of most other species are helpless when they hatch; their eyelids are closed, and they have few or no feathers. The naked chicks are very susceptible to heat loss and desiccation; the parent birds, however, reduce these hazards by brooding the chicks, covering the nest with body and wings. Both parents collect suitable food: worms, caterpillars, insects and other protein-rich materials. The sound or sight of the parents approaching the nest stimulates the nestlings to stretch their necks and gape their beaks (Fig. 18.13). The mouths of the chicks are bright orange in color inside, and the sight of this conspicuous feature stimulates the parents to thrust the food they are carrying into the open beaks.

When the young birds are a week or two old they begin to climb out of the nest and sit nearby in the bush or tree; the parents, however, continue to seek them out and feed them. The fledglings begin short practice flights as soon as their primary and secondary flight feathers are sufficiently

Fig. 18.13 *Dunnock chicks gaping for food*
(Heather Angel)

developed. This is one of the most dangerous periods of their lives, since they can feed themselves to only a limited extent and cannot escape from predators like cats and hawks. It has been estimated that only 25 per cent of the eggs laid in open nests reach the stage of independence from the parent birds.

18.4 QUESTIONS (Units Fifteen through Eighteen)

Apart from simple recall of facts, the most likely types of essay questions in examinations are those demanding a comparison of these organisms from the point of view of their methods of locomotion, breathing or reproduction, for example, the contrast between the lack of parental care in the frog and the elaborate parental behavior of birds, or between the use of the trachea in insects, and of gills in fish for gaseous exchange. The best practice for answering questions of this type is to select pairs of animals and draw up lists of features common to both in connection with breathing, locomotion or reproduction and then tabulate the differences relevant to those activities. The similarities will usually be concerned with fundamental principles; for example, oxygen and carbon dioxide are exchanged in both lungs and gills but the methods of ventilation are quite different.

To make the answers relevant, only points of similarity or difference should be mentioned. At least in practice answers, try to start all sentences with either "Both . . ." or "Whereas . . . ," e.g. "Both fish and frog can exchange oxygen and carbon dioxide with the water surrounding them," "Whereas the frog uses only its skin for this process, the fish has specialized structures, gills, with a greatly increased surface area for gaseous exchange."

Bear in mind that a "compare" question requires an account of both similarities and differences while a "contrast" question demands a description of the differences only. For either type of question, two separate accounts will not do.

UNIT 19 Food And Diet In Man

19.1 INTRODUCTION

The next eleven units of this book (Units Nineteen through Twenty-nine) are concerned with the biology of animals, in particular that of man. In this unit and the one which follows we begin with a discussion of human nutrition, and the subsequent three units deal with the ways in which food substances and oxygen reach the tissues where they are to be used, and the methods by which the waste products of their metabolism are removed from the body. Many of the principles and facts of these units apply to invertebrate animals as well.

Food eaten by animals is used in three ways:

(*a*) it may be oxidized in order to produce energy, which is expended in work and physical exercise, in maintaining the body's warmth, and in keeping going the many chemical reactions involved in metabolism;
(*b*) it may be incorporated into new cells and tissues to produce growth; or
(*c*) it may be used to renew or replace parts of cells of tissues that have been damaged or broken down by the chemical changes that occur during life.

An animal's body cannot function properly unless its diet provides

(*a*) a quantity of *energy* sufficient for its needs;
(*b*) appropriate amounts of *carbohydrates, proteins* and *fats;*
(*c*) *mineral salts;*
(*d*) *vitamins;*
(*e*) *water;*
(*f*) *roughage.*

19.2 ENERGY VALUE OF FOOD

A *joule* is a unit of energy; the same unit is used to measure all kinds of energy, whether it is apparent as heat or as mechanical, electrical or chemical energy. The number of joules of energy which can be obtained by oxidizing a food sample in the laboratory is a measure of the food's *energy value,* that is, its possible value as a source of energy to the body. The joule is a very small unit so the *kilojoule* (kJ), 1,000 joules, is a more convenient unit used in nutritional studies. (The *calorie* is still used as an energy unit by some writers; 1 calorie = 4.2 joules.)

The energy value of a foodstuff is found by burning a known weight of it completely to carbon dioxide and water in a special apparatus called a *bomb calorimeter*. This device is designed to ensure that all the heat given out during the combustion is transmitted to a known weight of water, and the rise in the water temperature is measured. Since 4.2 joules of heat energy raise the temperature of 1 g water by 1 °C, the number of joules given out by the burning food can be calculated. Whether the body can obtain the same quantity of energy when the food is eaten, however, depends on the efficiency of digestion and absorption, and on the chemical processes undergone by the food. For example, cellulose will burn and give out heat in the bomb calorimeter, but it cannot be digested by man and therefore cannot provide the human body with energy at all.

Only about 15 per cent of the energy in the food can be obtained as mechanical energy, but the additional heat energy set free is important in maintaining body temperature.

The number of kilojoules of energy needed each day by a human being varies very greatly according to age, sex, occupation and activity, but estimates can be made for particular groups of people. Thus, for example, a lumberjack doing eight hours of hard physical work each day requires from 23,000 to 25,000 kJ (5,500–6,000 Cal), a tailor or other sedentary worker 10,000 to 11,000 kJ (2,400–2,600 Cal), and a six-year-old child about 8,000 kJ (1,900 Cal). Even within a group, however, there are considerable variations: for instance, a large, heavily built man needs more energy to keep his body going than a small slight man, and an adolescent during a period of rapid growth needs more energy than an adult of the same height and build. People who are engaged in heavy manual work or other vigorous activity require an increased supply of energy; the following table indicates roughly how the body's energy requirements during the day depend on activity:

	kJ	*(Cal)*
8 hours asleep	2,400	(570)
8 hours awake: relatively inactive physically	3,000	(714)
8 hours awake and physically active	6,600	(1,571)
	12,000	(2,855)

If an individual's daily intake of food does not provide enough energy for his needs, he will lose weight as the existing food stores and tissues of his body are oxidized to release energy, and this capacity for work will fall off if the deficiency is prolonged and severe. A daily intake below about 6,300 kJ, if maintained for a long time, would probably produce wasting and eventually death, although symptoms of illness due to vitamin deficiency would be likely to appear in the first instance.

19.3 NUTRIENTS

(a) Carbohydrates

Carbohydrates are substances containing the elements carbon, hydrogen and oxygen in certain proportions. Examples are *glucose* ($C_6H_{12}O_6$), "table" sugar or *sucrose* ($C_{12}H_{22}O_{11}$), *starch* and *cellulose,* which can both be represented by the formula $(C_6H_{10}O_5)_n$, where n is equivalent to 300 or more for starch and at least 3,000 for cellulose; these large molecules are made up of long chains of glucose or similar units linked together. Foods relatively rich in carbohydrates are milk, fruit, jam and honey, all of which contain sugar and related compounds, and bread, potatoes and peas, all of which contain starch.

Carbohydrates are principally of value as energy-giving foods; 1 g carbohydrate can provide 17 kJ of energy. In mammals any excess of carbohydrate eaten is stored in the body, either as glycogen in the liver and the muscles, or converted into fat and stored in fat cells beneath the skin and elsewhere in the body.

(b) Proteins

Proteins contain the elements carbon, hydrogen, oxygen, nitrogen, usually sulphur and sometimes others according to their source. Examples of foods containing protein are lean meat, eggs, beans, fish such as herring and salmon, and milk and its products such as cheese.

Proteins are broken down by digestion to simpler, soluble substances called *amino acids* which are absorbed into the bloodstream and eventually reach the cells of the body. In the cells the amino acids are reassembled to form the proteins of the protoplasm. There are only about 20 different kinds of amino acid constituting proteins, but there are innumerable different proteins. The difference between one protein and the next depends on which amino acids are used to build it, how many of each there are, their sequence and their arrangement as shown in Fig. 19.1.

Plants can build all the amino acids they need from carbohydrates and salts but animals cannot. It follows that they must obtain their amino acids

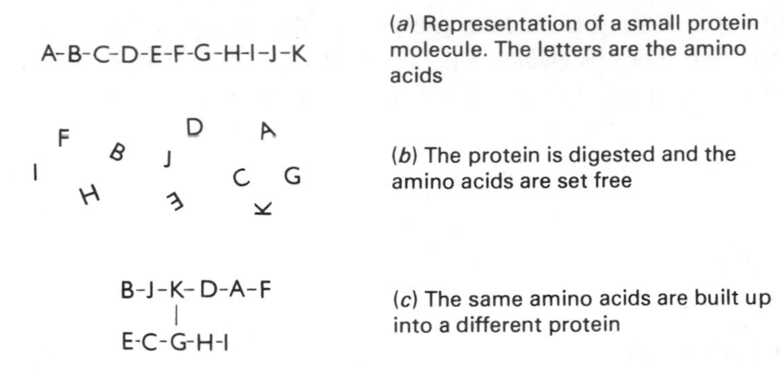

Fig. 19.1 *Diagram illustrating the breakdown and resynthesis of proteins*

from proteins already made by plants or present in other animals, and the diet must therefore include a minimum quantity of protein of one sort or another. A diet with enough fats and carbohydrates to provide the body with sufficient energy and rich in vitamins and salts will lead to illness and death unless it also provides proteins. Proteins are particularly important during periods of pregnancy and growth when new protoplasm, cells and tissues are being made.

Although animals cannot make amino acids they can, in some cases, convert one amino acid into another. There are, however, ten or more amino acids which animals cannot produce sufficiently quickly in this way and these *essential amino acids* must be obtained directly from proteins in the diet. Animal proteins generally contain more essential amino acids than do most plant proteins but since milk and eggs contain particularly high proportions, a vegetarian who includes plenty of these in his diet should not suffer from any deficiency.

If proteins are eaten in excess, there will be more amino acids in the body than are needed to produce or replace cells. The excess amino acids cannot be stored as such and are converted in the liver to carbohydrates which are then oxidized for energy, or converted to glycogen and stored. The energy value of protein is the same as that of carbohydrates, 17 kJ/g. The tendency for people in the industrialized countries to indulge heavily in meats is costly and wasteful.

(c) Fats

Like carbohydrates, fats contain only carbon, hydrogen and oxygen, but in different proportions. They are compounds of *glycerol* (glycerine) with *fatty acids*. Butter, margarine, lard and cooking oils such as olive oil or ground-nut oil consist almost entirely of fat, but milk, cheese, egg-yolk, some fish, nuts, and even "lean" meat and bacon all contain considerable amounts.

Although fats are less easily digested and absorbed than carbohydrates, they have more than double the energy value, providing 39 kJ/g. Fat can be stored in the body (see Section 20.11(*b*)), providing useful energy reserves. The layer of fatty tissue lying just below the skin may have a heat-insulating function; it is particularly thick in animals living in very cold climates.

(d) Mineral Salts

A wide varicty of salts is essential for the chemical activities in the body and for the construction of certain tissues. *Iron* is an essential constituent of hemoglobin, the red pigment in the blood; the hard tissues of bones and teeth contain *calcium, magnesium* and *phosphorus*; *sodium* and *potassium* are essential in nearly all cells, in the blood fluid and in nerves; *iodine* is necessary for the proper functioning of the thyroid gland. Minute traces of other elements, in particular *copper, cobalt* and *manganese,* are necessary to the body, even though larger amounts of these metals are actually poisonous.

Salts of these elements are present in small quantities in a normal diet, and the body can absorb and concentrate them.

(e) Vitamins

Vitamins are complex chemical compounds which, although they have no energy value, are essential in small quantities for the normal chemical activities of the body. Different species of animal vary greatly in the particular vitamins they require; for example, vitamin C is necessary to man but to very few other species. Plants can build up the vitamins they need from simple substances, but animals must obtain them "ready-made" directly or indirectly from plants.

It has long been known that certain diseases could be prevented or cured by making alterations in the diet. For centuries, sailors on long sea voyages, forced to live on a diet lacking in fresh foods, had suffered from the debilitating disease of scurvy; but in about 1750 a naval surgeon named James Lind found that the illness could be both avoided and treated by providing the seamen with citrus fruits. In 1896 Christiaan Eijkman, a Dutch doctor working in Java, was searching for the "germ" which he thought transmitted the disease beri-beri, but found that the disease could be induced in chickens by feeding them with polished (de-husked) rice and cured by feeding them unpolished rice. Eijkman concluded that the disease was caused by a poisonous substance in the polished rice, which was normally neutralized by something present in the husk of the unpolished grain, but over the next ten years he and other scientists, notably the English biochemist Gowland Hopkins, realized that the rice husk contained a so-called *accessory food factor,* the absence of which produced beri-beri.

Gowland Hopkins' experiments with rats between 1906 and 1912 demonstrated the importance of these accessory food factors. He fed young rats on a mixture of starch, sucrose, lard, salts and purified casein (a milk protein rich in essential amino acids), a diet which provided all the then known nutrients in the proportions believed to be necessary for health. The rats ceased growing and soon died. However, if a daily supplement of 3 cc of milk was added to their diet, the rats grew and remained healthy (Fig. 19.2). The energy value of this small amount of milk is very little, and it contains no more amino acids than the casein already being fed to the rats; the restorative properties of the milk must therefore have been due to a factor in it other than its energy value or its amino acid content. This and many other necessary food factors have since been identified, and although their chemical compositions are very different they are usually classed together under the collective term *vitamins.*

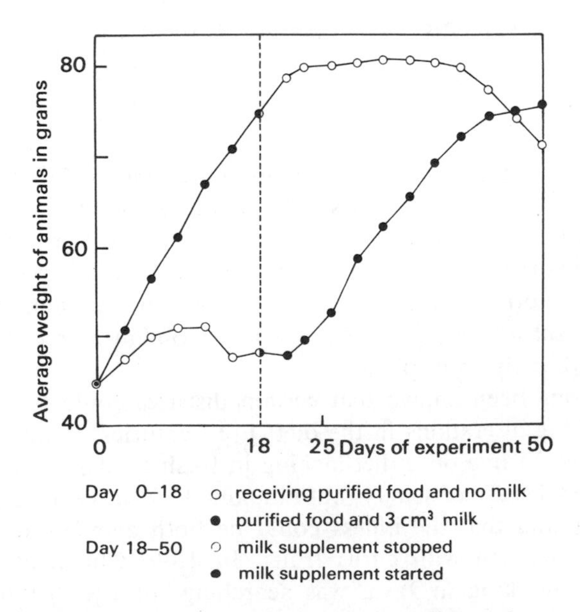

Fig. 19.2 *Gowland Hopkins' experiment with rats*
(From *J. Physiol.,* 44, 1912)

Fifteen or more vitamins have been isolated, and many of them can now be synthesized in the laboratory. Vitamins A, D, E and K are the *fat-soluble vitamins,* occurring mainly in animal fats and oils and absorbed into the body along with the products of fat digestion. Vitamin B (which is actually a group of compounds) and vitamin C are the *water-soluble vitamins.* Most vitamins seem to act in association with enzymes, biological catalysts which control the rate of all the essential chemical changes in the body, each vitamin influencing a number of vital reactions. If an individual's diet is deficient in one or more vitamins, some of the normal processes of his body cannot continue and symptoms of disease appear. Such diseases, *deficiency diseases,* can usually be effectively treated by adding the appropriate vitamin to the diet.

Table 19.1 *Vitamins and their characteristics*

Name	*Food rich in the vitamin*	*Diseases and symptoms caused by lack of the vitamin*
Vitamin A (*Retinol*)	Fresh green vegetables, milk and butter, cod-liver oil, liver.	Generally reduced resistance to diseases, particularly those which gain access by the skin. Poor night-vision. Skin becomes dry and scaly and so does cornea of eyes, causing *xerophthalmia.*
Note. Animals can obtain some vitamin A from *carotene,* the yellow pigment in vegetables, e.g. carrots, but most is taken in with butter and milk. Stored in the liver.		
Vitamin B_1 (*Thiamine*)	Wheat germ (embryo), yeast.	Loss of appetite, digestive disturbances, wasting. Other variable symptoms such as swelling of the feet and legs, leading to *beri-beri.*
At least 9 other B vitamins, including niacin	Yeast, milk, meat, eggs, cheese and green vegetables.	Niacin deficiency causes *pellagra,* characterized by loss of weight, diarrhea, dermatitis and mental disorders.
Note. White flour (72% extraction) is deficient in B_1 owing to removal of embryo. B_1 may be deficient in a slimming diet. Rice husks contain B_1. In rice-eating countries, a diet of polished rice (from which the husks are removed) leads to deficiency diseases. Maize is deficient in niacin.		

Table 19.1 *Vitamins and their characteristics* (continued)

Name	*Food rich in the vitamin*	*Diseases and symptoms caused by lack of vitamin*
Vitamin C (*Ascorbic acid*)	Oranges, lemons, tomatoes, black-currants, fresh vegetables.	Weakness, hemorrhages of the skin, especially in the mouth and gums, anemia, skin eruptions, poor healing of wounds and likelihood of infection: all are symptoms of *scurvy*.
Note. Scurvy is likely to occur only when fresh food is not available and preserved food constitutes the diet. The vitamin is destroyed by cooking in air.		
Vitamin D (*Cholecalciferol:* occurs in several forms)	Cod-liver oil, cream, egg-yolk.	Abnormal bone formation, soft bone with swollen ends, all symptoms of *rickets* in children.
Note. Natural fats below the skin will form a substance similar to vitamin D when exposed to *ultraviolet radiation,* either artificial or in sunlight.		
Vitamin E (*Tocopherol*)	Wheat embryo, e.g. in wholemeal bread, butter.	Affects reproduction. In the female the embryo dies and is reabsorbed; in the male the sperms die.
Note. Effects of deficiency have been demonstrated in rats but not in humans.		
Vitamin K	Vegetables, especially cabbage and spinach.	Symptoms rather like those of scurvy. Its absence increases the clotting time of the blood.
Note. Vitamin K is synthesized by intestinal bacteria; simple deficiency is unlikely. Bile is essential for its absorption.		

Table 19.1 lists some of the important vitamins, together with their sources and their properties. It must be emphasized that nearly all normal mixed diets include adequate amounts of vitamins without any supplementation, and deficiency diseases are most likely when the bulk of the diet consists of only one or two kinds of food, such as rice or maize, as in some of the developing countries.

19.4 WATER

Water makes up a large proportion of all the tissues in the body and is an essential constituent of normal protoplasm. Although not itself a nutrient, it plays an important part in the digestion and transport of food material and all the chemical reactions of the body take place in solution in water.

19.5 ROUGHAGE

Roughage is that part of the diet which cannot be digested; in man it consists largely of the cellulose in plant-cell walls. It adds bulk or fiber to the food and enables the muscles of the alimentary canal to grip it and keep it moving by peristalsis (see Section 20.2), particularly in the large intestine. Absence of roughage is likely to lead to constipation and its attendant disorders.

19.6 MILK

Milk is the sole article of diet during the first few weeks or months of a mammal's life and is an almost ideal food since it contains proteins, fats, carbohydrates, mineral salts (particularly those of calcium and magnesium) and vitamins. For adults, however, it is less satisfactory because of its high water content and lack of iron. Large volumes would have to be consumed if it were the principal article of diet for an adult, and serious blood deficiencies would result from the lack of iron. Iron deficiency does not arise in a young mammal living exclusively on milk because it has reserves of iron in its body, stored while it developed as an embryo inside its mother, which suffice for its needs until it is old enough to begin to eat solid food.

19.7 PRACTICAL WORK: FOOD TESTS

Test for starch. A little starch powder is shaken in a test-tube with some cold water and then boiled to make a clear solution. When the solution is cold, 3 or 4 drops of *iodine solution* are added. The dark blue color that results is characteristic of the reaction between starch and iodine. This test is very sensitive.

Test for glucose. A little glucose is heated to boiling-point with some *Benedict's solution* in a test-tube. The solution will change from clear blue to opaque green, then to yellow and finally a brick-red precipitate of copper(I) oxide will appear. The mouth of the test-tube should be directed away from people as the solution is likely to "jump" out of the tube if the liquid boils. Sucrose will not react with Benedict's solution until after it has been boiled with dilute hydrochloric acid and neutralized with sodium bicarbonate.

Test for protein (Biuret test.) To a one per cent solution of albumen is added 5 cm^3 dilute sodium hydroxide (CARE: this solution is caustic) and 5 cm^3 one per cent copper sulphate solution. A purple color indicates the presence of protein.

Test for fats. Two drops of cooking oil are thoroughly shaken with about 5 cm^3 ethanol in a dry test-tube until the fat dissolves. The alcoholic solution is poured into a test-tube containing a few cm^3 of water. A cloudy white emulsion is formed.

Application of the food tests. These tests can be applied to samples of food such as milk, raisins, potato, onion, beans, egg-yolk, ground almonds, to find out what food materials are present in them.

They can also be carried out in connection with work on seeds and plant storage organs. The seeds should be crushed in a mortar and then shaken with warm water to extract the soluble products. The solution and the suspension of crushed seed can be subjected to the tests given above.

19.8 QUESTIONS

1. In animals, which classes of food can be used to provide energy? Which ones provide the most energy?
2. What principles must be observed when working out a diet in order to lose weight? What dangers are there if such diets are not scientifically planned?
3. Kwashiorkor is a protein-deficiency disease which affects young children in the developing countries. Its onset is usually most severe when a child is weaned from the mother's milk to the starchy plant foods of the adults, such as yam or cassava. Why should weaning mark the onset of the disease?
4. It is a common fallacy that manual workers need much more protein than sedentary workers. Suggest why this idea is fallacious.
5. Eating a large amount of protein at one sitting is wasteful. It is better to take in a little protein at each meal. Why do you think this is the case?
6. What dietary problems confront (*a*) people living predominantly on rice or maize, (*b*) strict vegetarians?

UNIT 20
DIGESTION, ABSORPTION, AND METABOLISM OF FOOD

20.1 INTRODUCTION

If food taken into the mouth is to be of any value to the body, it must enter the bloodstream and be distributed to all the living cells. But a piece of meat or bread, for instance, cannot enter the blood unchanged: its insoluble constituents must be chemically altered to soluble compounds. *Digestion* is the breakdown of insoluble food materials, many of which have very large molecules, into soluble compounds having smaller molecules; these can pass in solution through the walls of the intestine and enter the bloodstream, the process of *absorption*. Digestion and absorption both take place in the *alimentary canal* or gut (Figs. 20.1 and 20.2), a muscular tube running from the mouth to the anus. Some of its regions have specific functions and accordingly specially adapted structures. In its lining or *epithelium* are glands * which produce some of the *digestive juices,* the fluids which bring about the breakdown of food; other juices are poured into the alimentary canal through ducts from glandular organs outside it. As the food passes through the alimentary canal it is broken down in stages until the digestible material is dissolved and absorbed. The undigested residue is expelled from the body through the anus and is called *stool* or *feces* (fecal matter).

Enzymes

Digestion of food is brought about by chemicals in the digestive juices called *enzymes.* Enzymes are catalysts, that is, substances which accelerate the rate of chemical reactions without altering the end-products. All

* A gland is a group of similar cells producing a substance which is used elsewhere in the body.

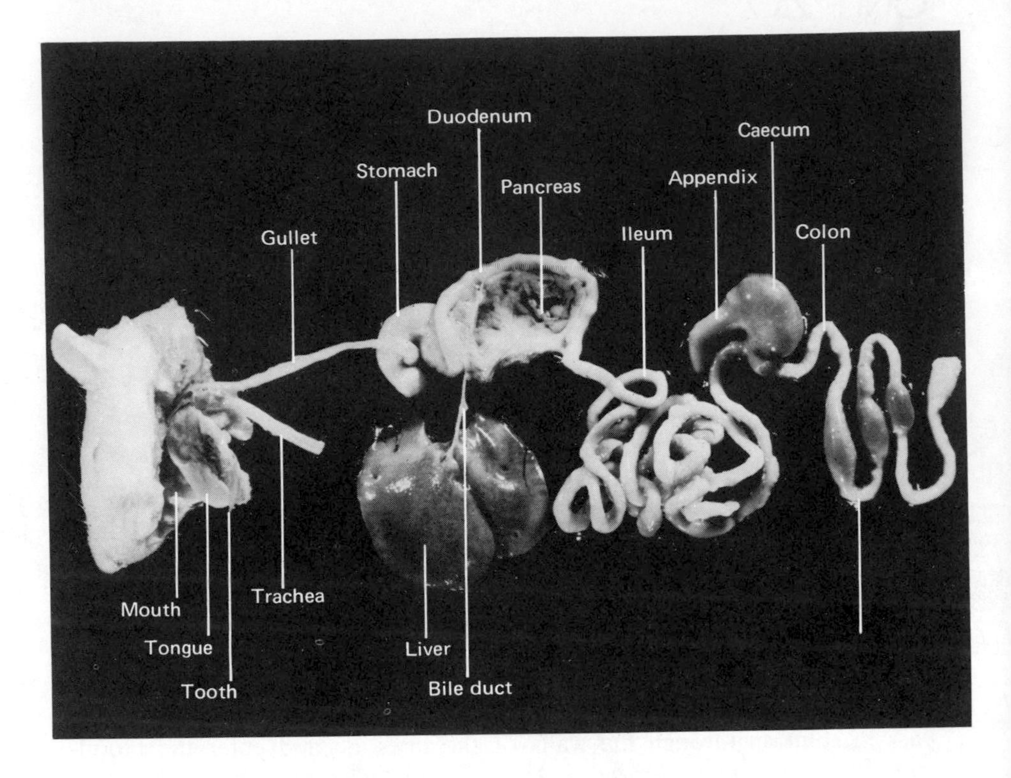

Fig. 20.1 *Alimentary canal of a rat unravelled*

(Dissection by Gerrar and Haig Ltd.)

enzymes are proteins, and they occur in great numbers and variety in all protoplasm, controlling virtually all the chemical processes which take place in living things; without them these processes would be too slow to maintain life. The vast majority of enzymes are *intracellular,* that is, they carry out their functions in the protoplasm of the cell in which they are made. Some enzymes, however, pass from the cells which have produced them to be used elsewhere, a process called *secretion.* These are called *extracellular* enzymes; bacteria (Unit 11) and fungi (Section 12.1) secrete such extracellular enzymes into the media in which they are growing in order to accelerate the conversion of their food materials into soluble substances that can enter the cells. The higher organisms secrete extracellular enzymes into the alimentary canal for the digestion of food taken into it.

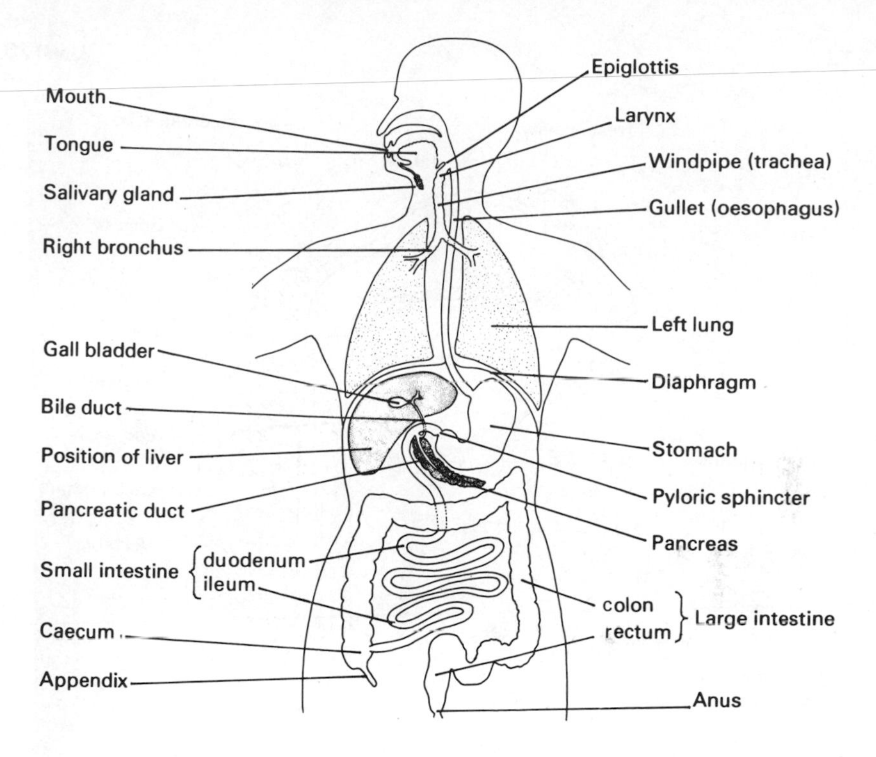

Fig. 20.2 *The alimentary canal*

Every enzyme has the following characteristics:

(*a*) like all proteins it is destroyed by heat;
(*b*) it acts best within a narrow temperature range (see Experiment 20.4 in Section 20.14);
(*c*) it acts most rapidly at a particular degree of acidity or alkalinity (pH) (see Experiment 20.3);
(*d*) it affects the rate of only one kind of reaction;
(*e*) since it affects only the *rate* of the reaction, it always forms the same end-product or products.

Digestive enzymes accelerate the rate at which insoluble compounds are converted into soluble ones. Enzymes which act on starch are called *amylases,* those acting on proteins are *proteinases,* and *lipases* act on fats.

20.2 MOVEMENT OF FOOD THROUGH THE ALIMENTARY CANAL

Ingestion. This is the act of taking food into the alimentary canal through the mouth.

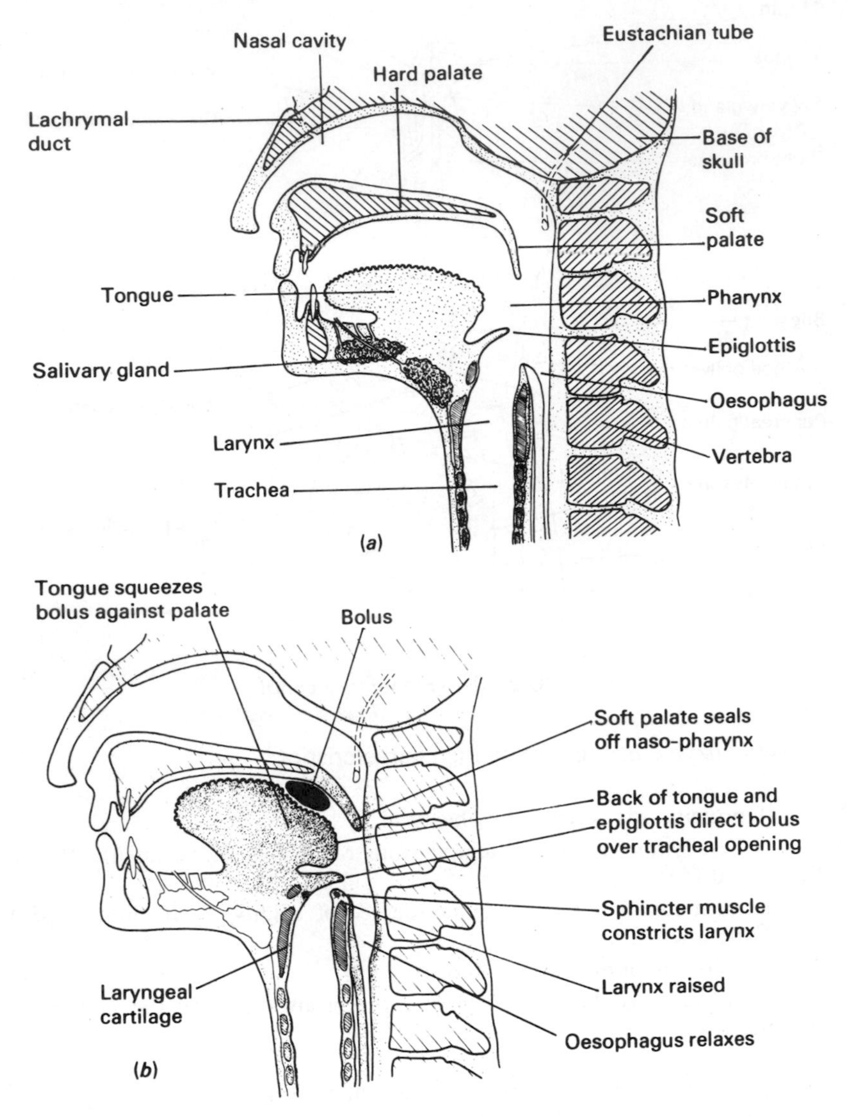

Fig. 20.3 *Sections through head to show swallowing action*

Swallowing (Fig. 20.3). In swallowing the following sequences of actions takes place:

(*a*) The tongue presses upward and back against the roof of the mouth, forcing the pellet of food, called a *bolus,* to the back of the mouth or *pharynx;*

(*b*) the *soft palate* rises and closes the opening between the nasal cavity and the pharynx;

(*c*) the *laryngeal cartilage,* a ring of tough gristle-like material around the top of the trachea or windpipe, is pulled upward by muscles so that the opening of the trachea lies beneath the back of the tongue: this opening is simultaneously constricted by the contraction of a circular *sphincter* muscle;

(*d*) the *epiglottis,* a flap of cartilage, directs food over the tracheal opening.

In this way food is able to pass over the entrance to the trachea without passing into it. The beginning of the action is voluntary, that is, it is under the conscious control of the will, but once the bolus of food reaches the pharynx swallowing becomes an automatic or reflex action (see Section 29.3).

Peristalsis. This is the way in which food is forced down the *esophagus* or gullet into the stomach and subsequently through the lower regions of the alimentary canal. The walls of the whole alimentary canal contain muscle fibers which run both circularly and longitudinally. The circular muscles contract and relax in such a way that waves of contraction pass along the alimentary canal, pushing the food steadily onward and preventing its movement in the reverse direction (Fig. 20.4).

Egestion. This is the expulsion from the body of the undigested remains of food.

20.3 DIGESTION IN THE MOUTH

In the mouth the food is mixed with *saliva* and chewed or *masticated* by the action of the teeth and tongue; this softens it and reduces it to pieces of a suitable size for swallowing, and also increases the surface available for

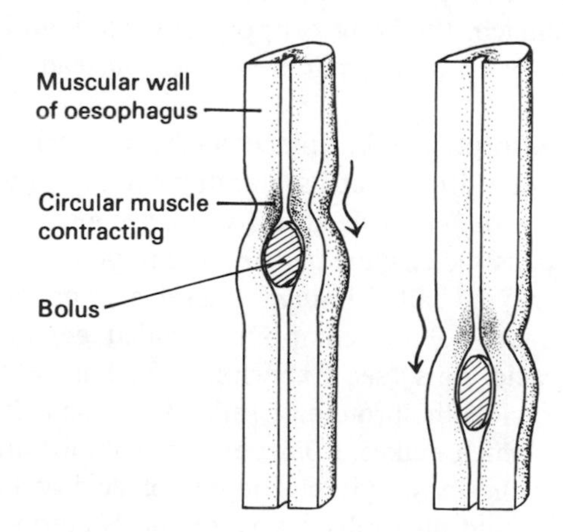

Fig. 20.4 *Diagram to illustrate peristalsis*

enzymes to act on. Saliva is a digestive juice secreted by three pairs of glands, the ducts of which lead into the mouth; an adult secretes 1 to 1.5 liters of saliva daily. The rate at which saliva is secreted increases when the body receives the stimuli of the taste, smell, sight or even the thought of food. It is a neutral or slightly acid fluid, consisting of water and a little *mucus,* a somewhat viscous, slimy substance which helps to lubricate the food and make the particles adhere to each other. Saliva also contains an enzyme, *salivary amylase* or *ptyalin,* which acts on cooked starch (for example, in bread or boiled potatoes) and begins its breakdown to a soluble sugar called *maltose,* the molecule of which consists of a pair of linked glucose units (see Experiment 20.2 in Section 20.14).

The longer food is retained in the mouth, the further this starch digestion proceeds and the more finely divided the food becomes as a result of chewing. In fact, even well-chewed food does not remain in the mouth long enough for much digestion of starch to take place, but saliva continues to act for a time even after the food has been swallowed. Relatively solid food takes about six seconds to pass from the mouth down the esophagus to the stomach; liquid food travels more quickly.

20.4 DIGESTION IN THE STOMACH

This part of the alimentary canal has flexible, elastic walls and so can contain a relatively large amount of food, which is retained in the stomach by the closure of a ring of muscle called the *pyloric sphincter* at its outlet. These characteristics enable food from a particular meal to be stored for some time and released at intervals to the rest of the alimentary canal. If there were no stomach, the body's supply of food would have to be eaten in small amounts every twenty minutes or so, instead of as three or four meals each day.

Very little absorption takes place in the stomach, except of ethanol (alcohol) and certain drugs, but its glandular lining (Fig. 20.5) produces a digestive juice, *gastric juice* (the adjective *gastric* means "of the stomach"). Gastric juice contains the enzyme *pepsin,* and in young children it may also contain another enzyme called *rennin.* Pepsin acts on proteins and breaks them down into more soluble compounds called *peptides,* each made up of several amino acid units (see Experiment 20.1 in Section 20.14). Rennin, when present, clots the protein of milk. The stomach wall also secretes *hydrochloric acid* which makes a 0.5 per cent solution in the gastric juice. The acid provides the most suitable degree of acidity (optimum pH) for pepsin to work in, and also kills many of the bacteria taken in with the food. The salivary amylase from the mouth cannot digest starch in such

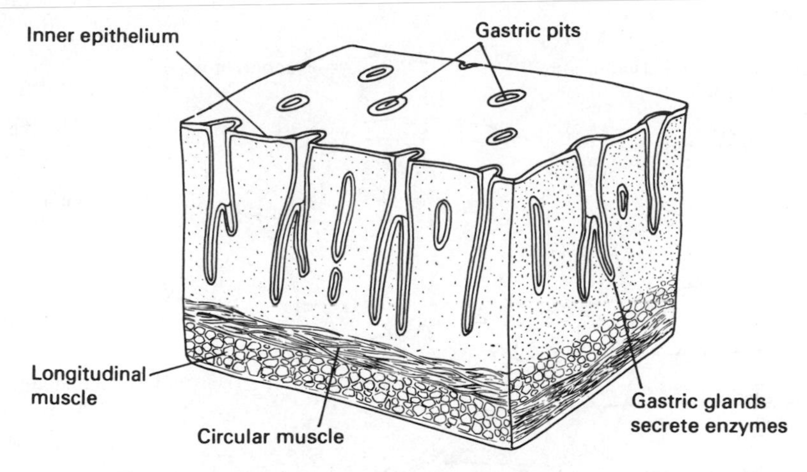

Fig. 20.5 *Stereogram of section through stomach wall*

an acid medium, but it seems likely that it continues to act within the bolus of food until this is broken up and the hydrochloric acid reaches all its contents.

The stomach walls contract about every twenty seconds with rhythmic peristaltic movements, churning the food and gastric juice together to form a creamy fluid called *chyme.* The length of time food is retained in the stomach depends to some extent on its nature. Water may pass through in a few minutes, a meal consisting largely of carbohydrate, such as porridge, may be retained for less than an hour, and a mixed meal containing protein and fat may be in the stomach for one or two hours.

When digestion in the stomach is complete the pyloric sphincter relaxes from time to time, each relaxation allowing a little chyme to pass through into the first part of the small intestine, called the *duodenum.*

20.5 DIGESTION IN THE DUODENUM

Two alkaline fluids are poured into the chyme in the duodenum, *pancreatic juice* from the pancreas and *bile* from the liver. Both juices contain sodium bicarbonate which partly neutralizes the strongly acid chyme, so creating the slightly acid medium favorable to the action of the pancreatic and intestinal enzymes.

Pancreas ("sweetbread"). This cream-colored gland lies below the stomach (Fig. 20.6). Its cells secrete several different enzymes. Three of these, including the powerful enzyme *trypsin,* break down proteins to pep-

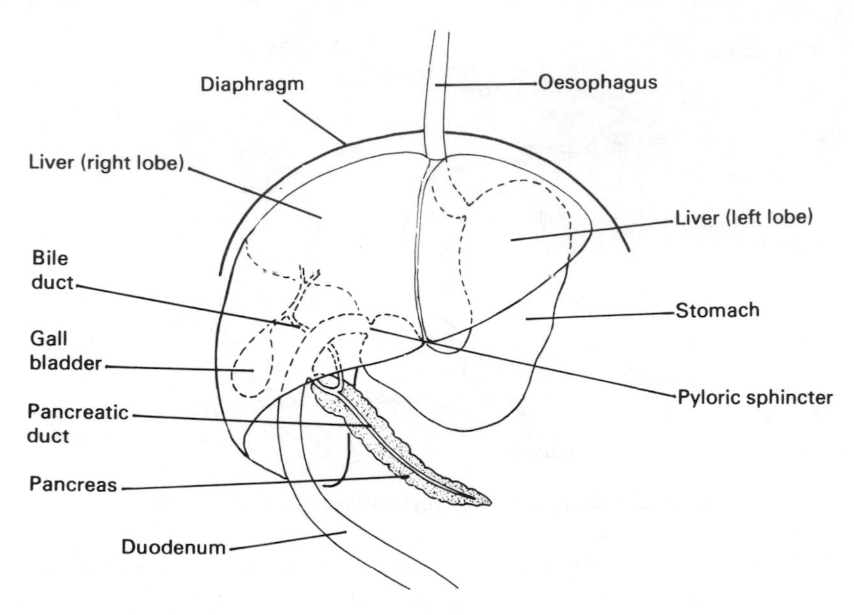

Fig. 20.6 *Diagram to show relation of stomach, liver, and pancreas*

tides, and peptides to soluble amino acids. Others break down starch to maltose, and fats to their constituent compounds, fatty acids and glycerol.

Bile. This green watery fluid is made in the liver, stored in the gall bladder and conducted to the duodenum by the bile duct. Its color is derived largely from breakdown products of the red pigment, hemoglobin, from decomposing blood cells. It contains sodium chloride, sodium bicarbonate and complex organic compounds called *bile salts,* but no enzymes.

Bile dilutes the contents of the intestine, and the bile salts reduce the surface tension of fats, so emulsifying them, that is, dispersing them as tiny droplets presenting an increased surface for enzyme action and allowing more rapid digestion. Many of the bile salts are reabsorbed in the *ileum,* the lower part of the small intestine.

20.6 DIGESTION IN THE ILEUM

The glands in the lining of the ileum (Fig. 20.8) secrete a digestive juice called the *succus entericus*; it contains at least five enzymes. These complete the breakdown to amino acids of any peptides still undigested, of maltose and other carbohydrates to glucose and related sugars like fructose and

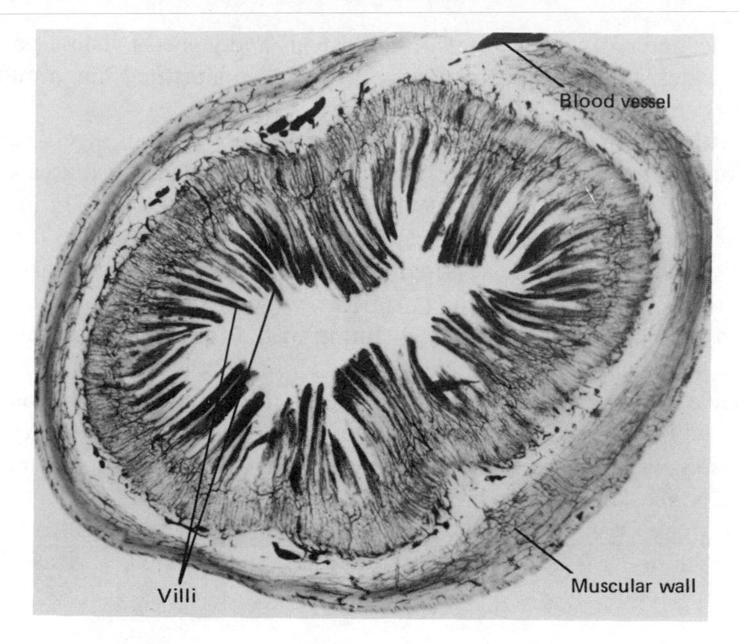

Fig. 20.7 *Transverse section through ileum of cat, showing villi*
(M. I. Walker)

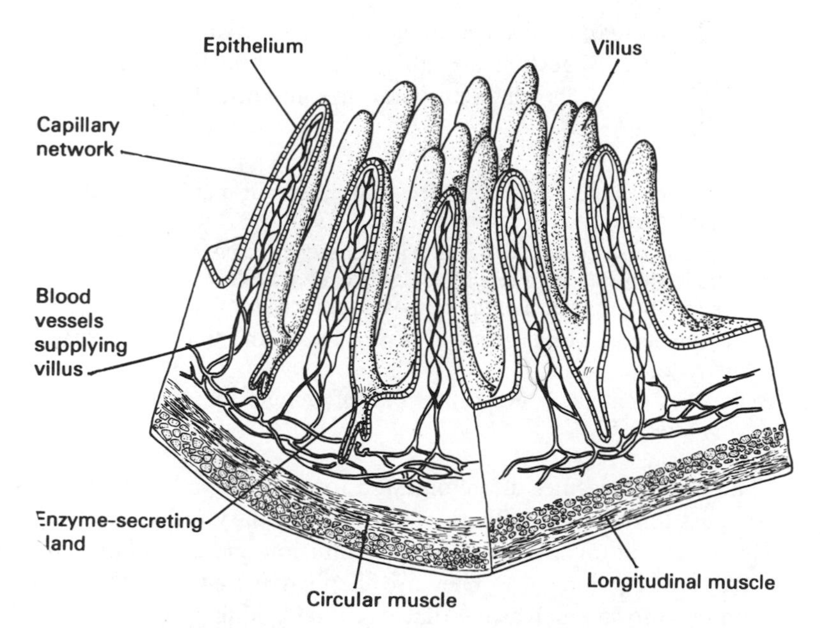

Fig. 20.8 *Stereogram to show structure of ileum*

galactose, and of unchanged fats to fatty acids and glycerol. These products are all water-soluble and so can pass through the intestinal lining and into the bloodstream.

Although the action of digestive enzymes dissolves all digestible food constituents, the cells which produce them are not themselves attacked by the enzymes, nor are the other cells lining the alimentary canal. This is partly because the enzymes are produced within the cells in an inactive form and cannot work until they are secreted into the cavity of the alimentary canal, where they are activated by the chemicals present. Pepsin, for example, is made by the glands of the stomach lining in the form of an inactive substance (or *precursor*) called *pepsinogen*; pepsinogen is only converted into the active enzyme when it is secreted into the stomach and comes into contact with the hydrochloric acid in the gastric juice. The alimentary canal is additionally protected from enzyme attack by the constant production of *mucus* by specialized cells in the epithelium. Mucus coats the epithelium with a slimy, viscous layer which helps to prevent the digestive juices from reaching it; it also lubricates the walls of the canal, and assists the passage of food along it.

20.7 ABSORPTION IN THE ILEUM

The ileum is an organ of absorption as well as an organ of digestion; nearly all the products of digestion are absorbed into the bloodstream through its walls. Certain of its characteristics are important adaptations to its absorbing function:

(*a*) in most species it is fairly long (several meters in man), presenting a large absorbing surface to the digested food;
(*b*) its internal surface is enormously increased by a covering of thousands of tiny finger-like projections called *villi,* each a millimeter or so long (Figs. 20.7 and 20.8), rather like the pile of a fabric like velvet;
(*c*) the epithelium covering the villi is very thin, and fluids can pass through it fairly easily;
(*d*) each villus contains a dense network of minute thin-walled blood vessels or *capillaries* (Fig. 20.9).

The small molecules of the digested food, principally amino acids and glucose, pass through the epithelium and the capillary walls; the mechanism of their movement involves both simple diffusion and some kind of "active transport" (see Section 3.2). They are then carried away in the capillaries, which unite to form small veins that eventually join up to form one large vein, the *hepatic portal vein.* This carries all the blood from the intestine

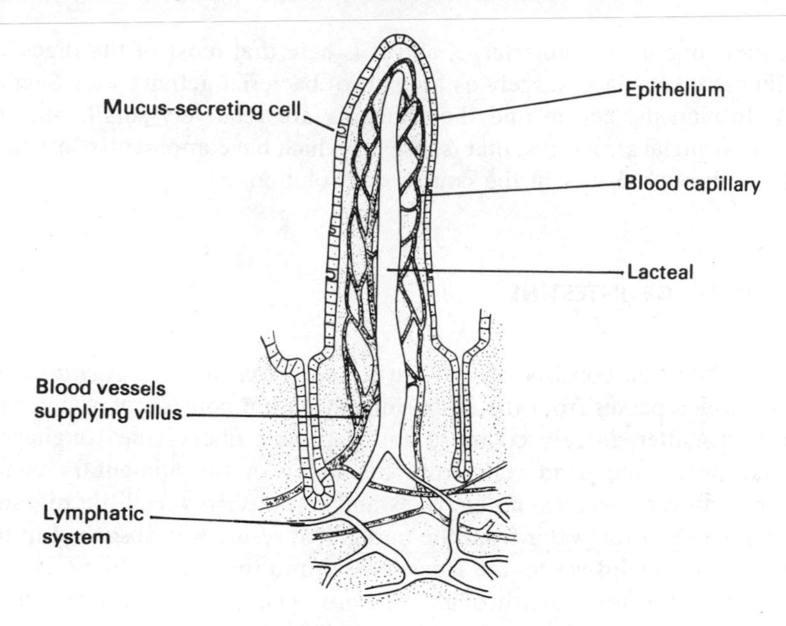

Fig. 20.9 *Villus structure*

to the liver, which may retain or alter any of the digestion products (see Section 20.12). The digested food is then carried in the bloodstream to all parts of the body.

Some of the fatty acids and glycerol from the digestion of fats enter the blood capillaries of the villi, but a large proportion recombine in the intestinal lining to form fats again. These fats pass into vessels in the villi called *lacteals* (Fig. 20.9), which derive their name from their milky-white appearance (the Latin word for milk is *lac*) due to the minute droplets of fat in the fluid they contain. Some of this fat may actually pass through the epithelium into the lacteals in finely emulsified form without any prior digestion.

The fluid in the lacteals drains into *lymph vessels,* which form a network all over the body called the *lymphatic system*; its contents are eventually emptied into the bloodstream (see Section 21.3).

20.8 THE CECUM AND APPENDIX

In many animals the *cecum* forms a branch of the alimentary canal at the point where the ileum meets the large intestine; it ends blindly in the *appendix* (see Fig. 20.2). In herbivorous animals like the rabbit and the horse these organs are quite large, and it is here that most of the digestion of cellulose takes place, largely as a result of bacterial activity (see Section 11.2). In man the cecum and the appendix are relatively small, and are probably vestigial structures, that is, organs which have apparently lost their functions through disuse in the course of evolution.

20.9 THE LARGE INTESTINE

The large intestine consists of two parts, the *colon* and the *rectum*. The material which passes from the ileum into the colon consists of water with undigested matter, largely cellulose and vegetable fibers (the roughage), bacteria, mucus and dead cells from the lining of the alimentary canal. The large intestine secretes no enzymes and can absorb very little digested food, but much of the water from the undigested residues is absorbed in the colon. The semi-solid waste, the *feces,* passes into the rectum by peristalsis and is expelled at intervals through the *anus*. The residues may spend 12 to 24 hours or more in the intestine, depending on the proportion of roughage in the diet.

20.10 HOW DIGESTED FOOD IS USED

The products of digestion are carried around the body in the blood, from which they are absorbed and metabolized by the cells.

(a) **Glucose.** Oxidized during respiration in the protoplasm (see Section 4.2), this reaction releases energy to drive the many chemical processes in the cell, and in specialized cells to produce, for example, contraction (muscle cells) and electrical changes (nerve cells).

(b) **Fats.** The fats are incorporated into cell membranes and other structures in cells. Those not used in this way for growth and maintenance are oxidized to carbon dioxide and water, releasing energy for the vital processes of the cells. More than twice as much energy is obtained in this way from a given weight of fat as from the same amount of glucose.

(c) **Amino acids.** These are reassembled in the cells to make proteins (see Section 19.3(*b*)); these proteins may go to make up the cell membrane, the nucleus, the cytoplasm or the structures contained in it, or they may be enzymes controlling and coordinating the chemical activity within the cell, or secreted by it for the control of extracellular reactions.

Table 20.1 *Digestive action*

Region of alimentary canal	*Digestive gland*	*Digestive juice produced*	*Enzymes in the juice*	*Class of food acted upon*	*Substances produced*
Mouth	Salivary glands	Saliva	Salivary amylase	Starch	Maltose
Note. Saliva is slightly acid or neutral. Mucus helps form bolus. Water lubricates food. Mastication increases surface available for digestion.					
Stomach	Gastric glands (in stomach lining)	Gastric juice	Pepsin	Proteins	Peptides
			(Rennin)	(Milk protein)	(Clots it)
Note. Hydrochloric acid is also secreted, providing an acid medium for pepsin action and killing most bacteria. No absorption except of alcohol and some drugs.					
Duodenum	Pancreas	Pancreatic juice	Trypsin and two other proteinases	Proteins and peptides	Amino acids
			Pancreatic amylase	Starch	Maltose
			Lipase	Fats	Fatty acids and glycerol
Note. Bile emulsifies fats and aids their absorption. Sodium bicarbonate in bile and pancreatic juice reduces acidity: duodenum contents are slightly acid.					
Ileum	Glands in ileum lining between villi	Succus enericus	Peptidase	Peptides	Amino acids
			Lipase	Fats	Fatty acids and glycerol
			Maltase	Maltose	Glucose
			Sucrase	Sucrose	Glucose and fructose
			Lactase	Lactose	Glucose and galactose
Note. Most absorption occurs in the ileum.					
Colon					
Note. Water is absorbed in the colon. No digestive juices are secreted.					

20.11 STORAGE OF DIGESTED FOOD

If the quantity of food taken in exceeds the body's needs for energy or for structural materials, the surplus must be stored.

(a) **Glucose** (Fig. 20.10). The concentration of glucose in the blood of a person who has not eaten for eight hours is usually between 90 and 100 mg per 100 cm^3 of blood. After a meal containing carbohydrate, the blood glucose level may rise to 140 mg per 100 cm^3, but within two hours the level returns to about 95 mg per 100 cm^3. Some of the glucose will have been oxidized to supply energy; the rest will have been removed from the blood as it passes through the liver, and converted into *glycogen,* the molecule of which consists of a long branching chain of glucose units, rather like that of starch. Glycogen is insoluble in water; about 100 g of the compound is stored in the liver and 300 g in the muscles. When the blood glucose level falls below about 80 mg per 100 cm^3, the glycogen in the liver is converted back into glucose which is released into the blood. The muscle glycogen, however, is not normally returned to the blood circula-

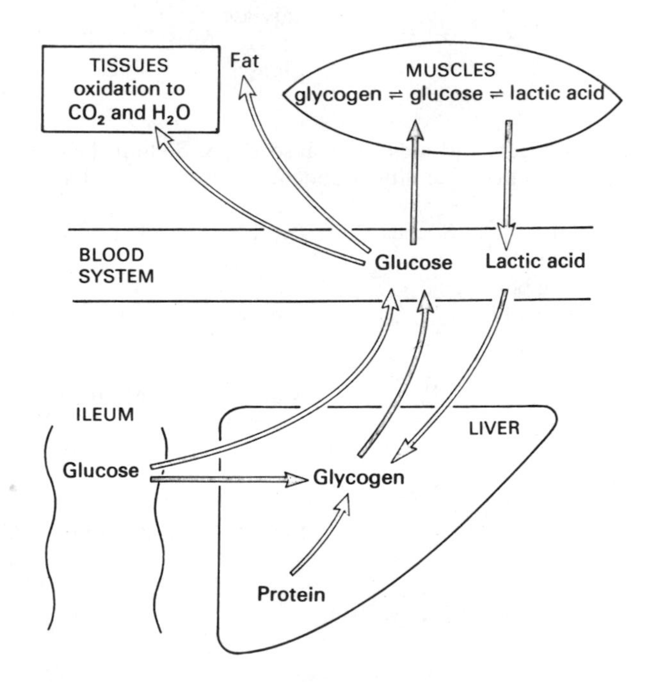

Fig. 20.10 *Carbohydrate metabolism*
(From Bell, Davidson, & Scarborough, *Textbook of Physiology and Biochemistry*, 8th ed, Livingstone, 1972, by permission)

tion but is reconverted to glucose for use as an energy source within the muscles themselves.

If the energy demand is high—for example, during vigorous exercise—insufficient oxygen may be available for complete oxidation of glucose to carbon dioxide and water. Lactic acid, the product of anaerobic respiration (see Section 4.3), may then accumulate in the muscle. It is carried away from the muscle in the blood and returned to the liver, where it may again be converted into glycogen, and stored.

The glycogen in the liver is a "short-term" store; if no other glucose supply is available it will last the body for about six hours. Excess glucose not stored as glycogen is converted into fat and stored in fat cells (see below).

(b) **Fats.** Certain cells can accumulate drops of fat in their cytoplasm. As these drops increase in size and number, they join together to form one large globule of fat in the middle of the cell, pushing the cytoplasm into a thin layer and the nucleus to one side (Figs. 20.11 and 20.12). Groups of fat cells form *adipose tissue* beneath the skin and in the connective tissue of most organs (Fig. 24.3).

Unlike glycogen, there is no limit to the amount of fat that can be stored in the body, and because of its high energy value it is an important food reserve.

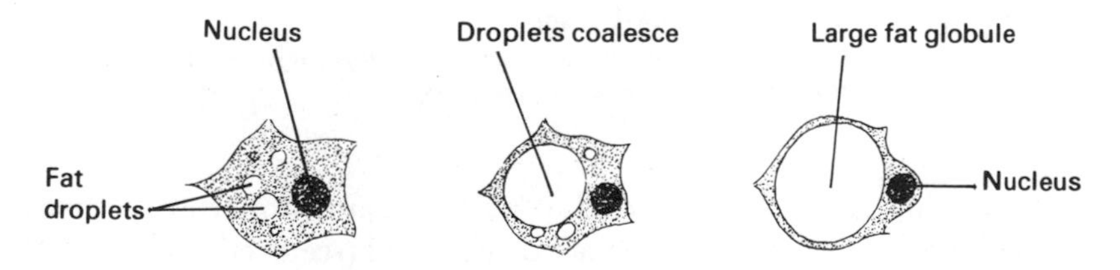

Fig. 20.11 *Accumulation of fat in a fat cell*

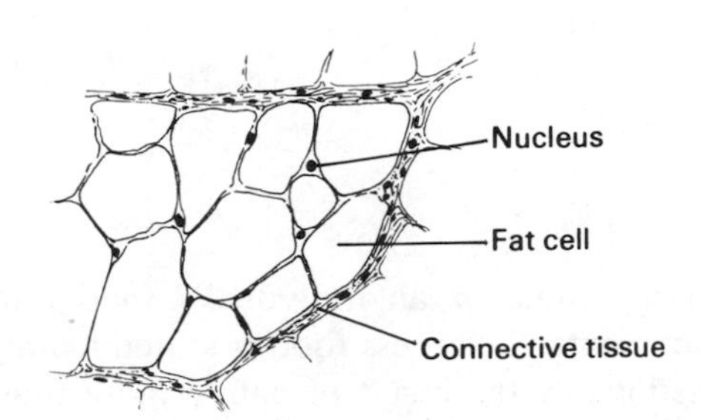

Fig. 20.12 *Small section of adipose tissue*

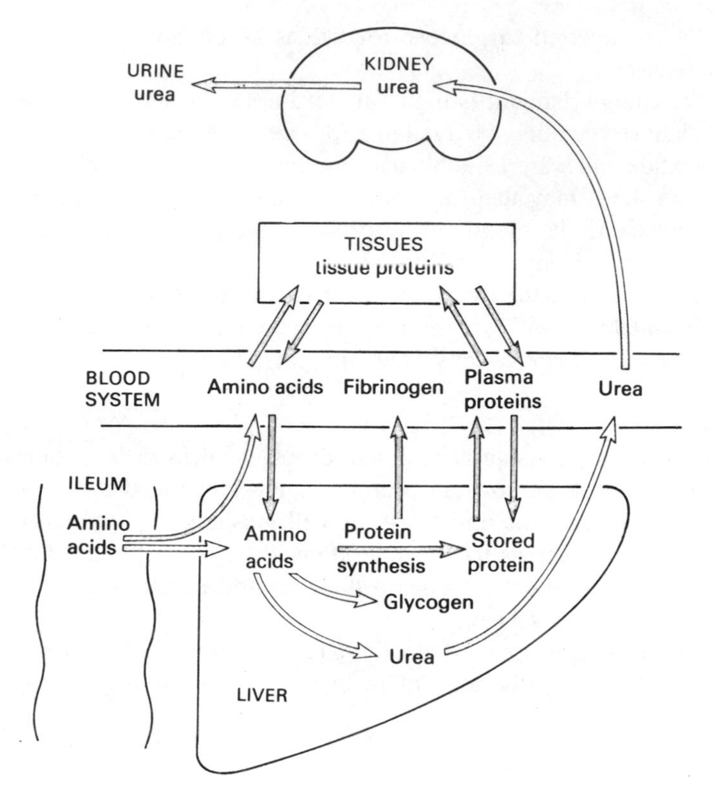

Fig. 20.13 *Protein metabolism*
(From Bell, Davidson, & Scarborough, *Textbook of Physiology and Biochemistry*, 8th ed, Livingstone, 1972, by permission)

(c) **Amino acids** (Fig. 20.13). Amino acids are not stored in the body. Those not used in protein formation are deaminated (see Section 20.12(*d*)). The protein of the liver and tissues can act as a kind of protein store to maintain the protein level in the blood but absence of protein in the diet soon leads to serious disorders.

Overweight

When the energy value of an indivdual's food intake exceeds his body's energy requirements, the excess food is stored mainly as fat: putting on weight is unquestionably the result of eating more food than the body needs. But it is still not fully understood why some people find it difficult to restrict their intake of food to the amount required to meet their needs, nor why some never seem to get fat however much they eat while others put on weight even when their intake only marginally exceeds their requirements. A partial explanation of people's differing reactions probably lies in the balance between the relative amounts of the various hormones produced by their bodies, a balance which is to some extent determined by hereditary factors.

An obese person can lose weight by eating food which supplies him with less energy than his body needs, so that his energy requirements must be met in part from his fat reserves. A "slimming diet" designed to reduce energy intake must, nevertheless, always include sufficient amounts of the essential amino acids, vitamins, minerals and certain essential fatty acids.

20.12 THE LIVER

The liver is a large, reddish-brown organ lying just below the diaphragm and partly overlapping the stomach (Fig. 20.6). It is supplied with oxygen by blood carried to it in the *hepatic artery* (*hepatic* means "of the liver"); in addition it receives all the blood which leaves the walls of the alimentary canal in the *hepatic portal vein* (see Section 20.7). It has a great many important functions.

(a) **Regulation of blood glucose levels.** Glucose enters the blood from the digested food in the ileum. If there is too much glucose in the blood some of it is removed by converting it into insoluble glycogen in the liver. If, on the other hand, the glucose level falls too low, some of the glycogen in the liver is converted by enzyme action into glucose which enters the blood and replenishes the supply (see also Section 20.11(*a*)). If the concentration of glucose rises above 160 mg per 100 cm^3 of blood, glucose is excreted by the kidneys in urine; if it falls below 40 mg per 100 cm^3, the brain cells may be affected by glucose deficiency, leading to convulsions and unconsciousness. By helping to keep the glucose concentration between 80 and 150 mg per 100 cm^3, the liver prevents these undesirable effects.

(b) **Formation of bile.** Bile is produced continuously by liver cells, but it is stored and concentrated in the gall bladder; it is poured into the duodenum when the acid chyme is discharged into it from the stomach. It contains *bile salts* produced by the liver, which are important in the emulsification and subsequent absorption of fats (see Section 20.5); they are reabsorbed with the fats they emulsify and eventually return in the

bloodstream to the liver. Bile owes its color to the green and yellow pigments formed in the daily breakdown of millions of red blood cells; the pigments are removed from the blood by the liver and excreted into the duodenum in the bile, eventually leaving the body in the feces.

(c) **Storage of iron.** The breakdown of red blood cells is completed in the liver. Their red pigment, hemoglobin, is a protein which contains iron; the iron from hemoglobin decomposition is stored in the liver and eventually incorporated into new red blood cells.

(d) **Deamination.** Amino acids over and above the body's needs are not stored as such. Those which are not built up into proteins for the manufacture of new or replacement protoplasm or for enzyme production are converted into carbohydrate by *deamination,* that is, the removal of the nitrogen-containing amino group,—NH_2, from their molecules. The nitrogen of the amino group is converted in the liver into the soluble compound *urea,* $NH_2.CO.NH_2$, which is carried in the bloodstream to the kidneys where it is removed from the blood and excreted in the urine. The carbohydrate residue is converted into glycogen and either stored or oxidized to release energy.

(e) **Manufacture of plasma proteins.** The fluid part of the blood, the *blood plasma,* contains a number of proteins in solution. The liver makes most of these proteins, including fibrinogen, which plays an important part in the clotting of blood (see Section 21.2(*c*)).

(f) **Metabolism of fats.** When the fats stored in the body are needed for the supply of energy, they are carried in the blood from the fat depots to the liver. They are there converted into substances which can be readily oxidized by other tissues to release energy.

(g) **Maintenance of body temperature.** Chemical changes, including those listed above and a great many others, are constantly going on in the liver; many of these release energy in the form of heat. This heat is carried throughout the body by the blood and helps to maintain body temperature.

(h) **Detoxification.** The action of the bacteria on amino acids in the large intestine produces certain poisonous substances which are absorbed by the blood, but which on reaching the liver are converted into harmless compounds that can be safely carried in the bloodstream to the kidneys and are excreted in the urine. Many other substances are similarly modified by the liver before being excreted. Hormones, made by the body itself, and

certain drugs are converted into inactive compounds in the liver, so limiting the period during which they can influence the body's metabolism.

(i) **Storage of vitamins.** The fat-soluble vitamins A and D are stored in the liver. This is the reason why liver, especially fish liver, is a valuable source of these vitamins in the diet. The liver also stores a product of the vitamin B_{12} which is necessary for the normal production of red cells in the bone marrow.

20.13 HOMEOSTASIS

If a mobile, single-celled animal such as *Ameba* or *Paramecium* (Section 14.2) finds itself in conditions which are unfavorable—for example, too acid, too warm or too light—it is capable of moving until it encounters more suitable conditions.

The cells in a multicellular organism cannot move to a fresh environment but are no less dependent on a suitable temperature and pH for the chemical reactions which maintain life. It is therefore essential to their efficient functioning that the medium around them does not alter its composition very much. In fact, the composition and concentration of the body fluids remain remarkably constant, with only minor fluctuations. Even though the food eaten may vary from day to day in its amount and chemical nature, the composition of the blood, and of the fluid which bathes all the cells of the body, changes comparatively little. This constancy of the *internal environment,* as it is called, is of great importance to the smooth functioning of the cells of the body. If it were not maintained, the chemical changes in their protoplasm would become so erratic and unpredictable that quite a slight change of diet or activity might bring about a complete breakdown of the body's biochemistry. Enzymes, for example, can only function properly within a certain narrow range of acidity or alkalinity; if the acidity of the body's fluids were allowed to rise beyond normal levels, the activity of essential enzymes would be inhibited and life could not continue. A rise in the concentration of the body fluids might produce a disastrous withdrawal of water from the cells by osmosis; a fall in body temperature could slow down vital chemical reactions.

The regulation of the internal environment is called *homeostasis.* Section 20.12 makes it clear that the liver plays a vital part in the process, in the several ways in which it maintains the composition of the blood. The homeostatic functions of the blood and the kidneys are discussed respectively in Sections 21.2(*a*) and 23.5, the role of the skin in the maintenance of body temperature in Section 24.2, and the control of many homeostatic processes by hormones in Section 29.4.

20.14 PRACTICAL WORK

Experiment 20.1 *To investigate the action of pepsin on egg-white*

Egg-white is a solution of protein in water, the protein coagulates on heating. The white of one egg is stirred into 500 cm^3 tap-water. The mixture is boiled and filtered through glass wool to remove large particles. Four test-tubes are labeled *A, B, C* and *D,* and each is filled to a depth of about 2 cm with the cloudy protein suspension. A 1 per cent solution of commercial pepsin is prepared. 1 cm^3 pepsin solution and three drops bench "dilute" hydrochloric acid are added to test-tube *A*. 1 cm^3 pepsin solution only is added to *B,* and three drops hydrochloric acid only to *C*. Some pepsin solution is boiled (this inactivates the enzyme) and 1 cm^3 of the boiled solution and three drops hydrochloric acid are added to test-tube *D*. All four test-tubes are placed in a beaker of warm water (35–40 °C) for five or ten minutes.

Result. The cloudy liquid in test-tube *A* becomes clear. No change can be seen in the other three tubes.

Interpretation. The change from a cloudy suspension to a clear solution suggests that the solid egg-white particles have been digested to soluble products. This change is not brought about by the action of pepsin alone (test-tube *B*) nor by the action of the acid in the absence of any enzyme (test-tube *C*). The experiment with boiled pepsin solution shows that commercial pepsin contains no digestive substance other than the enzyme. The clearing of the liquid in test-tube *A* supports the idea that pepsin brings about the digestion of egg-white in acid conditions.

Experiment 20.2 *To investigate the action of saliva on starch*

Saliva is collected in a test-tube, after rinsing the mouth to remove traces of food. About 2 cm^3 5 per cent starch solution are placed in each of four test-tubes labeled *A, B, C* and *D*. About 1 cm^3 of saliva is added to each of tubes *A* and *B* and about 1 cm^3 boiled saliva is added to each of tubes *C* and *D*. After five minutes, the contents of *A* and *C* are tested with iodine, and those of *B* and *D* with Benedict's solution (see Section 19.7).

Result. Iodine gives the blue color characteristic of starch with the contents of the control test-tube *C*, but not with those of *A*. The contents of tube *B* give a red precipitate with Benedict's solution, indicating the presence of a sugar, but those of *D* do not react.

Interpretation. The results of the experiment support the theory that saliva contains a substance that brings about the conversion of starch to a sugar; the substance is inactivated on boiling, indicating that it is probably an enzyme.

Experiment 20.3 *To investigate the effect of pH on an enzyme*

Approximately 0.1M solutions of acid and alkali are required: these are respectively prepared (*a*) by making 1 cm^3 bench "concentrated" hydrochloric acid up to 100 cm^3 with water, and (*b*) by dissolving 0.8 g sodium bicarbonate in a little water, and making up to 100 cm^3. Six test-tubes are labeled *A* to *F*, and 5 cm^3 1 per cent starch solution are placed in each. The pH (degree of acidity or alkalinity) is then adjusted by adding alkali or acid as follows: ten drops and four drops of 0.1M sodium bicarbonate solution respectively to *A* and *B;* six drops, seven drops and eight drops of 0.1M hydrochloric acid respectively to *D, E* and *F*. The pH of *C* is left unaltered. Rows of drops of iodine solution are placed on a white glazed tile. Saliva is collected as before, and 1 cm^3 is added to each tube. A drop of solution is withdrawn on the end of a glass rod from each test-tube in turn, and each is added to one of the iodine drops; the testing is repeated at intervals. Initially all the samples show a blue color on testing with iodine, but when a sample just *fails* to give a blue color it may be assumed that all its starch has been digested. The time taken for the starch to disappear from each solution is noted.

Result. Digestion is most rapid in test-tube *C*, that is, in neutral conditions.

Interpretation. The digestive enzyme of saliva (salivary amylase) acts most effectively in neutral or nearly neutral solutions.

Experiment 20.4 *To investigate the effect of temperature on an enzyme reaction*

Place 10 cm^3 2 per cent starch solution in each of three test-tubes, *A, B* and *C*. *A* is placed in a beaker of ice-water, *B* in cold tap-water and *C* in warm water (about 40 °C). All three are left to stand for five minutes to allow their contents to reach the temperature of their surroundings; meanwhile, iodine drops are arranged on a tile as in Experiment 20.3. 1 cm^3 saliva is then added to each test-tube, and samples are tested at intervals as before. The time taken for starch to disappear from each solution is noted.

Result. The starch disappears most quickly from the solution in tube *C*.

Interpretation. Of the temperatures tried, that most favorable to the digestion of starch by salivary amylase is approximately body temperature (35–40 °C).

Note. These experiments and several others are fully detailed in the laboratory manual *Enzymes;* see *Further Reading* at the end of this book.

20.15 QUESTIONS

1. List the chemical changes undergone by (*a*) a molecule of starch from the time it is placed in the mouth to its ultimate use in providing energy, (*b*) a molecule of protein from the time it is swallowed to the time when its components are used in a cell (other than in the liver).
   In each case, state where the changes are taking place.
2. Write down the menu for your breakfast and lunch (or supper); indicate the principal food substances present in each component of the meal and state the final digestion product of each and the use your body is likely to have made of them.
3. What advantages is it to an animal to take food into its alimentary canal for digestion rather than digest it externally as do the fungi?
4. Herbivorous animals have very long intestines with a large cecum and appendix, but carnivorous animals have a relatively shorter intestine with small cecum and appendix. In what ways are these differences related to the differences in diet?

Unit 21
Blood And the Circulatory System

21.1 THE COMPOSITION OF BLOOD

The body of an adult man contains between five and six liters of blood. It consists of a watery fluid, *plasma,* in which are suspended large numbers of cells of several different kinds.

(a) Red Cells (erythrocytes)

Red blood cells (Figs. 21.1(*a*) and 21.2) are minute discs, concave on both sides, consisting of spongy cytoplasm contained in a thin elastic "skin." They have no nuclei. Their cytoplasm is colored by a red pigment,

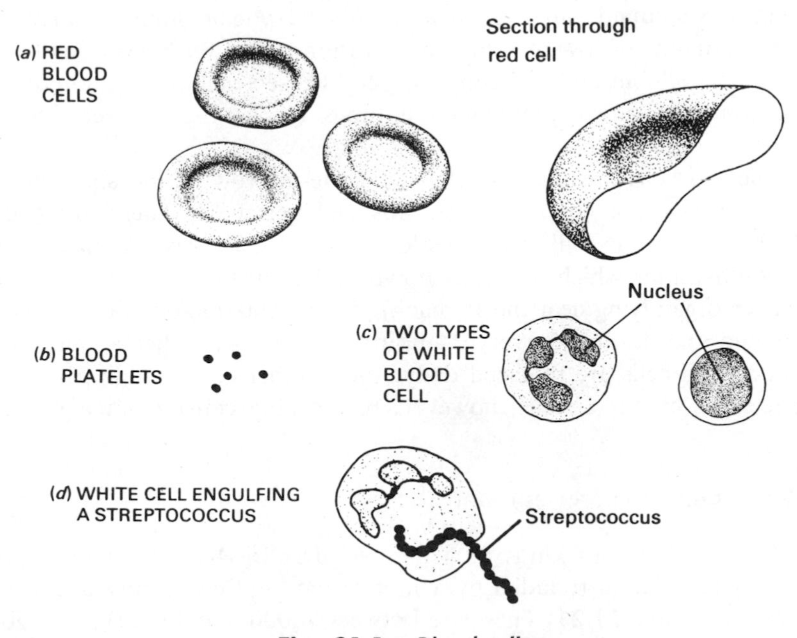

Fig. 21.1 *Blood cells*

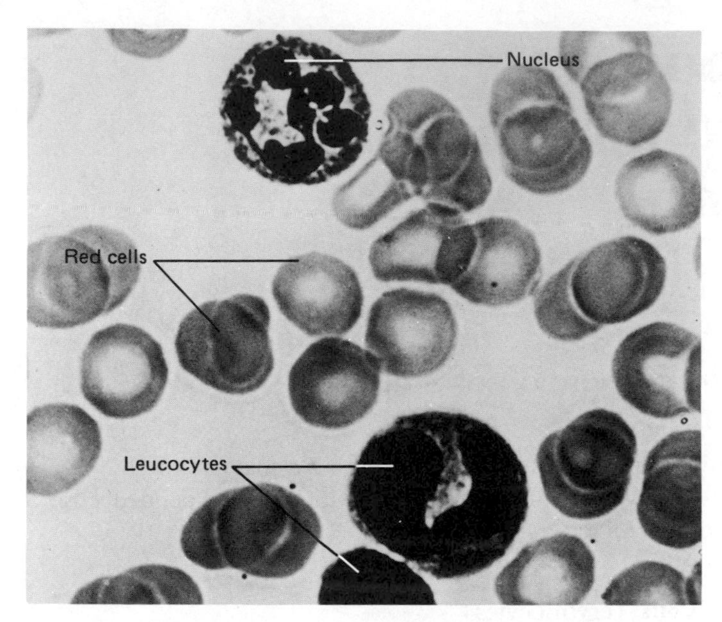

Fig. 21.2 *Red and white cells from human blood*
(Gene Cox)

hemoglobin, which is a protein containing iron atoms in its molecule. It readily combines with oxygen in conditions of high oxygen concentration, forming a compound called *oxyhemoglobin.* Oxyhemoglobin is unstable, and in conditions of low oxygen concentration it rapidly breaks down, reforming hemoglobin and releasing oxygen. These reactions are important in the transport of oxygen from the lungs to the tissues (see Section 21.2(*b*)).

The red cells are made in the red bone-marrow of the short bones such as the sternum (breast-bone), ribs and vertebrae. There are about 5,500,000 in a cubic millimeter of blood. A red cell lives for only about four months, after which it breaks down and disintegrates in the liver or spleen (an organ lying near the stomach). About 200,000,000,000 red cells are formed and destroyed every 24 hours, which means that about 1 per cent of all red cells are replaced daily; the number of red cells per cubic millimeter of blood remains, however, remarkably constant in a healthy person.

(*b*) **White Cells** (leucocytes)

There are various kinds of white blood cells. All consist of unpigmented cytoplasm, surrounded by a membrane, and containing a nucleus (Figs. 21.1(*c*) and 21.2). There are between 4,000 and 13,000 in a cubic millimeter of blood, that is, there are about 600 red cells to every white cell. White cells are made in the bone-marrow, the lymph nodes (see Section 21.3(*c*)) and the spleen. Most of them are of a type called *phagocytes* which can move by a flowing action of their cytoplasm, rather like an *Ameba.* They can ingest and destroy bacteria and dead cells by flowing around them and engulfing them in cytoplasm, and then digesting them by enzyme action (Fig. 21.1(*d*)). Phagocytes can pass out of blood capillaries by squeezing between the cells of the capillary wall; they move to the site of an injury of infection and accumulate there, destroying invading bacteria and damaged tissue, and preventing the spread of harmful bacteria to other parts of the body as well as accelerating the healing of the injured region.

(*c*) Platelets

These round or oval structures are very much smaller than red or white cells (Fig. 21.1(*b*)). There about 400,000 of them in a cubic millimeter of blood; they are cells without nuclei budded off from special, very large cells in the red bone-marrow, and they play an important part in the clotting action of the blood.

(*d*) Plasma

Plasma is the liquid part of the blood. It is a solution in water of many compounds; they include *mineral salts,* especially sodium chloride and sodium bicarbonate, and *soluble proteins,* such as albumin, fibrinogen (see Section 21.2(*c*)) and the globulin antibodies (see Section 21.2(*d*)). It also contains *hormones* (see Section 29.4) and the products of the digestion of food, mainly *glucose* and *amino acids,* together with the waste products of metabolism, *carbon dioxide* and *urea.*

Blood serum is blood plasma from which the fibrinogen has been removed.

21.2 THE FUNCTIONS OF BLOOD

(*a*) Homeostatic Functions of Blood

Every living cell in the body is bathed with a fluid, the *tissue fluid,* which supplies it with the food and oxygen necessary for its metabolism and removes the waste products that could poison the cell if they accumulated. Tissue fluid is derived from blood plasma (see Section 21.3(*c*)); it contains a lower concentration of plasma proteins, but similar amounts of minerals, glucose and amino acids. Since the levels of these compounds in blood plasma are kept very precisely within certain limits by the liver (see Section 20.12) and the kidneys (see Section 23.3), the body's cells are in contact with a liquid of almost constant composition. This provides them with the environment they need and enables them to live and grow in the most favorable conditions. By delivering oxygen and nutrients to the tissue fluid and removing the waste products of metabolism, the blood is important in maintaining the constancy of the internal environment.

In addition the blood has an important function in the maintenance of the body at an even temperature; this is discussed in Section 21.2(*b*).

(*b*) The Blood as a Transport System

Blood circulates through every part of the body in the vessels of the *circulatory system* (see Section 21.3); on the average, a particular red cell completes the circulation of the body in 45 seconds. This movement constantly brings fresh supplies of oxygen and nutrients to the tissue fluid as fast as they are used up, and removes the poisonous waste products before they can accumulate to harmful levels.

(*i*) Transport of oxygen from the lungs to the tissues. Oxygen is carried in the blood as oxyhemoglobin, formed when blood is exposed to the relatively high concentrations of oxygen in the lungs. Oxygen is appreciably soluble in water. It dissolves in the film of moisture lining the lungs, diffuses into the plasma in the capillaries there, and hence enters the red cells, where it reacts immediately with the hemoglobin in their cytoplasm to form oxyhemoglobin (see also Section 22.3). This prompt removal of oxygen from solution maintains a steep diffusion gradient between the source of oxygen and the red cell, and thus ensures rapid diffusion of oxygen into the cell.

Oxyhemoglobin is a bright red compound, while hemoglobin itself is dark red; oxygenated blood is thus a much brighter red in color than deoxygenated blood. When blood containing oxyhemoglobin reaches a tissue where oxygen is being used up, the pigment breaks down into hemoglobin and free oxygen; the oxygen diffuses through the walls of the capillary blood vessels in the tissue, into the tissue fluid and so to the cells.

(*ii*) Transport of carbon dioxide from the tissues to the lungs. All living cells produce carbon dioxide as a waste product of their respiration (see Section 4.1). It diffuses from the cells into the tissue fluid and thence into the blood plasma. Some of it is carried in solution in the plasma as sodium bicarbonate, and some in the cytoplasm of the red cells. It is released in the lungs (see Section 22.3) where it diffuses through the lining of the lungs into the lung cavity, and is expelled from the body in the expired air.

(iii) **Transport of excretory material from the tissues to the kidneys.** The waste products of living cells must be removed from the body, or *excreted,* before they can accumulate to harmful levels; carbon dioxide is excreted by the lungs and most other excretory material by the kidneys. These substances diffuse from the cells into the tissue fluid and thence into the blood plasma, which carries them away in solution. When they eventually reach the kidneys a large proportion is removed and excreted (see Section 23.3).

(iv) **Transport of digested food from the ileum to the tissues.** The soluble products of digestion pass into the capillaries of the villi lining the ileum (Section 20.7). They are carried in solution by the plasma and after passing through the liver enter the general circulation. Glucose and amino acids diffuse out of the capillaries and into the cells of the body. Glucose may be oxidized in a muscle, for example, and provide the energy for contraction; amino acids will be built up into new proteins and make new cells and fresh tissues.

(v) **Distribution of hormones.** Hormones are chemical compounds produced by certain glands, the *endocrine glands* (see Section 29.4). Each hormone affects the rate of one or more specific vital processes in the body. They are carried away from the endocrine glands in solution in the blood plasma to every part of the body, although each hormone exerts its effect on only one organ or group of organs. For example, *adrenaline* secreted by the adrenal glands near the kidneys speeds up the heart-beat and the breathing rate, while *antidiuretic hormone* is secreted by the pituitary gland at the base of the brain, and controls the amount of water excreted by the kidneys.

(vi) **Distribution of heat.** Many of the chemical reactions of the body release energy in the form of heat. Such processes occur more rapidly in some parts of the body than others: for example, considerable amounts of heat are generated by the chemical activities of the liver and by the functioning of muscular organs such as the limbs or the heart. The heat so produced locally is distributed all around the body by the blood, so that local overheating is avoided and an even temperature is maintained in all parts of the body.

The temperature of the body is partly controlled by the regulation of heat loss through the skin. The diversion of blood to or from the skin is important in the process, and is discussed in Section 24.2.

(c) Formation of Clots

When a blood vessel is cut open or injured, the blood platelets and damaged tissue produce chemicals which help to convert the soluble protein *fibrinogen* into insoluble *fibrin*. The blood platelets adhere to the damaged area, and threads of fibrin "crystallize out" onto them, forming a network fiber across the wound. Red cells become entangled in the network and a clot is formed, which stops further loss of blood, and prevents the entry of bacteria and poisons. The dried clot evenutally becomes a scab, which protects the injured area while new skin is forming.

(d) Prevention of Infection

(i) **Infected wounds.** Normally, the skin provides a barrier to the entry of any bacteria. The layer of dead cells on the skin provides a mechanical barrier while the mucus and digestive juices of the alimentary canal offer a chemical defense. If the skin is broken, however, the bacteria enter the cut, certain of the white cells migrate through the capillaries in that region and begin to engulf and digest any bacteria that have invaded the tissues. Many dead white cells and self-digested, dead tissues may accumulate at the site of infection and form pus. In this way, and as a result of clot formation which prevents free circulation, the site of the infection is localized and most of the bacteria are destroyed before they can enter the general circulation. Those which escape into the lymphatic system (see Section 21.3(*c*)) are trapped by the large numbers of white cells in the lymph nodes or in the spleen and liver.

(ii) **Disease and immunity.** Many diseases are caused by the presence of bacteria or viruses which may have entered the body through wounds, in contaminated food, water or air, or by other means (see Section 11.3). Some virulent strains of bacteria cannot be ingested by the white cells until they have been acted upon by *antibodies,* proteins released into the blood plasma by certain specialized white cells when stimulated by substances called *antigens* produced by the bacteria themselves. If the appropriate antibodies are not already present in the blood, or if they cannot be made quickly enough, the harmful bacteria may invade the whole body, and give rise to disease symptoms. These symptoms may be due to one or more of the following:

foreign proteins of the bacteria themselves;
poisonous chemicals or *toxins* (usually proteins) secreted by the bacteria;
breakdown products of infected tissue.

Recovery from the disease depends to a large extent on the production of antibodies in the blood. There are several kinds of antibodies, with differing kinds of protective action:

Opsonins adhere to the outer surface of bacteria and so make it easier for the phagocytic white cells to ingest them;

Agglutinins cause bacteria to stick together in clumps, and in this condition the bacteria cannot invade the tissues;

Lysins destroy bacteria by dissolving their outer coats; and

Antitoxins combine with and so neutralize the poisonous toxins produced by bacteria.

When a person recovers from a disease, the antibodies he has produced remain in his blood for a short time only. However, the white cells' ability to produce them is greatly increased, so that a further invasion by the same species of bacteria or virus is likely to be stopped at once. The individual is then said to be "immune" to the particular disease. People may possess this immunity from birth, or they may acquire it after recovering from an attack of the disease; immunity may also be acquired when disease bacteria are present in the body even though they may not be sufficiently numerous or suitably placed to produce symptoms of illness.

Immunity to a disease may also be induced artificially by *vaccination* or *inoculation.* A *vaccine* is a preparation of bacteria or viruses which have been killed or treated in such a way as to prevent their reproduction. When these are injected into the body the organism undergoes, in effect, a mild form of the disease, often without any apparent symptoms. Its white cells are stimulated to manufacture antibodies, and subsequently retain the ability to produce them rapidly. In this way the body has artificially acquired immunity, although the period for which the immunity is retained varies with the infection concerned from a few months to many years.

The blood of a person or an animal which has recently recovered from a disease contains considerable amounts of antibodies, especially antitoxins. A serum may be prepared from this blood (by removing the cells and fibrinogen from it) which if injected into a person who has never had the disease may confer a temporary immunity on the recipient because of the antibodies it contains. If he already has the disease the antibodies in the serum will fight the infection and assist his recovery. *Snake-bite* is treated with serum prepared in this way from horse's blood. The horse is injected with a very dilute preparation of snake venom, which stimulates the production of the appropriate antitoxins. Samples of blood are later taken from the horse, and serum prepared from these samples is used to treat snake-bite cases. *Tetanus* is treated in a similar way, using an anti-tetanus serum prepared from the blood of horses that have been injected with diluted tetanus toxin. The study of the immune responses of animals and humans is called *immunology*.

21.3 THE CIRCULATORY SYSTEM

(a) The Blood Vessels

The blood is distributed around the body in blood vessels, most but not all of them tubular, and varying in diameter from about 1 cm to 0.001 mm. They branch and rejoin to form a continuous system, communicating with every living part of the body (Fig. 21.3). Blood circulates

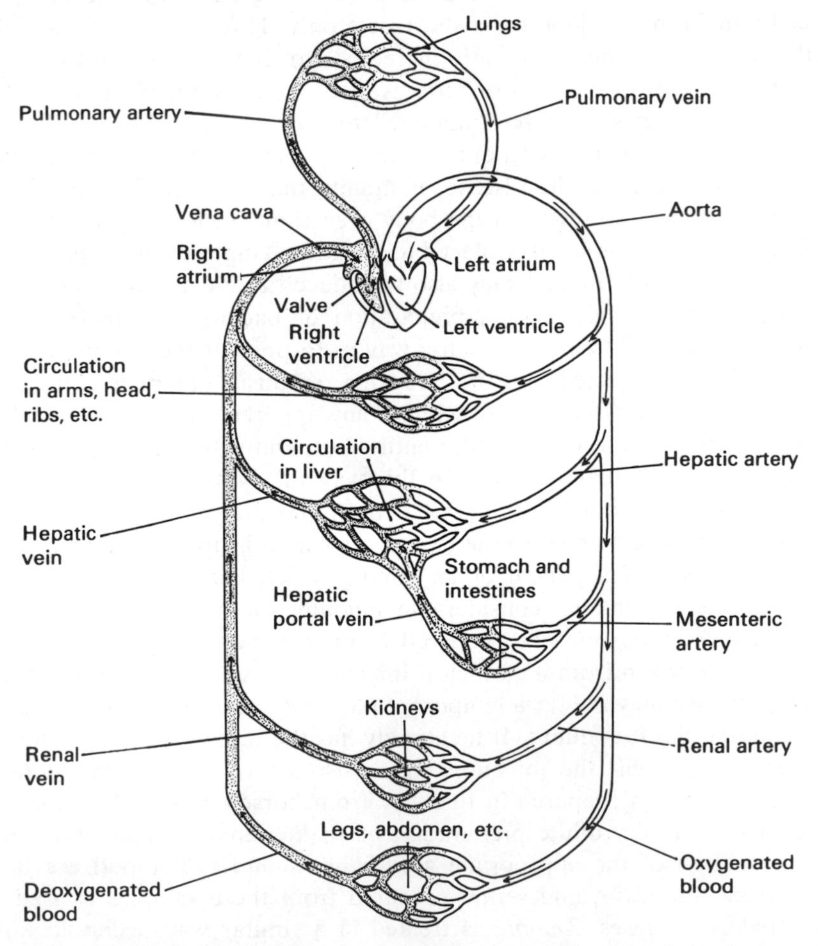

Fig. 21.3 *Diagram of human and other mammalian blood circulation*

around the body, flowing always in the same direction and passing repeatedly through the heart, which maintains the circulation by the pumping action of its muscular contractions.

There are three kinds of blood vessel: *arteries, veins* and *capillaries,* different in both structure and function.

(i) **Arteries** (Figs. 21.4(*a*) and 21.5) are blood vessels which carry blood from the heart to the organs of the body (Fig. 21.6). They are fairly wide and have thick muscular walls that can stand up to the pressure surges caused by the heart's pumping action. The arteries divide and subdivide into smaller vessels called *arterioles,* which themselves divide repeatedly until they form a dense network of microscopic vessels penetrating between

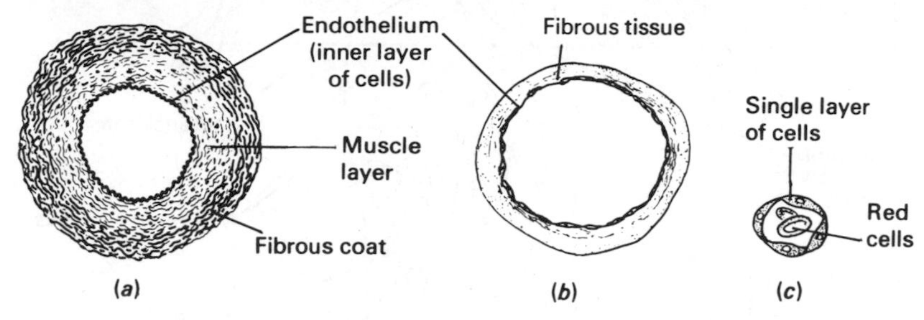

Fig. 21.4 *Transverse sections of blood vessels: (a) an artery; (b) a vein; (c) a capillary (not drawn to scale)*

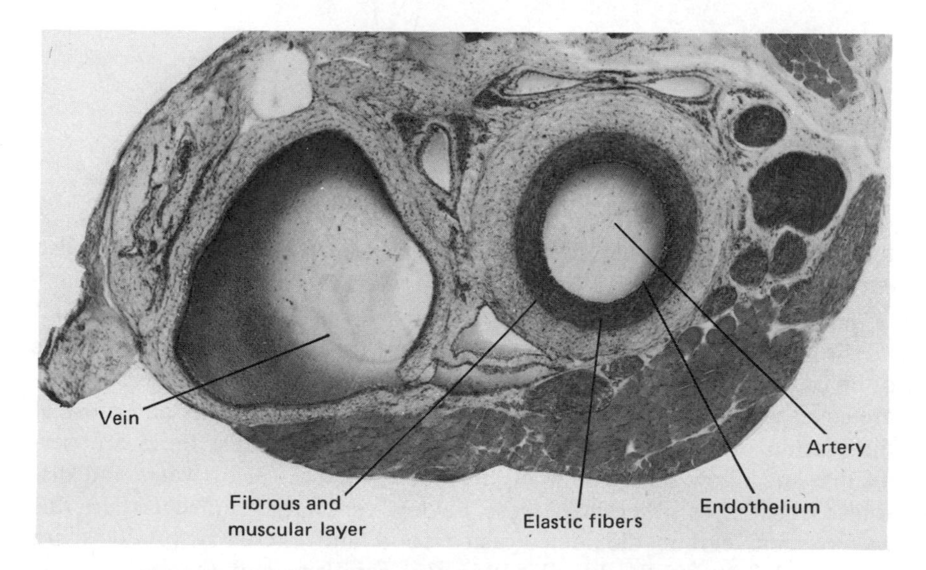

Fig. 21.5 *Transverse section through an artery and a vein* (GBI Laboratories Ltd.)

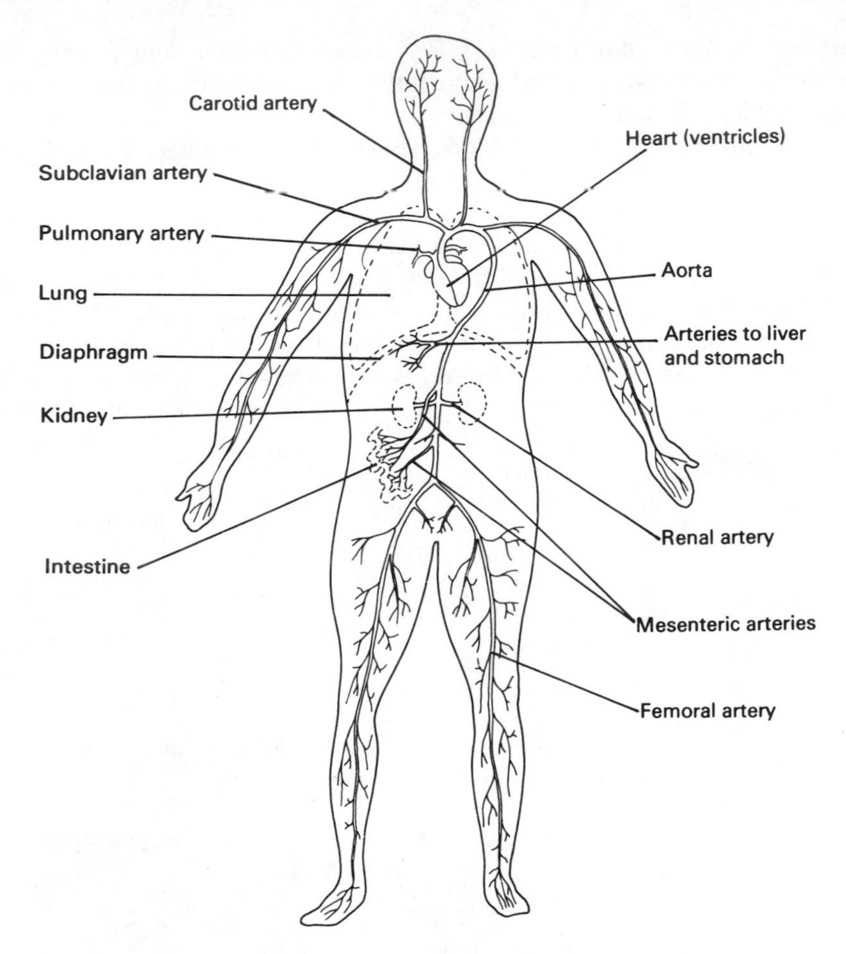

Fig. 21.6 *Diagram of human arterial system*

the cells of every living tissue; these minute blood vessels are called *capillaries,* a name derived from the Latin word meaning "a hair."

(ii) **Capillaries** (Figs. 21.4(*c*) and 21.7) are tiny blood vessels, with walls often only one cell thick. Some capillaries are so narrow that the red cells are squeezed flat in passing through them. Although red cells and most of the plasma proteins are contained by the capillary walls, these are permeable and allow the passage in and out of the vessel of water and dissolved substances other than those having very large molecules. In this way oxygen, carbon dioxide, digested food and excretory products are exchanged between the blood within the capillary and the cells of the tissues outside it (see Section 21.3(*c*)). The capillary network is so dense that no living cell is far from a supply of oxygen and food. In the liver

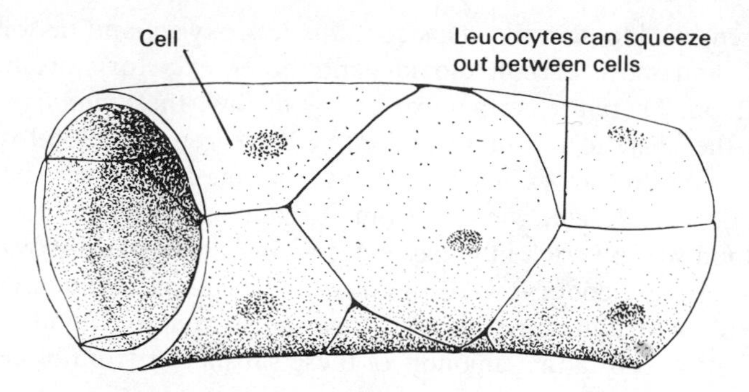

Fig. 21.7 *Stereogram of blood capillary*

every cell is in direct contact with a capillary. Eventually, the capillaries unite together to form larger vessels, *venules,* which join to form still larger vessels, the *veins.*

(iii) **Veins** (Figs. 21.4(*b*) and 21.6) return blood from the tissues to the heart. They are wider and have thinner walls than the arteries; the pressure of the blood in them is steady and lower than in the arteries. The movement of blood toward the heart is assisted by the contraction of body muscles adjacent to the veins during the body's normal activities. Blood movement away from the heart is prevented by *valves,* flaps of tissuc attached to the walls of the vein which open readily to admit blood passing toward the heart but close if there is any tendency for blood to move in the reverse direction (Fig. 21.8). Sedentary people have poor circulation, especially in their extremities, due to blood accumulating in their veins and capillaries.

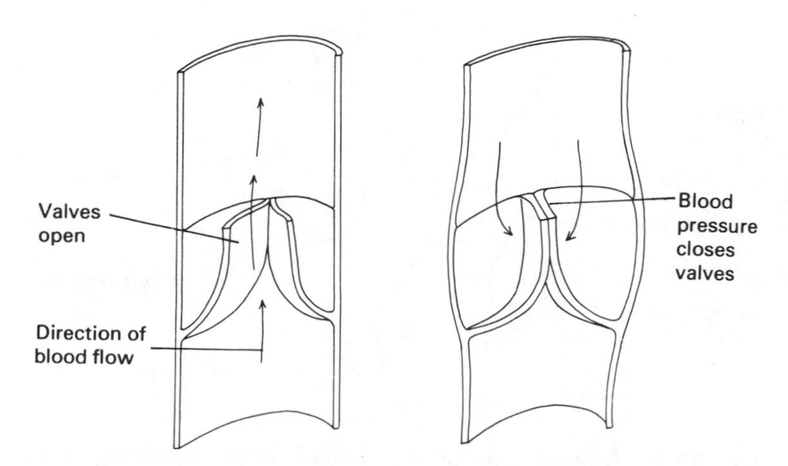

Fig. 21.8 *Diagram to show the action of valves in a vein*

In general, blood in the veins contains less oxygen and dissolved food materials, and more carbon dioxide and other excretory products than arterial blood. There are certain exceptions to this: the *pulmonary arteries* run from the heart to the lungs and carry deoxygenated blood, while the *pulmonary veins* return oxygenated blood from the lungs back to the heart. The *hepatic portal vein,* running from the alimentary canal to the liver, carries blood which is rich in glucose and amino acids, and the *renal veins,* leading from the kidneys toward the heart, carry blood containing smaller amounts of urea and mineral salts than most venous blood, since the kidneys remove substantial amounts of these substances from blood passing through them.

(b) The Heart

The heart is a muscular pumping organ. It is divided into four chambers, the left and right *atria* and the left and right *ventricles;* there is no direct communication between the left and right sides. The upper chambers, the atria, are relatively thin-walled; the walls of the ventricles are thicker, especially that of the left ventricle, which pumps blood around the entire body. The muscles of the ventricles require constant and ample supplies of food and oxygen to maintain their activity, and they are provided with a plentiful blood supply by the *coronary arteries* (Fig. 21.9). Heart attacks involve a blockage of these arteries usually.

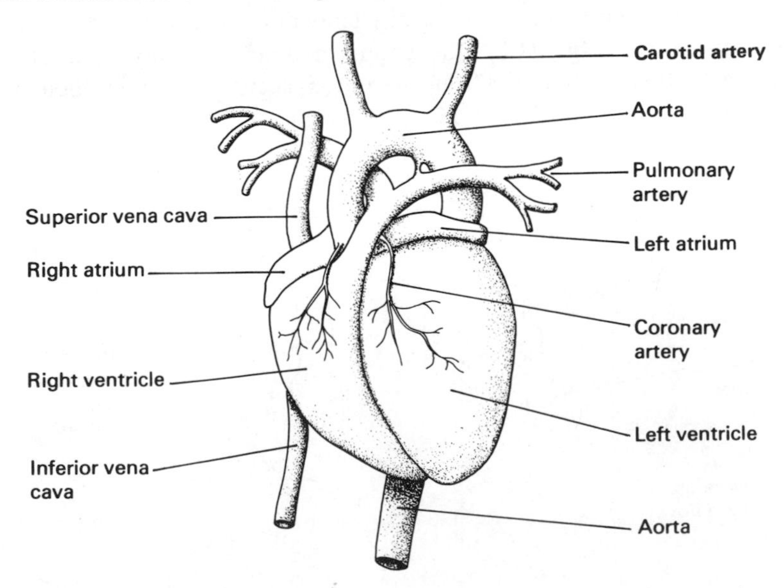

Fig. 21.9 *External view of mammalian heart (pulmonary veins not shown)*

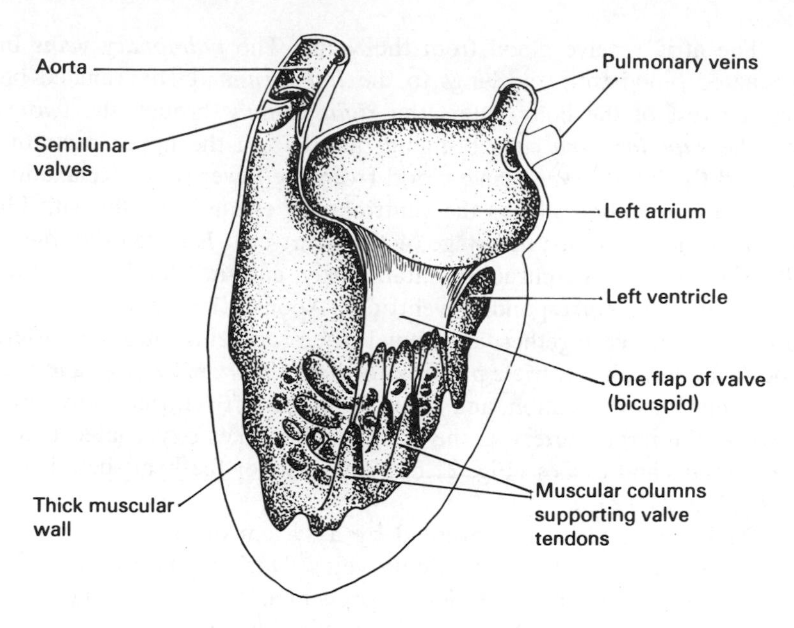

Fig. 21.10 *View of heart cut open*

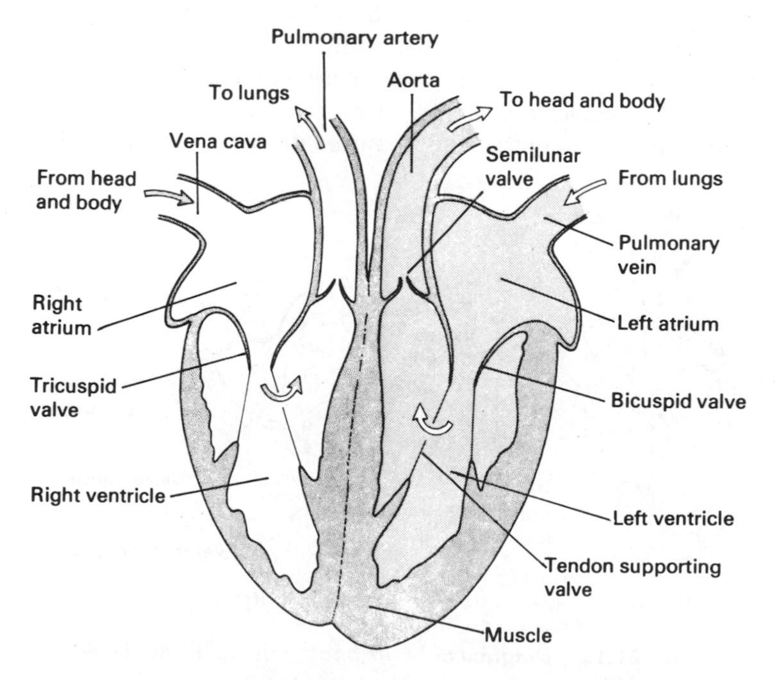

Fig. 21.11 *Diagram of heart, longitudinal section*

The atria receive blood from the veins. The *pulmonary veins* bring oxygenated blood from the lungs to the *left atrium.* Deoxygenated blood from the rest of the body enters the *right atrium* through the two *vena cava,* the *superior vena cava* bringing blood from the upper parts of the body and the *inferior vena cava* blood from the lower parts. Relaxation of the ventricular muscle allows the ventricles to expand and fill with blood from the atria and veins; this stage of the heart-beat is called *diastole.* The walls of the two atria contract simultaneously, and the blood each contains is forced into the corresponding ventricle. About 0.1 second later the two ventricles contract together and expel their contents into the arteries: blood from the right ventricle passes into the *pulmonary arteries* and thence to the lungs for oxygenation, and blood from the left ventricle is forced into the *aorta,* the largest artery in the body, which takes oxygenated blood to all the organs and tissues (Fig. 21.6). This part of the heart-beat is called *systole* (Fig. 21.12).

Backflow of blood is prevented by a system of valves, which act in a similar way to the valves in the main veins. During diastole, blood in the aorta and the pulmonary arteries is prevented from returning into the relaxed ventricles by the closure of the pocket-like *semilunar valves* (Figs. 21.10 and 21.11), so called because the flaps of tissue of which they consist are roughly half-moonshaped. During systole these valves open to allow blood to enter the arteries, while blood is prevented from passing back into the atria by the closure of the parachute-like *bicuspid valve* in the left ventricle and the *tricuspid valve* in the right.

In an adult person at rest, the heart contracts about 70 times a minute, but the rate increases to 100 or more during activity or excitement. The heart's rhythmic contraction is automatic and can proceed without

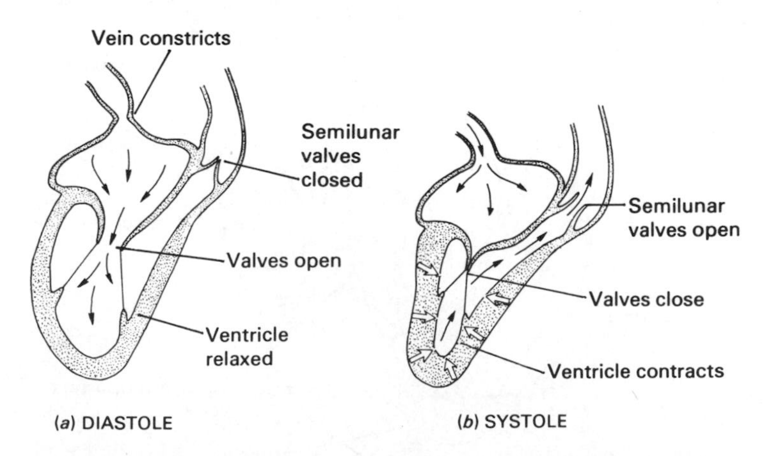

Fig. 21.12 *Diagram of heart beat (only right side shown)*

any stimulation from the nervous system; if a mammalian heart is removed from the body and provided with an artificial circulation, it will continue beating for several hours. Nervous stimulation is, however, superimposed on the heart's natural rhythm and helps to maintain and control its rate. An increased heart-rate increases the speed at which blood is supplied to the tissues, and so allows a greater rate of activity.

Blood pressure. The pressure of the atmosphere on the outer surface of the body tends to flatten the blood vessels, and a fairly high pressure must be developed by the heart in order to overcome this effect and to force blood through the narrow vessels of the capillary system. This pressure varies according to the part of the body considered and the age of the individual, but an average pressure (over atmospheric) produced in the ventricle when it contracts is equal to 130 mm of mercury (about one-sixth of an atmosphere).

For convenience, blood pressure is normally measured in the main artery of the arm. An instrument called a *sphygmomanometer* is used: this consists of a hollow rubber tube connected to a pressure-measuring device. The tube is wrapped firmly around the upper arm and inflated with a pump. When the pressure exerted by the tube against the tissues of the arm just equals the pressure in the artery, the blood flow is cut off and the pulse at the wrist disappears. In practice this point is not easy to detect precisely; the pressure is therefore first increased beyond the point where blood ceases to flow and then allowed to fall slowly. The pressure in the tube is read at the point at which the pulse again becomes perceptible. This reading is a measure of the maximum blood pressure in the artery, that is, the pressure at systole.

(c) Exchange between Capillaries, Cells and Lymphatics

At the arterial end of the capillary network the pressure of blood in the capillaries is high, and fluid is forced out of the vessels through their thin walls (Fig. 21.13). The liquid so expelled has a composition similar to that of plasma, containing dissolved glucose, amino acids and salts, but it has a much lower concentration of proteins than plasma. This exuded fluid bathes the surfaces of the cells of all living tissues, and is called *tissue fluid*. From it the cells extract the oxygen and nutrients they need for the biochemical processes of living, and into it they excrete their carbon dioxide and waste products.

The narrow capillaries offer considerable resistance to the flow of blood. This slows down the movement of blood, so facilitating the exchange of substances by diffusion between the plasma and tissue fluid (Fig. 21.13).

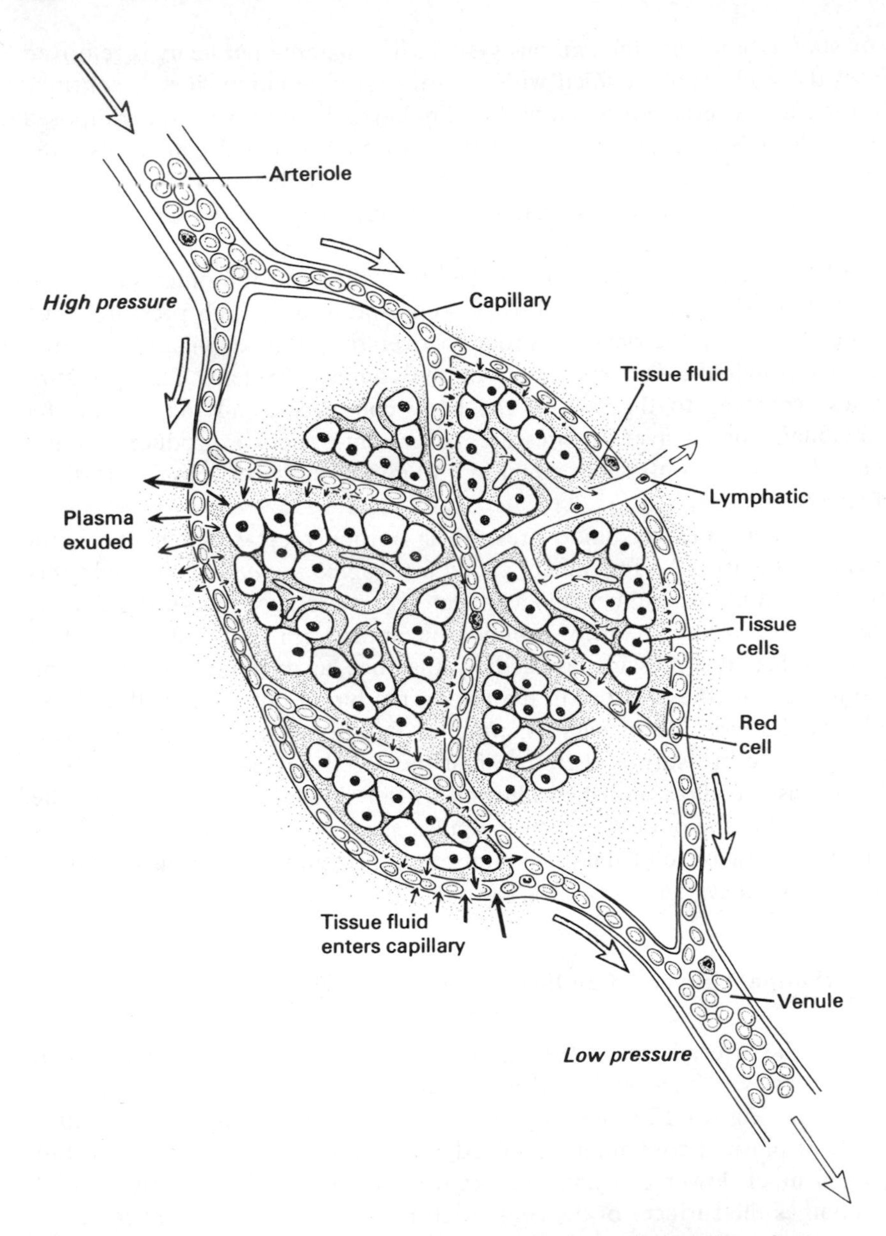

Fig. 21.13 *Relationship between capillaries, cells, and lymphatics*

The capillary resistance also results in a drop of pressure so that at the venous end of a capillary bed the blood pressure is less than that of the tissue fluid and much of the latter passes back into the capillaries.

The fact that the plasma contains more proteins than the tissue fluid gives the blood a low osmotic potential (Section 3.3) which tends to cause water to pass from the tissue fluid into the capillary. At the arterial end of the capillary network the blood pressure is greater than this osmotic pressure, so forcing water out, but at the venous end water from the tissue fluid enters the capillary by osmosis.

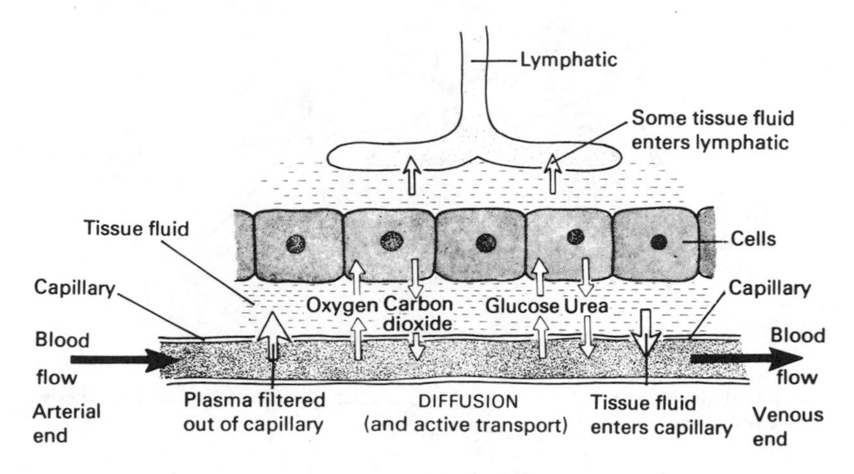

Fig. 21.14 *Blood, tissue fluid, and lymph*

Lymphatic system. Part of the tissue fluid returns to the main circulation by the capillaries, as described above. Some of it, however, drains into blindly ending thin-walled vessels, *lymphatic vessels,* lying between the cells of the tissues. The lymphatic vessels join up to form larger vessels which eventually unite into two main ducts, *lymphatic ducts,* which empty their contents into the large veins entering the right atrium. At various points along the lymph vessels are small organs called *lymph nodes* (Fig. 21.15), which have an important part in protecting the body against infection: they destroy bacteria and manufacture antibodies, and also make new white blood cells. Body infections often lead to a swelling and tenderness of lymph nodes. Tonsils are the body's largest lymph glands located at the back of the mouth on either side of the throat.

The fluid in the lymphatic vessels is called *lymph.* Its composition is similar to that of blood plasma, but it contains less protein. No red blood cells are present, but it contains special white cells, *lymphocytes,* made in the lymph nodes and especially concerned with the production and transport of antibodies. After a meal containing fats the lymph is a milky-white color due to the fat droplets absorbed in the lacteals of the intestinal lining (see Section 20.7), which drain into the lymphatic system.

Lymph flows in one direction only, from the tissues to the heart.

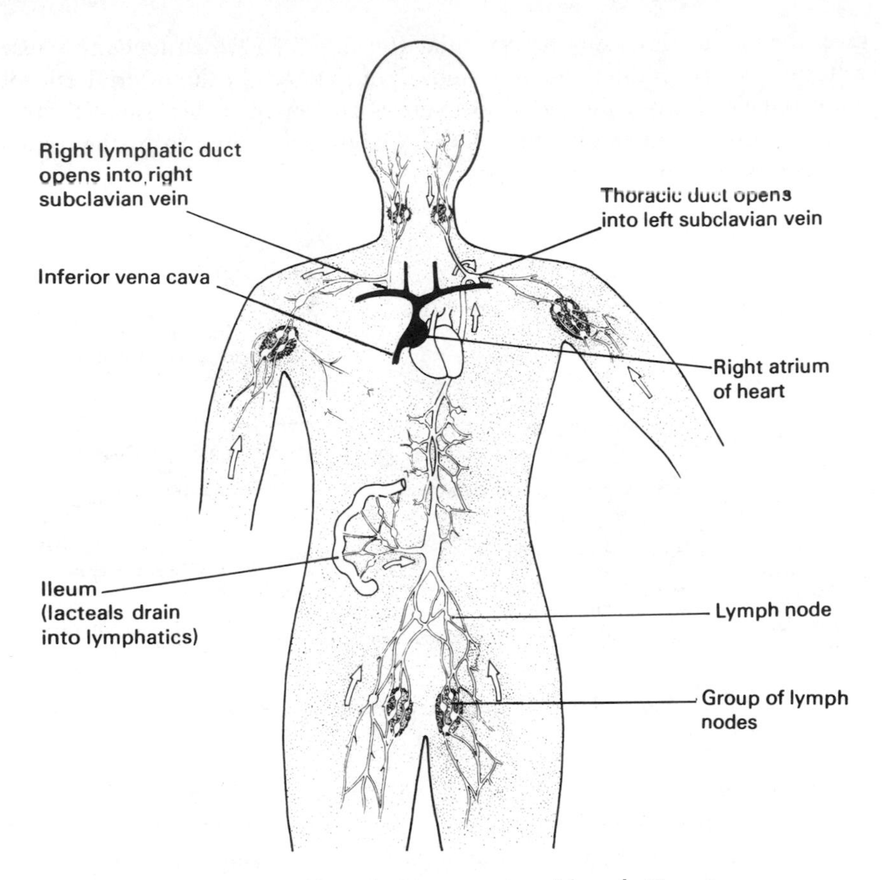

Fig. 21.15 *Main drainage routes of lymphatic system*

There is no specialized pumping organ; the flow is brought about partly by the pressure of the fluid accumulated in the tissues and partly by muscular exercise, the pressure of the contracting muscles around the lymphatics assisting the fluid through the vessels. Some of the lymphatics contain valves (Fig. 21.16), which allow lymph to flow in one direction only.

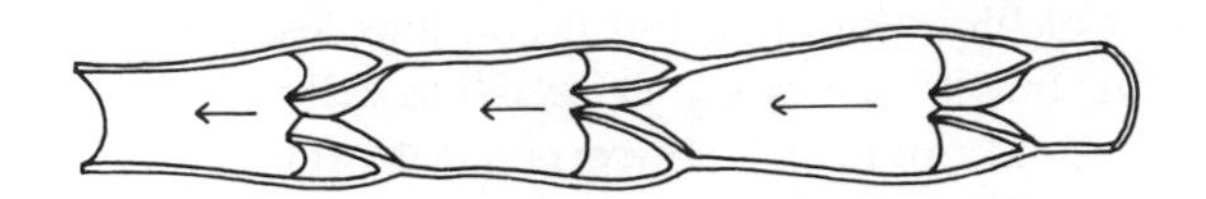

Fig. 21.16 *Deep lymphatic vessel cut open to show valves*

21.4 QUESTIONS

1. Although the walls of the left ventricle are thicker than those of the right ventricle, the volumes of the two ventricles are the same. Why is this necessarily so?
2. State in detail what happens to a glucose molecule and a fat molecule in the ileum, and an oxygen molecule in a lung, during their absorption and thereafter, until all three reach a muscle cell in the leg.
3. Why is a person whose heart valves are damaged by disease unable to participate in active sport?
4. A system for transporting substances in solution might as well be filled with water. What advantages has the blood circulatory system over a water circulatory system?
5. What is the advantage to an animal of having capillaries which are (*a*) very narrow, (*b*) repeatedly branched and (*c*) very thin-walled?
6. How do you think microscopic animals can survive without having a circulatory system?

UNIT 22
BREATHING

22.1 INTRODUCTION

The movement, growth, reproduction and other vital processes of a living animal's body all require the expenditure of *energy* which it can only obtain from the food it eats. The process of liberating energy is called *respiration* (see Unit Four); in animals and many plants it involves the uptake of oxygen and the production of carbon dioxide.

Oxygen enters the animal's body from the air or water surrounding it: oxygen makes up about 20 per cent of the composition of air, and is appreciably soluble in water. In the less complex animals the entire exposed surface of the body absorbs oxygen, but in the higher animals there are special respiratory organs such as lungs or gills where oxygen is taken up; carbon dioxide is usually excreted from the same organ. In the respiratory organ of a vertebrate oxygen is taken up into the blood, which transports it to all the living parts of the body where it is used in tissue respiration.

An efficient respiratory organ has a large surface area for rapid oxygen uptake and a plentiful supply of blood in a dense network of capillary blood vessels, which are separated from the surrounding air or water by a very thin layer of cells or *epithelium*. In land-dwelling animals the whole absorbing surface of the organ is covered by a layer of moisture. In many animals there is also a mechanism which constantly renews the air or water in contact with, or near, the respiratory surface, a process called *ventilation*.

In man and other mammals the respiratory organs are the lungs. The sections that follow describe the respiratory system in man and its ventilation by breathing movements, and the way in which oxygen is taken up and carbon dioxide excreted, a process known as *gaseous exchange*.

22.2 THE LUNGS

The lungs (Fig. 22.1) are two thin-walled elastic sacs lying inside the rib-cage in the upper part of the trunk, the *thorax*. The outer surface of the

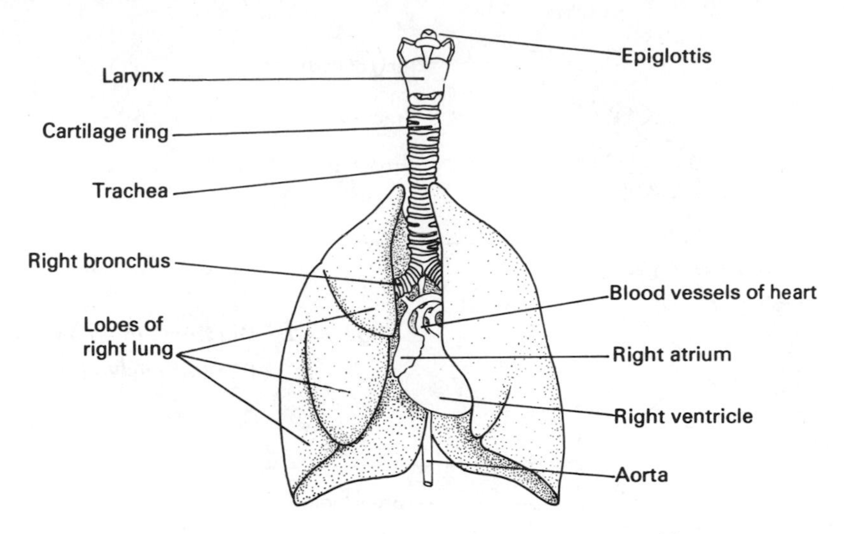

Fig. 22.1 *Diagram of lungs showing position of heart*

lungs and the inner surface of the thorax are both covered with a smooth elastic membrane, the *pleural membrane.* It secretes a liquid, *pleural fluid,* which lubricates these surfaces so that during the movements of breathing the regions of contact between the lungs and thorax slide over one another with very little friction (Fig. 22.10).

The lungs communicate with the atmosphere through the windpipe or *trachea.* The upper part of the trachea, the *larynx,* opens into the *pharynx* at the back of the mouth (see Fig. 20.3). The movements of the thorax alternately expand and compress the lungs repeatedly in such a way that air is taken into them during their expansion and expelled as they are compressed. Gaseous exchange takes place while the air is in the lungs: some of the atmospheric oxygen is absorbed, and carbon dioxide is released from the blood into the lung cavities.

Lung Structure

The trachea divides into two *bronchi* which enter the lungs and divide into smaller branches or *bronchioles* (Fig. 22.2). These divide further and terminate in a mass of little thin-walled pouch-like air sacs or *alveoli* (Figs. 22.3, 22.4 and 22.5), which make up the greater part of lung tissue.

(a) **Air passages.** The walls of the trachea and bronchi contain rings of cartilage (gristle-like material), which keep the air passages open and prevent them from collapsing when the air pressure inside them falls during inspiration (that is, breathing in; see Section 22.4). The lining of the air

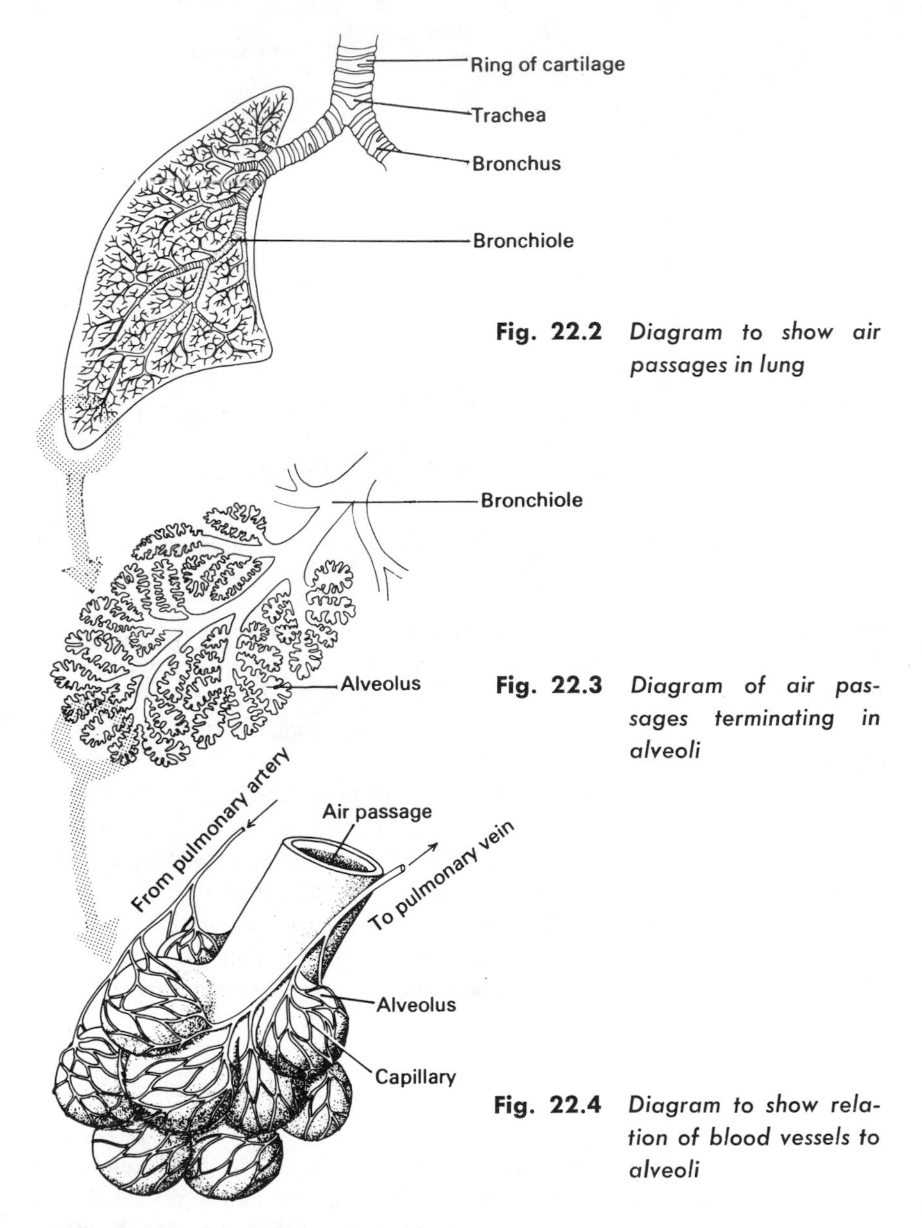

Fig. 22.2 *Diagram to show air passages in lung*

Fig. 22.3 *Diagram of air passages terminating in alveoli*

Fig. 22.4 *Diagram to show relation of blood vessels to alveoli*

passages is covered with numerous *cilia,* minute cytoplasmic hairs which constantly flick to and fro; it also contains glandular cells which secrete mucus. If any fragments of foreign material such as dust particles or bacteria are swept into the lungs with the inspired air, they become trapped in the mucus film. Eventually they are carried by the movements of the cilia up to the larynx and into the pharynx, where they are swallowed.

The epiglottis and other structures at the top of the trachea (see Section 20.2) prevent large particles from entering the air passages, particularly during swallowing. Choking and coughing are reflex actions (see Section 29.3), which tend to remove any foreign particles which accidentally enter the trachea or bronchi.

(b) **Alveoli.** There are about 350 million alveoli in a human lung, having a total absorbing surface of about 90 square meters. The walls of the alveoli are very thin and elastic, consisting of a single layer of cells or *epithelium*. Beneath the epithelium is a dense network of capillaries (Fig. 22.4); these carry deoxygenated blood which has been brought to the heart from all parts of the body and pumped to the lungs from the right ventricle through the pulmonary arteries.

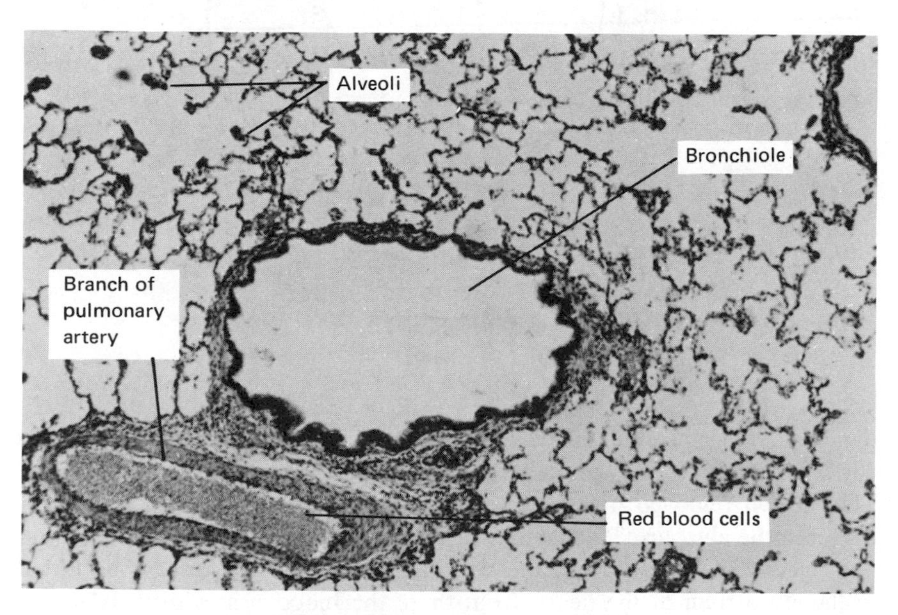

Fig. 22.5 *Microscopic structure of lung tissue seen in section*
(Gene Cox)

22.3 GASEOUS EXCHANGE IN THE LUNGS

The oxygen concentration in the blood in the lung capillaries is low, and an oxygen diffusion gradient therefore exists between the air in the alveolus and the blood. Thus oxygen diffuses from the alveolus, through the moisture film covering the alveolar lining, the epithelium and the capillary wall into

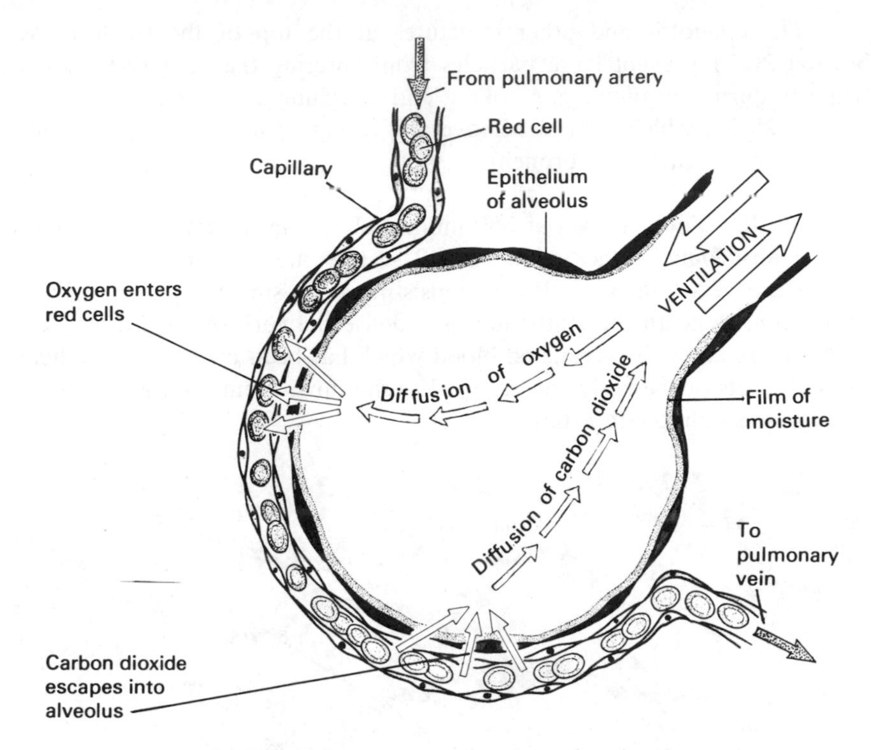

Fig. 22.6 *Gaseous exchange in the alveolus*

the blood plasma, and thence into the cytoplasm of the red blood cells where it combines immediately with hemoglobin (see Section 21.1(*a*)). The oxygen removed in this way is quickly replaced from the oxygen in the air in the alveolus. The capillaries carrying the freshly oxygenated blood join up and eventually form the pulmonary veins which return the blood to the left atrium of the heart; from there the blood passes directly to the left ventricle and the aorta, and thence to all parts of the body.

The concentration of carbon dioxide in the alveolar air is low; in these conditions an enzyme in the blood, *carbonic anhydrase,* breaks down the bicarbonates in solution in the plasma, releasing carbon dioxide. The gas diffuses in solution from the plasma across the capillary wall and the epithelium and is released into the alveoli, from which it is expelled with the expired air.

Because of the gaseous exchange in the alveoli, the composition of the air leaving the lungs differs markedly from that of the atmosphere: it contains considerably less oxygen and more carbon dioxide (compare Experiment 4.5).

Table 22.1 *Approximate percentage composition of inspired and expired air*

	Inspired	*Expired*
Oxygen	21	16
Carbon dioxide	0.04	4
Nitrogen	79	79
Water vapor	varies	saturated

The proportions of nitrogen in inspired and expired air are the same because the gas is chemically unreactive. Although nitrogen dissolves in blood plasma it does not take part in chemical reactions within it, and the rates of nitrogen diffusion into and out of the blood are therefore the same.

Diffusion gradients

The rates of diffusion of oxygen and carbon dioxide in solution depend in part on the diffusion gradients of the gases (see Section 3.2): the steeper the diffusion gradient, the more rapid is the diffusion.

A steep *oxygen diffusion gradient,* and therefore relatively rapid oxygen diffusion, is maintained between the alveoli and the blood in the capillaries by several factors:

(*a*) the constant replacement of oxygenated blood by deoxygenated blood;
(*b*) the very short distance between the alveolar lining and the blood;
(*c*) the rapid removal of oxygen from solution by reaction with hemoglobin;
(*d*) the constant replenishing of the oxygen in the alveoli by ventilation.

Factors (*a*) and (*b*) are similarly significant in maintaining the *carbon dioxide diffusion gradient* in the reverse direction to the oxygen gradient. Ventilation plays a part as well in that it constantly removes carbon dioxide from the alveoli. The action of carbonic anhydrase in releasing carbon dioxide from bicarbonates, so keeping the carbon dioxide level in the blood above that in the alveoli, is also important.

22.4 VENTILATION OF THE LUNGS

The lungs cannot expand or contract except by passive movements: ventilation is brought about by movements of the thorax.

The thorax is an airtight cavity enclosed at the sides by the ribs and the tissues covering them, and by a sheet of muscle extending across the body cavity between the thorax and the abdomen, the *diaphragm*. When it is relaxed the diaphragm is dome-shaped, extending upward into the thoracic cavity, with the liver and stomach lying immediately below it.

(a) **Inspiration** (breathing in). During inspiration the volume of the thorax is increased by two movements:

(*i*) the diaphragm contracts and flattens out;
(*ii*) the *intercostal muscles,* which run obliquely from one rib to the next, contract and pull the lower ribs upward and outward (Figs. 22.7, 22.8 and 22.9).

Any change in the volume of the thorax is followed by the elastic tissue of the lungs; thus as the thorax expands the capacity of the lungs increases, and there is a consequent fall in the air pressure within them. As a result the pressure of the atmosphere forces air into the lungs through the trachea via the nose and mouth.

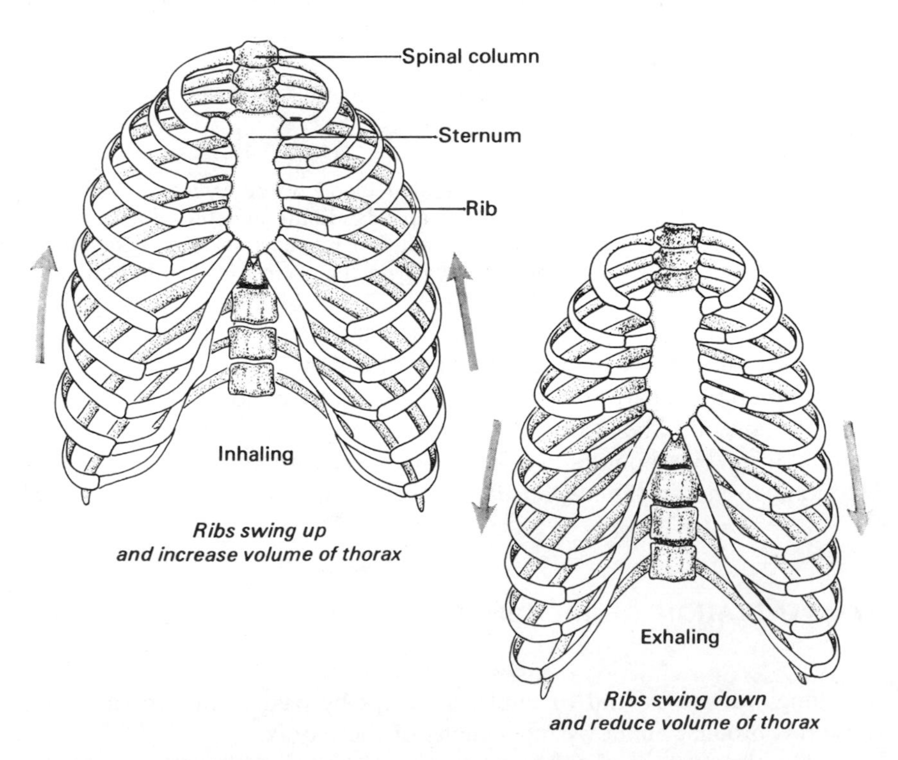

Fig. 22.7 *Movement of rib-cage during breathing*

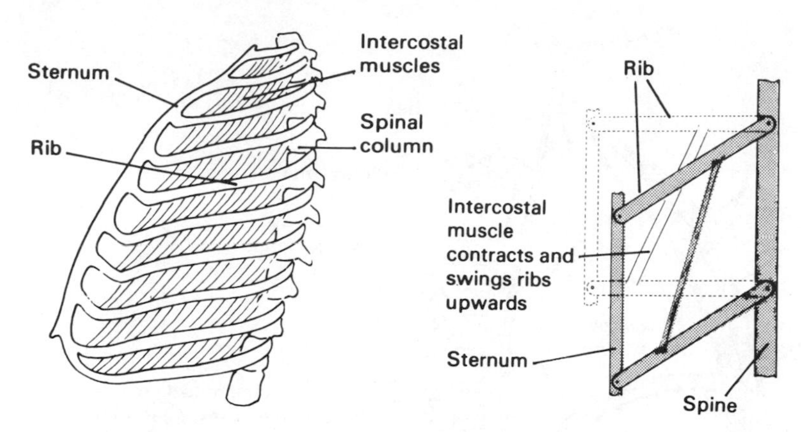

Fig. 22.8 *Rib-cage seen from left side, showing intercostal muscles*

Fig. 22.9 *Model to show action of intercostal muscles*

(b) **Expiration** (breathing out). During expiration, the muscles of the ribs and diaphragm relax. The ribs move down under their own weight, and the organs lying just below the diaphragm, under pressure from the muscular abdominal wall, push the diaphragm back into its domed shape again. As a result of these movements the lungs return to their original volume, expelling part of the air they contain.

In quiet breathing the movements of the diaphragm alone are responsible for the ventilation of the lungs; during vigorous activity the thoracic movements are more extensive.

The lungs of an adult man contain between 3,000 and 3,500 cm^3 of air during quiet breathing; however, only about 500 cm^3 of this air, the *tidal air,* is exchanged at each breath. Deep inspiration can take in another 2,000 cm^3 of air, and vigorous expiration can expel an additional 1,500 cm^3. The thorax cannot collapse completely, because of the rigidity of the surrounding rib-cage, so that the remaining 1,500 cm^3 of air can never be expelled. This *residual air* remains stationary in the alveoli, and exchanges carbon dioxide and oxygen by diffusion with the tidal air sweeping into and out of the air passages.

Rate of Breathing

The rhythmical breathing movements are usually carried out quite unconsciously about sixteen times a minute. They are controlled by a region of the brain which is very sensitive to the level of carbon dioxide in the blood. If there is a rise in the carbon dioxide concentration of the blood reaching this region of the brain, nerve impulses are automatically sent

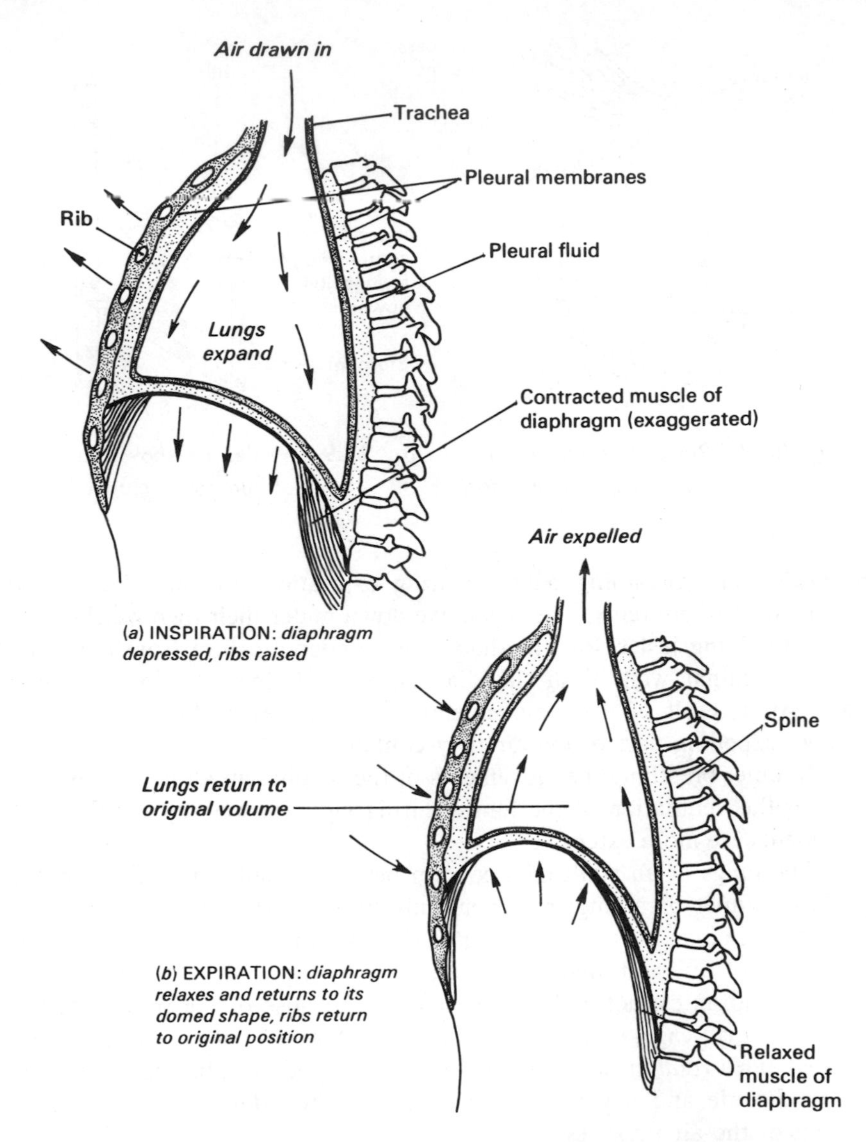

Fig. 22.10 *Diagrams of thorax to show mechanism of breathing*

to the diaphragm and rib muscles which increase the rate and depth of breathing. The concentration of carbon dioxide in the blood is most likely to rise during vigorous activity, and the accelerated rate of breathing helps to expel the rapidly accumulating carbon dioxide and to increase the amount of oxygen in the blood, so meeting the demands of increased tissue respiration. By regulating the oxygen and carbon dioxide levels in the blood, the lungs are fulfilling a homeostatic function (Section 20.13).

At most times the rate of breathing can be controlled voluntarily within certain limits, as in singing or when playing a wind instrument.

22.5 THE NOSE

Air passing through the nose is warmed slightly before it passes into the lungs, and some of the dust particles and bacteria carried in it are removed by being trapped in the film of mucus which lines the nasal cavities. In addition the lining of the nose contains sensory cells which respond to certain chemicals in the air, producing a sensation of smell (see Section 28.3).

22.6 VOICE

The *vocal cords* are a pair of membranes, containing ligaments (tough strands of fibrous material), protruding from the lining of the larynx. When air passes over the vocal cords in a particular way they vibrate, producing sounds. They are controlled by muscles, which can alter the tension in the cords and the distance between them, thus varying the pitch and the quality of the sounds produced.

22.7 PRACTICAL WORK

Experiment 22.1 *To investigate the composition of exhaled air*

(a) **Carbon dioxide concentration.** Experiment 4.5 demonstrates the increased proportion of carbon dioxide in exhaled air as compared with inhaled air.

(b) **Oxygen concentration.** Exhaled air is collected in a gas jar (or any jar with a lid) by downward displacement of water (Fig. 22.11(*a*)). A candle stub on a deflagrating spoon is lit, placed in the jar (Fig. 22.11(*b*)) and the length of time for which the candle burns is noted. The more oxygen there is in the air sample, the longer will the candle stay lit. The air exhaled during the first part of an expiration (tracheal and bronchial air) and that from the last part (alveolar air) should be collected in separate gas jars and compared with each other and with a jar of atmospheric air.

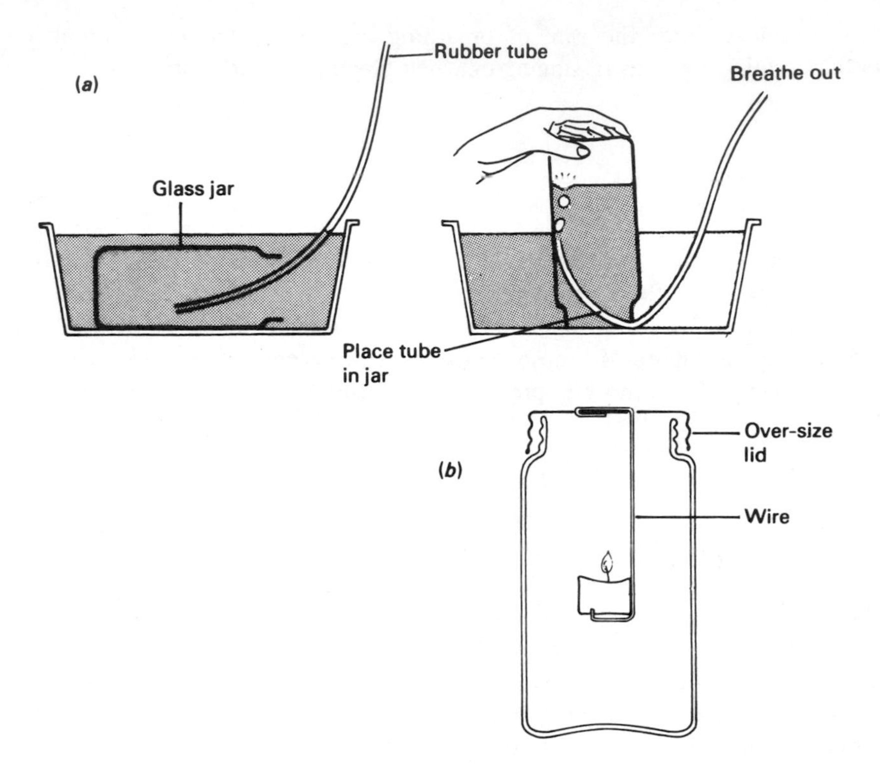

Fig. 22.11 *Experiment to investigate the oxygen concentration in exhaled air: (a) collection of exhaled air; (b) testing the sample*

Experiment 22.2 *To investigate lung capacity*

A large (5-liter) bottle is calibrated in liters by filling it with water 1 liter at a time and marking the levels. The bottle, full of water, is inverted in a trough or bowl of water, the stopper removed, and a rubber tube inserted through the neck. The experimenter takes a deep breath and exhales through the tube so that the exhaled air collects in the bottle, displacing the water. The amount of water displaced will give a measure of the lung capacity.

22.9 QUESTIONS

1. Outline the events which take place in the course of vigorous exercise which leads to a change in the rate and depth of breathing both during and after the activity. (See also Unit Four.)

2. The lungs and ileum are adapted for absorption. Point out the features they have in common which facilitate absorption.
3. The table below gives the approximate percentage volume composition of air inhaled, exhaled or retained in the lungs. Explain how these differences in composition are brought about by events in the lungs.

	Inhaled air	*Exhaled air*	*Alveolar air*
% Oxygen	21	16	14
% Carbon dioxide	0.03	4	5.5

UNIT 23
EXCRETION

23.1 INTRODUCTION

A living cell is in a condition of constant change; chemical reactions are continually taking place within it. Reactions involving the release of energy by respiration go on in everything which is alive. Other reactions concern the alteration of one kind of material into another. Even the apparently permanent structures of the body like muscles, blood or skin are, in fact, changing from day to day. The chemical units of living protoplasm are constantly being renewed. New molecules are being added, and defective molecules or entire cells are being digested away. All these reactions give rise to end-products, some of which are poisonous above a certain concentration and which could affect the normal functioning of the body if they were allowed to accumulate.

Excretion is the process by which such toxic products are removed from the body before they can accumulate to harmful levels.

The main *excretory products* in animals are carbon dioxide and water produced as end-products of respiration (see Section 4.2) and nitrogenous compounds from the breakdown of amino acids. Part of this nitrogenous material arises from the deamination in the liver of amino acids absorbed after a meal containing more protein than the body needs (see Section 20.12(*d*)), and part from the breakdown of proteins in the renewal and repair of the protoplasm of the body. The products of amino acid breakdown are ammonia and other nitrogen-containing compounds which could, if allowed to accumulate in the body, cause death in a matter of days or weeks. They are converted in the liver to *urea* and *uric acid,* which are less poisonous.

In man and other mammals the excretory products of the body's cells pass into the tissue fluid surrounding them and thence into the blood, which carries them to the *excretory organs*, the lungs, the liver and the kidneys. The lungs excrete carbon dioxide and water vapor in exhaled air, the liver excretes bile pigments derived from hemoglobin decomposition into the

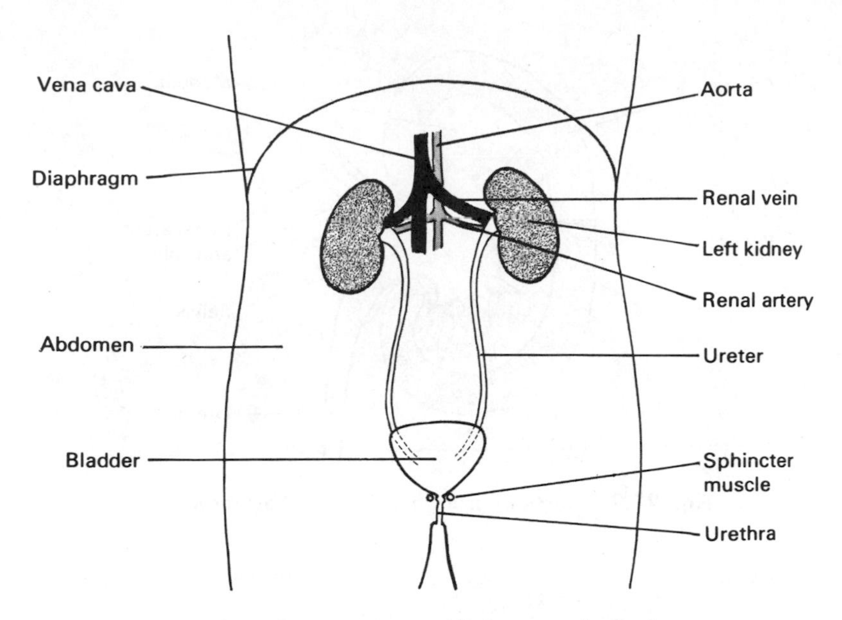

Fig. 23.1 *Position of kidneys in the body*

alimentary canal, where they are incorporated eventually into the feces, and the kidneys remove nitrogenous material and excess mineral salts and water from the blood and eliminate them in the urine.

23.2 STRUCTURE OF THE KIDNEYS

The two kidneys are fairly solid structures attached to the back of the abdominal cavity (Fig. 23.1). They are oval in shape, with an indentation on their innermost sides, and red–brown in color. Each kidney weighs about 250 g, and is enclosed in a transparent membrane. The kidneys receive oxygenated blood through a branch of the aorta, the *renal artery* (*renal* means "of the kidney"), and deoxygenated blood is carried away from them in the *renal vein* to the inferior vena cava. A tube, the *ureter,* runs from each kidney to the base of the *bladder* in the lower abdomen.

A section through a kidney (Fig. 23.2) shows a darker, outer region, the *cortex,* an inner zone which is lighter in color, the *medulla,* and a hollow space called the *pelvis* from which the ureter leads. Cones or *pyramids* of kidney tissue project into the pelvis.

The kidney tissue consists of many minute tubes, *renal tubules,* and

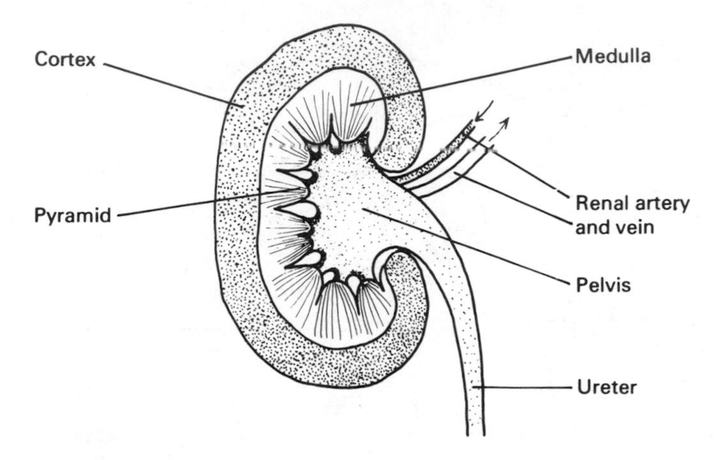

Fig. 23.2 *Section through kidney to show regions*

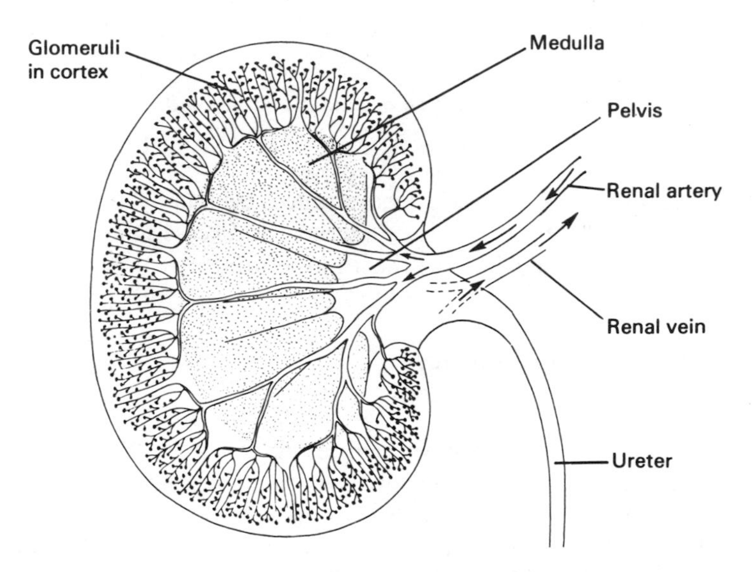

Fig. 23.3 *Section through kidney to show distribution of glomeruli*

capillary blood vessels, all held together by connective tissue. The renal artery divides up into a great many arterioles, mostly in the cortex (Fig. 23.3). Each arteriole leads to a *glomerulus,* which is a little knot of capillary blood vessels, repeatedly divided and coiled. Each glomerulus is almost entirely surrounded by a cup-shaped organ, called a *Bowman's capsule,* about 0.1 to 0.2 mm in diameter, which leads to a renal tubule. This tubule,

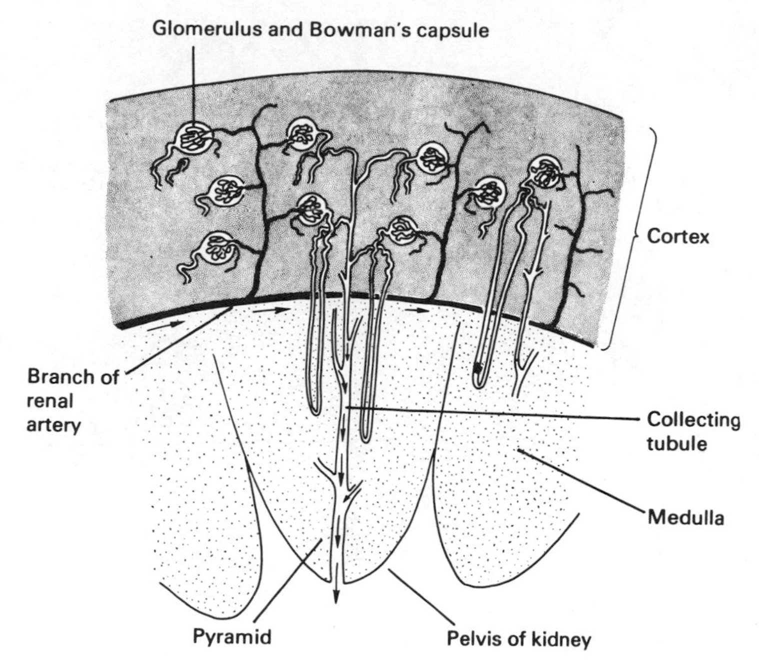

Fig. 23.4 *Section through cortex and medulla*

after a series of coils and loops, joins up with other tubules to form a *collecting tubule* which passes through the medulla to open into the pelvis at the apex of a pyramid. The capillaries from the glomeruli and the renal tubules unite to form the renal vein (Fig. 23.4).

23.3 MECHANISM OF KIDNEY EXCRETION

The narrow and tortuous capillaries of the glomerulus offer considerable resistance to the flow of blood, so that a high pressure is set up. This is sufficient to overcome the osmotic pressure of the blood plasma (see Section 21.3(*c*)) so that fluid filters out through the capillary walls in the glomerulus and collects in the Bowman's capsule. The filtered fluid, *glomerular filtrate,* has a composition similar to that of blood serum, containing glucose and other products of digestion, salts and nitrogenous waste products; fibrinogen and other plasma proteins remain in the blood. The filtration process is continuous; in man, 180 liters of the filtrate, carrying 145 g glucose and 1,100 g sodium chloride, pass each day into the

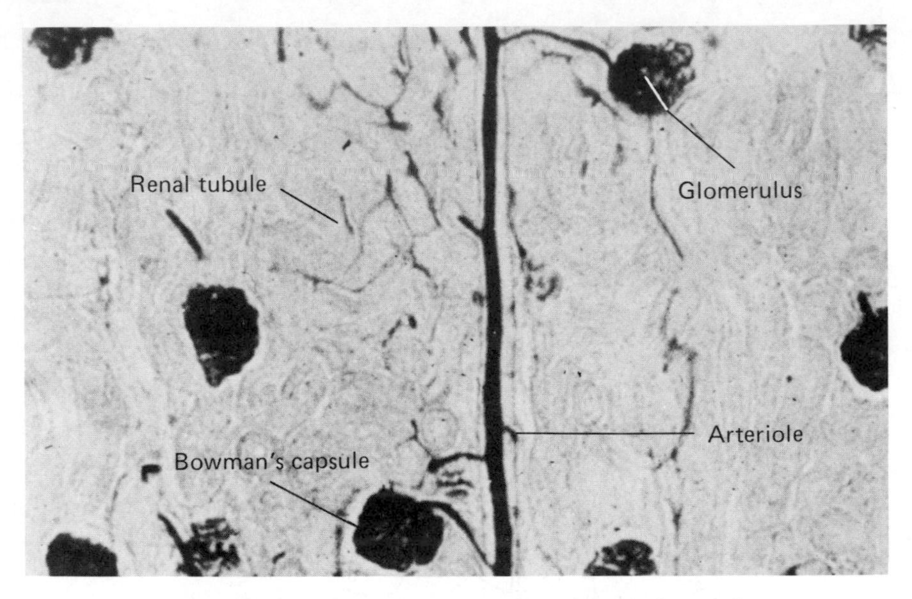

Fig. 23.5 *Section through cortex to show glomeruli*
(Brian Bracegirdle)

Bowman's capsules. As the filtrate passes down the renal tubule, all the nutrients, some of the salts and much of the water are absorbed back into the network of capillaries that surrounds the tubule (Fig. 23.6). This selective reabsorption prevents the loss of useful substances from the body. The remaining liquid, now called *urine,* contains only the waste products such as urea and related substances, hormones which have been inactivated in the liver, excess salts and water.

The urine passes down the collecting tubule, where more of the water is reabsorbed. The amount of water removed from the urine is controlled very precisely by hormone action (see Section 23.5) in such a way that the concentration of the blood is regulated. If the blood is dilute, as it is when a great deal of liquid has been drunk, comparatively little water is reabsorbed, the urine is therefore dilute and the excess water is eliminated. If, on the other hand, the blood is concentrated, which may happen after loss of water by profuse sweating, more water is absorbed in the collecting tubule and less water is lost in the urine. From the collecting tubule the urine enters the pelvis of the kidney, whence it passes down the ureter to the bladder as a result of waves of contraction moving along the ureter walls.

The selective reabsorption from the glomerular filtrate is performed by the cells of the walls of the kidney tubules. It often proceeds against a diffusion gradient by mechanisms which are not fully understood but which certainly need a supply of energy, obtained by respiration within the cells.

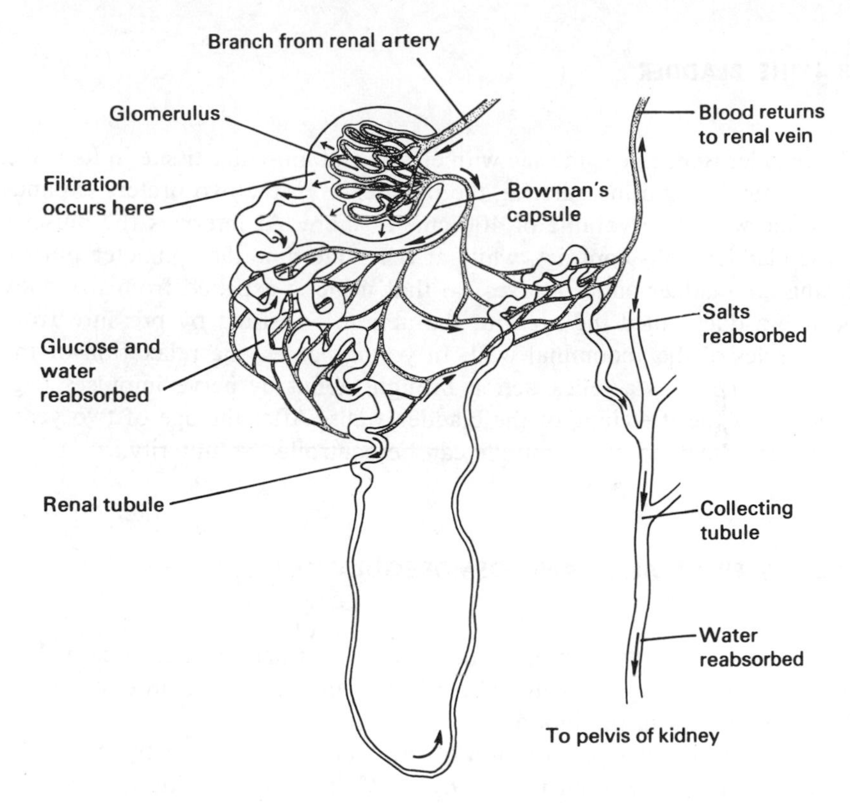

Fig. 23.6 *Diagram of glomerulus and Bowman's capsule*

The blood leaving the kidneys in the renal vein therefore contains less oxygen and glucose and more carbon dioxide, as a result of tissue respiration, and less water, salts and nitrogenous waste, as a result of excretion, than the blood in the renal artery.

Table 23.1 lists the main nitrogenous compounds removed from the blood by the kidneys.

Table 23.1 *Approximate percentage of nitrogenous compounds in blood plasma and urine*

	Blood plasma	*Urine*
Proteins	7–9%	0%
Urea	0.03	2
Uric acid	0.004	0.06
Ammonium compounds	0.0001	0.04
[*Water*	*90–93*	*95*]

23.4 THE BLADDER

The bladder is an extensible sac with elastic and muscular tissue in its walls. The accumulating urine entering the bladder from the two ureters expands its elastic walls to a volume of 400 cm^3 or more. At intervals the muscles in the bladder walls contract, while at the same time the sphincter muscle closing the bladder outlet relaxes, so that urine is expelled from the body through a duct called the *urethra*; the action is assisted by pressure from the muscles of the abdominal wall. In young babies, the relaxation of the sphincter muscle is a reflex action brought about by nerve impulses triggered off by the stretching of the bladder walls. After the age of two years or so, the relaxation of the muscle can be controlled voluntarily.

23.5 WATER BALANCE AND OSMOREGULATION

Water is lost from the body in urine, feces, sweat and exhaled breath. It is gained by eating and drinking. These losses and gains will produce corresponding changes in the blood.

Changes in the concentration of the blood are detected by the *hypothalamus,* part of the brain (Fig. 29.11). If the blood passing through the brain is too concentrated, the hypothalamus stimulates the *pituitary gland,* which lies just beneath it, to secrete a hormone called *antidiuretic hormone* (ADH). When this hormone is carried in the blood to the kidneys it stimulates the reabsorption of water from the glomerular filtrate by the cells in the walls of the kidney tubules. Thus the urine becomes more concentrated and the further loss of water from the blood is reduced. If, on the other hand, the blood passing through the hypothalamus is too dilute, production of ADH by the pituitary gland is suppressed and less water is absorbed from the glomerular filtrate.

The intake of water by the body is controlled by the sensation of thirst, which stimulates the individual to take in fluid. The mechanism producing feelings of thirst is not well understood, but it undoubtedly serves to regulate the intake of water and so to maintain the concentration of the blood within certain limits.

The kidneys have an important part in the homeostasis of the body (see Section 20.13) in the three ways in which they control the composition of the blood:

(*a*) they remove urea and other harmful substances;
(*b*) they remove excess water;
(*c*) they remove excess mineral salts.

These activities are both excretory, in that they remove the unwanted products of metabolism, and osmoregulatory, in that they keep the osmotic potential of the blood more or less constant.

23.6 QUESTIONS

1. In an experiment, a man drank a liter of water. His urine output increased so that after two hours he had eliminated the extra water. When he drank a liter of 0.9 per cent sodium chloride solution, there was little or no immediate increase in urine production. Explain the difference in these results.
2. In cold weather one may need to urinate frequently, producing a fairly colorless urine. In hot weather, urination is infrequent and the urine is often colored. Explain these observations.
3. Consult Section 20.7 and 20.12 and then explain briefly why glucose does not normally appear in the urine.
4. Re-read Sections 3.2 and 3.3. In the artificial kidney a patient's blood is circulated through dialysis tubing immersed in a warm solution of glucose and salts. Explain how this results in the elimination from his blood of nitrogenous wastes without loss of essential glucose and salts.
5. Explain why the elimination of water by the kidneys may be considered to be both excretion and osmoregulation.

Unit 24
Skin And Temperature Control

24.1 SKIN FUNCTION AND STRUCTURE

The skin is a continuous layer of tissue covering the surface of the body. It has three principal functions:

(*i*) it protects the tissues beneath from mechanical injury, from the injurious effects of ultraviolet radiation in direct sunlight, from bacterial invasion and from damage by desiccation;
(*ii*) it contains numerous sense organs which are sensitive to temperature, touch, pressure and pain, and so make the organism aware of changes in its surroundings (see also Section 28.1(*d*));
(*iii*) it helps to keep the body temperature constant.

The skin consists of two main layers, an outer *epidermis* and an inner dermis (Figs. 24.1 and 24.3). The relative thickness of the layers and abundance of the structures within them varies with its position on the body. For example, the skin on the soles of the feet has a very thick epidermis and no hair follicles. The description which follows is therefore given in general terms.

(*a*) Epidermis

The epidermis is made up of three layers of cells.

(*i*) **Malpighian.** This is the innermost layer of the epidermis. It is a continuous layer of cells which divide actively and so are constantly renewing the epidermis from beneath. The cells may contain granules of dark-brown pigment, *melanin,* the concentration of which determines the color of the skin. Melanin is opaque to ultraviolet light, and protects the tissues beneath it against injury from this radiation.

(*ii*) **Granular.** This layer contains some living cells, but toward the surface it gives way gradually to the *cornified layer.*

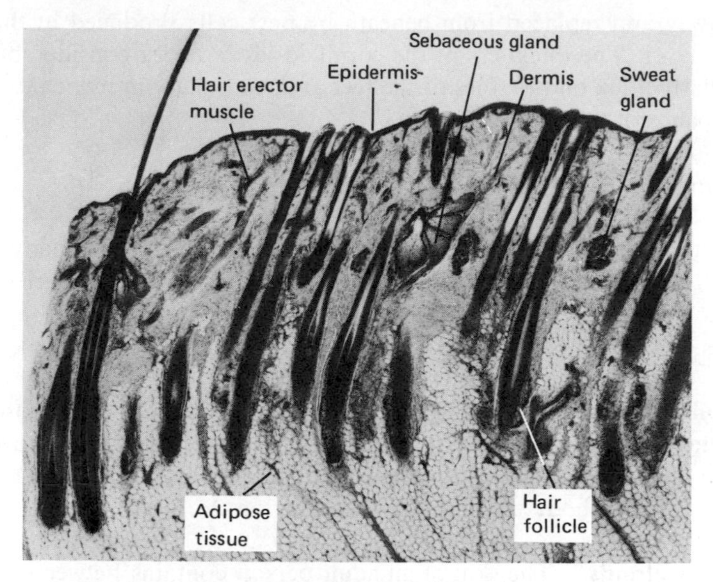

Fig. 24.1 *Section through hairy skin (×15)*
(Brian Bracegirdle)

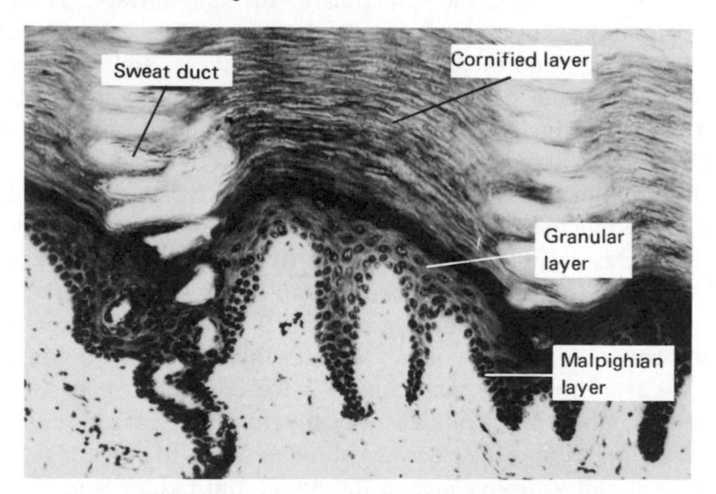

Fig. 24.2 *Section through non-hairy skin (×200); the sweat ducts are contorted in passing through the cornified layer*
(Brian Bracegirdle)

(iii) **Cornified.** This layer contains dead cells only. They become impregnated with *keratin,* a fibrous protein, and form a tough outer coat which resists mechanical damage and bacterial invasion and reduces the loss of water by evaporation. The cells of the cornified layer are continually being

worn away and replaced from beneath by new cells produced in the Malpighian layer. The thickness of the cornified layer varies considerably; it is particularly thick on the soles of the feet and, especially in manual workers, on the palms of the hands (Fig. 24.2).

(b) Dermis

The dermis is a layer of connective tissue containing many elastic fibers. It is thicker than the epidermis, and contains blood capillaries, nerve endings and sensory organs, lymphatic vessels, sweat glands and hair follicles.

(i) **Capillaries.** The capillaries in the skin supply its cells, and those of its associated structures, with the necessary food and oxygen and remove their excretory products. Those lying close to the surface have an important part in the control of body temperature (see Section 24.2(*a*)).

(ii) **Sweat glands.** The skin of an adult person contains between two and three million sweat glands. Each is a coiled tube of secretory cells; one end of the tube passes through the epidermis to the skin surface. The gland is supplied with blood by a network of capillaries. The cells absorb fluid from the capillaries and the tissues surrounding them, and pass it into the duct of the gland and thence to the exterior. The fluid consists mainly of water, with salts, principally sodium chloride, and small quantities of urea and lactic acid dissolved in it.

Small amounts of water are constantly lost by evaporation through the skin at normal temperatures. If the body temperature rises by 0.2 to 0.5 °C the glands begin to secrete sweat on to the surface of the skin, where it rapidly evaporates. In a hot climate a man doing manual work may lose about 1 kg of sweat in an hour. Since about 0.5 per cent of sweat consists of mineral salts, especially sodium chloride, this represents a considerable loss of salts from the body, particularly if the conditions persist for some time; workers in high temperatures must therefore take care to replace these salts in their food and drink. If water alone is taken to replace that lost in sweat, the salt and water balance of the blood and tissues is upset, leading to the symptoms of "heat cramp."

(iii) **Hair follicles.** The hair follicle is a deep pit in the dermis, lined with cells of the granular and Malpighian layers, which multiply and build up a hair inside the follicle. The cells of the hair become impregnated with a horny deposit of the protein keratin, and die. The constant adding of new cells to the base of the hair causes it to grow. Growth continues for about four years; the hair then falls out and a new period of growth begins. Where

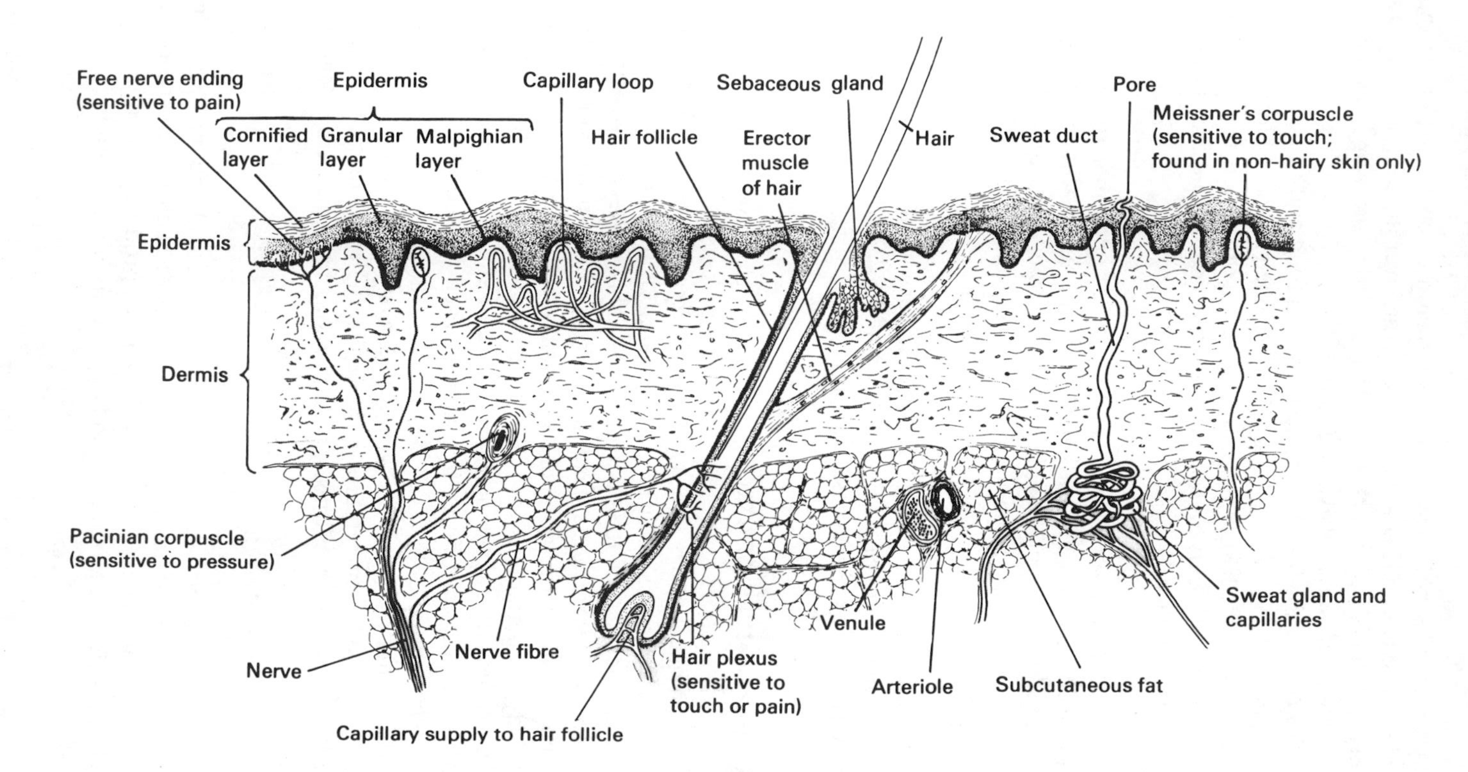

Fig. 24.3 *Generalized section through skin*

hairs grow thickly they form a covering which protects the skin from injury, and which also has an insulating function: the layer of stationary air trapped between the hairs reduces the loss of heat from the skin, and also cuts down the loss of water by evaporation.

The cells of the follicle are supplied with food and oxygen by capillary blood vessels. A network of sensory nerve fibers, a *hair plexus,* responds to movements of the hair. This sensory function of the hair is well developed in the whiskers or *vibrissae* that grow on the sides of the face in mammals such as cats and mice.

(iv) **Sebaceous glands.** The sebaceous glands open into the hair follicles and produce an oily secretion which gives the hairs water-repelling properties, keeps the epidermis supple, and reduces the tendency for it to become too dry as a result of evaporation. It also has antiseptic properties against certain bacteria.

(v) **Subcutaneous fat.** The layers of adipose tissue beneath the dermis contain numerous fat cells (Figs. 20.11 and 20.12) where fat is stored. The fat may also act as a heat-insulating layer.

24.2 TEMPERATURE CONTROL

Fish, amphibia, reptiles and all the invertebrates are *poikilothermic* (see Section 16.1), that is, their body temperature is the same as, or only a few degrees above, that of their surroundings. This makes their activities very dependent on temperature conditions. When their surroundings are warm their bodies are warm too and are normally active, but in cold conditions their body temperature is low, all the chemical changes in their protoplasm are slowed down, and the organism may be reduced to a state of complete inactivity. Insects can be entirely immobilized by the effects of a sudden fall in temperature.

Birds and mammals are *homoiothermic* or constant-temperature animals. Their body temperature is normally higher than that of their surroundings, and is not affected by fluctuations in the external temperature.

(a) Heat Loss and Gain

The body *gains* heat mainly from the chemical activities which go on in all living protoplasm. The chemical changes such as respiration taking place rapidly in the liver and in active muscle produce a good deal of heat, which is distributed around the body by the circulatory system.

The body *loses* heat to the atmosphere, mainly by convection and by radiation, as long as the air temperature is below that of the body. The conversion of water from liquid to a vapor requires considerable amounts of heat, the *latent heat of vaporization,* so the evaporation of water from the surface of the body and the lungs has a marked cooling effect. Heat losses are reduced by the insulating effects of an outer layer of hair, feathers or clothing.

Normally a balance is maintained between gains and losses of heat, so that the body temperature remains constant. In man this temperature, as measured with a thermometer placed under the tongue, is kept at about 36.8 °C, although it is slightly different (but still steady) in other parts of the body. External conditions may, however, require that the loss of heat to the air must be accelerated (when the body is overheated) or slowed down (in overcooling).

(i) **Overheating.** Vigorous activity, disease, hot weather and many other causes may bring about overheating. If the blood reaching the brain is a fraction of a degree warmer than normal, nerve impulses are sent from the brain to the skin, and produce two marked effects.

Vasodilation is the dilation (widening) of the arterioles which supply the capillary network beneath the epidermis. This causes an increase in the amount of blood flowing near the surface, and consequently an increase in the loss of heat to the surrounding air by convection and radiation (Fig. 24.4(*a*)). The increased flow of blood just beneath the epidermis can be seen as a flush on the skin.

Sweating is accelerated by nerve impulses, starting mostly in the brain, which increase the rate of sweat production so that a continuous layer of moisture may be produced on the skin surface. The latent heat absorbed by the sweat as it evaporates is taken from the body, so reducing the body's temperature.

Any air movement over the body helps to speed up the evaporation of sweat, which is why fans have a cooling effect on the body, even though they do not lower the temperature of a room.

In humid conditions, the air contains so much water vapor that the sweat may not evaporate rapidly enough to produce an adequate cooling effect, and may lead to *heat stagnation* in which the body temperature rises to over 41 °C, causing collapse and sometimes death. *Heat stroke* is a similar result of extreme overheating when, after prolonged sweating due to vigorous activity at high temperatures, sweat production ceases and the body temperature rises to a dangerously high level. Both conditions are sometimes called "sun-stroke," but it is the high temperature, rather than the effect of direct sunlight, which produces the symptoms of illness.

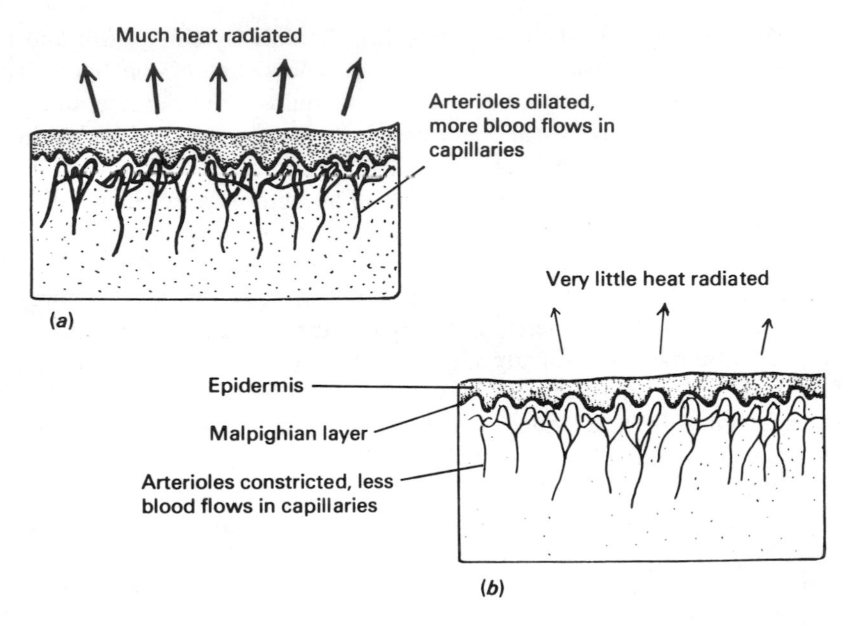

Fig. 24.4 *Diagram of section through skin: (a) vasodilation; (b) vasoconstriction*

(ii) **Overcooling.** In very cold conditions, the body tends to lose more heat than it is generating. Several different mechanisms may operate to reduce heat losses.

Vasoconstriction is the narrowing of the arterioles which supply blood to the capillaries lying just beneath the epidermis, thus reducing the flow of blood near the surface of the skin, and consequently slowing down the heat loss through the skin (Fig. 24.4(*b*)). Vasoconstriction makes the skin look pale or bluish.

Sweat production is reduced, and heat loss through evaporation is thus minimized.

Increased metabolism, for example, in the liver, releases more heat, which is carried to all parts of the body in the circulatory system. Muscular activity may be increased: movements like swinging the arms or stamping the feet release heat, while shivering, a reflex action (see Section 29.3) operating when the body temperature falls, is a spasmodic contraction of the muscles which has the same effect.

Fur or feathers can be fluffed out by contraction of the erector muscles in the skin (Fig. 24.1), thus increasing the volume of air trapped in them, and so improving the heat-insulation effect. In man, a similar contraction of the muscles of the hairs only produces "goose-pimples."

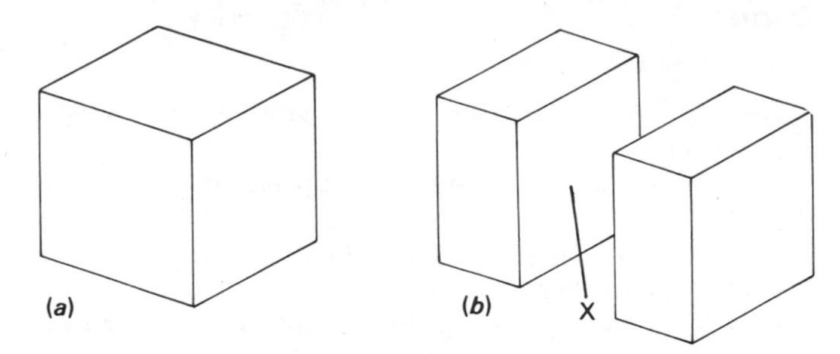

Fig. 24.5 *Diagram illustrating the relation between volume and surface area*

(b) Surface Area and Heat Loss

Consider the cube drawn in Fig. 24.5(*a*). If it is cut in half as shown in Fig. 24.5(*b*) each portion has half the volume of the original cube but more than half its surface area because an extra surface (marked X in the diagram) has been added to each half. Each time the solid is cut into smaller parts, the ratio surface area/volume increases. This means that small animals have a relatively larger surface area per unit of volume than have larger animals and so the former lose heat more rapidly to the surroundings. This is thought to be one reason why small homoiothermic animals like humming birds are restricted to areas with warm climates, and why the mammals and birds living in polar regions are relatively large.

(c) Hibernation

During very cold conditions many animals *hibernate,* that is, they cease feeding and pass into a state of dormancy which lasts throughout the winter. Mammals like hedgehogs and chipmunks fall asleep in specially prepared burrows or nests. Their breathing slows down and often becomes imperceptible. All the chemical activities in their bodies go on very slowly, using their food stores of fat and glycogen as sources of energy. The body temperature falls well below normal, and may be only a few degrees above that of the surroundings.

Hibernating animals are quite insensible and cannot be roused by touching them; in fact they are likely to die if such attempts are made. They awaken quite spontaneously at the end of the hibernation period; the body temperature returns gradually to its normal level, starting from the innermost organs, the animals begin to feed and their usual activities are resumed.

24.3 QUESTIONS

1. Why is it more accurate to describe fish as "variable-temperatured" animals rather than "cold-blooded"?
2. When a dog is hot, it hangs its tongue out and pants. Why should this have a cooling effect?
3. Why do you think we experience more discomfort in hot humid weather than we do in hot dry weather?
4. You may "feel hot" after exercise or "feel cold" without your overcoat and yet your body temperature is not likely to differ by more than 0.5 °C on the two occasions. Explain this apparent contradiction. (See also Section 28.1.)
5. Draw up a balance sheet showing all the possible ways in which the human body can gain and lose heat.

Unit 25
Sexual Reproduction

25.1 INTRODUCTION

Sexual reproduction is the formation of a new individual by a process involving the joining or fusing together of two reproductive cells or *gametes* into a single composite cell or *zygote,* from which the new organism develops. The fusion of the gametes is called *fertilization:* its most important aspect is the fusion of the two nuclei, since the nuclei carry the factors which will determine the characteristics of the new individual.

Usually one gamete comes from a male animal and the other from a female, although in some species such as earthworms and snails each individual animal produces both male and female gametes, that is, they are *hermaphrodite.*

The male gametes are known as *sperms* and the female gametes as *ova* (singular, *ovum*).

Internal and External Fertilization

In most *fish* and *amphibia,* fertilization is external. The female lays the eggs first and the male fertilizes them by placing sperms on them afterward. A behavior pattern which brings the sexes into proximity usually ensures that sperm is shed near the eggs and so increases the chances that they will be fertilized.

In *reptiles* and *birds,* the eggs are fertilized inside the body of the female by sperms passed by the male into the egg ducts. A sperm meets the ovum and fertilizes it before it is laid. Very little development of the egg takes place before laying, however, and the embryo grows in the egg *after* it has left its mother's body.

In *mammals,* fertilization is also internal, but after fertilization the eggs are retained in the female's body until the young animals are more or less fully formed. After birth the young are fed on milk secreted by the *mammary glands* of the female, and they are protected by their parents until they become fully independent. The description which follows concerns sexual reproduction in man.

25.2 THE REPRODUCTIVE ORGANS

(a) Female Reproductive Organs

The ova are produced in the *ovaries,* two cream-colored oval structures attached to the body wall at the back of the abdominal cavity below the kidneys. Close to each ovary is the funnel-shaped opening of the *oviduct,* the narrow tube down which the ova pass when they are released from the ovary. The oviducts open into a wider tube, the *uterus* or *womb,* lying lower down in the abdomen. Normally the uterus is only about 8 cm

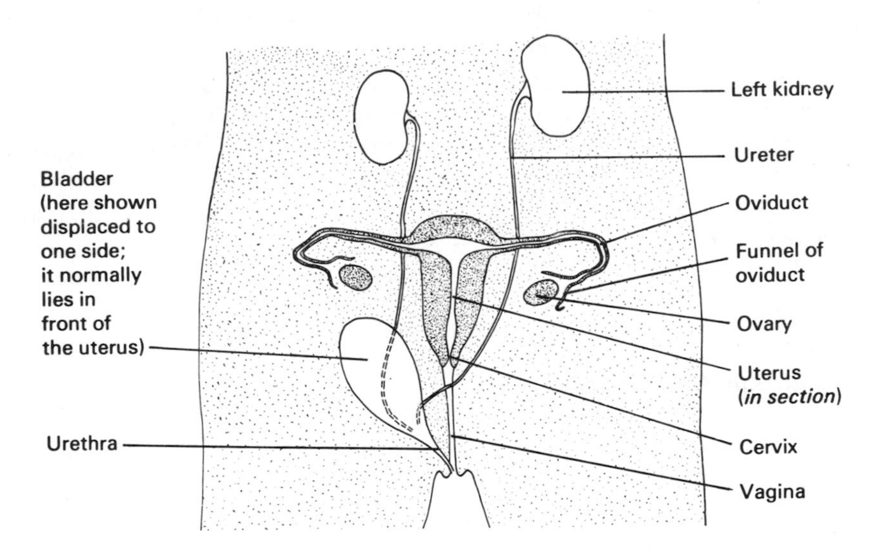

Fig. 25.1 *Female reproductive organs*

long, but it becomes greatly enlarged during pregnancy. The lower end of the uterus, usually almost completely closed by a ring of muscle called the *cervix,* leads into a muscular tube called the *vagina,* which opens to the outside of the body: the lower opening of the vagina is called the *vulva.* The opening of the urethra leading from the bladder lies close to the vulva (Figs. 25.1 and 25.2).

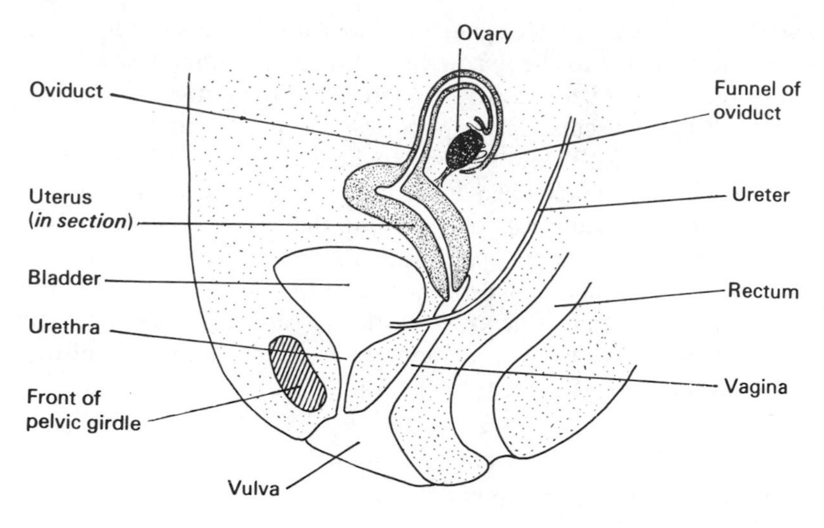

Fig. 25.2 *Female reproductive organs (vertical section, side view)*

(b) Male Reproductive Organs

The sperms are produced in the *testes.* In man the testes lie outside the abdominal cavity, suspended by the *spermatic cords* in a sac of skin called the *scrotum:* this results in the temperature of the testes being lower than that of the rest of the body, a condition favorable to sperm production. Each testis is an oval structure consisting of a mass of narrow tubules

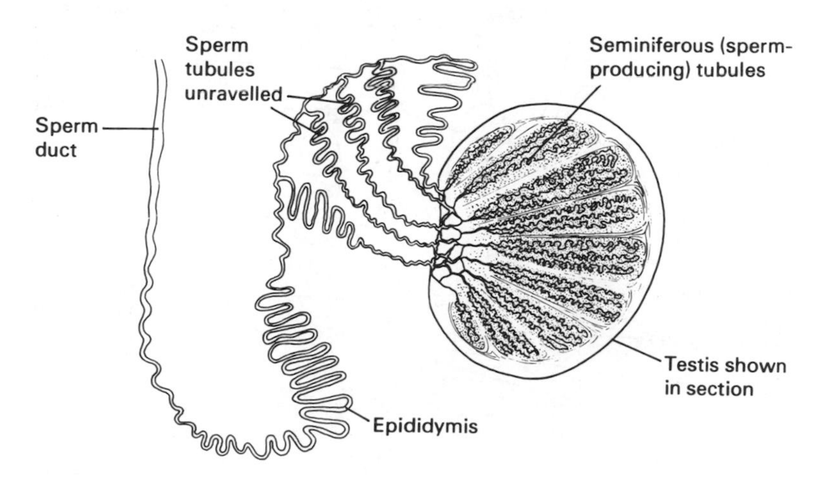

Fig. 25.3 *Diagram to show relation of sperm ducts and testis*

(*seminiferous tubules*), which meet and join up at one side of the organ to form ducts leading to the *epididymis* (Fig. 25.3), a narrow tube about 6 m long lying in a coiled mass on the outside of the testis. The epididymis in turn leads to a muscular *sperm duct;* the two sperm ducts open into the top of the urethra just below the point where it leaves the bladder (Figs. 25.4 and 25.5). A short coiled tube, the *seminal vesicle,* branches from each sperm duct just above its opening into the urethra. Two glands open into the urethra: the *prostate gland,* which surrounds it at the point where it leaves the bladder, and *Cowper's gland,* a little lower down. The urethra in males is prolonged into a *penis,* consisting of connective tissue containing numerous small spaces which are normally empty but which are filled with blood when the penis is erect (see Section 25.4).

25.3 PRODUCTION OF GAMETES

(*a*) Ovulation

The ovary consists of connective tissues, blood vessels and potential ova (Fig. 25.6). It is thought that the ovaries of a newborn girl already contain all her potential ova, about 70,000 in number; but of these only about 500 will mature during her lifetime.

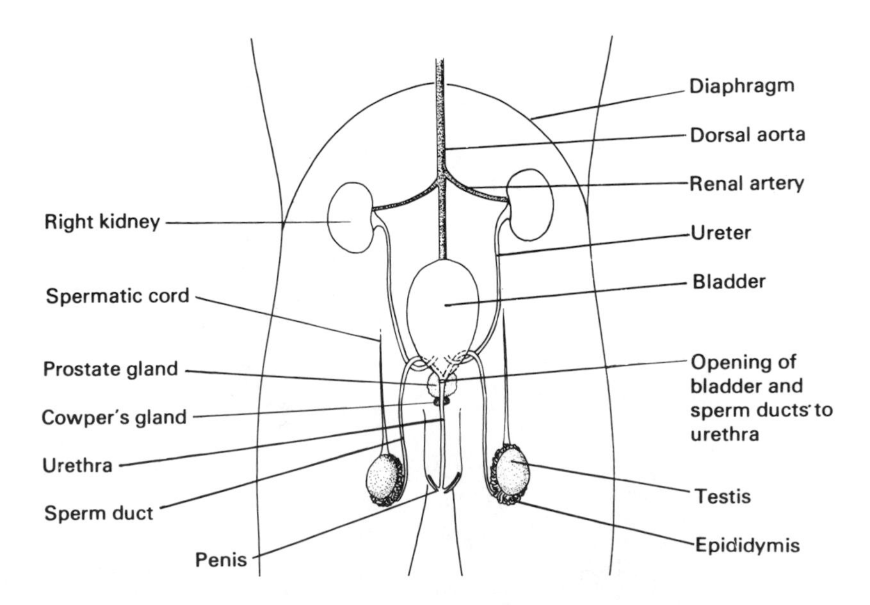

Fig. 25.4 *Male reproductive organs*

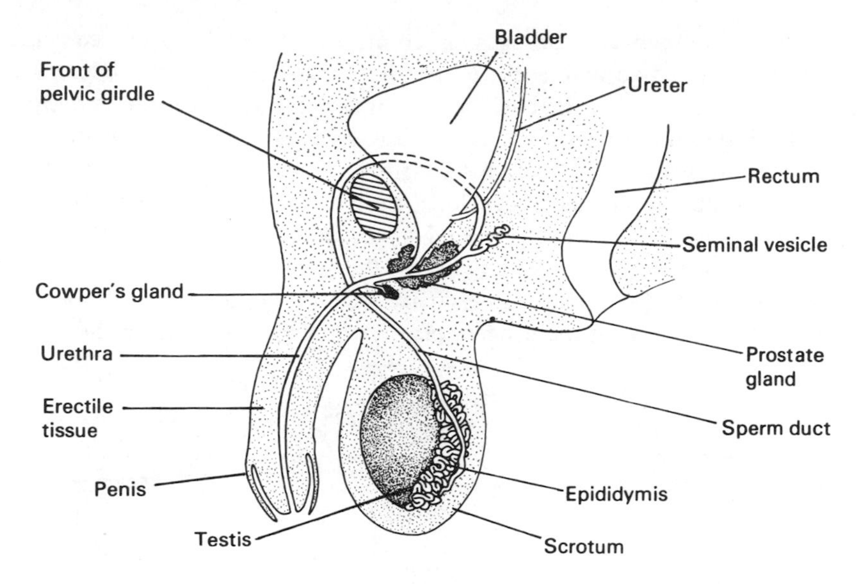

Fig. 25.5 *Male reproductive organs (vertical section, side view)*

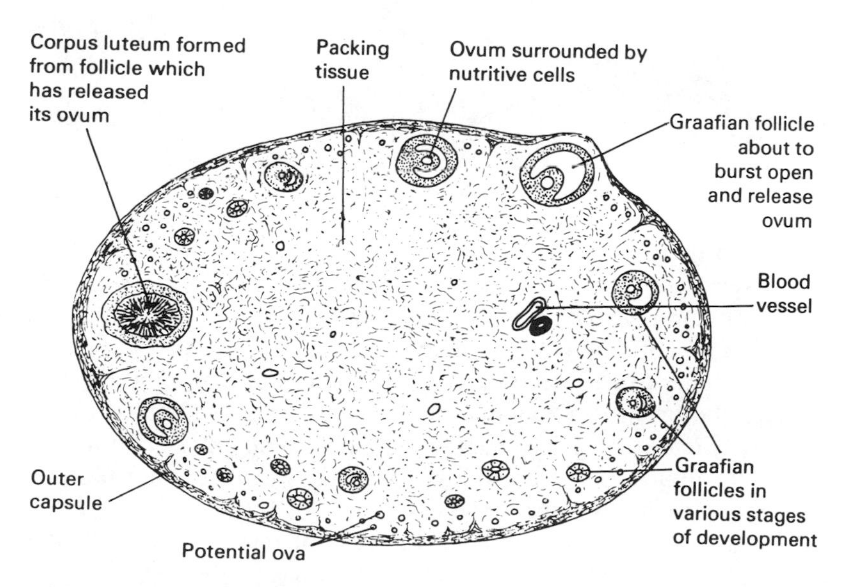

Fig. 25.6 *Section through an ovary (as seen under the low power of a microscope)*

Some time between the ages of about eleven and sixteen years the ovaries become active and begin to produce mature eggs. The beginning of this period of life is called *puberty*. Some of the ova begin to enlarge; the cells around them start dividing rapidly, until each developing ovum is surrounded by a layer of nutritive cells richly supplied with blood vessels. Eventually a fluid-filled cavity develops, partly encircling the ovum and its surrounding cells; the region is called a *Graafian follicle* (Fig. 25.7). A fully developed Graafian follicle is about the size of a pea and protrudes above the surface of the ovary. Every four weeks a Graafian follicle bursts, releasing its ovum; this is the process of *ovulation*. Occasionally in humans,

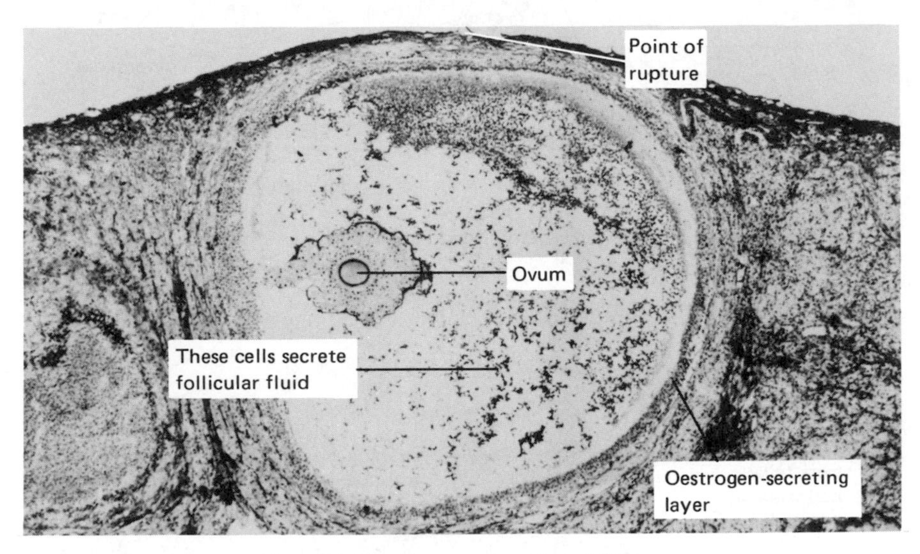

Fig. 25.7 *Mature Graafian follicle*
(Brian Bracegirdle)

but frequently in other animals such as rabbits and mice, more than one ovum is released from the ovary at one time.

The ovum released from the Graafian follicle passes into the funnel-shaped opening of the oviduct, which is lined with cilia; the movements of the cilia flicking to and fro waft the ovum into and along the oviduct. At this stage the ovum is a spherical mass of protoplasm about 0.13 mm in diameter, bounded by a thin membrane and containing a central nucleus. Some of the follicle cells may still be adhering to it. If it is not fertilized, it is thought to live for 24 hours or less.

After the release of the ovum the cells of the Graafian follicle continue to grow and divide, and a small solid structure, the *corpus luteum,* is formed. If the ovum is not fertilized the corpus luteum degenerates after about two weeks and is gradually replaced by ordinary ovary tissue.

(b) Sperm Production

The seminiferous tubules making up the testis are lined with actively multiplying cells which ultimately give rise to sperms (Fig. 25.8). Each sperm is a nucleus surrounded by a thin layer of cytoplasm which extends into a long tail (Figs. 25.9 and 25.10). From the testis they pass into the

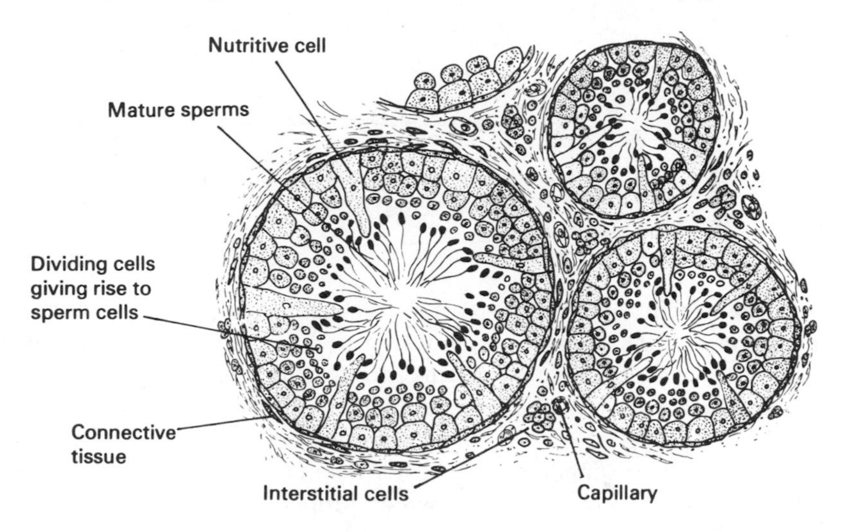

Fig. 25.8 *Section through seminiferous tubules of mammalian testis (greatly enlarged)*

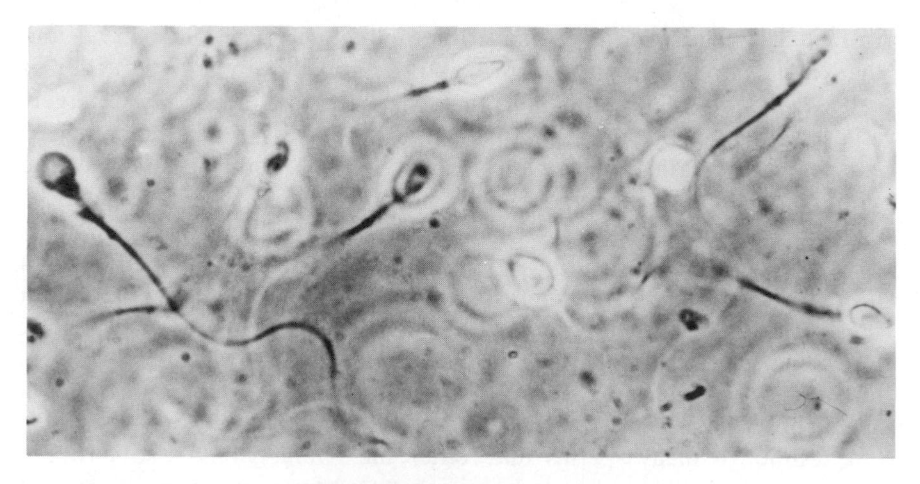

Fig. 25.9 *Human sperms*
(Brian Bracegirdle)

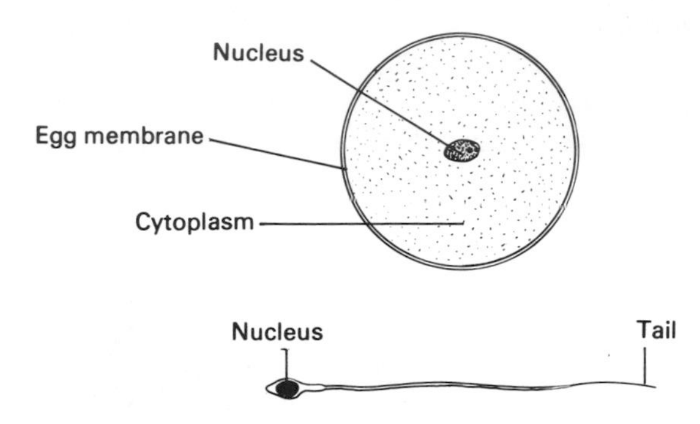

Fig. 25.10 *Sperm and ovum of a mammal (not to scale)*

epididymis where they are stored. During mating (*copulation*), muscular contractions of the epididymis and sperm duct and of their accessory muscles force the accumulated sperms through the urethra. Here secretions from the seminal vesicles, the prostate gland and Cowper's gland are added to the fluid now called *semen.* The presence of the nutrients and enzymes in these secretions stimulates the hitherto immobile sperms into action; their tails begin lashing movements which propel them along.

25.4 FERTILIZATION

In humans, fertilization takes place internally. Sperms are introduced into the female through the penis, which is placed in the vagina: the action is called *copulation.* To facilitate this the penis becomes erect and firm, largely as a result of blood flowing into the blood spaces more rapidly than it escapes, so increasing the turgidity of the tissues around the urethra. The stimulation of the sensory organs in the penis sets off a reflex action, *ejaculation,* which results in some hundreds of millions of sperms accumulated in the semen being expelled into the vagina. The sperms swim through the cervix, into the uterus and up the oviduct. If there is an ovum present in the oviduct, the head of a sperm will probably eventually stick to it and penetrate the surrounding membrane. The sperm's nucleus passes into the cytoplasm and fuses with the female nucleus there to form a zygote or fertilized egg cell.

Although many sperms may enter the oviduct, only one fertilizes the ovum; once the ovum has been penetrated by a sperm, no more will usually be admitted. In some animals the remaining sperms produce an

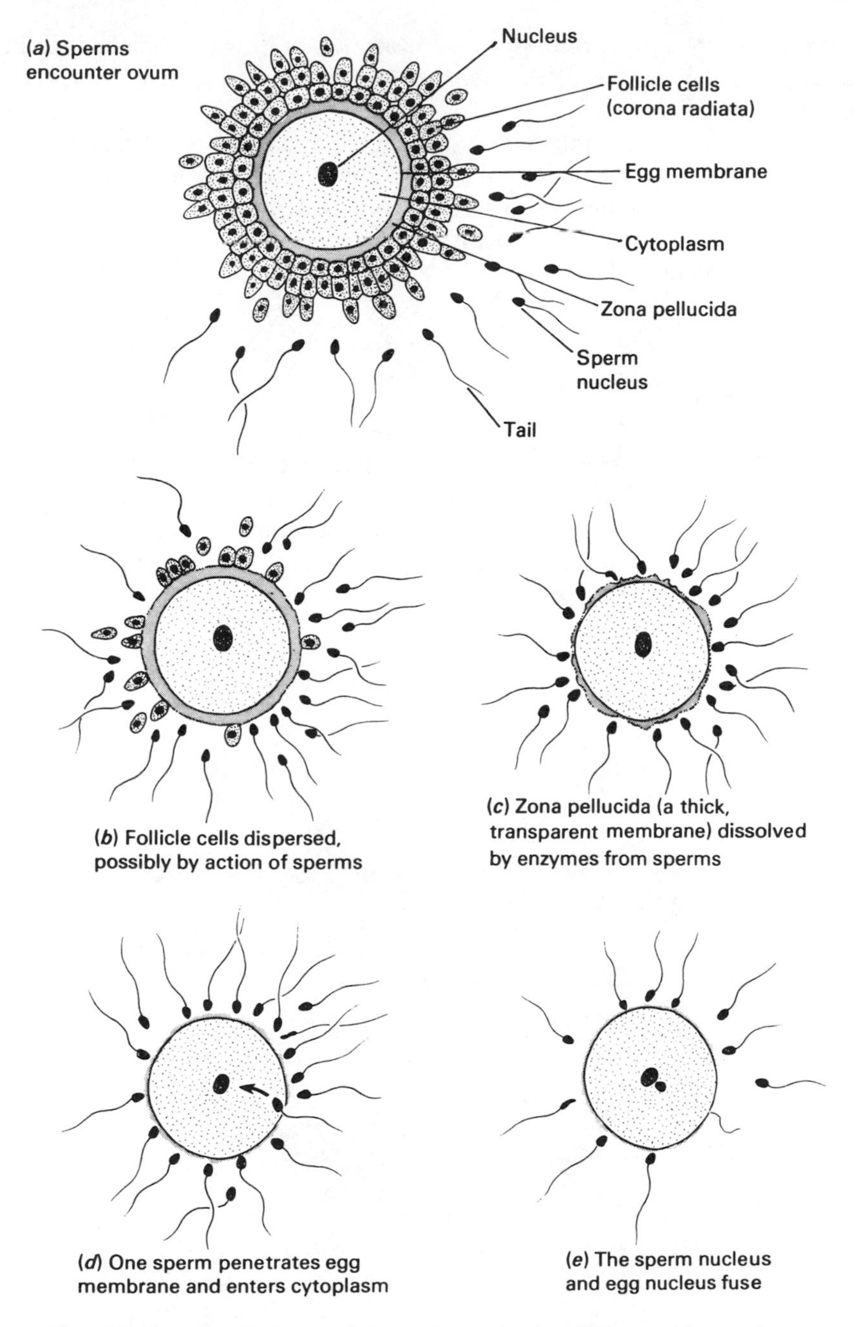

Fig. 25.11 *Fertilization of human ovum (the diagrams show what is thought to happen but the course of events is not known for certain)*

enzyme which helps to disperse the *corona radiata* (the follicle cells still adhering to the surface of the ovum), but this action has not yet been confirmed in humans.

A sperm can probably live for two or three days in the female reproductive organs. Since only one ovum is produced each month, and since its life if unfertilized is thought to be no more than a day or so, there is a relatively short period of three or four days each month during which fertilization can take place (Fig. 25.18).

25.5 PREGNANCY AND DEVELOPMENT

The fertilized egg undergoes rapid cell division (Fig. 25.12) and is referred to as an *embryo*. It passes down the oviduct and into the uterus where it first adheres to and then sinks into the uterine lining. When an ovum is fertilized, the corpus luteum in the ovary does not degenerate but enlarges and persists, secreting a hormone called *progesterone*. Progesterone brings

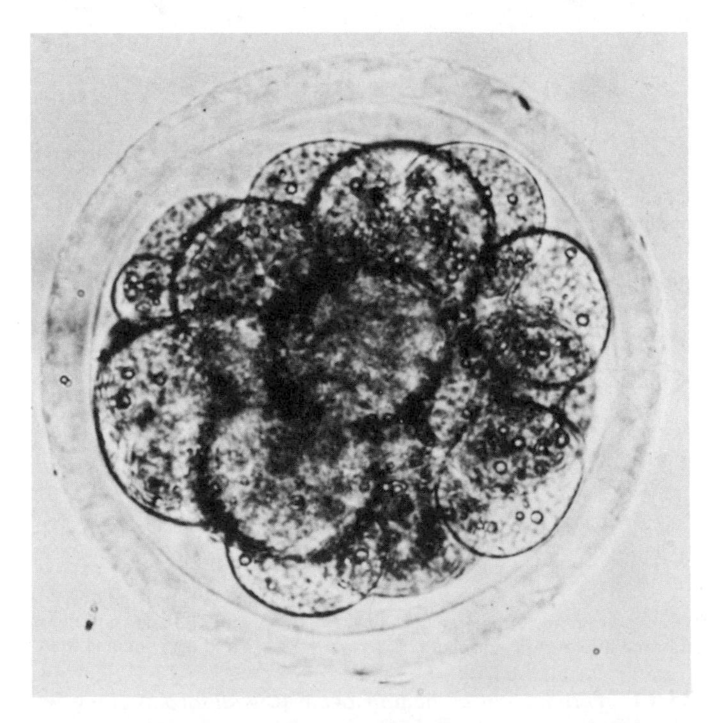

Fig. 25.12 *Early stages of cell division in zygote of sheep* (W. J. Hamilton)

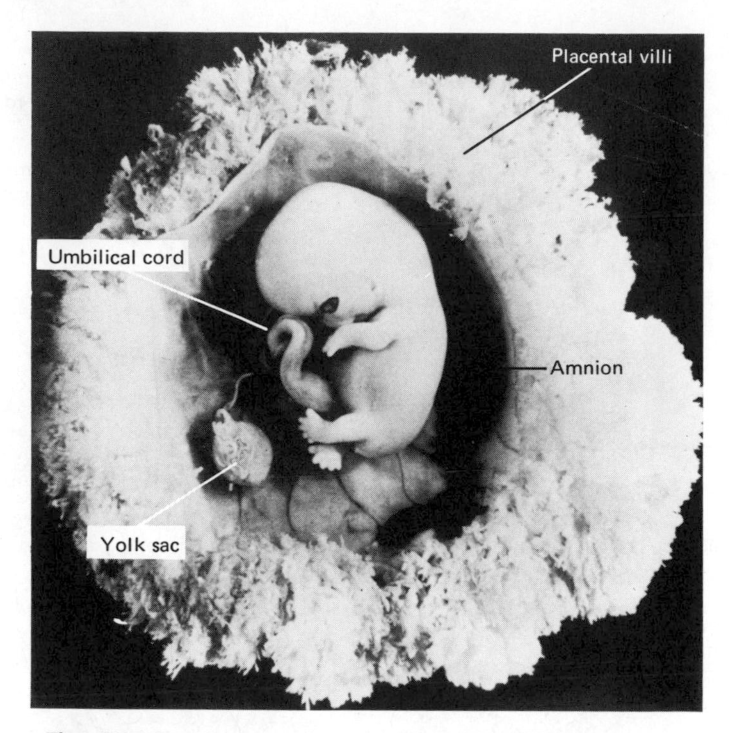

Fig. 25.13 *Human embryo, seven weeks after conception*
(W. J. Hamilton)

about changes in the uterus which enable it to accept the developing embryo and to accommodate it during its development. Initially the uterus has a volume of 2 to 5 cm^3, but it extends with the growth of the embryo to 5 000 to 7 000 cm^3, enlarging the abdomen and displacing the organs in it to some extent. Under the influence of progesterone and of other hormones produced in the ovary, called *estrogens,* the uterine lining becomes thicker and it develops a rich supply of blood vessels.

The cells of the embryo grow and divide repeatedly; initially they all appear alike, but they soon develop differing structures and functions, and eventually form the various specialized tissues of the new individual. The tissues grow and extend in relation to each other, and give rise to the organs of the body, which are recognizable quite early in pregnancy (Fig. 25.13): the embryo's heart and circulatory system, for example, are formed within four weeks of fertilization.

During the rapid growth that takes place in pregnancy, the embryo obtains all the oxygen and nutrients it needs through the *placenta.* This is a disc of tissue adhering to the lining of the uterus (Figs. 25.14(*b*) and (*c*)); minute finger-like processes called *villi* containing blood capillaries

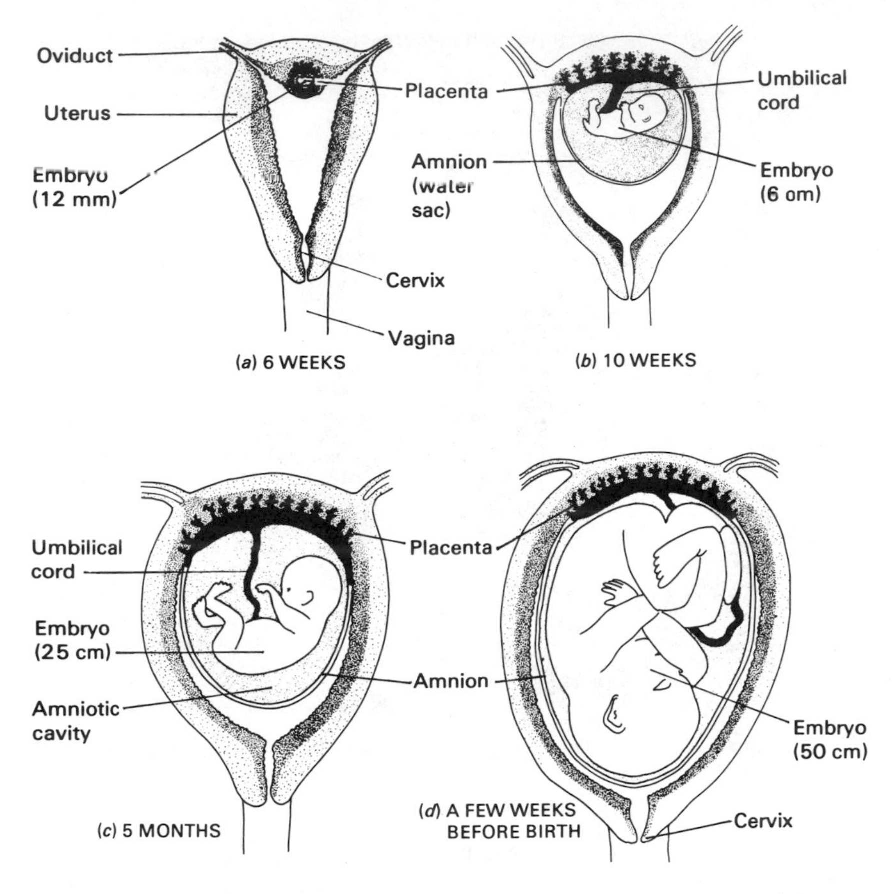

Fig. 25.14 *Growth and development in the uterus (not to scale)*

protrude from it into the uterine lining. Blood from the embryo's circulatory system is carried to and from the capillaries in the villi by an artery and a vein running in the *umbilical cord* which connects the embryo's abdomen to the placenta. The capillaries of the placenta are separated from blood spaces or *sinuses* in the uterine lining by very thin membranes: hence dissolved substances can pass between the two circulatory systems in either direction (Fig. 25.15). Dissolved oxygen, glucose, amino acids and salts in the mother's blood pass from the uterine blood vessels into the capillaries of the embryo, while the waste products of the embryo's metabolism, mainly carbon dioxide and urea, diffuse across in the opposite direction and are carried away in the mother's blood and excreted through her lungs and kidneys (Fig. 25.16).

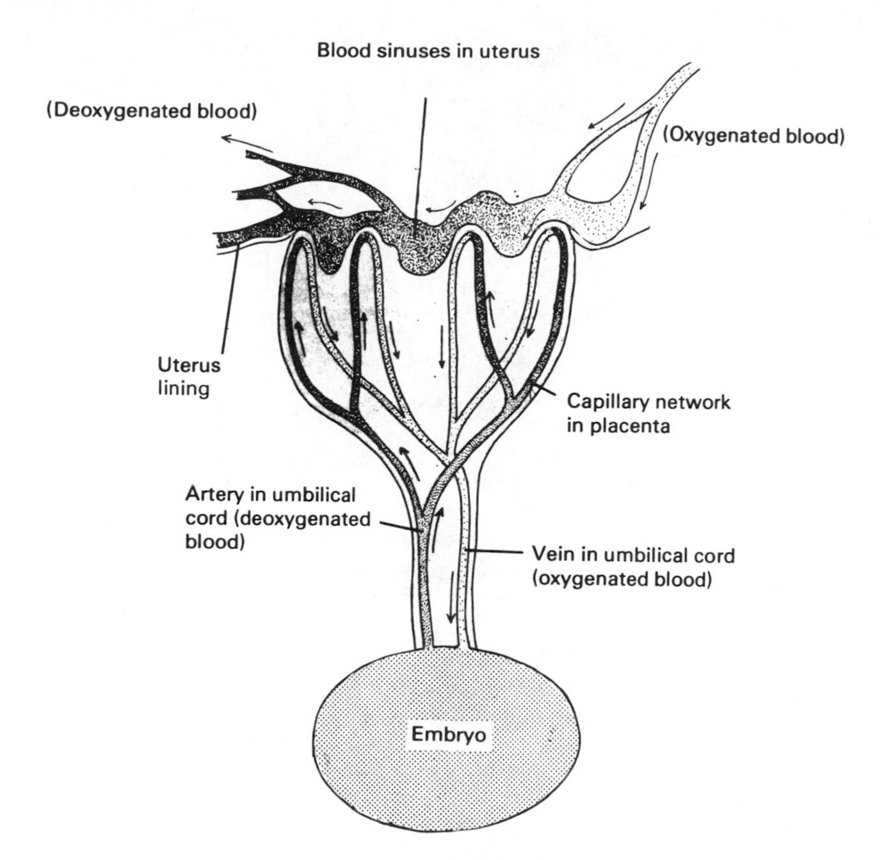

Fig. 25.15 *Diagram to show relationship between blood supply of embryo, placenta, and uterus*

Although the embryo depends on the mother's blood for its food and oxygen, its circulatory system is never directly connected with the maternal blood vessels. If it were, the adult's blood pressure would burst the delicate capillaries forming in the embryo. In addition, the placental membranes separating the two circulatory systems have the ability to prevent some harmful substances from reaching the embryo.

The placenta also secretes hormones, both estrogens (see Section 25.8) and progesterone, throughout pregnancy. While it does so, the Graafian follicles cease to develop and ripen, and ovulation stops.

Throughout its development the embryo is surrounding by fluid, *amniotic fluid,* held in a *water sac* or *amnion* (Fig. 25.14(*c*)), which protects it from injury or distortion by equalizing the pressures acting upon it. Quite early in pregnancy the embryo begins to move its limbs inside the water sac; after four or five months the movements are fairly vigorous.

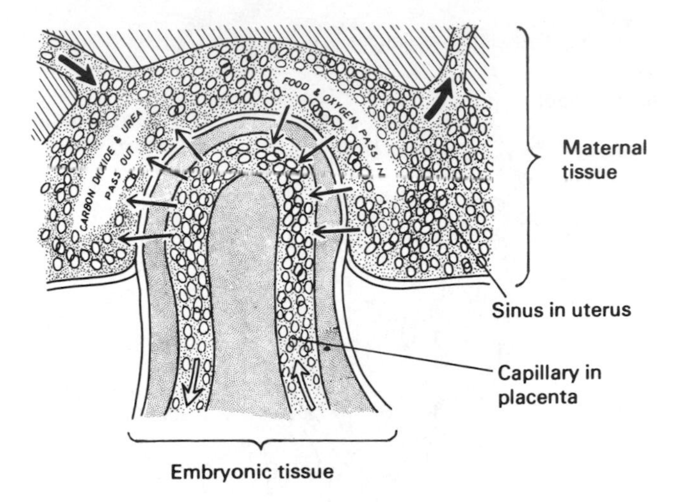

Fig. 25.16 *Diagram to show exchange of oxygen and food from uterus to placenta*

The period from the fertilization of the ovum to birth, the *gestation period,* is about nine months in humans. It varies greatly from one species to another, but is longest in the higher animals. In elephants it is two years.

Twins

Sometimes a fertilized ovum separates completely into two parts at an early stage of development, and each part grows separately into a normal embryo. Such "one-egg" or *identical twins* are indeed identical in nearly every physical respect, although differences in position and blood supply while in the uterus may produce differences in weight and vigor at birth. Twins also result if two ova are released from the ovaries simultaneously, and if both are fertilized. These *fraternal twins* may be of different sexes, and are not necessarily any more alike than other brothers and sisters of the same family.

25.6 BIRTH

A few days before birth the embryo turns head downward in the uterus so that its head comes to lie just above the cervix (Figs. 25.14(*d*) and 25.17).

The process of birth begins with regular contractions of the muscles in the uterine wall; this is the onset of what is called "labor." These con-

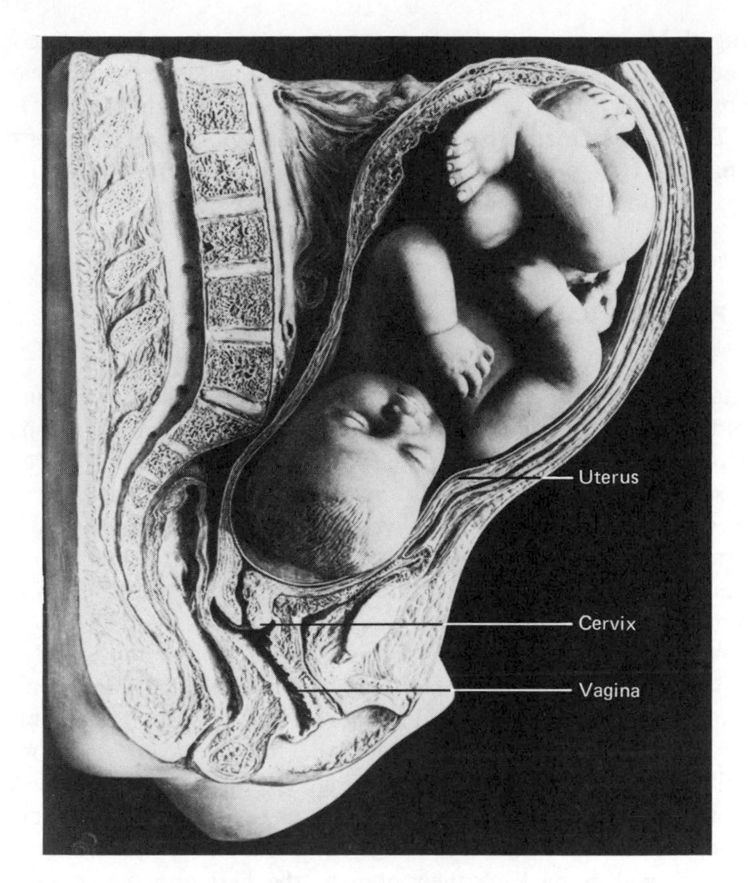

Fig. 25.17 *Model of human fetus just before birth*

(Reproduced with permission from the *Birth Atlas,* published by the Maternity Center Association, New York)

tractions become gradually stronger and more frequent. The opening of the cervix dilates little by little to let the child's head pass through, and the uterine contractions are reinforced by pressure from the muscles of the diaphragm and abdominal wall. At some point during the contractions the water sac breaks and the amniotic fluid escapes through the vagina. Finally the muscular contractions of the uterus and abdomen expel the child head first through the dilated cervix and vagina. The sudden fall in temperature experienced by the newly born baby stimulates it to take its first breath, usually accompanied by crying.

The umbilical cord which still connects the child to the placenta is tied tightly to prevent bleeding, and cut. Later, the placenta breaks away from the wall of the uterus and is expelled separately as the "afterbirth."

During the days after the birth the remains of the umbilical cord still attached to the baby's abdomen shrivel and fall away, leaving a scar in the abdominal wall, called the *navel*. The average birth weight of babies is 3 kg.

The baby is ready to suck at the breast shortly after its birth. The mammary glands in the breast, which have become enlarged during pregnancy, are stimulated by the baby's sucking to secrete milk.

25.7 PARENTAL CARE

All mammals protect and care for their young in various ways until they are old enough to move about efficiently and obtain their own food. Most prepare a nest in which to bear their young, and in this way the babies are protected from predatory animals and from temperature changes. The nest reduces the chance that they may wander away from their parents and get lost, and prevents them from injuring themselves. Young mammals of many species are born without fur, but the shelter of the nest and the warmth of the mother's body protect them from overcooling.

All mammals suckle their young. The milk contains nearly all the food, vitamins and salts that the young animals need for their energy requirements and tissue building, but there is no iron in it for hemoglobin manufacture. All the iron needed for the first weeks or months of the animal's life is stored in its body during its gestation in the uterus. The mother's milk is secreted more plentifully as the young animals grow and their needs increase. Later, the milk is supplemented by solid food; carnivorous mammals bring back their prey to the nest and tear it into pieces small enough for the young to swallow.

Often parent animals are particularly aggressive when they are caring for young, and they may react violently to intruders.

When the young animals are old enough to obtain their own food and escape from predators, they leave the nest and disperse. The period of dependence on the parents varies from one species to another; in humans it is lengthy.

25.8 SECONDARY SEXUAL CHARACTERS

Both male and female reproductive organs secrete a number of hormones (see Section 29.4) into the bloodstream: the hormones secreted by the testes are mainly substances called *androgens* and those produced by the ovaries are mainly *estrogens*. At puberty these hormones are released into the bloodstream and circulate around the body; they stimulate the development of certain tissues, producing the features characteristic of masculinity and femininity, the *secondary sexual characters.*

In males, the voice becomes deeper, hair begins to grow on the face, in the armpits and in the region of the external reproductive organs, and the body becomes more muscular.

In females the breasts grow and the pelvic girdle enlarges, and body hair begins to grow in the armpits and in the region of the vulva.

25.9 MENSTRUATION

Whenever an ovum is released from an ovary, the lining of the uterus thickens and its blood supply increases. If the ovum is fertilized and embryological processes begin, the embryo sinks into the thickened lining and continues its development as described in Section 25.5. However, the great majority of the 500 or so ova produced during a woman's lifetime are not fertilized, and die shortly after their release. The new uterine lining then disintegrates and its fragments, together with some blood and mucus, are discharged from the body through the cervix and the vagina, 12 to 14 days after ovulation (Fig. 25.18). This period of *menstruation* lasts for a few days, and takes place about every four weeks. Menstruation is only known in certain monkeys and apes and in humans.

25.10 BIRTH CONTROL

When people wish to limit the size of their families they usually make use of one or other form of birth control. This may involve restricting copulation to a period during which fertilization is unlikely to occur that is, 5–10 days or 18–28 days after the onset of menstruation (Fig. 25.18). This is known as the *rhythm method.* However, this is not a very reliable method because the time of ovulation and the length and regularity of the menstrual periods vary considerably, not only between individuals but in the same individual from time to time.

The other principal methods of contraception are:

(a) **The sheath or diaphragm.** The sperms are kept from entering the cervix and reaching the ovum by the interposition of a physical barrier. This may be a thin rubber sheath worn on the penis of the male (called a condom) or a small rubber diaphragm inserted in the vagina of the female.

(b) **The intra-uterine device (IUD).** A small plastic strip bent into a loop or coil is inserted through the cervix into the uterus. The device may be

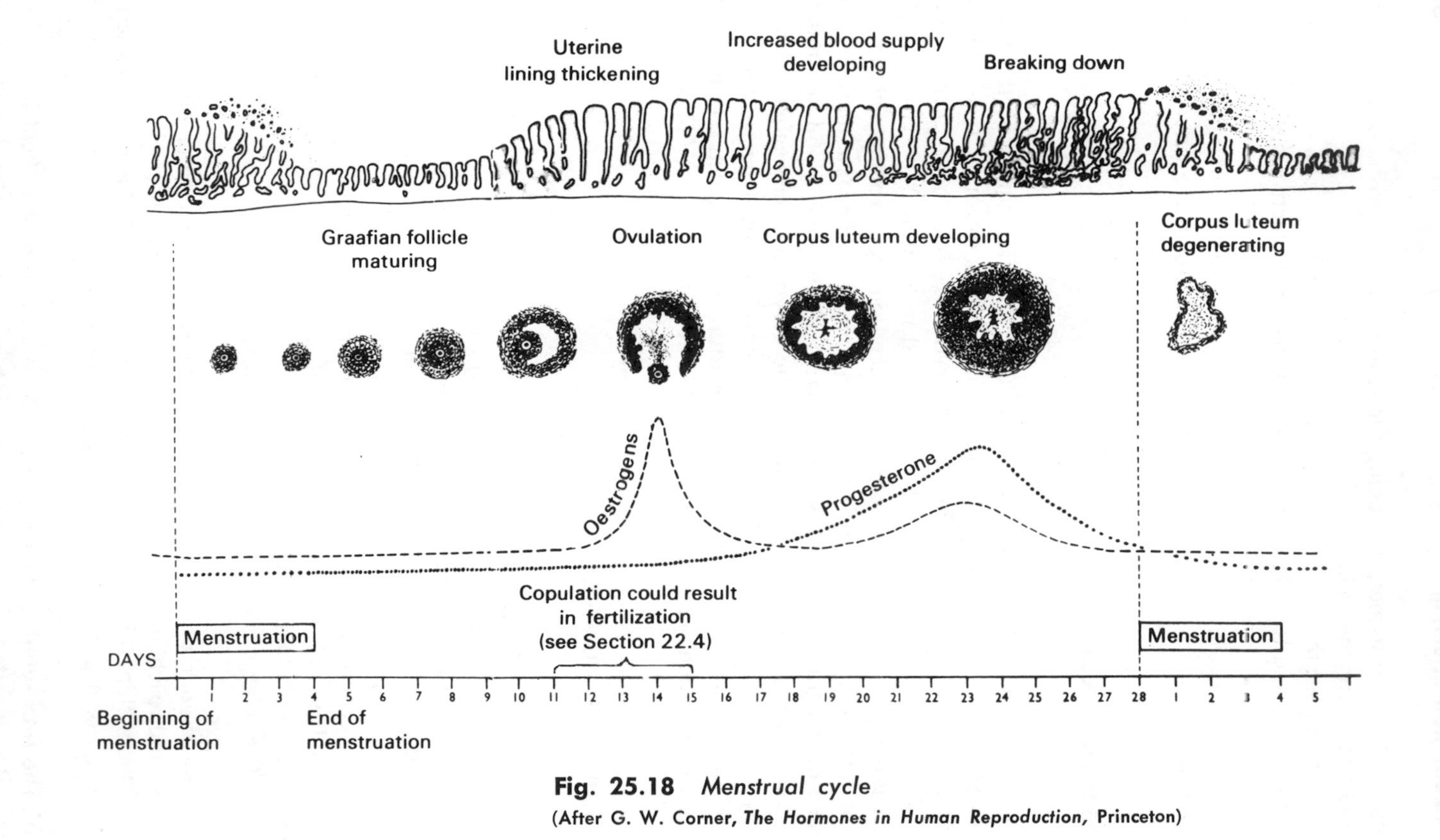

Fig. 25.18 *Menstrual cycle*

(After G. W. Corner, *The Hormones in Human Reproduction,* Princeton)

retained there for several years and while it is in place it is very effective in preventing pregnancy. It is not certain whether its action is due to interference with the fertilization of the ovum, or to prevention of the implantation of the fertilized ovum in the uterine wall.

(c) **The contraceptive pill.** The "pill" contains synthetic chemicals which have biochemical effects similar to those of the estrogens and progesterone. When taken in suitable proportions these hormones suppress ovulation and so prevent conception. The pills must be taken daily for the 21 days between menstrual periods, and are then almost 100 per cent effective.

25.11 WORLD POPULATION

With the increasing extent and application of medical knowledge, fewer people, and especially fewer children, die from infectious diseases year by year, and in most parts of the world the average lifespan has increased considerably during the last few decades. The birth rate, however, has not decreased in proportion to the falling death rate, and the world population is rising with increasing speed: it is now doubling in number every 35 years or less. In countries with limited natural resources, this leads to shortages of food and living space, while in the industrialized countries the population increase has led to pressing problems in the disposal of industrial wastes and sewage, with the attendant dangers of pollution, and to environmental damage by intensive methods of food production (see Unit 33).

There is clearly a physical limit to the number of people who can live on the Earth, although authorities differ in their estimates of this number. Therefore it seems essential, at the very least, to educate people into accepting the need to limit their families and in methods of doing so. The alternative to voluntary population control is involuntary control by famine and disaster.

25.12 QUESTIONS

1. In what ways does a zygote differ from any other cell in the body?
2. What is the advantage to the embryo of the early development of its heart and circulatory system?
3. What differences are there in the numbers, structure and activity of the male and female gametes in humans?
4. What do you consider are the advantages of (*a*) internal fertilization over external fertilization and (*b*) development of the embryo in the uterus rather than in the egg?

5. In what ways is parental care in humans similar to and different from parental care in other mammals?
6. List the changes in the composition of the maternal blood which are likely to occur when it passes through the placenta.
7. Explain why there are only a few days in each menstrual cycle when fertilization is likely to occur.

Unit 26
The Skeleton, Muscles, And Movement

26.1 INTRODUCTION

The hard parts of an animal's body, such as the bones and teeth of a human or the shell of a snail, are produced by living cells. Frequently they contain non-living material, usually calcium salts and especially calcium phosphate and carbonate, and they may consist almost entirely of such substances. Such structures may, however, be able to grow and change as a result of the activities of living cells which dissolve away and replace the hard materials.

Exoskeletons. Where the hard material is formed mainly on the outside of the body it is often called an *exoskeleton.* Insects such as beetles or dragonflies and crustaceans like crabs or lobsters have a hard covering to their bodies called a *cuticle;* projections extend from the exoskeleton into the body cavity for muscle attachment (Fig. 15.12). Exoskeletons cannot grow: animals with exoskeletons increase their size by periodically replacing their cuticles, a process called *ecdysis.* The cells lying just beneath the cuticle dissolve and absorb its inner layers; the outer layers split and are shed from the body, while a new cuticle forms on the exposed surface (for a more detailed discussion see Section 15.3).

Endoskeletons. All vertebrate animals have *endoskeletons,* rigid structures within their bodies. These are made of bone and cartilage. The *bones* consist mostly of mineral material, but also contain living cells and protein fibers. Although the bones are hard and rigid, they can grow by a steady increase in size. Their mineral content is constantly being exchanged with mineral salts in the blood. For example, if the calcium content of the diet is low, calcium from the bones will go into solution in the blood, and will be redeposited when the supply again becomes adequate. *Cartilage* is a tissue which is widely distributed in the body and is not restricted to the skeleton. Its structure varies according to its position and function. The

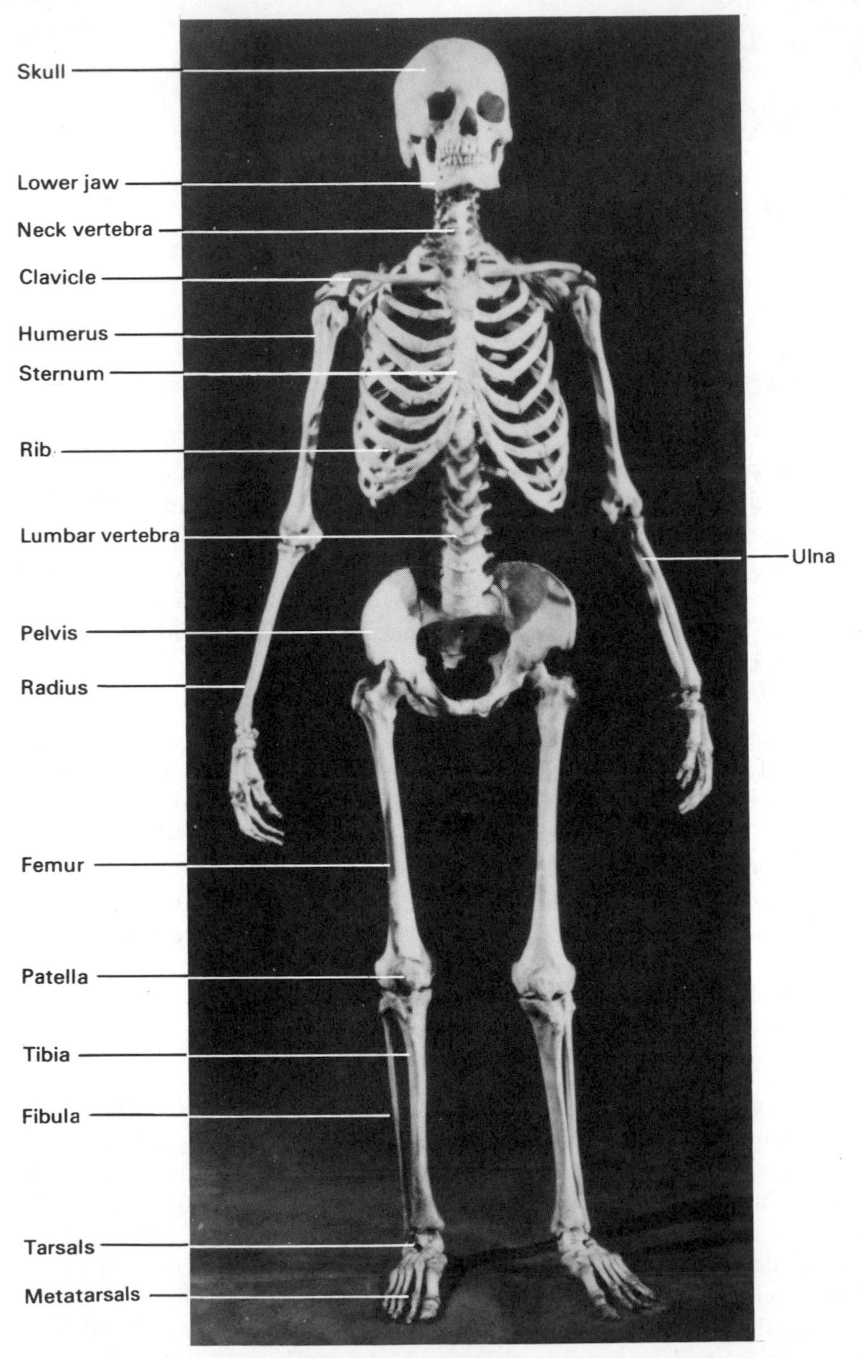

Fig. 26.1 *Skeleton of man*
(Rank Organization)

cartilage of the ear pinna is tough and flexible, containing many elastic fibers. That on the articular surfaces of joints (see Section 26.3) is translucent and smooth, with few fibers. The cartilaginous discs between the spinal vertebrae (Fig. 26.3(*b*)) contain a high proportion of tough inelastic collagen fibers.

26.2 FUNCTIONS OF THE SKELETON

(*a*) Support

Many invertebrate animals have no skeleton at all. The weight of the tissues of jellyfish and similar creatures is supported and buoyed up by the water surrounding them. In others, like earthworms, the organs are supported by the pressure of their body fluids acting against a strong muscular body wall. In animals like insects or crustaceans, the rigidity of the exoskeleton prevents the soft inner tissues of the animal from collapsing.

The bony framework of a vertebrate maintains the shape of its body even during vigorous muscular activity, and also supports some of the vital organs and prevents them from pressing against and distorting each other. In most land-dwelling animals the skeleton permits the body to be raised up on its limbs; this allows more rapid movement than can be achieved with the body dragging along the surface of the ground. The backbone or *spine* of the rabbit (Fig 26.2), for example, takes the form of a bridge-like arch or span from which the organs of the body are suspended.

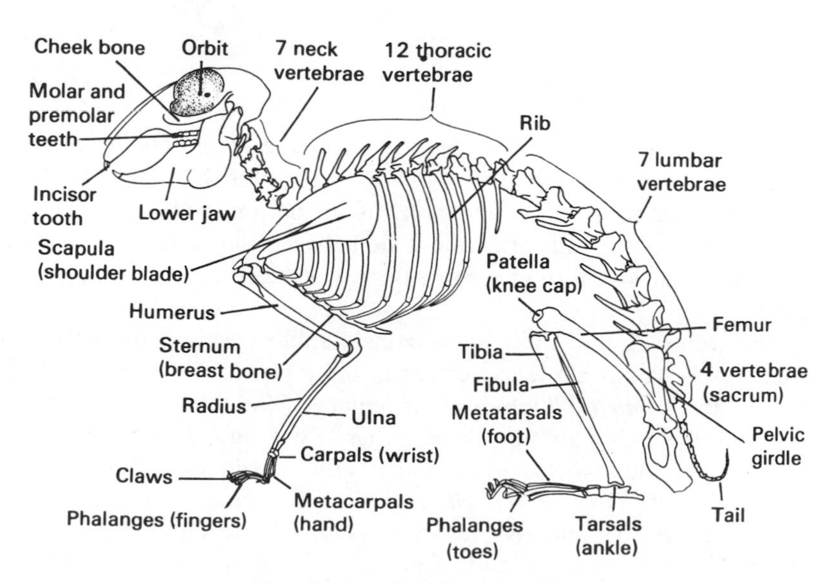

Fig. 26.2 *Skeleton of rabbit*

(b) Protection

In animals with exoskeletons, the entire body is protected by the hard, horny cuticle. The surface of a vertebrate's body is more vulnerable, but many of the most important and delicate organs are protected by a casing of bone, which has great mechanical strength both in compression and in tension. The brain is completely enclosed by the skull, and the spinal cord by the backbone, while the heart and lungs are surrounded by a cage of ribs. The organs are thus protected from distortion resulting from pressure and from injury resulting from impact. The rib-cage is important in the breathing mechanism (see Section 22.4), in addition to protecting the organs of the thorax.

(c) Movement

The internal projections of the exoskeleton and the bones of the endoskeleton provide fixed points for the attachment of muscles (see Figs. 26.8 and 26.9 and Section 26.4). The contraction of these muscles produce movements, such as limb and jaw action. *Locomotion* is the result of the coordinated action of muscles on the limb bones; locomotion of mammals is discussed in Section 26.6, and that of insects, fish, amphibians and birds in Section 15.7, 16.3, 17.3 and 18.2 respectively.

26.3 JOINTS

A joint is formed wherever two bones meet. Some are immobile, as at the joints or sutures between the bones of the skull or of the pelvis. In others only very limited movement can occur, as in the joints of the spine (Fig. 26.3) where disc-like pads of cartilage lie between the vertebrae and allow just sufficient movement to give some degree of flexibility to the spine as a whole.

The joints of the limbs allow considerable freedom of movement. *Ball-and-socket* joints allow movement in three planes: for example, at the hip where the *femur* or thigh bone articulates (that is, makes a movable joint) with the *pelvis* or hip girdle (Figs. 26.4 and 26.5), and at the shoulder where the *humerus* in the upper arm articulates with the *scapula* or shoulder blade (Fig. 26.6). *Hinge joints,* like those in the elbow and fingers (Fig. 26.10) and in the knee (Fig. 26.7), allow the rounded processes or *condyles* at the end of one bone to move against the hollow in the other in one plane only.

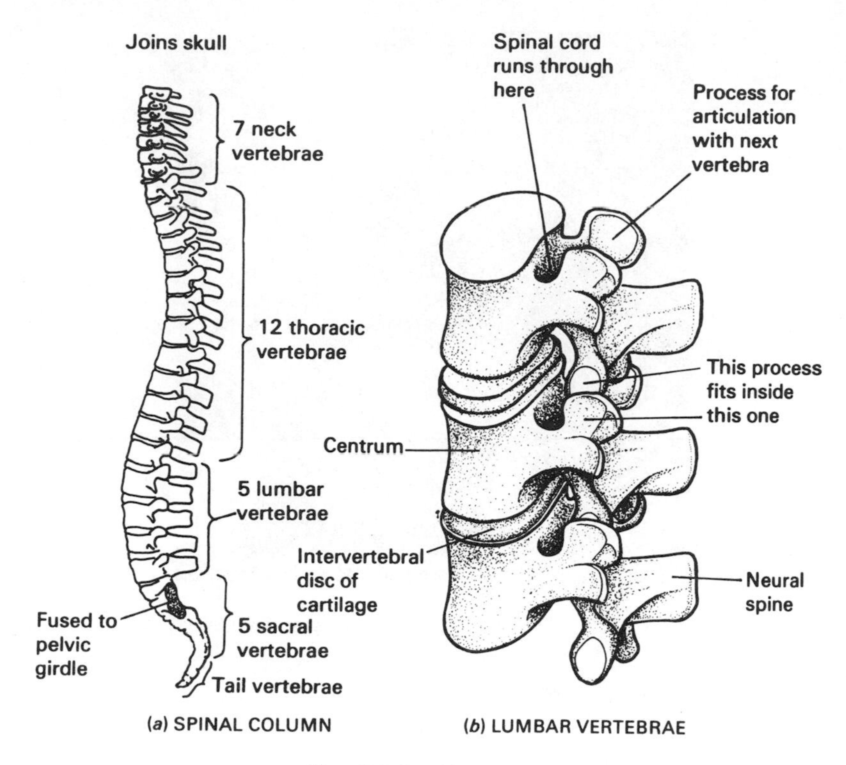

Fig. 26.3 *Human spine*

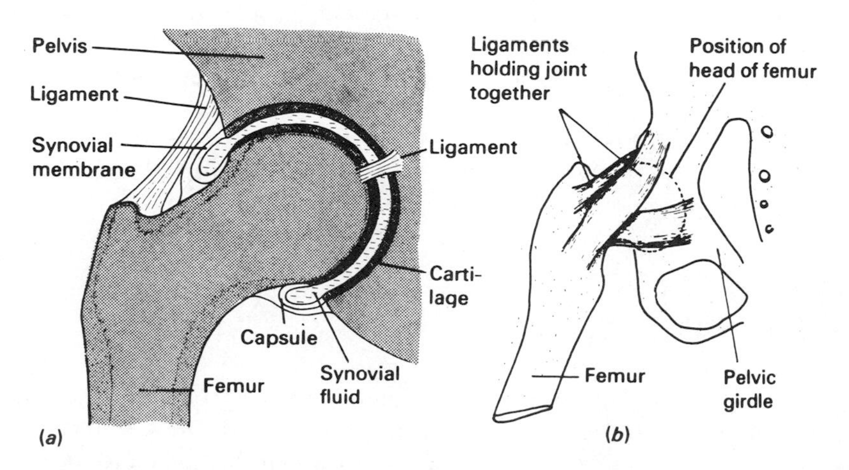

Fig. 26.4 *Ball-and-socket joint of hip: (a) vertical section; (b) external view*

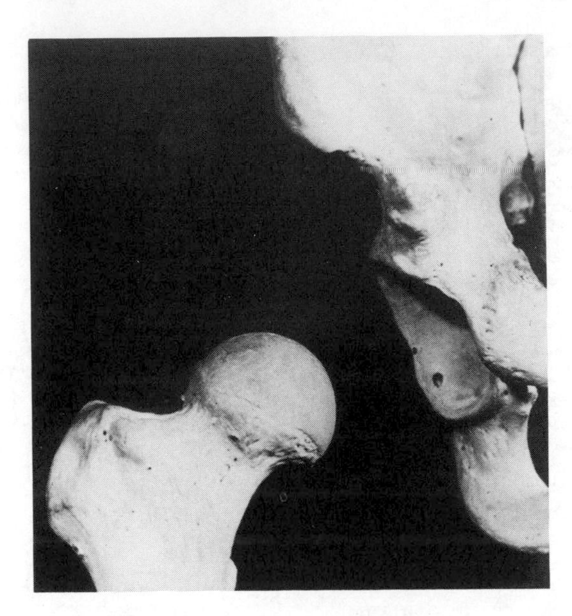

Fig. 26.5 *The hip joint (disclosed to show socket)*
(Rank Organization)

Where the surfaces of bones move over each other they are covered with a smooth, slippery layer of cartilage. This layer is lubricated with a viscous fluid, the *synovial fluid,* held within a capsule of fibrous material surrounding the entire joint; the lining of the capsule, the *synovial membrane* (Figs. 26.4 and 26.7), secretes the synovial fluid. Joints of this kind are called *synovial joints,* and movement of the bones within them is virtually friction-free.

The bones of the joint are held together with strong fibrous strands called *ligaments* (Fig. 26.4), which prevent dislocation of the joint during normal movement.

26.4 MUSCLES

A muscle is made up of bundles of elongated cells lying side by side, sheathed in connective tissue. At each end it narrows to a strong, inextensible strand of connective tissue, a *tendon,* which is attached to the tough membrane or *periosteum* which covers all the bones of the skeleton (Fig.

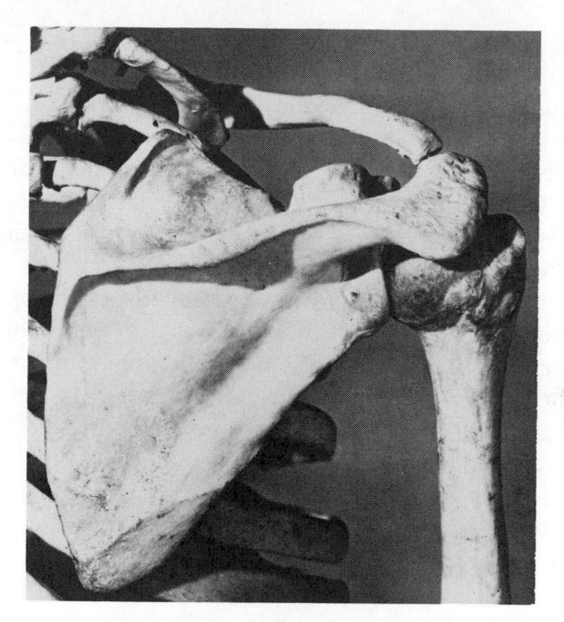

Fig. 26.6 *The shoulder joint*
(Rank Organization)

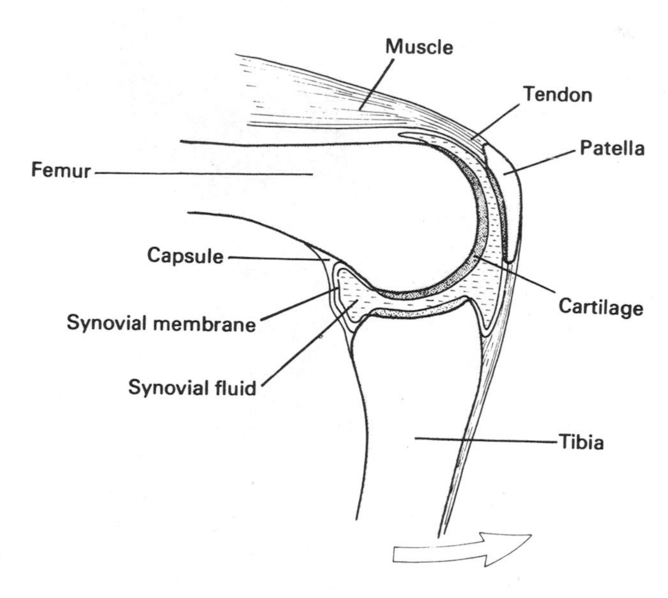

Fig. 26.7 *Hinge joint of knee (vertical section)*

26.9). Bones frequently have projections or ridges on their surfaces at the points where muscles are attached (Fig. 26.8).

If a muscle cell is stimulated by a nervous impulse, it contracts to two-thirds or less of its resting length, so that when the muscle as a whole contracts it becomes shorter and thicker. Since each end of the muscle is attached to a bone, usually across a joint, the contraction produces movement. If one end is attached to a bone which is fixed in position, then only one part of the limb can move when the muscle contracts. For example, the biceps muscle in the upper arm is attached at its upper end to the scapula in the trunk and at its lower end to the radius of the forearm; when it contracts only the forearm is normally able to move (Fig. 26.12). Sometimes the "fixed" end of the muscle is attached to the upper part of the limb; for example, the upper end of one of the muscles which extends the foot is attached to the femur in the thigh.

The muscles of the limbs act across the joints in such a way that the bones are moved like levers, with the joint itself as the fulcrum of the lever and the applied force acting at the point of attachment of the muscle

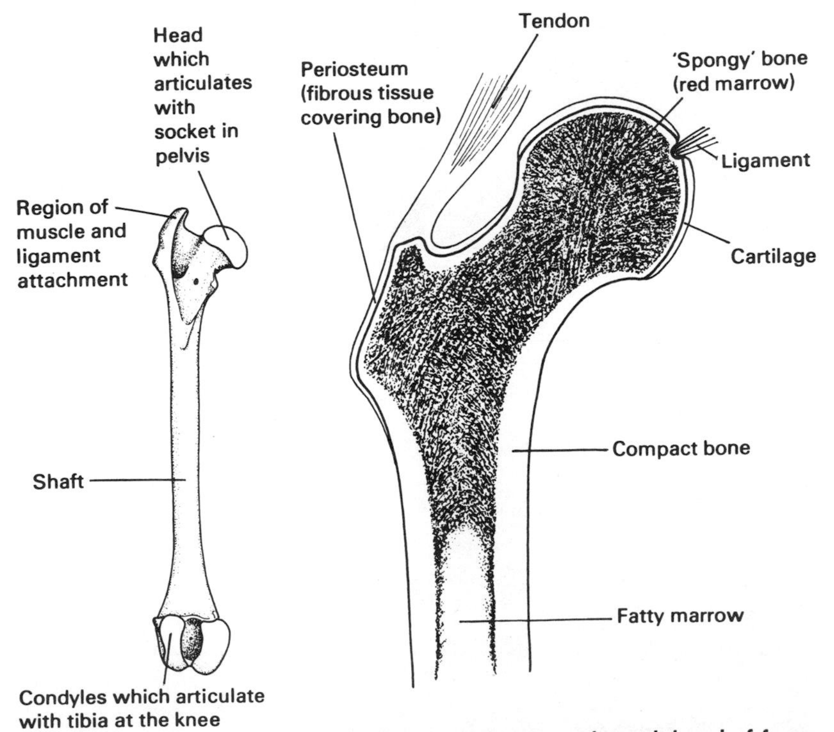

Fig. 26.8 *Femur of rabbit*

Fig. 26.9 *Section through head of femur*
(After Koch, *Amer. J. Anat.*, 21, 1917)

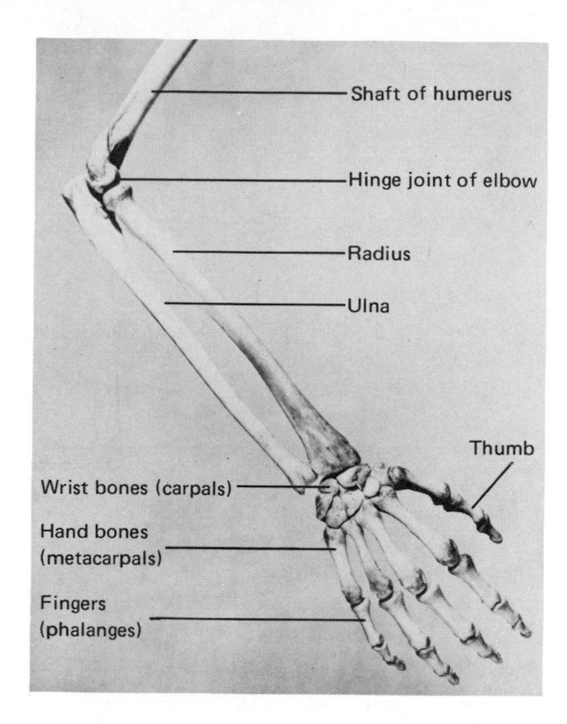

Fig. 26.10 *Skeleton of the forearm*
(Rank Organization)

(Fig. 26.11 is a generalized diagram of such a system). The muscle insertion is close to the joint, so that even when the muscle contracts through a short distance only, the movement of the limb is considerably greater. For

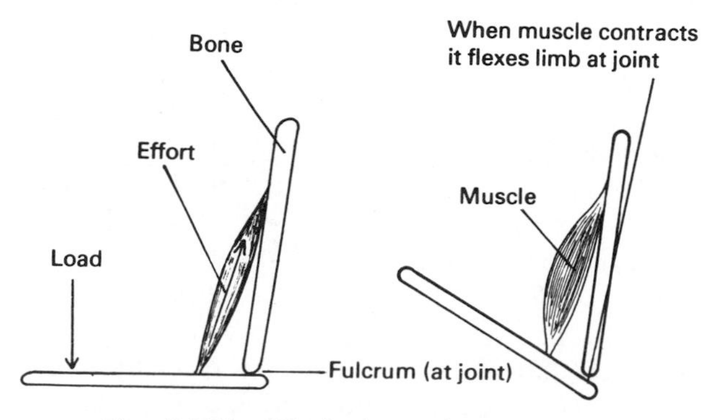

Fig. 26.11 *The limb as a lever*

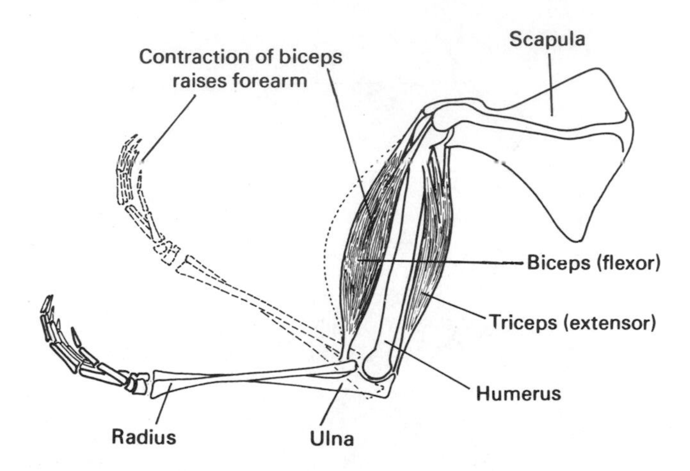

Fig. 26.12 *Antagonistic muscles of the forearm*

example, the biceps muscle (Figure 26.12) only shortens by about 10 cm when it contracts fully, but this produces a hand movement of about 60 cm.

Muscles can only contract and relax; they cannot lengthen of their own accord. After its contraction a muscle must be pulled back to its original length by a second muscle. Consequently most muscles are paired, the members of the pair acting in opposite directions. Where such *antagonistic pairs* act across a hinge joint they are called *extensor* and *flexor* muscles (Fig 26.12); the contraction of the extensor muscle straightens or extends the limb, while contraction of the flexor bends or flexes it. One member of an antagonistic pair of muscles is usually much stronger than the other: of the pair of muscles in the upper arm, the biceps which flexes the arm is visibly larger and better developed than the triceps which extends it. Similarly, the frog's hind-leg extensor muscles which make it leap are stronger than the flexors which return the limb to rest.

Walking is brought about by movements of the limbs produced by sets of antagonistic muscles contracting and relaxing alternately. When the animal stands still or sits down, however, the muscles do not relax completely but remain slightly contracted in a state of tension or *tone,* so holding the body in position.

The skeletal muscles and others such as those in the tongue are called *voluntary muscles* and can be contracted at will. The muscle tissue of the alimentary canal and in the walls of the arteries is quite different in its structure and is called *involuntary muscle,* because it cannot be consciously controlled. The muscles of the heart are made up of a tissue, *cardiac muscle,* which is different from either; it is specially adapted to its function of constantly alternating contraction and relaxation throughout the animal's lifetime.

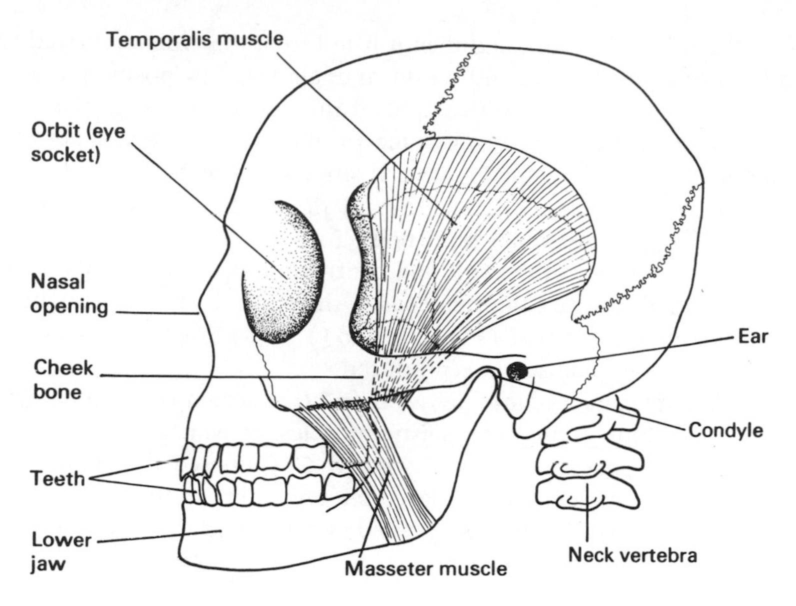

Fig. 26.13 *Human skull and chewing muscles*

26.5 GIRDLES

When an animal walks along, the backward thrust of its feet against the ground is transmitted to the body as a whole through a bony structure called a *girdle*. The girdles link the limbs with the spine, the central support of the body.

The *pelvic girdle* or hip girdle (clearly seen in Fig. 26.1) is rigidly joined to the base of the spine, so that the weight of the body is supported when standing, while in walking the force of the leg-thrust is transmitted directly and very effectively to the spine.

In mammals the *pectoral girdle* (Fig. 26.6) is made up of the shoulder blades and clavicles; the shoulder blades are bound by muscles to the back of the thorax. Although this muscular attachment is less effective than fusion of the bones in transmitting force from the arms to the body, it allows greater freedom of movement to the arms. In man the shoulder is more mobile than in most other mammals, but the movement of the scapula is limited to some extent by the clavicle (Fig. 26.1).

26.6 LOCOMOTION

Walking or leaping takes place by movement of the limbs in a coordinated sequence. Each limb in turn thrusts backward against the ground, so propelling the animal forward, and then is lifted from the ground, moved forward into its original attitude, and set down again in a new position.

Fig. 26.14 is a greatly simplified diagrammatic representation of the hind-leg of a rabbit, showing some of the muscles only. Contraction of muscle *A* pulls the femur backward, since the foot cannot slide backward because of the friction between it and the ground, the effect of the contraction is to move the whole body forward. Contraction of muscle *B* straightens the leg at the knee, and contraction of *C* extends the foot. Thus if *A, B* and *C* contract simultaneously, the leg extends and straightens, and the body moves forward (Fig. 26.14(*b*)); if both hind-limbs act together in this way, a leaping action is produced.

The relaxation of muscles *A, B* and *C* is accompanied by the contraction of the appropriate antagonistic muscles, of which only *A′* (the action of which is opposed to that of *A*) and *C′* (opposed to *C*) are shown in Fig. 26.14. Contraction of *C′* flexes the ankle, lifting the foot to prevent it from being dragged along the ground, while contraction of *A′* returns the limb to its original position.

Coordination

It is essential that the complex series of muscular contractions described above are smoothly coordinated if movement is to be effective, so that the legs move precisely in the correct sequence, and the pairs of antagonistic muscles do not contract simultaneously. This is partly achieved through a system of sensory organs in the muscles. These stretch-receptors or *proprioceptors* (Fig. 28.5) are sensitive to the degree of tension in the muscle, and send nervous impulses to the spinal cord and thence to the brain when the muscle is being stretched. The brain can coordinate the impulses reaching it from proprioceptors in many different muscles and so obtain information as to the body's posture, thus enabling it to compute a pattern of muscular activity which will produce effective movement.

26.7 QUESTIONS

1. Construct a diagram similar to Fig. 26.7 to show a section through the elbow joint, using Fig. 26.10 for guidance. Show the attachment of the biceps and triceps tendons and state where you would expect to find the principal ligaments.
2. From Fig. 26.2 make an enlarged drawing of the scapula and fore-limb of the rabbit. Draw in muscles which you think would help to push the animal forward. Show the points of attachment of the muscles very clearly and state what each muscle would do when it contracted.

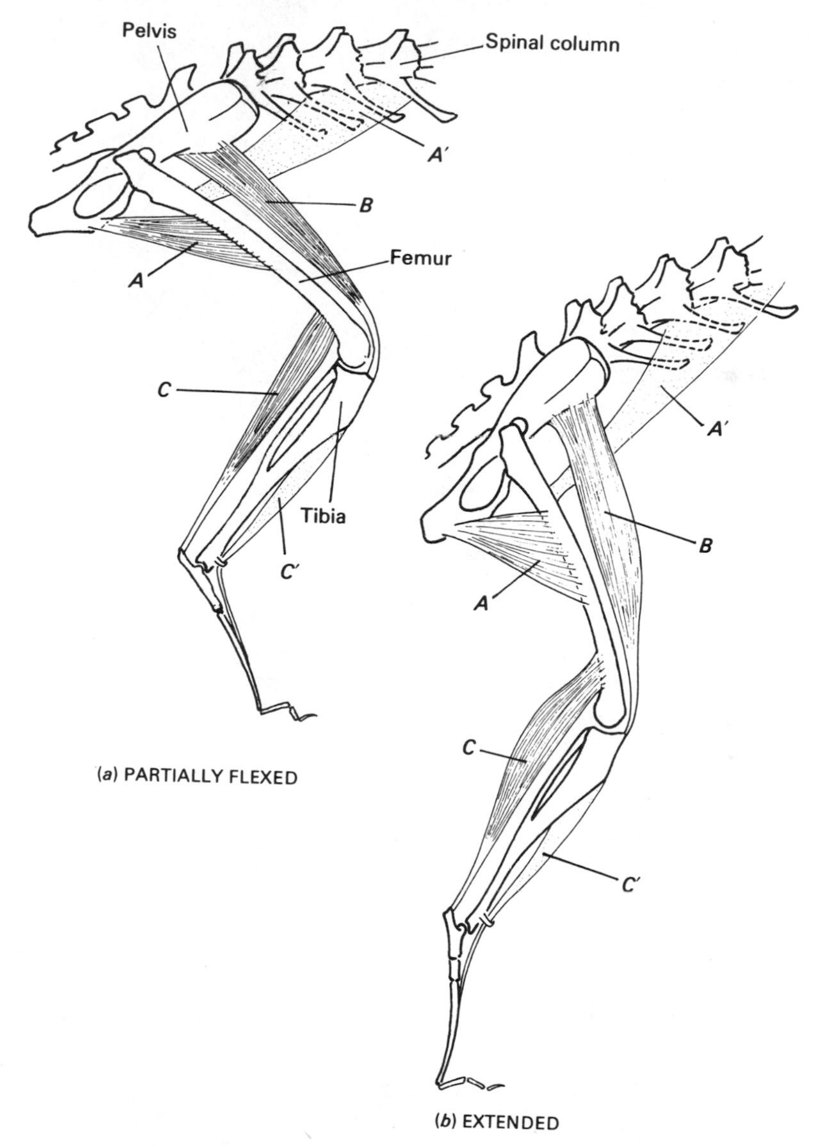

Fig. 26.14 *Action of a rabbit's hind-leg in leaping*

3. What is the principal action of (*a*) your calf muscle, (*b*) the muscle in the front of your thigh and (*c*) the muscles in your forearm? If you don't already know the answer, try making the muscles contract and feel where the tendons are pulling.
4. Unlike most mammals, man stands upright on his hind-legs. What differences do you think this has made to his skeleton and musculature?

UNIT 27
TEETH

27.1 INTRODUCTION

Teeth develop from specialized cells in the skin covering the jaws. The production of a tooth begins quite early in the development of an embryo. In time the root of the tooth becomes enclosed in the developing jaw bone, while the crown breaks through the skin into the mouth cavity.

The numbers and shapes of the teeth of mammals vary from one species to another. In general, the *incisor* teeth at the front of the jaw are adapted to grip or to gnaw food. The *canine* teeth lie on either side of the incisors. The *premolar* and *molar* teeth lie in the cheeks, at the side of the jaws, and grind or crush the food.

Most mammals have two sets of teeth during their lives. A human two-year-old child has 20 "milk teeth"; at the age of about five these begin to fall out as a result of their roots being dissolved away in the jaw, and they are gradually replaced by the 32 permanent teeth. These may not all appear until the age of seventeen years or more.

27.2 TOOTH STRUCTURE

Enamel. The outermost layer of the crown of the tooth, enamel is deposited by cells in the gum before the tooth reaches the surface. Enamel is a non-living substance containing calcium salts; it is the hardest substance made by animals, and forms a smooth, hard-wearing biting surface.

Dentine. This forms the greater part of the bulk of the tooth. It is nearly as hard as enamel, but less brittle. Its structure resembles that of bone to some extent. Strands of cytoplasm from cells in the pulp run through it; these cells are able to add more dentine to the inside of the tooth. Ivory consists of dentine from the tusks (incisor teeth) of elephants.

Pulp. This is soft connective tissue filling the cavity in the middle of the tooth. It contains blood capillaries and sensory nerve endings. The blood in the capillaries brings supplies of oxygen and nutrients to the cells of the pulp and to the cytoplasmic strands in the dentine. The nerve endings in dental pulp are particularly sensitive to heat and cold but produce only the sensation of pain.

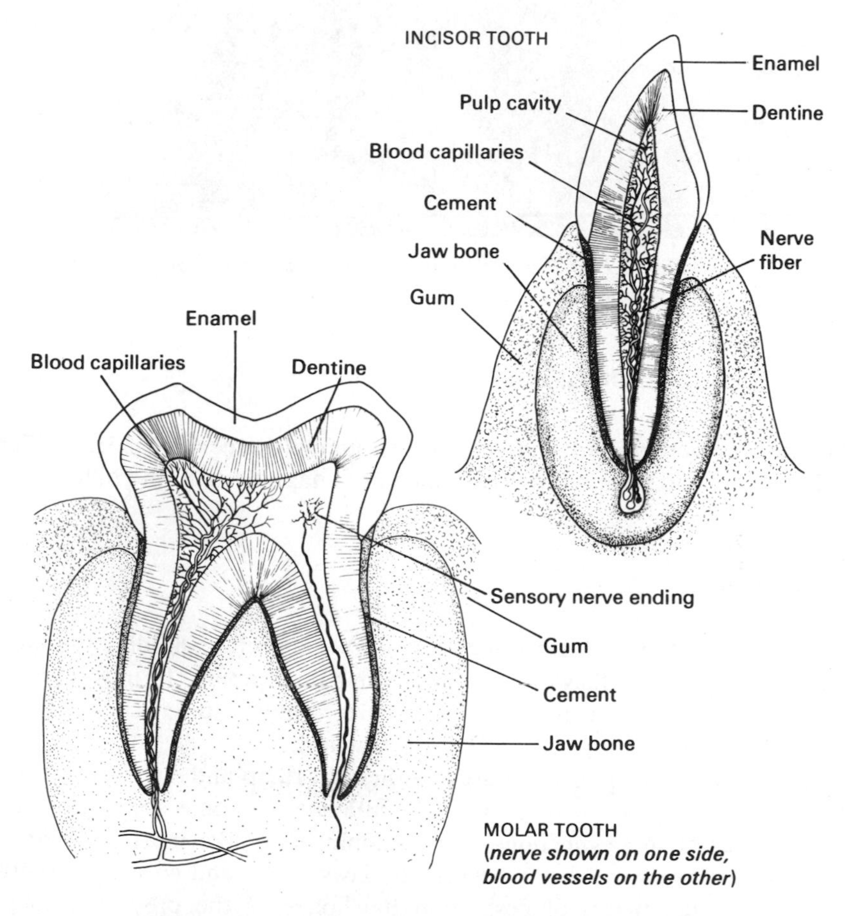

Fig. 27.1 *Sections through incisor and molar teeth*

Cement. This is a thin layer of bone-like material covering the dentine of the root of the tooth. The root is not held rigidly in the jaw but is anchored by tough fibers embedded in the cement at one end and in the jaw bone at the other; the tooth is therefore able to move slightly in its socket as a result of chewing and biting movements.

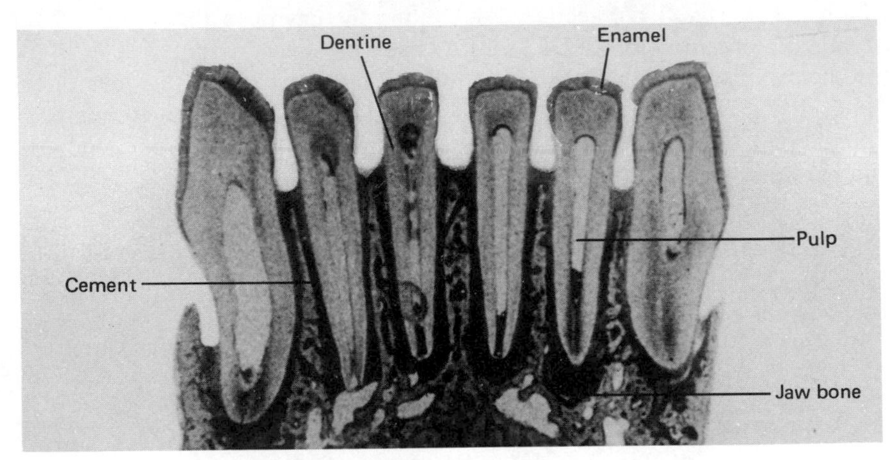

Fig. 27.2 *Section through teeth and jaw of cat*
(GBI Laboratories Ltd.)

27.3 SPECIALIZATION OF TEETH

The teeth of herbivorous and carnivorous animals (see Unit 33.1) differ in size, shape and position in ways which are adaptations to the differences in their diet.

(a) Carnivores

Where all the teeth serve the same purpose, such as holding the prey to prevent its escape, they are all very similar in structure. In most carnivorous fish, many of which swallow their prey whole without chewing, the teeth are simply sharp pegs projecting backward.

In animals which first capture and then break up and chew their food, the teeth in difference regions of the mouth may become specialized for one or other of these functions. For example dogs, like most carnivores, have sharp *incisors* which meet when the jaws close, and so can grip and strip away small pieces of flesh from the bones of the prey. The long, pointed *canine* teeth near the front of the jaw can penetrate the prey, preventing its escape and often killing it. The massive *carnassial* teeth (specially adapted molars in the lower jaw and premolars in the upper) have sharp cutting edges which pass each other like the blades of scissors or shears, slicing off pieces of flesh and cracking bones (Fig. 27.4). The surfaces of the *molars* are flatter and meet when the jaw closes, so crushing the bones and flesh to smaller pieces before swallowing. These crushing teeth develop near the back of the jaw where the mechanical advantage is greater.

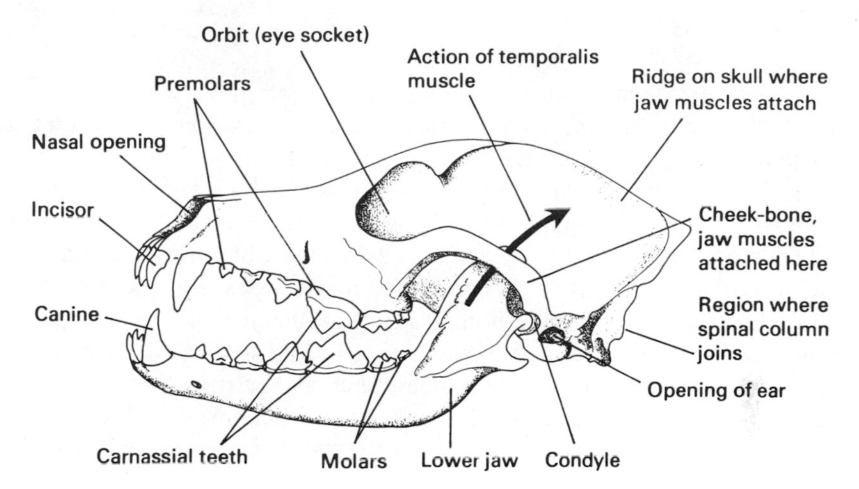

Fig. 27.3 *Skull of dog (carnivore)*

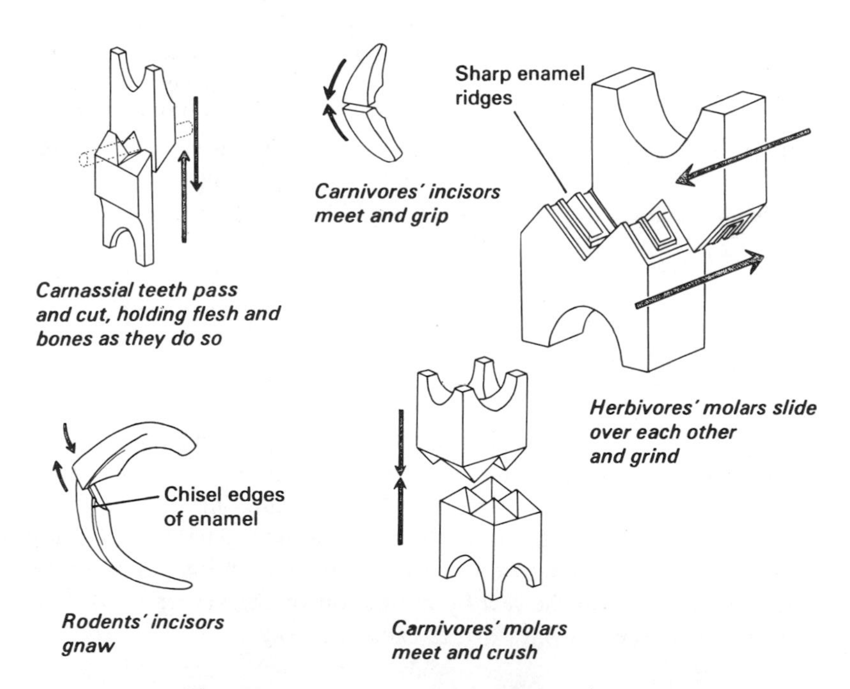

Fig. 27.4 *Action of specialized teeth*

The teeth of young carnivorous and omnivorous mammals, including man, continue to grow after they are first formed up to a certain size only; they then cease growing. This may be due to the gradual closing up of the hole in the root which admits the blood vessels to the pulp cavity; the consequent reduction of the blood supply is thought to prevent further growth.

(b) Herbivores

The permanent teeth of herbivorous animals have wide openings at their roots and continue to grow throughout life. As they grow they are constantly worn down by friction against the teeth just above or below them. Mastication is particularly important in herbivores; the animals' diet of grass and plants requires prolonged chewing in order to break down the cellulose cell walls and release the easily digestible protoplasm and cell sap from the plant cells. Crushing the cellulose walls also provides an increased surface area for digestion. During mastication, the molar and premolar teeth of the upper and lower jaws grind against each other, moving backward and forward in some species, such as squirrels, and sideways in others, like sheep; they gradually wear each other away and come to fit together exactly. The hardest layer of the tooth is the enamel, which is therefore the slowest to wear down and in time comes to form sharp-edged ridges projecting above the rest of the tooth (Fig. 27.6).

Herbivores' canine teeth are often absent; if present they are usually very similar in shape and form to the incisors. The toothless gap between the incisors and premolars (Fig. 27.5) allows the tongue to manipulate the food in grazing. The premolars and molars are almost identical in shape and size, as might be expected from the similarity of their functions. In many herbivores the incisors in the top jaw are also missing; the grass or other vegetation is gripped between the bottom incisors and the gum of the upper jaw (Fig. 27.7).

27.4 JAW ACTION

In the herbivores, the joints between the lower jaw and the skull are held together quite loosely, so that the lower jaw can move laterally as well as up and down, the action seen in cows and sheep "chewing the cud" (masticating food returned from the stomach to the mouth). The jaws of carnivores, however, can move up and down only. This adaptation probably helps to prevent dislocation of the jaw by the action of the very powerful chewing muscles, or by the struggles of the captured prey.

27.5 TEETH IN MAN

Man is an omnivorous mammal, and human teeth do not show the same degree of specialization as those of purely carnivorous or herbivorous mammals (Fig. 27.8). When the jaws close, the top and bottom incisors can

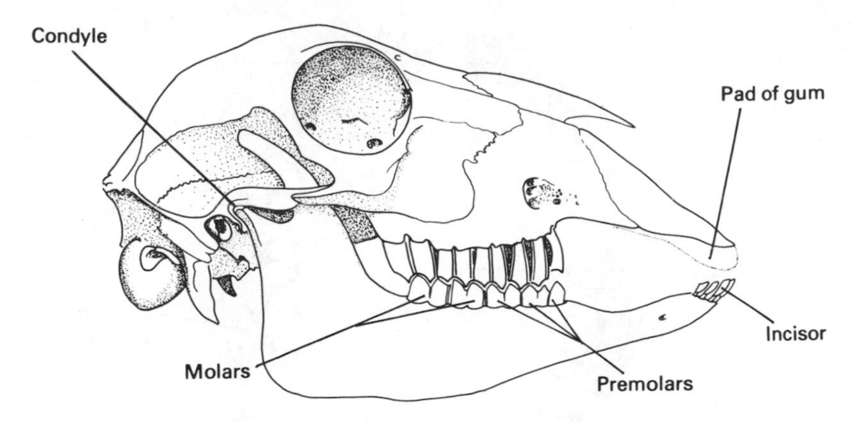

Fig. 27.5 *Skull of sheep (herbivore)*

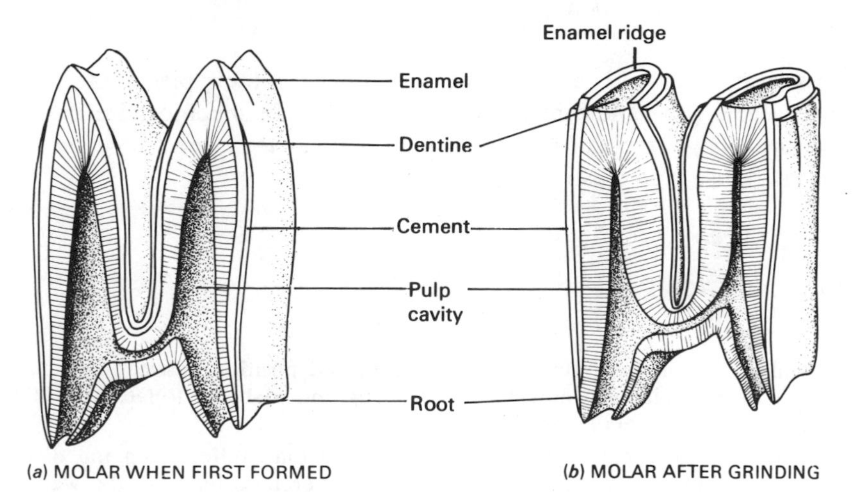

Fig. 27.6 *Sections through herbivore's molar tooth, to show how it is worn down*

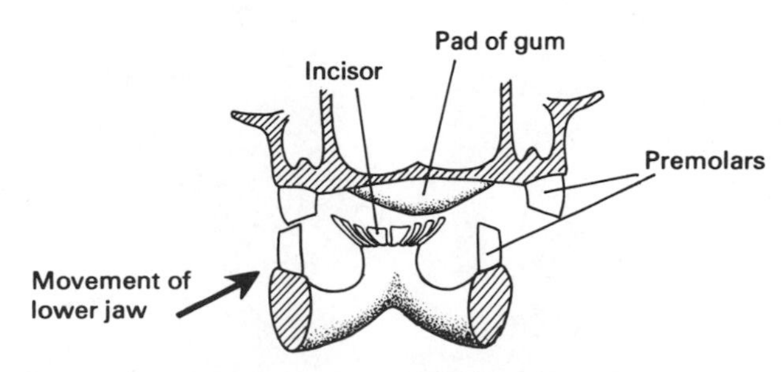

Fig. 27.7 *Diagram of section through herbivore's skull, to show action of teeth*

(After J. Maynard Smith)

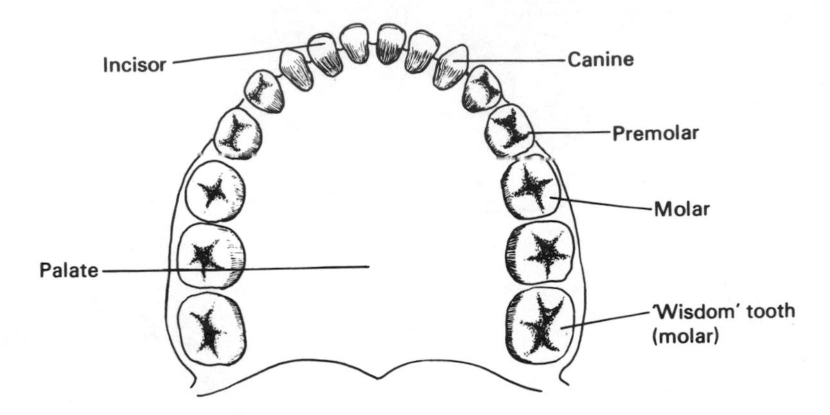

Fig. 27.8 *Arrangement of teeth in human upper jaw*

pass each other with a scissor action, and so cut off pieces of food of manageable size. The canines are little different in shape or size from the incisors. The surfaces of the molar and premolar teeth meet together when the jaws close, and crush food between them.

27.6 QUESTIONS

1. Make a list of the differences in structure and position between the incisor, canine, premolar and molar teeth of a dog and a sheep. Relate each difference to the normal diet of each animal.
2. What general aspects of the diet of civilized man differ from the diets of other mammals? What effects might these differences have on the way man uses his teeth?
3. What problem would result if a rodent were not allowed to gnaw on hard seeds or other hard objects?

Unit 28
The Sensory Organs

28.1 THE SENSORY SYSTEM

All animals respond in certain definite ways to changes in their environment; for example, a sparrow responds to a sudden sound by flying away, and an *Ameba* responds to the presence of food materials by flowing toward them. In every case the animal is made aware (although not necessarily in the sense of "conscious") of its surroundings through a sensory system. The sensory systems of the simpler animals can only perceive stimuli of a very general nature, such as the change from darkness to light, or the difference between heat and cold. In higher animals the sensory system is capable of obtaining detailed information about surroundings, such as the distance, size and color of objects, as a result of the development of specialized sensory cells, often as part of elaborate sensory organs, and of a complex nervous system.

(a) Stimulation and Conduction of Impulses

In general, a particular sensory cell can respond to only one kind of stimulus. Thus, for example, a light-sensitive cell in the eye does not respond to sound waves, and a sense organ in the skin which is sensitive to pressure is not affected by the stimulus of heat. However, some sense organs are less specific in their action than this; the sensory nerve endings in the skin which produce the sensation of pain respond to a variety of stimuli, including pressure, heat and cold.

Sense organs and sensory cells are all connected to the brain or spinal cord by nerve fibers. When the sensory cell receives the particular stimulus to which it can respond, it causes an electrical impulse to travel along the nerve fiber to the brain or spinal cord. When the impulse reaches one of these centers the animal receives an impression of the nature of the stimulus and of where it was received; the impulse may also initiate an automatic or *reflex* action (see Section 29.3).

All the impulses transmitted along the nerves are fundamentaly exactly alike, whether they come from a cell in the skin responding to heat or a cell in the inner ear responding to a sound. Sensations themselves are not transmitted: only an electrical impulse travels along the nerve. All the sense organs of the same kind in a particular part of the body are connected with the same region of the brain. The stimulus and the part of the body affected by it are identified according to the part of the brain which receives the impulse. For example, if the nerves from an arm and a leg could be interchanged at a point just below the brain, stubbing one's toe would produce a sensation of pain in the arm or the hand, since the impulse from the toe would enter the part of the brain specialized to receive impulses from the arm. Similarly, any kind of stimulation of the part of the brain which receives impulses from sensory cells in the leg produces sensations which seem to be in the leg; even if the leg has been amputated, the nerves from the stump may still send impulses to the brain which produce sensations as if the leg were still there.

Whether or not the animal has sensations arising from stimulation of a sensory organ thus depends on the functioning of the appropriate region of the brain; if there is an injury to the part of the brain receiving impulses from, say, an eye, vision in the eye is damaged or destroyed even though the sensory cells in the eye are healthy and functioning normally.

(b) Intensity of Sensation

In general, a strong stimulus produces a more pronounced sensation than a weak stimulus. But vigorous stimulation does not affect the quality or the intensity of the electrical impulses emitted by the sensory cells: all such impulses are alike. However, a strong stimulus does increase the number of impulses reaching the brain. This is due in part to the stimulation of a greater number of sensory organs in the area, and in part to the proportion of sensory cells which only respond to intense stimuli. Many sensory organs are made up of different kinds of cells, some of which are activated by the slightest stimulus, while others only emit impulses on powerful stimulation. The activation of all these cells together means that the stimulus can be recognized as being stronger than usual.

(c) Pain

Although we tend to regard pain as inconvenient and alarming, it has important biological advantages. By making animals respond quickly, sometimes automatically by reflex action, pain tends to remove the animal or the affected part from danger. For example, a finger which touches something unexpectedly hot is quickly and automatically withdrawn the instant pain is felt, usually before a serious burn can ensue: without the sensation of pain considerable tissue damage could result before one was aware of it. A sensation of pain is not essential for an effective reflex action, but it probably helps the animal to learn to avoid a recurrence of the painful situation.

Pain does not always produce a reflex action but sometimes has a protective function nevertheless. In man, pain such as a toothache serves as a warning that all is not well and gives an opportunity to seek advice or treatment.

(d) Skin Senses

Small sensory organs are fairly evenly distributed throughout the dermis of the skin (Fig. 28.1). They consist of branched nerve endings, some encapsulated in connective tissue (such as the *Meissner's corpuscles* and *Pacinian corpuscles,* sensitive to touch and pressure respectively), some entwined around the roots of hairs (*hair plexuses,* touch- and pain-sensitive) and some bare nerve endings in the epidermis (pain-sensitive). Other sensory nerve endings respond to changes in temperature. However, the responses of the skin sensory nerve endings are not always very specific; in man, for example, the skin of the ear lobe contains only hair plexuses

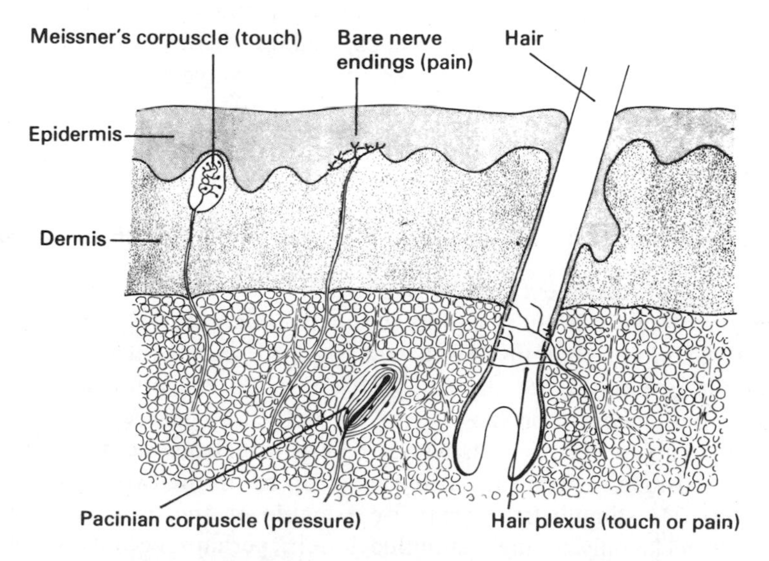

Fig. 28.1 *Sense organs of the skin (generalized diagram)*

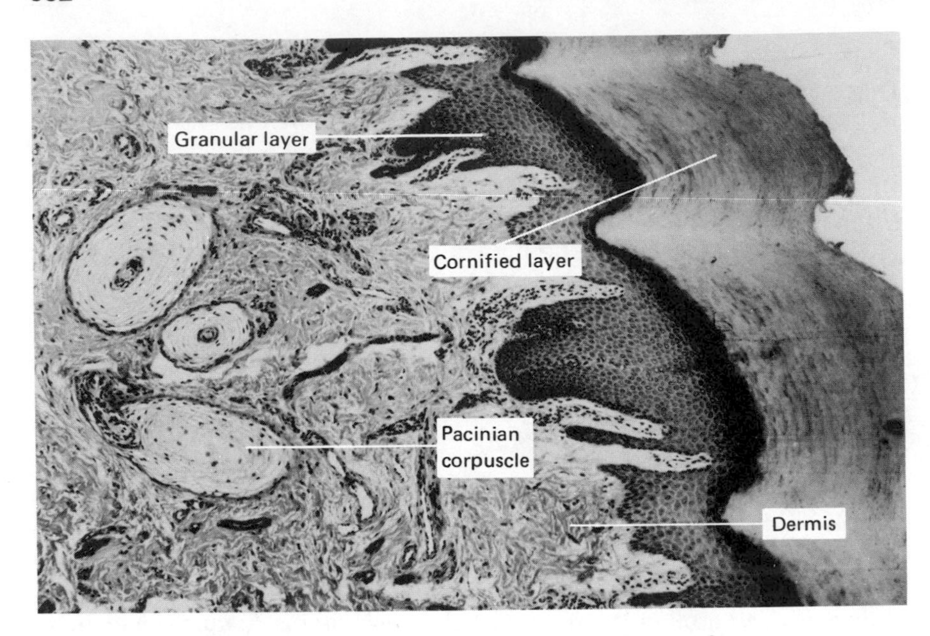

Fig. 28.2 *Pacinian corpuscle in the human skin*
(Brian Bracegirdle)

and free nerve endings, but can detect pressure and temperature changes as well as touch and pain.

Certain regions of the skin have a greater concentration of sensory nerve endings than others. The skin of the fingertips, for example, contains a large number of Meissner's corpuscles, making them particularly sensitive to touch. The front of the upper arm is sensitive to heat and cold. Some areas of the skin have relatively few sensory nerve endings, and in certain regions it can be pricked or lightly burned without any sensation being felt.

Experiment 28.1 *To investigate the varying sensitivity of the skin to touch*

(*i*) This experiment measures the ability of the skin to detect the separation of two points. A pair of compasses is used; if these are not available a simple apparatus like that in Fig. 28.3 can be made. The compass points or pins are set initially 2 cm apart. The experimenter touches the two simultaneously on the skin of the subject, whose eyes should be closed; the subject then reports whether he can feel one point or two. In the more sensitive regions such as the fingertip or lip the two pinpricks are felt as separate stimuli. Elsewhere, for example on the back of the hand or on the neck, only a single stimulus is felt, perhaps because there are fewer touch-sensitive nerve endings in the skin in these regions. By reducing

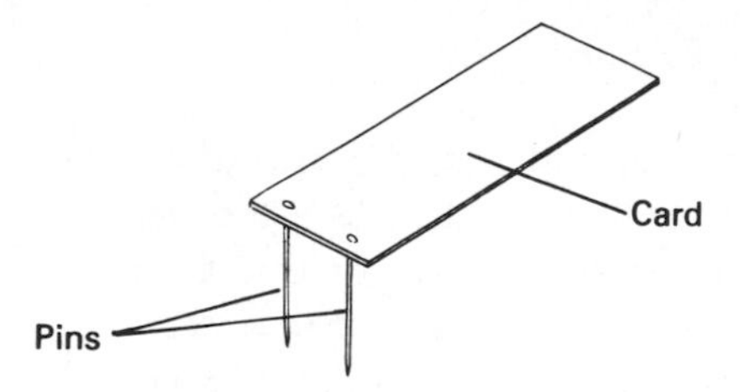

Fig. 28.3 *Apparatus for testing skin's sensitivity*

the distance between the points from time to time, the degree of sensitivity of different regions of the skin can be mapped out.

It is best to vary the stimulus, using sometimes one point and sometimes two so that the subject does not know in advance which is to be used.

(*ii*) A rubber stamp like that in Fig. 28.4(*a*) is first prepared, by making transverse and longitudinal cuts on the surface of an ordinary eraser. This is used to mark a pattern of regularly spaced dots on the skin of the wrist (Fig. 28.4(*b*)) and a second pattern, for recording purposes, on a sheet of paper. A stiff bristle or length of horse hair, held in forceps, is pressed against each dot of the pattern on the skin in turn, using enough force just to bend the bristle. The subject, who must not watch the experiment, reports when he can feel the stimulus. The positive and negative results of each test are entered against the set of marks on the paper, and finally can be expressed in terms of the percentage of the total number of stimuli which produce a sensation of touch.

The technique can be used to compare the relative concentrations of touch-sensitive nerve endings in different parts of the skin, although the sensitive "spots" do not necessarily correspond to single nerve endings.

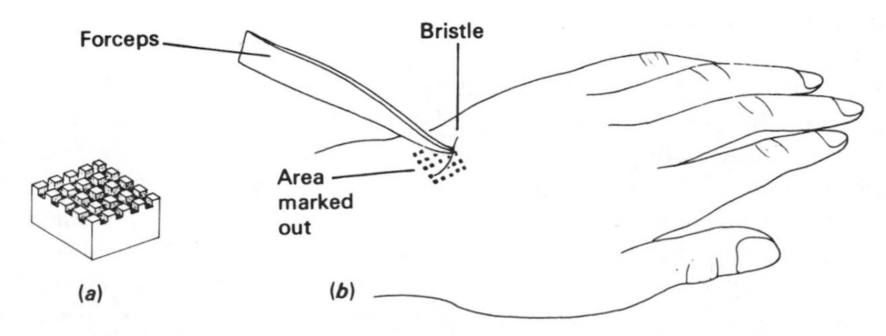

Fig. 28.4 *Testing skin sensitivity to touch: (a) rubber stamp; (b) applying the stimulus*

(e) Internal Sense Organs

Sensory organs lie in the inner parts of the body as well as in the skin. One kind is found in the muscles: the nerve endings are entwined around the muscle fibers and are sensitive to the degree of tension in them (Fig. 28.5). These sensory organs or *proprioceptors* enable an animal to learn to place its limbs accurately as it walks or runs, and to know their exact position without having to look at them.

Pain-sensitive nerve endings are also found in many internal organs, but there are none in the brain.

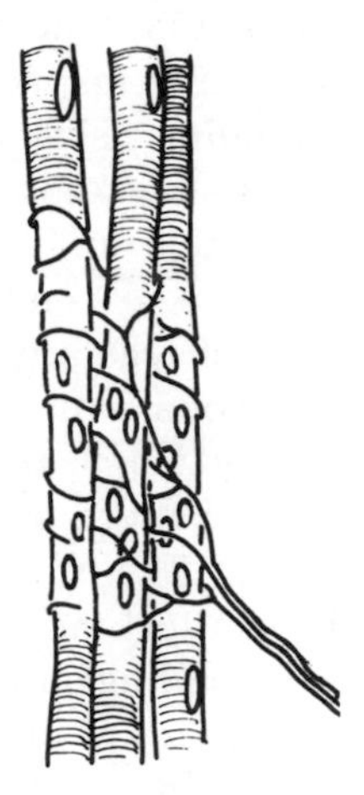

Fig. 28.5 *Stretch-receptor (proprioceptor) sensitive to tension in muscle fiber*

(f) Special Senses

The sensory nerve endings in the dermis and in the internal tissues make up the *general sensory system,* by which the animal is made aware of its own body's activities and immediate environment. In addition, the *special senses* of taste, smell, sight, hearing and balance provide it with much more detailed information concerning its relation to its surroundings; these senses are discussed in the sections which follow. The sense organs concerned with the special senses each consist of a great concentration of cells which are sensitive to one kind of stimulus. These sensory cells may be associated with structures that direct the stimulus on to the sensitive region.

28.2 TASTE

The upper surface of the tongue contains numbers of groups of sensory cells which respond to the presence of certain chemicals dissolved in the moisture overlying them. These groups are called *taste-buds;* they lie mostly in the grooves between the little projections on the tongue's surface. Each taste-bud contains several sensory cells, and nerve endings connected to the brain (Fig. 28.6).

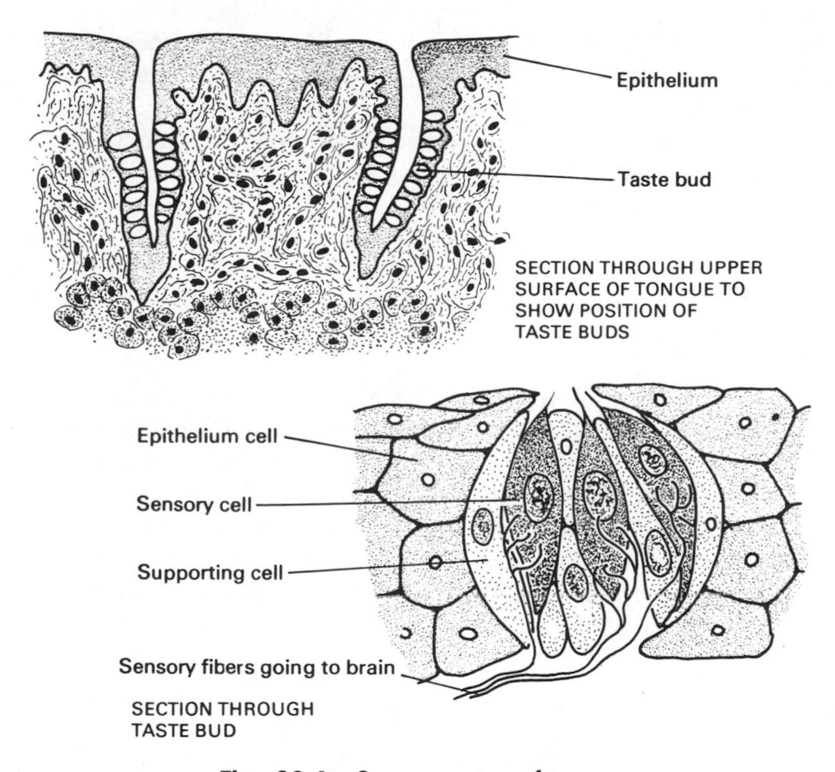

Fig. 28.6 *Sensory system of tongue*

There are four kinds of taste-buds: those sensitive to sweet, sour, salty and bitter chemicals. Each kind is specific in its response: those responding to sweet-tasting substances are not activated by sour compounds, and so on. However, it is not known why, for instance, some chemicals stimulate the taste-buds responding to sweetness while other closely related compounds have no effect on them at all. The different kinds of taste-buds are unevenly distributed on the tongue so that the back of the tongue is more sensitive to bitter chemicals, and the tip and sides to sweet and sour compounds.

The sense of taste is limited to the four factors of sweetness, sourness, saltiness and bitterness, and probably only serves to distinguish between foods suitable and unsuitable for eating.

28.3 SMELL

The epithelium (the outermost layer of cells) lining the upper parts of the nasal cavity contains numbers of spindle-shaped cells which have long strands of cytoplasm extending out into the mucus film covering the epithelium. From these cells nerve fibers pass into the brain (Fig. 28.7). Like the taste-sensitive cells of the tongue, these sensory cells respond to the presence of chemicals, in this case to volatile compounds in the air entering the nose, which dissolve in the mucus film; but unlike the taste-sensitive cells they are stimulated by a very large number of different substances and produce many different sensations of smell. No satisfactory classification of smells, or explanation of how they are distinguished, has yet been made.

The sensation of flavor, which has a much greater range than that of taste, is closely associated with the sense of smell. If volatile substances cannot reach the nasal cavity—for example, if the nose is held firmly or if one has a heavy cold—it is difficult to differentiate by taste alone between many kinds of food which normally have quite distinctive flavors.

The sense of smell is easily fatigued. If one remains for some time in air contaminated by, for example, hydrogen sulphide (a vile-smelling toxic

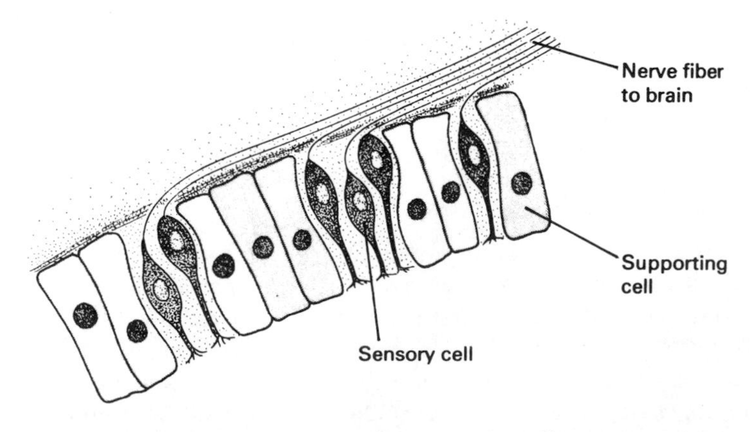

Fig. 28.7 *Sensory epithelium of smell receptors*

gas) one's awareness of the smell gradually disappears, although a newcomer might detect its presence at once. Similarly, the pleasant fragrances of a bakery go unnoticed by anyone who is in the bakery for long.

28.4 STRUCTURE OF THE EYE

The eyes are the organs of sight. They are spherical structures housed in deep depressions in the skull called the *orbits,* and are each attached to the wall of the orbit by six muscles, which by their contractions can make the eye move from side to side or up and down (Fig. 28.8). The structure is best studied in a horizontal section (Fig. 28.9).

The *eyelids* can cover and so protect the eye. They can be closed voluntarily, but also close rapidly by reflex action (see Section 29.3) if the eyeball is touched or appears likely to be touched, or at a flash of bright light. They also rapidly close and open periodically throughout the waking hours; this regular blinking action prevents the surface of the eye drying out, by distributing fluid over it from the tear glands (see below).

The *conjunctiva* is a thin epithelium (layer of cells) lining the inside of the eyelids, and also covering the front of the eyeball, where it is transparent.

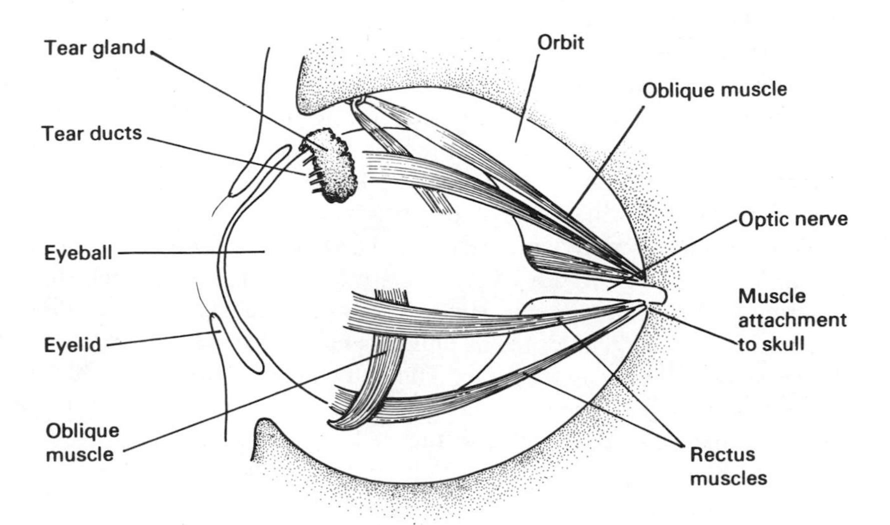

Fig. 28.8 *Diagram to show eye muscle attachment (left eye seen from side)*

The *tear glands* open under the top eyelids. They secrete a solution of sodium bicarbonate and sodium chloride, which keeps the exposed surface of the eye moist. It also washes away dust and other particles, and contains an enzyme, *lysozyme,* which can destroy bacteria. Excess fluid drains away into the nasal cavity through the *lachrymal duct* which opens at the inside corner of the eye.

The *sclera* is a tough, non-elastic fibrous coat around the eyeball. At the front of the eye it is continuous with a disc of transparent tissue, the *cornea,* through which light enters the eye. Since the surface of the cornea is curved, the light is bent or *refracted* as it passes through, and the rays begin to converge (see Section 28.5(*a*)).

The *choroid* is a layer of tissue lining the inside of the sclera. It contains a network of blood vessels supplying food and oxygen to the eye. It is deeply pigmented, the black pigment reducing the reflection of light within the eye. At its rim the choroid is thickened to form a ring-shaped structure, the *ciliary body,* lying just below the edge of the cornea, containing blood vessels and muscle fibers.

The transparent *crystalline lens* lies within the ring formed by the ciliary body, held in position by the fibers of the *suspensory ligament* radiating between the two structures. The shape of the lens, and hence the degree of refraction of the light passing through it, can be altered by the contraction or relaxation of the muscle fibers in the ciliary body (see Section 28.5(*b*)).

The *iris* is a ring of opaque tissue, continuous with the choroid; it lies just over the ciliary body and extends in front of the lens. In the center is a hole, the *pupil,* which admits light to the eye (see Section 28.5(*c*)). The iris contains blood vessels, and sometimes a pigment layer that determines what is usually called the "color" of the eyes. "Blue" eyes have no pigment in the iris, the color being produced by the combined effects of the black pigmentation of the choroid, the blood capillaries in the inner layers of the iris, and the layer of white tissue on its outer surface.

The *aqueous humor* lying between the cornea and the lens and the *vitreous humor* filling the rest of the eye are both solutions of salts, sugars and proteins in water (*humor* is the term that was used by medieval physicians for the body fluids). The aqueous humor has a fluid consistency, but the vitreous humor is jelly-like. The outward pressure of the humors on the sclera maintains the shape of the eye. The aqueous humor serves as a source of oxygen and nutrients to the cells of the transparent tissues of the cornea and the crystalline lens, which contain no blood vessels.

The *retina* is the layer of cells lining the eye which contains the sensory cells. There are two kinds of light-sensitive cells, called *rods* and *cones* according to their shape. The rods are particularly sensitive to low-intensity light, whereas the cones are concerned with the perception of light of different colors. The *fovea* is a small depression in the center of the

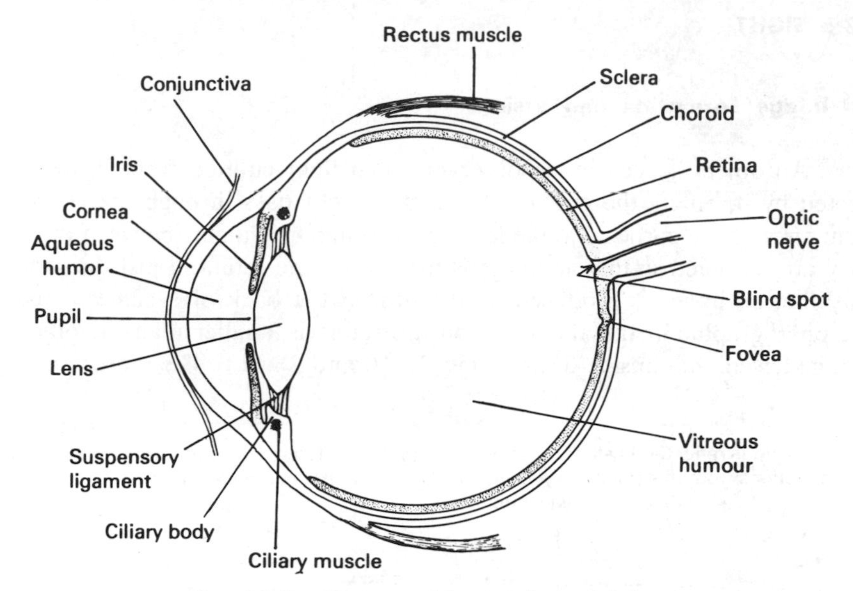

Fig. 28.9 *Horizontal section through left eye*

retina which contains only cones. This region has a greater concentration of sensory cells than any other part of the retina (see Section 28.5(*a*)). The fovea lies exactly on the optical axis of the eye so that an image of an object under close observation falls directly onto it.

Nerve fibers from the rods and cones pass across the front of the retina and joint to form the *optic nerve* which passes through the skull into the brain. At the point where the nerve fibers leave the eye there are no light-sensitive cells, so that even if light falls on this region no impulses are received by the brain. We are not normally aware of this blank or *blind spot* in our field of vision, partly because it is compensated by the use of two eyes scanning the same field, and partly because it never coincides with the image of an object which we are examining closely; the image of such an object always falls on the fovea.

Experiment 28.2 *To appreciate the presence of the "blind spot"*

Hold the book about 60 cm away from the eyes. Close the left eye and concentrate on the cross with the right eye. Slowly bring the book closer to the face. When the image of the dot falls on the blind spot it will seem to disappear.

28.5 SIGHT

(a) Image Formation and Vision

An object is seen by an observer when light emitted from it, or reflected by it, enters the eye. As the rays of light pass through the curved structures of the cornea and the lens and through the humor between them, they are refracted so that an image is formed on the retina in just the same way that an image is produced on the film at the back of a camera: like the photographic image, the image on the retina is smaller than the object being viewed, and upside-down (Figs. 28.10 and 28.11). The sensory cells

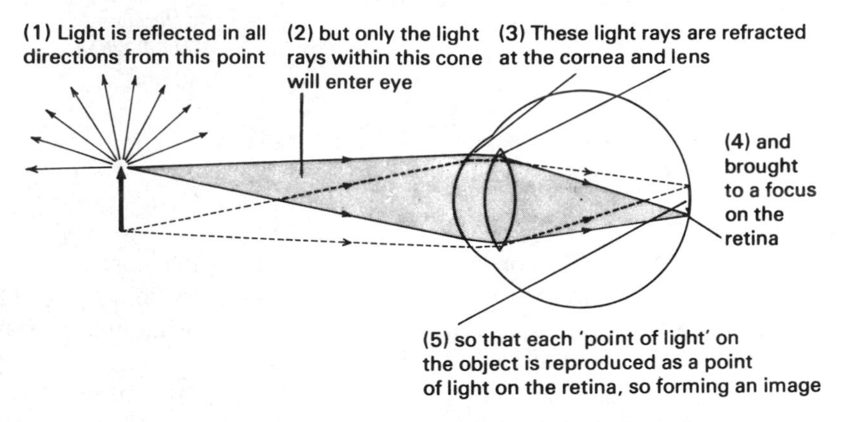

Fig. 28.10 *Diagram of image formation on the retina*

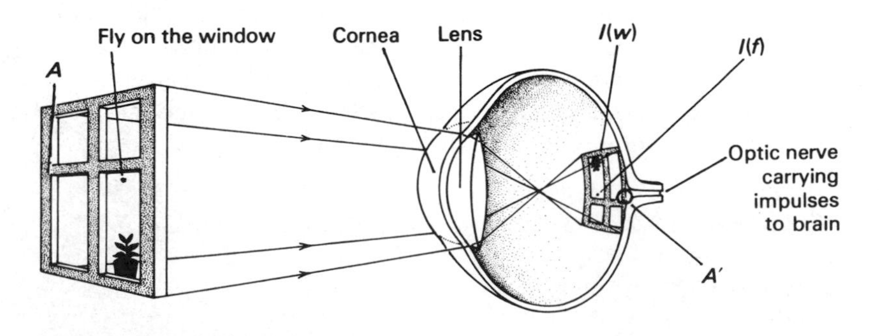

I(w) Image of window focused on retina
I(f) Image of fly falls on fovea; the fly is the only object seen in detail
A′ Part *A* of window forms image on blind spot', and so cannot be seen

Fig. 28.11 *Image formation in the eye*

on which the image falls respond to the light stimulus, and impulses are transmitted along the nerve fibers. These impulses pass along the optic nerve to the brain, where, as a result, an impression is formed of the size, color, nature and distance of the object. The inversion of the image on the retina is corrected during the interpretation of the images received in the brain, so that the object is "seen" the right way up.

The accuracy of the impression that the brain gains of the image depends on how numerous and how closely packed are the light-sensitive cells in the part of the retina which receives the image, since each cell can only respond to the presence of a point of light and (in the case of cones only) its color. The fovea, with its high concentration of sensory cells, is in fact the only region that can give rise to a precise and detailed appreciation of the form and color of an object. In a hawk's fovea there are about a million cones per square millimeter.

(b) Accommodation

A glass lens has a definite, constant focal length; when it is at a certain distance from a screen it is possible to obtain a sharply focused image of an object on the screen only when the object is at a certain fixed distance from the lens. However, the lens in the eye can change its shape, so altering its focal length and enabling it to produce sharp images on the retina of objects lying anywhere from 25 cm or so away from the eye up to the limits of visibility. This ability of the eye to adjust its lens system for viewing either near or distant objects is called *accommodation*.

The focusing power of the lens is determined by its thickness: a thick lens refracts more strongly than a thin one, and has a shorter focal length. The lens of the eye consists of semi-solid material held in an elastic capsule, which tends to shrink and thicken the lens at the center. When the eye is at rest, this tendency is counteracted by the tension in the suspensory ligament radiating between the lens and the ciliary body, maintained by the outward pressure of the aqueous humors against the sclera. The lens in the resting eye is therefore thin and has a long focal length, adapted for focusing light from distant objects onto the retina. When a nearer object is to be observed, the lens must thicken and refract the light more strongly to produce a sharp image. The change in shape is brought about by contraction of the muscles running around the ciliary body parallel to the edge of the lens, reducing the diameter of the opening in the ciliary body and so relaxing the tension in the suspensory ligament: the lens is thus allowed to shrink and become thicker.

When the ciliary muscles relax, the pressure of the humors against the sclera pulls the ciliary body back to its original size, restoring the tension in the suspensory ligament and pulling the lens back to its thin shape.

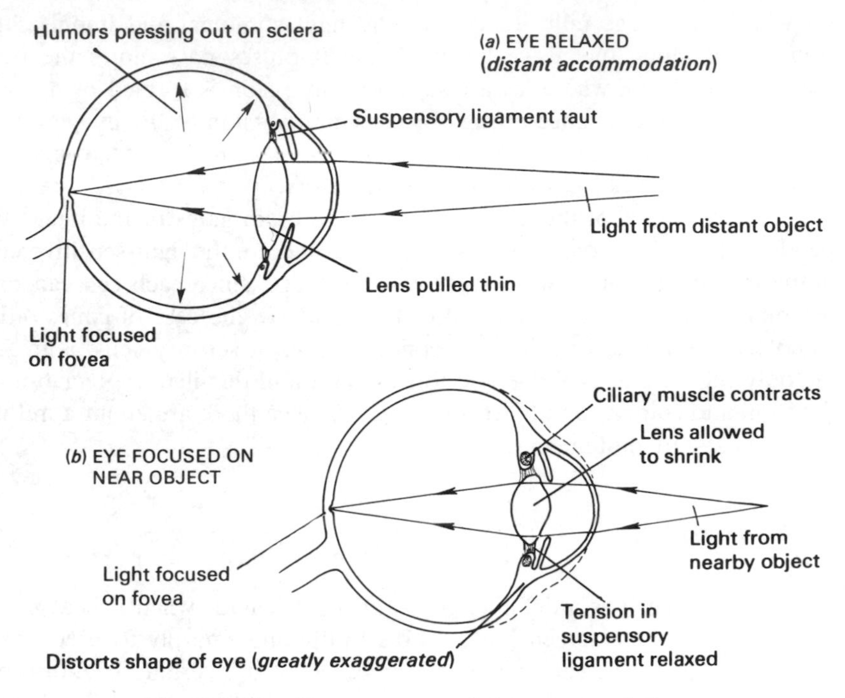

Fig. 28.12 *Diagrams to explain accommodation*

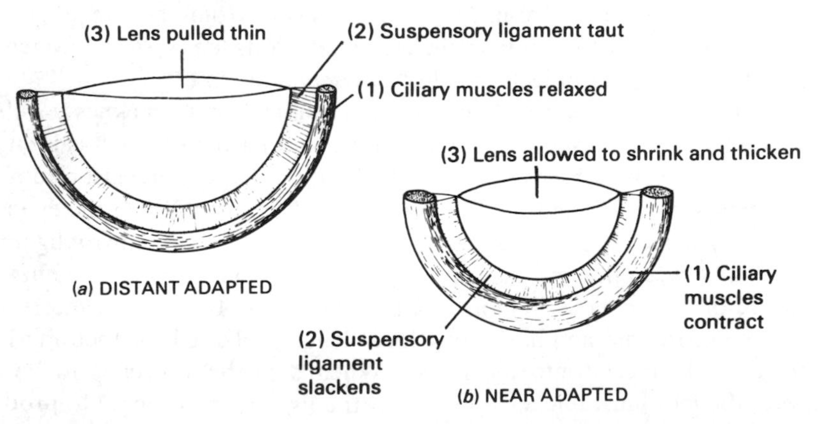

Fig. 28.13 *Diagram to explain how accommodation is brought about*

(c) Adaptation to Varying Light Intensity

The amount of light admitted to the eye is controlled by the adjustment of the size of the pupil by the muscles of the iris. The iris contains two sets of muscle fibers, running radially and circularly. Contraction of

the circular muscles reduces the size of the pupil and less light is admitted to the eye. Contraction of the radial muscles widens the pupil and more light is admitted. The adjustment of the pupil diameter is a reflex action (see Section 29.3), triggered by changes in the intensity of the light. In dim light the pupils are wide open, so that the maximum amount of light is admitted to the eye, helping to increase the brightness of the image. In bright light the pupils are contracted, so that the retina is to some extent protected from damage by light of high intensity.

(d) Color Vision

The mechanism by which color is appreciated is not yet fully understood. One of the most straightforward theories, and one which fits many of the observed facts, suggests that there are three types of cone in the retina, sensitive to red, blue and green light respectively. Each cone produces its maximum response when it receives light of one particular color, and the brain's impression of color depends on the relative number of each kind of cell which are stimulated.

It is believed that the primates (lemurs, monkeys, apes and man) are the only mammals which can perceive color; others see only black, white and shades of gray.

(e) Stereoscopic Vision

Each eye forms its own image of an object under observation, so that two sets of images are sent to the brain. The two sets are not quite identical, since each eye "sees" a slightly different aspect of the object (Fig. 28.14), but, they are correlated in the brain so that the observer gains a single impression of the object. The combination of the two images produces an appreciation of the solidity and three-dimensional properties of the object which is not obtained when it is viewed with only one eye.

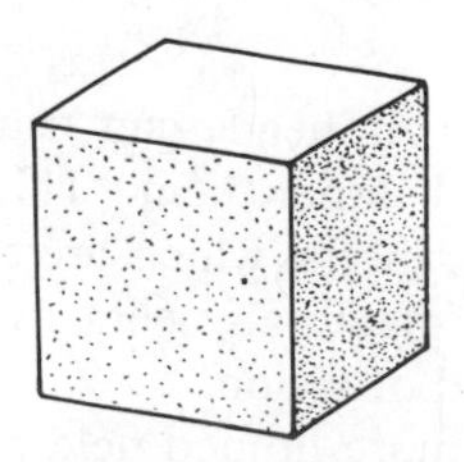

Left eye sees this view

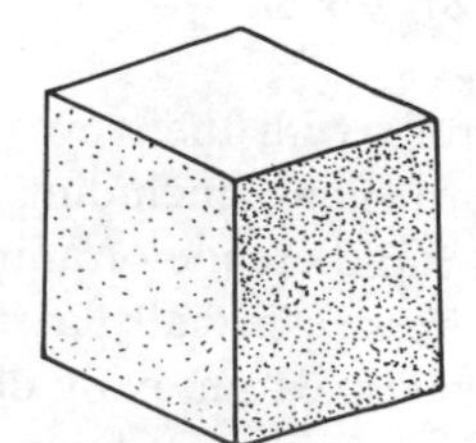

Right eye sees this view

Fig. 28.14 *Different views of a cube seen by left and right eyes*

If the eyes are not aligned normally, or if the brain centers dealing with sight impressions are injured or dulled (for example, by alcohol), the two sets of sensory impulses from the eyes are not properly correlated, and we "see double" (see Experiment 28.3).

(f) Judgment of Distance

For the eyes to focus an image from a nearby object they must be turned slightly inward, directed toward the object. The stretching of the eye muscles controlling this movement stimulates the sensory receptors in their tissues. Impulses reaching the brain from these receptors indicate the extent to which the eyes are converging and so give an impression of the distance of the object.

The stereoscopic vision described in Section 28.5(*e*) also helps one to judge distance; it is very difficult to estimate distance accurately using only one eye. Stereoscopic vision is best developed in animals whose eyes are set in the front of their heads and directed forward. These include most predatory animals, to whom stereoscopic vision is an advantage in judging the distance of their prey before they leap or dive. Examples are lions and tigers among the mammals, hawks, owls and gannets among the birds, and pike among fish. The stereoscopic vision of apes is probably advantageous to them as tree-dwelling animals.

Animals whose eyes are set in the sides of their heads have stereoscopic vision only in the limited zone where their two fields of vision overlap; in most birds, for example, the overlapping of the fields of vision could give stereoscopic vision of objects within an angle of 6–10° (compare Fig. 28.15(*a*)). These animals can otherwise judge distance only by the apparent size of objects and by *parallax,* that is, the apparent movement of nearby objects against a distant background when the head is turned from side to side.

(g) Field of Vision

Animals with their eyes in the sides of their heads can usually see nearly all around, including objects directly behind them. This ability favors the rapid escape of animals like deer or rabbits, which are often the prey of others, especially as the eyes of most mammals seem to be particularly sensitive to moving objects in their field of vision.

Animals with eyes facing forward have a more limited field of vision, but even a man is aware of objects within an angle of about 200°, although only those included in an angle of 2° from the eye will form an image on the fovea and so be observed accurately. This is considerably less than most

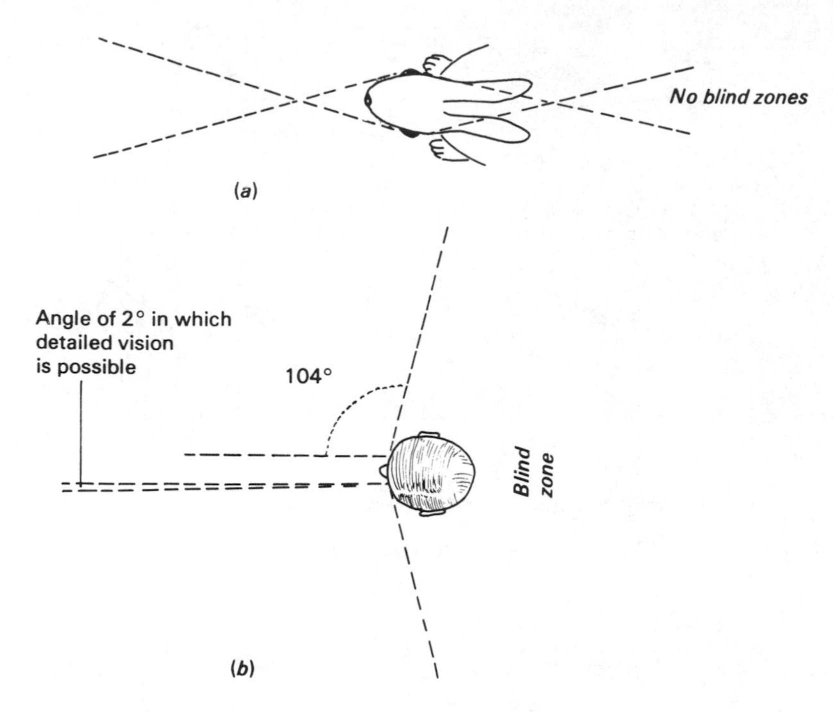

Fig. 28.15 *Fields of vision: (a) hare; (b) man*

people imagine and means, for example, that only about two letters in any word on this page can be studied in detail.

(h) Long and Short Sight

Long sight and short sight are both defects of vision. A *long-sighted* person is unable to see nearby objects clearly, either because his eyeballs are small, or because the lenses of his eyes are unable to thicken sufficiently. In either case light from a nearby object cannot be refracted sufficiently strongly to form a sharp image on the retina (Fig. 28.16). Wearing eye glasses with converging lenses, which bring forward the focused image, corrects the condition.

A *short-sighted* person cannot see distant objects clearly, usually because his eyeballs are too large, and light from a distant object is focused in front of, rather than on, the retina (Fig. 28.17). Spectacles with diverging lenses shift the image backward to the correct position and enable distant objects to be seen clearly.

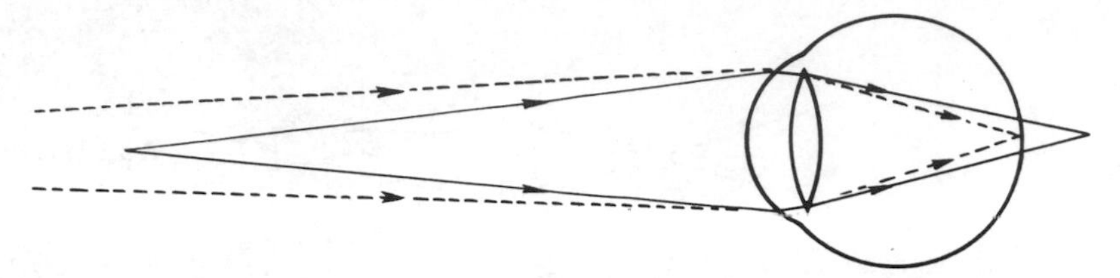

Long sight is caused by small eyeballs or weak lenses. Light from a distant object is brought to a focus on the retina but from a close object its focus is behind the retina

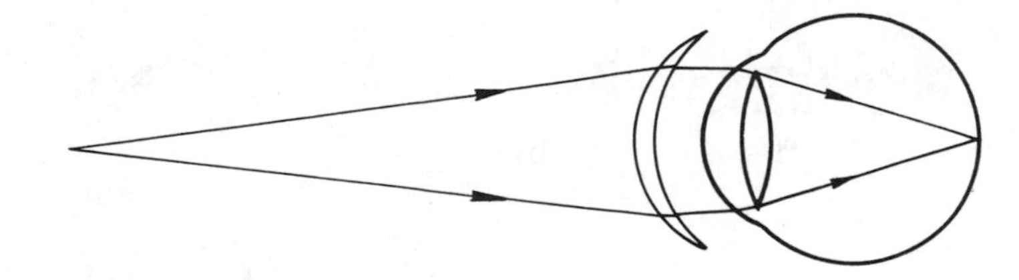

Long sight can be corrected by wearing converging lenses

Fig. 28.16 *Long sight*

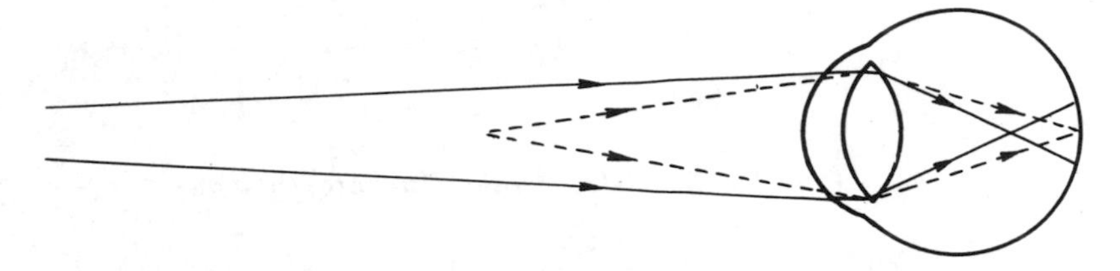

Short sight is usually caused by large eyeballs. Light from a distant object is focused in front of the retina

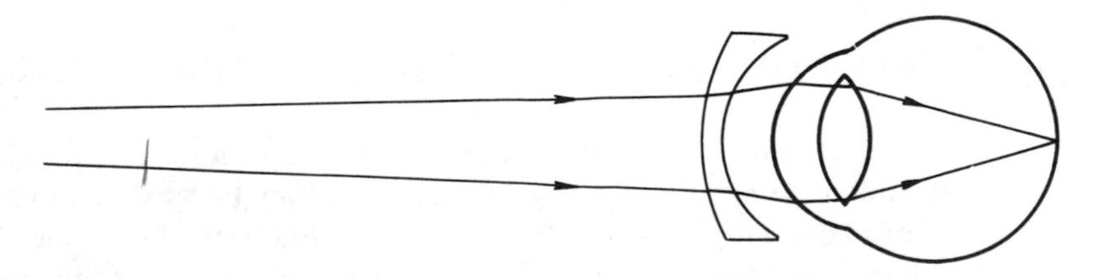

Short sight can be corrected by wearing diverging lenses

Fig. 28.17 *Short sight*

Experiment 28.3 *To experience double vision*

If a nearby object is observed, and a finger pressed against the lower lid of one eye so as to displace the eyeball slightly, an impression of two separate images will result, one formed in each eye.

Experiment 28.4 *To find which eye is used more*

A pencil is held at arm's length in line with a distance object. First one eye is closed and opened and then the other. With one the pencil will seem to jump sideways. This shows which eye was used to line up the pencil in the first place.

28.6 STRUCTURE OF THE EAR

The *outer ear* is a short tube opening at the side of the head, leading into the skull. The *ear-drum,* a membrane of skin and fine fibers, closes off its innermost end completely. At the outer end there is, in most mammals, an outgrowth of skin and cartilage, the *pinna;* this may help to concentrate and direct sound vibrations into the ear, and assist in judging the direction they have come from.

The *middle ear* is an air-filled cavity in the skull, communicating with the back of the nasal cavity through a narrow tube, the *Eustachian tube* (pronounced u-stay-key-un). Three small bones or *ossicles,* the *malleus,*

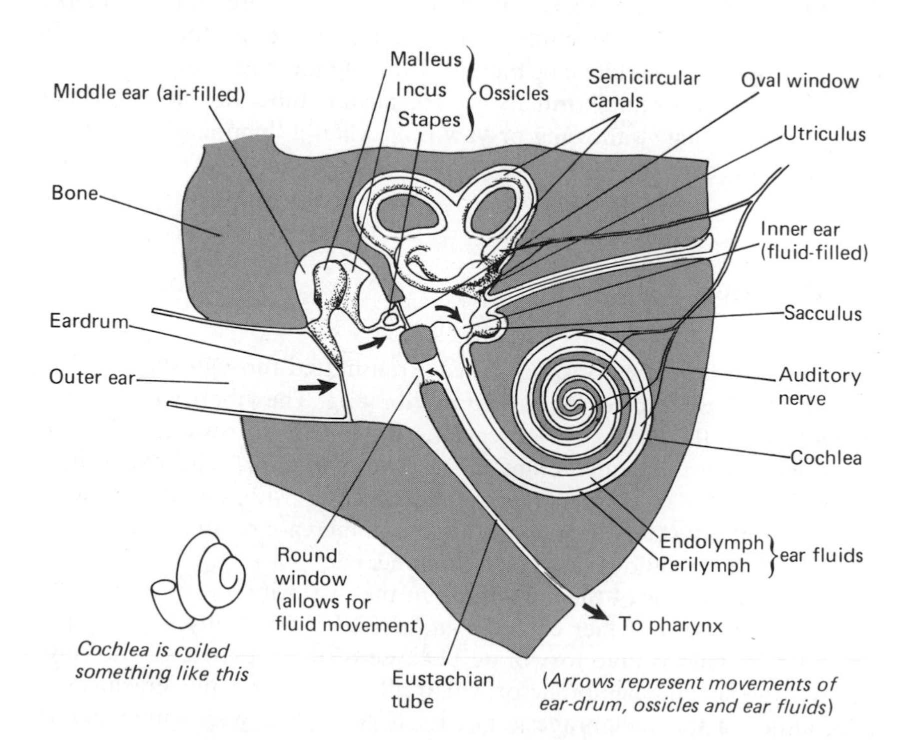

Fig. 28.18 *Diagram of the ear in section*

the *incus* and the *stapes,* link the ear-drum with a small opening in the skull, the *oval window,* which leads to the inner ear. The oval window, and a second opening between the middle ear and the inner ear, the *round window,* are both closed by membranes.

The *inner ear* consists of a complex network of fluid-filled membranous tubes, lying in a liquid called *perilymph.* The longest tube, which is coiled up rather like a snail-shell, is called the *cochlea;* it contains the sensory nerve endings which respond to sound vibrations by sending nervous impulses to the brain. Three semicircular tubes, the *semicircular canals,* lying at right angles to each other, lead from a sac called the *utriculus,* which is connected by a smaller sac, the *sacculus,* to the cochlea. They are of importance in the sense of balance (see Section 28.8). The middle ear and the inner ear are surrounded and protected by the bones of the skull.

Pressure Equalization

Air pressure in the middle ear is usually the same as atmospheric pressure. If the pressure outside the ear-drum changes, for example in a climbing aircraft or even in a rapidly ascending lift, the pressure is equalized by the Eustachian tube opening and admitting more air to, or releasing excess air from, the middle ear, thereby avoiding the possibility of pressure damage to the ear-drum. Normally the Eustachian tubes are closed and are opened only during swallowing or yawning, when a "popping" sound may be heard in the ears.

28.7 HEARING

Sounds reach the ear as pressure waves transmitted through the air. They enter the outer ear and set the ear-drum vibrating. The vibrations are transmitted through the three ossicles, causing the innermost ossicle, the stapes, to vibrate against the membrane in the oval window. The ear-drum is larger than the oval window, and this, together with the fact that the ossicles act as a system of levers, results in an increase of about 22 times in the force of the vibrations that reach the inner ear.

The oscillations of the membrane in the oval window are transmitted to the fluid inside the inner ear (the *endolymph*), especially to that in the cochlea. The fluid is able to vibrate because of the presence of the round window, the flexible membrane of which allows for some movement of the fluid, although this membrane is not itself concerned with sound perception. The cochlea is lined with a membrane consisting of transverse fibers with a layer of sensory cells resting on them. Although details of the mechanism have not been fully worked out, it is thought that the short fibers in the first part of the cochlea respond only to high-frequency vibrations, and the long fibers in the last part to low-frequency vibrations, the fibers of intermediate length lying between them each responding to an intermediate frequency; the receptors are sensitive to sound vibrations with frequencies between 30 and 20,000 per second. When the transverse fibers vibrate, they stimulate the sensory cells resting upon them, and impulses are transmitted from these along the auditory nerve to the brain. According to this theory, the brain determines the pitch of a note by detecting which sensory cells have been stimulated and which are unaffected.

Sense of Direction of Sound

In general, the source is slightly closer to one ear than the other, so that the sound is heard more loudly and a fraction of a second sooner in one ear than the other. The differing extent to which the two ears are stimulated enables animals to estimate the direction from which the sound comes. Most mammals can also move their ear pinnae to a favorable position for receiving the sound, and so obtain a more accurate bearing.

An invisible source which is equidistant from both ears is difficult to locate because both ears are stimulated equally whether it is below eye-level, directly in front, directly above or behind the head. A dog in such a situation will cock its head on one side so that one of its ears may be stimulated more than the other.

A blindfolded man can locate the position of a sound in one of eight possible positions all around him, but a dog's perception is far more precise: it can recognize the source of a sound in any of 32 possible positions. Cats have been shown to be capable of distinguishing between the positions of two sources only half a meter apart and 18 meters away.

28.8 THE SEMICIRCULAR CANALS AND THE SENSE OF BALANCE

The *utriculus, sacculus* and *semicircular canals* are organs of balance, controlling the posture of the animal (Fig. 28.19). The linings of the utriculus and sacculus contain patches of sensory cells, connected by nerve fibers to the brain. Fine strands of cytoplasm arising from these sensory cells are embedded in minute chalky granules called *otoliths* contained in gelatinous plates floating in the endolymph. When the head is tilted the floating plates

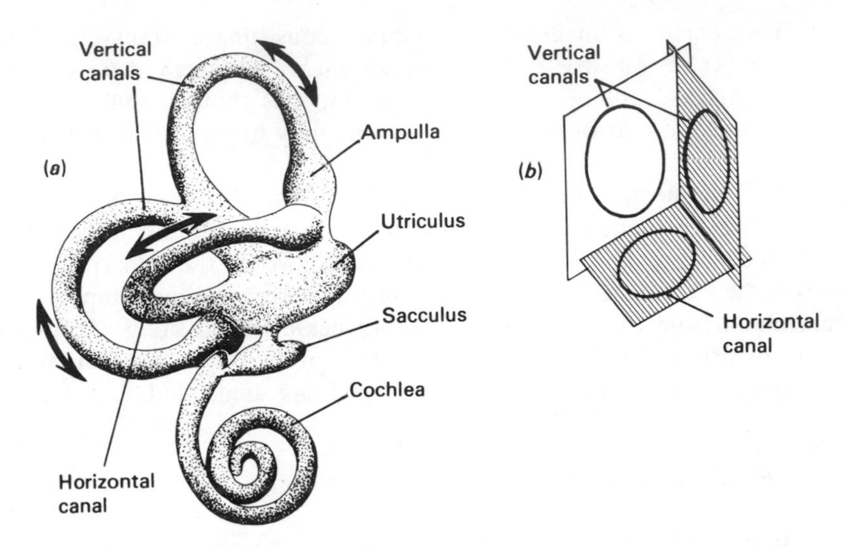

Fig. 28.19 *(a) The semicircular canals (the arrows show the direction of rotation that stimulates each canal); (b) the relative positions of the semicircular canals*

move within the fluid, and the otoliths pull on the sensory fibers. The sensory cells send nervous impulses to the brain, which set off reflex movements (see Section 29.3) of the muscles which tend to return the body to its normal posture.

Each of the fluid-filled semicircular canals carries a swelling called an *ampulla* close to one end. Within each ampulla is a conical gelatinous structure or *cupula,* attached to a patch of sensory cells below it, called a *crista,* by a number of fibers; nerve fibers lead from each crista to the brain (Figs. 28.20 and 28.21). When the head turns in a horizontal plane, the fluid in the semicircular canal in that plane moves more slowly than the walls of the canal (compare the movement of fluid in a cup when the cup is rotated). The cupula in that canal is thus displaced to one side by the drag of the fluid, although those in the other two canals are relatively unaffected. The displacement of the cupula stimulates the cells in the crista below it and impulses are sent to the brain which coordinates the impulses from the three semicircular canals, each mainly transmitting information concerning rotation in one plane.

The sensory cells in the utriculi and sacculi respond to tilting movements of the head or body, while those in the semicircular canals respond principally to rotating movements in their particular plane. If these sensory organs did not function properly, animals would keep falling over unless they constantly relied on their eyes. Without the semicircular canals one

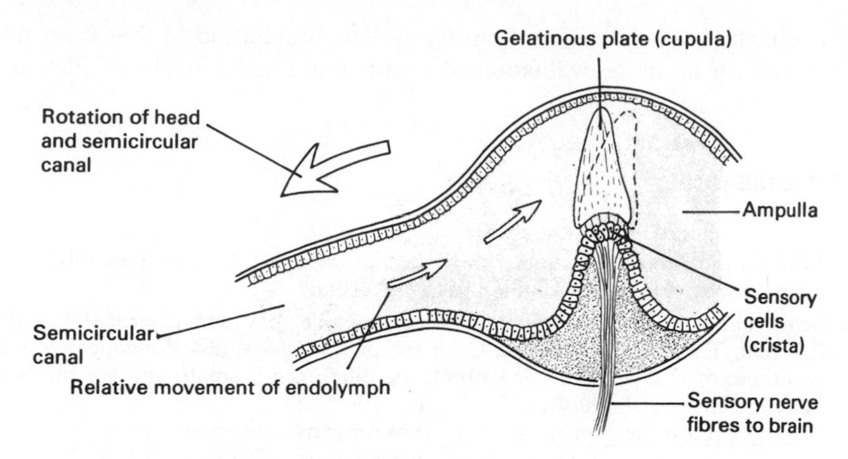

Fig. 28.20 *Diagrammatic section through ampulla*

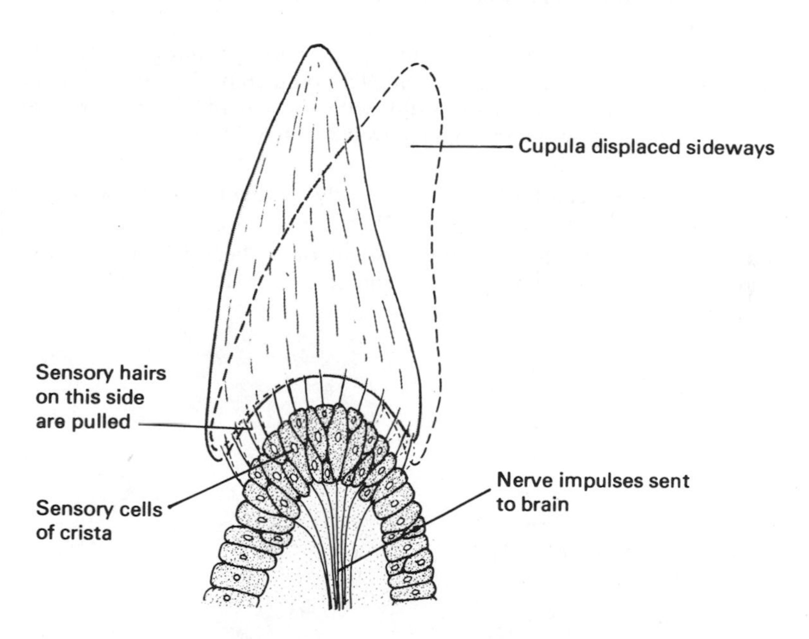

Fig. 28.21 *Detail of crista and cupula*

could probably stand upright if quite stationary, but it would be very difficult to maintain one's balance while walking or changing direction.

Some people experience vertigo (dizziness) during night flying, when they are unable to determine their position by looking at their surroundings. It thus seems possible that we normally rely on information from the eyes and probably also the general sensory system, in addition to that from the inner ear, for accurate evaluation of position and maintenance of posture.

28.9 QUESTIONS

1. Most animals have a distinct head end and tail end. Why do you think the main sensory organs are confined to the head end?
2. Sugar and saccharin both taste sweet and yet they are chemically quite different. The strings of middle C on the piano vibrate 264 times per second and those of D 297 times per second: the difference is small, and yet the two notes are easily distinguished.

   What are the properties of the sense organs concerned which make for poor discrimination of chemicals and precise discrimination of sounds?
3. In what functional way does a sensory cell in the retina differ from a sensory cell in the cochlea?
4. An eye defect known as "cataract" results in the lens becoming opaque. To relieve the condition, the lens can be removed completely. Make a diagram to show how an eye without a lens could, with the aid of glasses, form an image on the retina. What disadvantages would result from such an operation?
5. In poor light, an object can be seen more clearly in silhouette by looking to one side of it than by looking at it directly. Why?
6. A person whose ear ossicles are defective can often hear a ticking watch pressed against his head better than he can if it is held close to his ear. Explain this effect.

UNIT 29
COORDINATION

29.1 INTRODUCTION

The various physiological processes in living animals have been described in the preceding units as if they were quite separate functions of the body, the total result of which constitutes a living organism. Although these processes can usefully be considered individually, they are in fact all very closely linked and dependent on each other. The digestion of food, for example, would be of little value without a bloodstream to absorb and distribute the products.

These systems do not work together haphazardly: the timing and location of one set of activities is closely related to that of the others. For example, a person can walk down the street without having to think consciously about the way his legs move alternately, flexing and extending knees and ankles, or, if he begins to hurry, about the increased rate of breathing and heart-beat needed to send a greater volume of oxygenated blood to his muscles. Similarly, when a person is eating a meal, the position of the food on his plate as recorded by his eyes is used to place his arms and hands correctly to take it up, not laboriously by trial and error, but with precision and accuracy. As the food is raised to the mouth the jaws open to receive it at just the right moment, chewing movements begin and saliva is secreted. When the food is swallowed, several different muscles contract in sequence so that the bolus passes into the esophagus and not into the windpipe; peristaltic movements in the esophagus propel the food into the stomach, where the gastric glands begin to secrete the enzyme for its digestion.

In both these examples, a number of bodily functions come into action in sequence, each at just the right moment, with the result that energy is not wasted in unnecessary movements or in the secretion of enzymes when no food is present. Some of the functions—like walking and chewing—are under the conscious control of the will, while others, such as heart-rate regulation and enzyme secretion, proceed quite automatically. The linking together in time and space of these and other activities of many different kinds is called *coordination*. Without coordination the bodily activities would be thrown into chaos and disorder: food might pass undigested through the alimentary canal for lack of enzyme secretion, even assuming it negotiated the hazard of the windpipe; both legs might be bent simultaneously in an attempt at walking; a runner would collapse after a few meters from lack of an increased oxygen supply, and so on.

Coordination is effected partly by the *nervous system* (discussed in Section 29.2), a network of specially adapted tissues leading to all parts of the body, and partly by the *endocrine system* (Section 29.4), made up of a number of glands which secrete chemicals into the bloodstream.

29.2 THE NERVOUS SYSTEM

(a) Neurons

The nervous system is made up of specialized nerve cells or *neurons*. These are small masses of cytoplasm containing a nucleus. A neuron has the property of being able to transmit an electrical impulse, usually in one direction only. Branching cytoplasmic filaments or *dendrites* conduct impulses toward the *cell body* (the part of the neuron containing the nucleus), while a single long fiber called an *axon* conducts impulses away.

There are three kinds of neuron. *Motor neurons* (Fig. 29.1(*a*)) conduct impulses from the brain or spinal cord along an axon to a muscle fiber or a gland. Where it reaches a muscle fiber, the axon terminates in a branched structure called a *motor end plate,* lying within the sheath of the muscle fiber: impulses are transmitted through this structure to the fiber, causing it to contract. A *sensory neuron* (Fig. 29.1(*b*)) receives impulses from sensory cells or receptors via a single elongated fiber or *dendron,* and transmits them to the next cell in the network in its axon. *Multipolar neurons* (Fig. 29.1(*c*)) occur in the brain and spinal cord. They have many dendrites but no associated long fibers. Dendrons and axons, collectively called *nerve fibers,* consist of fluid-filled tubes of cytoplasm, often surrounded by a sheath of fatty material. In mammals the cell body is usually in the brain or spinal cord, while the nerve fiber extends the whole distance to the organ concerned, often for considerable lengths, for example, from the base of the spine down to the big toe.

Nerve fibers transmit electrical impulses down their whole length very rapidly, passing the impulses on to the next cell in line. The mechanism of transmission is not the same as that of electrical conduction in metals, where the current flow depends on the voltage applied. The fiber builds up

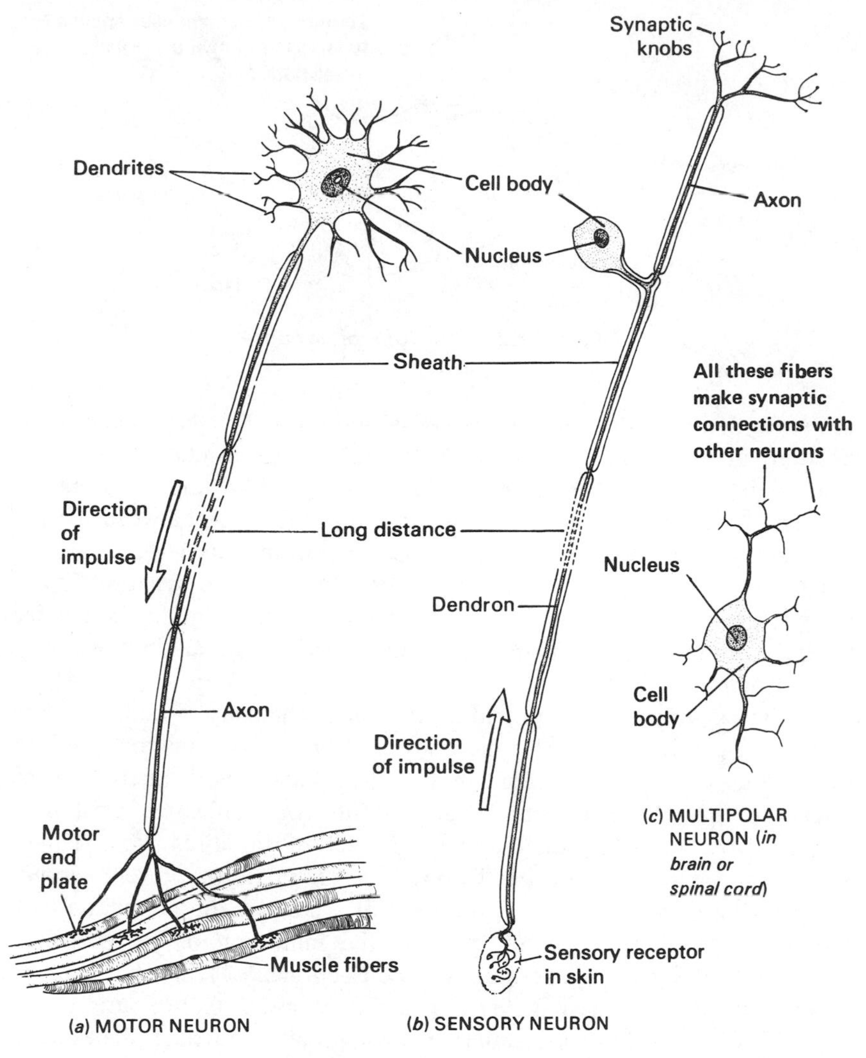

Fig. 29.1 *Nerve cells (neurons)*

an electrical charge within itself which is released when the cell is stimulated, and which has to be built up again before the next impulse can pass.

(b) The Synapse

Neurons can only transmit impulses from one to another at special junctions or *synapses,* where the branched ending of an axon is applied to the dendrites or cell body of another neuron (Fig. 29.2). The two neurons

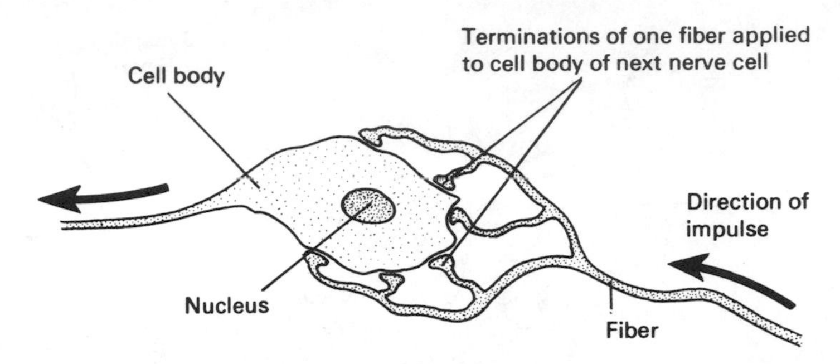

Fig. 29.2 *Diagram of synapse*

are separated by a microscopic space, without any direct cytoplasmic connection, and it is believed that the impulses are conducted across the synapse by a chemical mechanism. The passage of an impulse in the first neuron stimulates it to secrete a chemical into the space between the cells. This substance stimulates the second cell to transmit the impulse along its length. A single impulse does not necessarily cross the synapse. Two or three impulses, arriving in rapid succession, or possibly simultaneously from several transmitting fibers, may be needed to set off an impulse in the second neuron.

The nervous system, providing the main channels of communication between the different parts of the body and brain, contains many millions of neurons, each one of which may have synapses with many incoming fibers, so that enormous possibilities of intercommunication exist in the system. Moreover, the simultaneous arrival of impulses from different sources at the synapses of a cell body may stimulate or prevent (*inhibit*) the relaying of a subsequent impulse. For example, the stimulation of a muscle to contract is accompanied by the inhibition of the passage of impulses to the antagonistic muscle (see Section 26.4), so that the latter relaxes and the muscular movement is not opposed. Coordination of the activities of the body's various parts is brought about through the establishment of a system of such stimulatory and inhibitory patterns at the synapses; the process by which an animal learns how to perform a certain activity can be regarded as the setting up of this kind of system.

(*c*) **Nervous Systems**

The nervous system of a relatively simple animal like a sea-anemone or a *Hydra* consists of a network of nerve cells spread fairly evenly throughout the animal (Fig. 29.3), so that an impulse started at one point spreads out gradually in all directions through many synapses. The nerve cells of

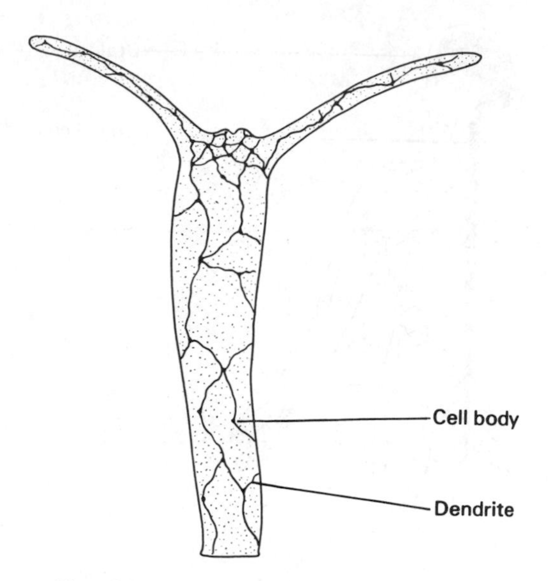

Fig. 29.3 *Nerve-cell network of a* Hydra *(X8)*

higher animals, however, are grouped together to form specialized tissues and organs. In vertebrates these form a *central nervous system* consisting of the brain and spinal cord, and a *peripheral nervous system* consisting of nerves running from the central nervous system to all the parts of the body (Fig. 29.4).

Nerves are bundles of nerve fibers, sheathed in connective tissue (Fig. 29.5). Most nerves contain both motor and sensory fibers, although one or other type may predominate. The cell bodies of sensory fibers sometimes form a bulge or *ganglion* on the nerve (Fig. 29.6).

Most of the cell bodies and their synapses lie in the central nervous system: every nervous impulse traveling to or from the various parts of the body passes through it. Because of the enormous number of synapses in the central nervous system, there are far more possible nervous interconnections and linkages than there would be if the nerves simply ran from one organ to another. In this, the central nervous system is a little like a telephone exchange (although, of course, its mechanism is totally different). Direct connection of your telephone to each of the people you might want to call up could need hundreds of wires; if a separate system had to be installed for every subscriber the numbers and confusion of wires would be overwhelming, and yet each would still be limited to communication with a few hundred people. A telephone exchange makes it possible for you to call up any other subscriber in the country, rapidly and efficiently. In a small and very uncomplicated society (analogous to a very simple

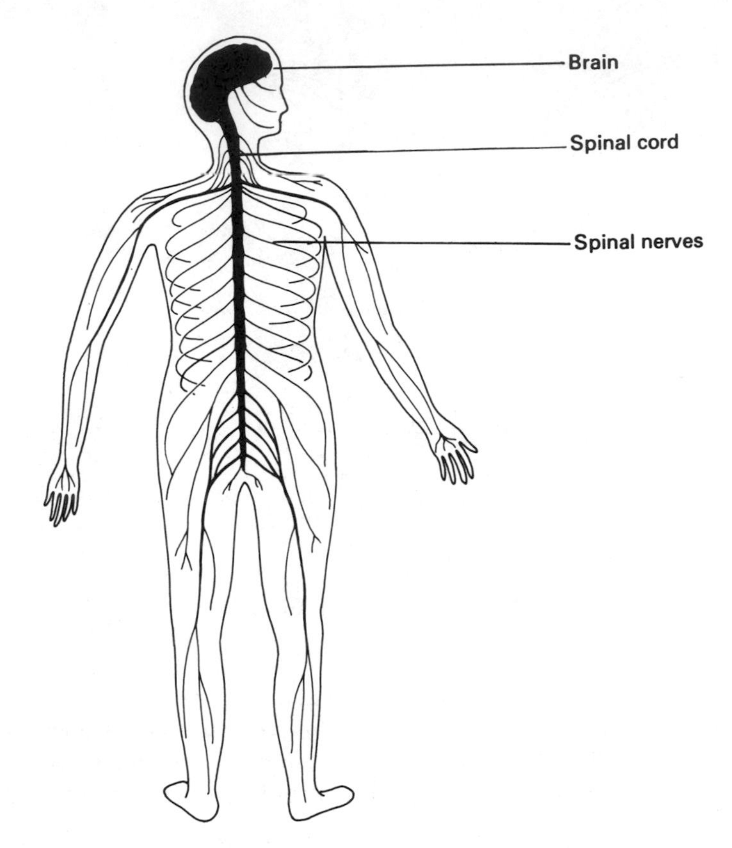

Fig. 29.4 *Nervous system of a man (X1/20)*

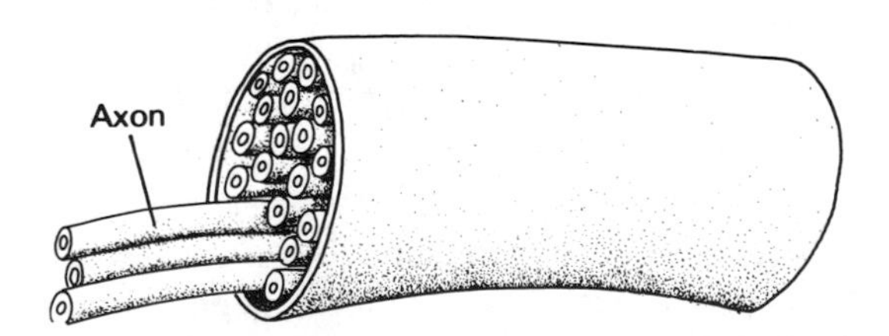

Fig. 29.5 *Diagram to show nerve fibers grouped into a nerve*

organism like *Hydra*) the direct-connection telephone system might be adequate; in a larger community, a central telephone exchange is essential for the swift and easy coordination of the activities of its members—acting in a similar way to the central nervous system of one of the higher animals.

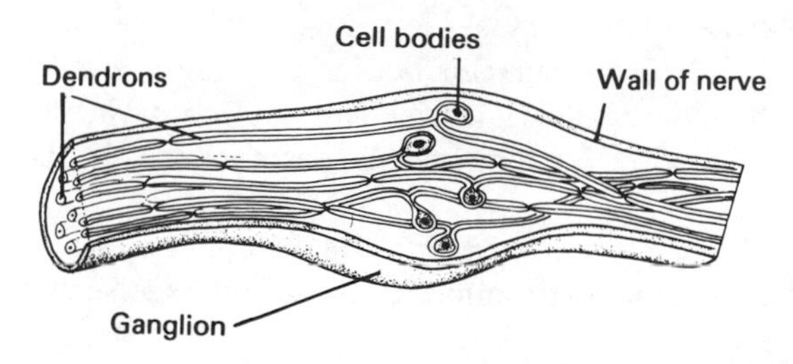

Fig. 29.6 *Diagram of cell bodies forming a ganglion*

(d) The Spinal Cord

The spinal cord consists of a great number of neurons grouped into a cylindrical mass running from the brain of the animal to the tail. It is surrounded and protected by the bones of the spinal column. The neuron cell bodies and their synapses lie in the center of the cord, making a region of *gray matter* which is roughly H-shaped in cross-section (Fig. 29.7). Outside this is the *white matter,* made up of nerve fibers; its whitish appearance is due to the fatty material of the fiber sheaths. The *spinal nerves* emerge from between the vertebrae and run to all parts of the body (Fig. 29.4). Each nerve consists of sensory fibers, bringing impulses from the sensory organs of a particular part of the body to the brain, together with motor fibers, carrying impulses from the brain to the muscles and other

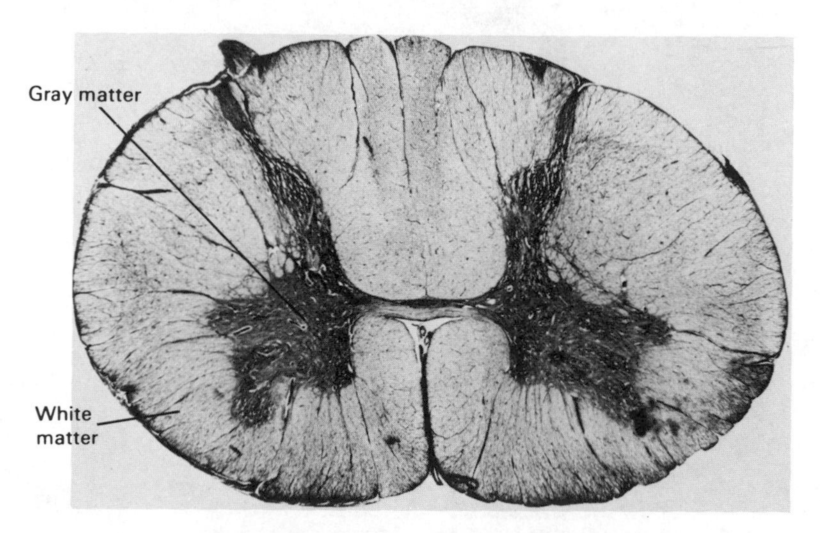

Fig. 29.7 *Section through spinal cord*
(Brian Bracegirdle)

organs of the same region. The sensory fibers enter the spinal cord at the back via the *dorsal root,* their cell bodies forming a swelling, the *dorsal root ganglion.* The motor fibers leave the spinal cord via the *ventral root,* and run beside the corresponding set of sensory fibers in the spinal nerve (Fig. 29.14).

The spinal cord is concerned with the conduction of nervous impulses to and from the brain and with spinal reflex actions (see Section 29.3).

(e) The Brain

Like the spinal cord, the brain consists of neurons with a great concentration of cell bodies. During evolution, the organs of the head have become increasingly specialized and well developed, and the number of

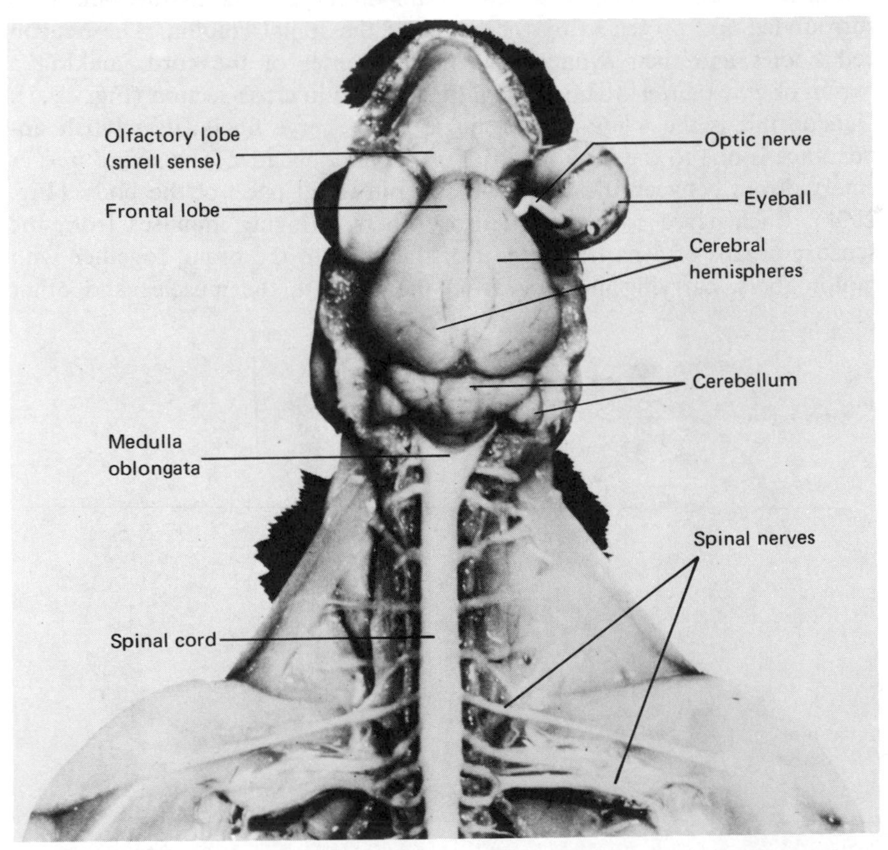

Fig. 29.8 *Dissection of the brain and spinal cord of a rabbit (seen from above)*
(Dissection by Gerrard and Haig Ltd.)

sensory cells at the head (and consequently of sensory fibers entering the front of the central nervous system) has increased. As a result this part of the nervous system has enlarged and developed to form a distinct organ, the brain of vertebrate animals, at the top of the spinal cord (Fig. 29.8).

In simple vertebrates, three regions of the brain can be distinguished: the fore-brain, mid-brain and hind-brain. The fore-brain receives the impulses transmitted from the nasal organs, the mid-brain those from the eyes, and the hind-brain those from the ears and semicircular canals and from the sensory organs of the skin. These three regions can also be recognized in the early stages of development of higher animals. As the embryo develops, the roof of the hind-brain thickens and enlarges to form the *cerebellum,* which coordinates the balancing organs and the muscles, making precise and accurate movements possible. The floor of the hind-brain thickens to form the *medulla oblongata,* the region which controls the "involuntary"

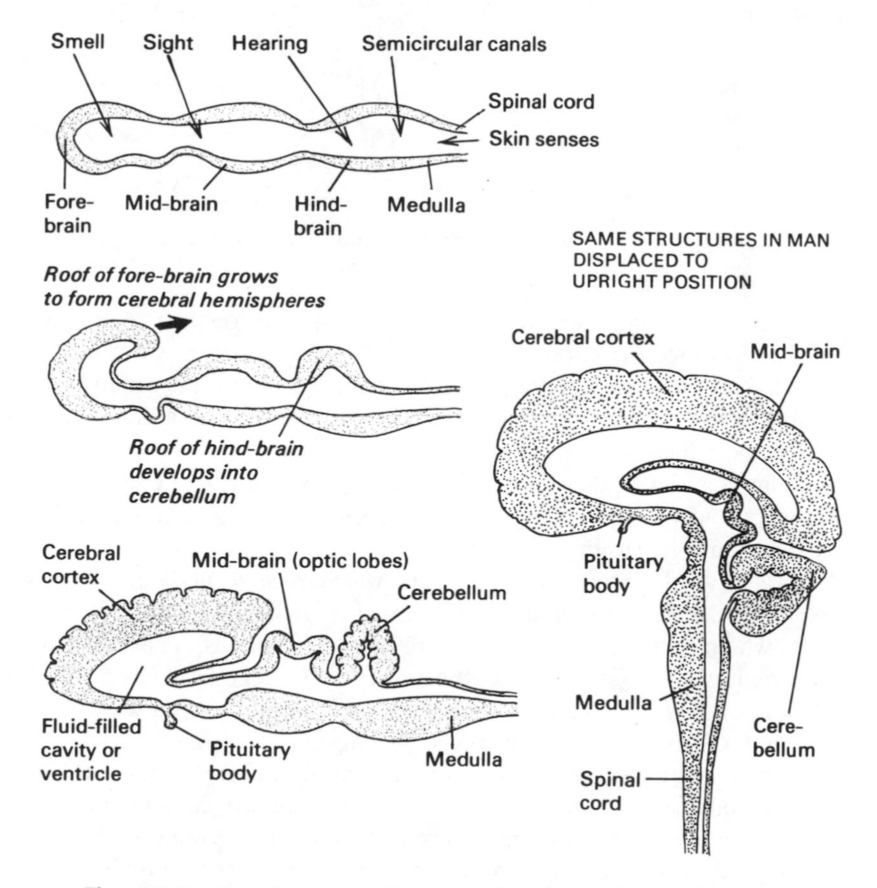

Fig. 29.9 *Development of mammalian brain (vertical sections)*

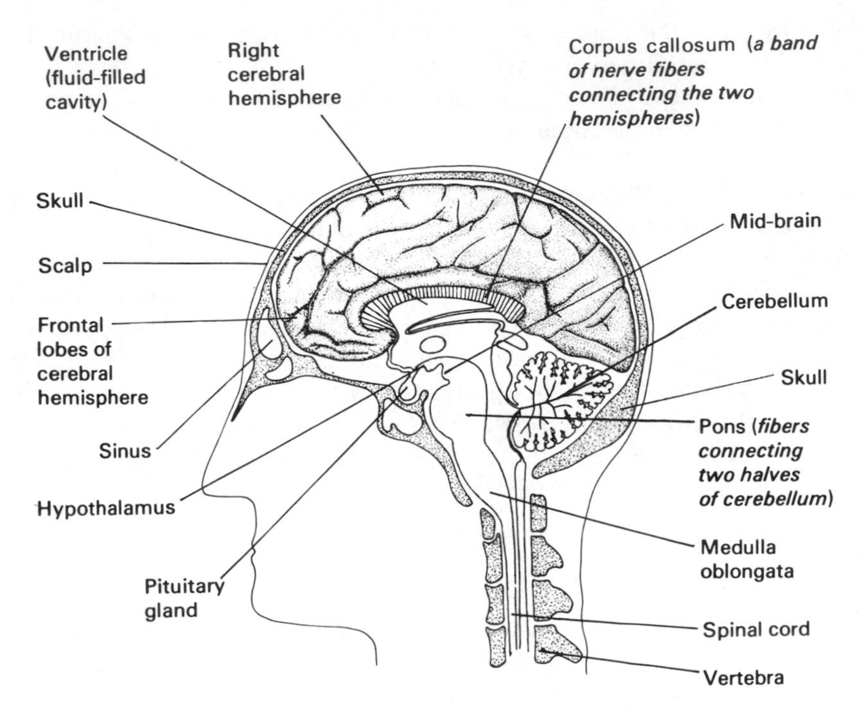

Fig. 29.10 *Section through head to show brain*

actions like breathing, heart-beat, and body-temperature control by vasodilation and vasoconstriction.

The relative sizes of these principal regions of the brain usually bears a relation to the respective importance of the senses of the animal. In the dogfish shark, which hunts its prey by smell, the front lobes of the fore-brain are very large and well developed. The salmon, by contrast, depends on its sight for capturing food, and the optic lobes of its mid-brain are much larger than the fore-brain.

As well as the regions of the brain which receive and interpret impulses from the sensory organs, the *motor areas* of the brain transmit impulses which initiate activities in the muscles and glands. These activities may be relatively simple, such as the secretion of digestive juices or the withdrawal of a hand from a pinprick, or more complex responses, like the search for food initiated by sensations of hunger.

Nerve fibers pass from the sensory centers in the fore-, mid- and hind-brain to certain regions of the brain called *association centers,* in which impulses from different sense organs are correlated. For example, the smell of meat may stimulate the nose of a dog and cause sensory impulses to be sent to the fore-brain. These would normally be relayed to the motor areas and produce coordinated movement toward the food, but the sight of another dog in possession of the meat and the sound of its ferocious growling will also stimulate sensory cells in the eyes and ears. All these impulses are relayed to the association centers, where they can be correlated with information from past experience "stored" in the brain and the relative strengths of the stimuli assessed; the motor areas of the brain will then receive impulses from the association center which will activate the dog's muscles either to retreat, or to fight its rival for the meat.

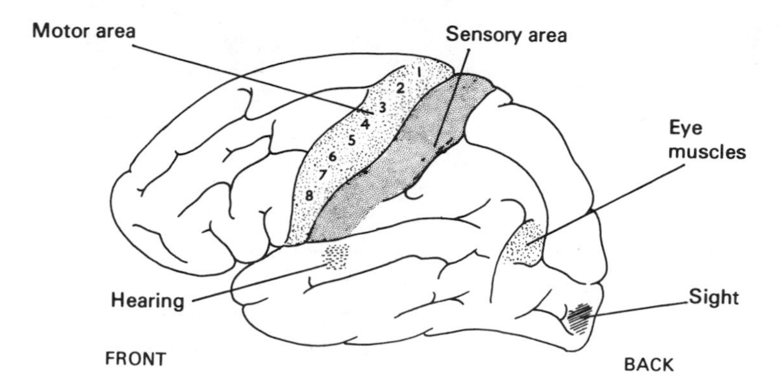

Fig. 29.11 *Localization of areas in the left cerebral hemisphere*
Numbered regions of the motor area are respectively concerned with impulses from (1) abdomen, (2) thorax, (3) arm, (4) hand, (5) finger, (6) thumb, (7) neck, (8) tongue

Without association centers and their "information-storage" or memory systems, learning and the conditioned reflexes described in Section 29.3 would be impossible. In man and other mammals there are important association centers in the two lobes or *cerebral hemispheres* (together forming the *cerebrum*), large outgrowths from the fore-brain spreading backward over the rest of the brain. In these centers the complex systems of synapses between huge numbers of neurons allow intelligent behavior, memory and, in man at least, consciousness of his own activities. The behavior of animals without cerebral hemispheres is a matter of simple reflex actions and of inborn behavior patterns called *instinct*. Certain regions of the cerebral hemispheres have been shown to be concerned with impulses from particular regions of the body or sense organs (Fig. 29.11).

Summary of Brain Functions

To sum up:

(*i*) The brain receives impulses from all the sensory organs of the body.

(*ii*) As a result of these sensory impulses, it sends off motor impulses which cause muscles and glands to function appropriately.

(*iii*) It correlates impulses from the sense organs in its association centers.

(*iv*) The association centers and motor areas co-ordinate bodily activities so that the mechanisms and chemical reactions of the body work efficiently together.

(*v*) It "stores" information so that behavior can be modified according to past experience.

29.3 REFLEX ACTION

Reflex action is a rapid, automatic response to a stimulus, by an organ or system of organs. For example, the iris of the eye contracts or dilates the pupil in response to changing light intensity without our being aware that it is happening. More commonly we know that a reflex action is occurring but are unable to control it. Blinking when a foreign particle touches the cornea is a reflex action which protects the eyes: we know it is happening

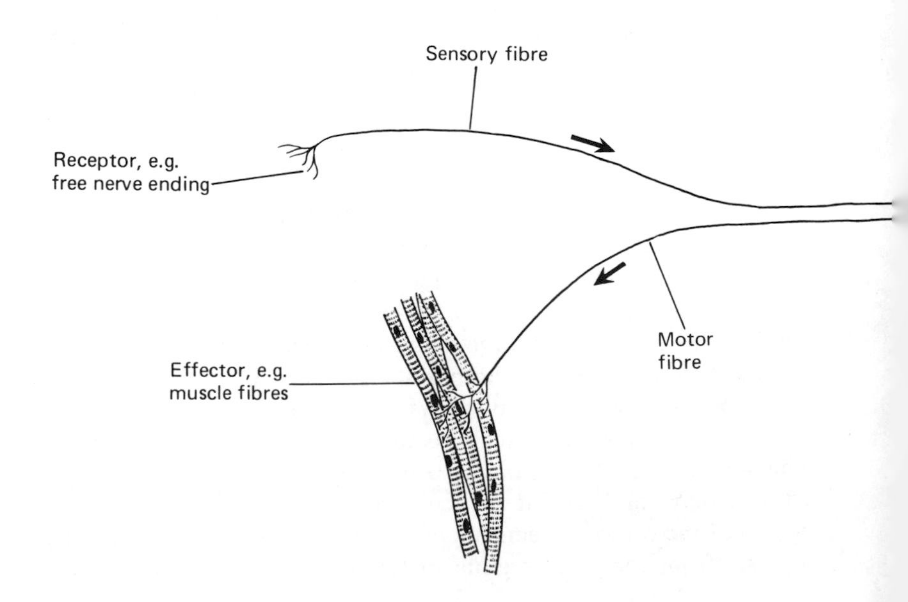

but can do nothing to prevent it or modify it. Sneezing is a reflex response to a stimulus in the nose. The knee-jerk is another example. If the right leg is crossed over the left and struck sharply just above or below the knee cap, the lower leg jerks outward by reflex action.

The Reflex Arc

It is possible to trace the path taken by the impulses in a reflex action (the *reflex arc*), although the description which follows is an over-simplification. Consider, for example, the movement of a hand that has touched an unexpectedly hot object (Fig. 29.13). Heat- and pain-sensitive sensory cells in the skin are stimulated, and set off impulses in the sensory fibers running from them. These fibers form part of the nerve running in the arm from the hand to the spinal cord, and enter the spinal cord via the dorsal root. The axon of each sensory cell makes a synapse with a relay neuron, which in turn makes a synapse with one or more motor fibers. The impulses are thus transmitted from the sensory fibers to motor fibers which leave the spinal cord through the ventral root, and pass in a nerve (probably the same one in which the sensory impulse traveled) to the biceps muscle. The impulse causes the muscle to contract, so removing the hand from the painful stimulus and preventing damage to the tissues.

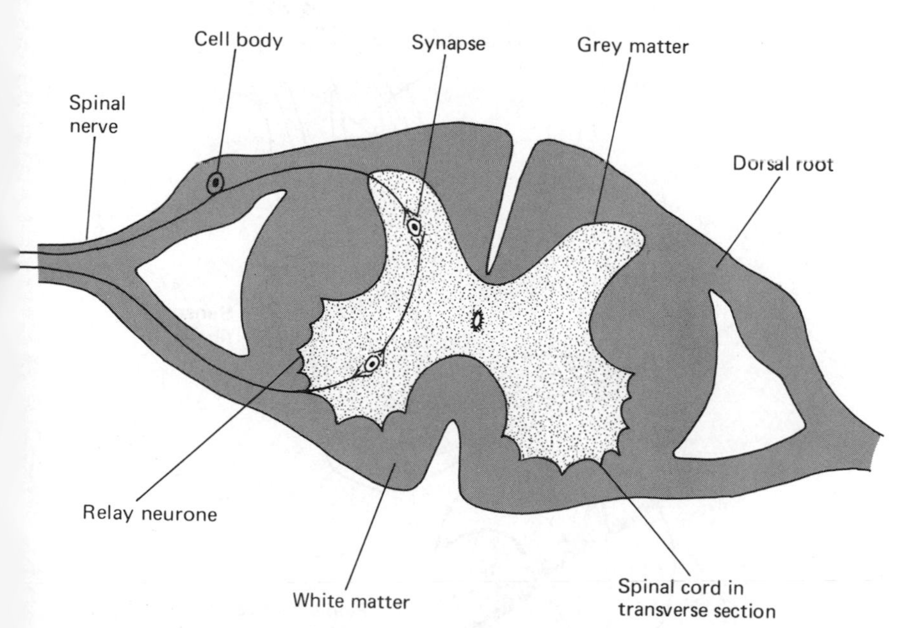

Fig. 29.12 *One of the simplest connections for a reflex action*

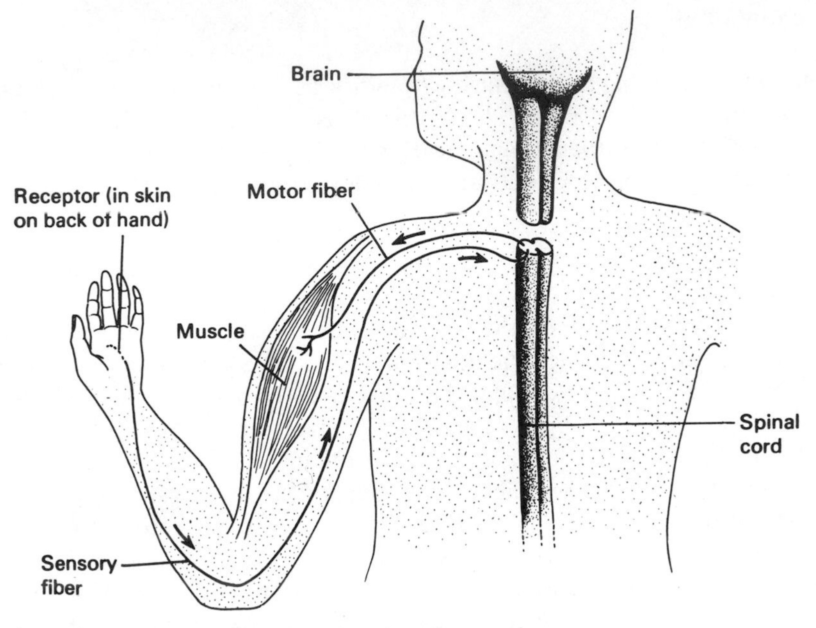

Fig. 29.13 *A reflex pathway*

Fig. 29.14 *Diagram to show reflex arc (withdrawal reflex)*

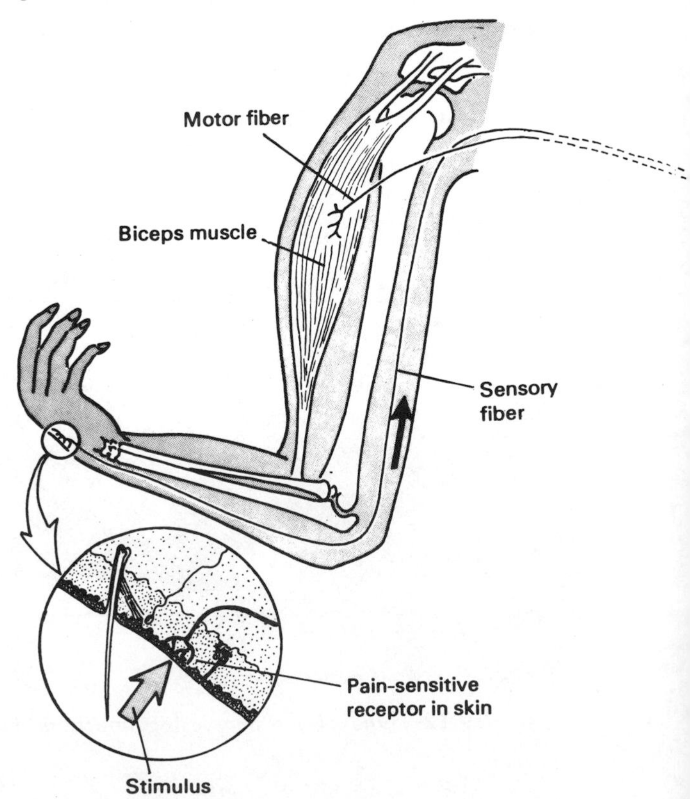

A similar reflex arc produces the knee-jerk, but in this case the sensory fibers make a synapse directly with the motor fiber, and no relay neuron is involved. Striking the tendon below the knee cap stimulates stretch receptors in the leg extensor muscle. The impulse travels to the spinal cord and is transmitted directly to the motor fiber, which stimulates the same muscle to contract.

The two examples described above are *spinal reflexes,* that is, reflex actions which do not involve the brain; in fact, spinal reflexes can be demonstrated in a frog even when its brain has been destroyed. However, since one is usually aware that a reflex action is taking place, the impulses passing in the reflex arc must be relayed by nerve fibers in the spinal cord to the brain. Reflex actions concerning organs of the head, like blinking or sneezing, take place in the brain.

Reflex actions are in fact less simple than described here. Several receptors of different kinds may be stimulated at once and many sets of muscles or glands may be brought into action, involving many more neurons

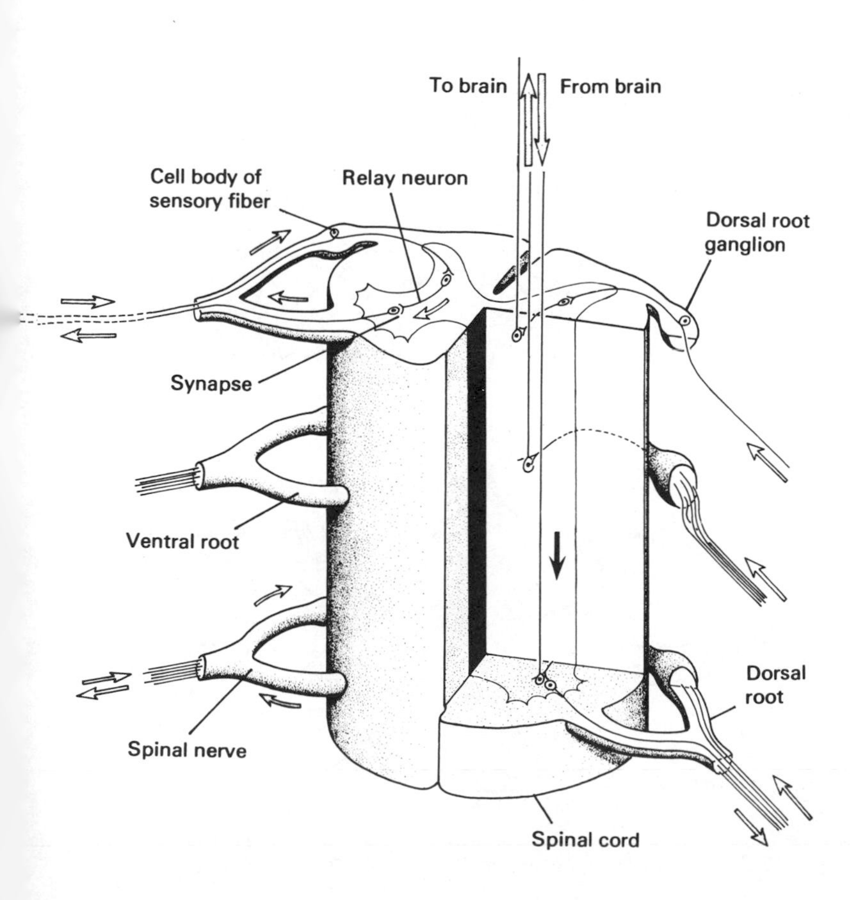

than the two or three mentioned in the preceding descriptions of reflex arcs. Additionally, the relay neuron usually has synapses with many other motor fibers. Thus, for example, a reflex muscle contraction is accompanied by the simultaneous inhibition of the action of the antagonistic muscle.

Conditioned Reflexes

In most simple reflexes, the stimulus produces a related response: for example, the stimulus of the presence of food in the mouth produces the reflex of salivation. After a period of learning or training, however, it is possible for a different and often irrelevant stimulus to produce the same response. A *conditioned reflex* has then been established, and the animal is said to be *conditioned* to respond to this stimulus.

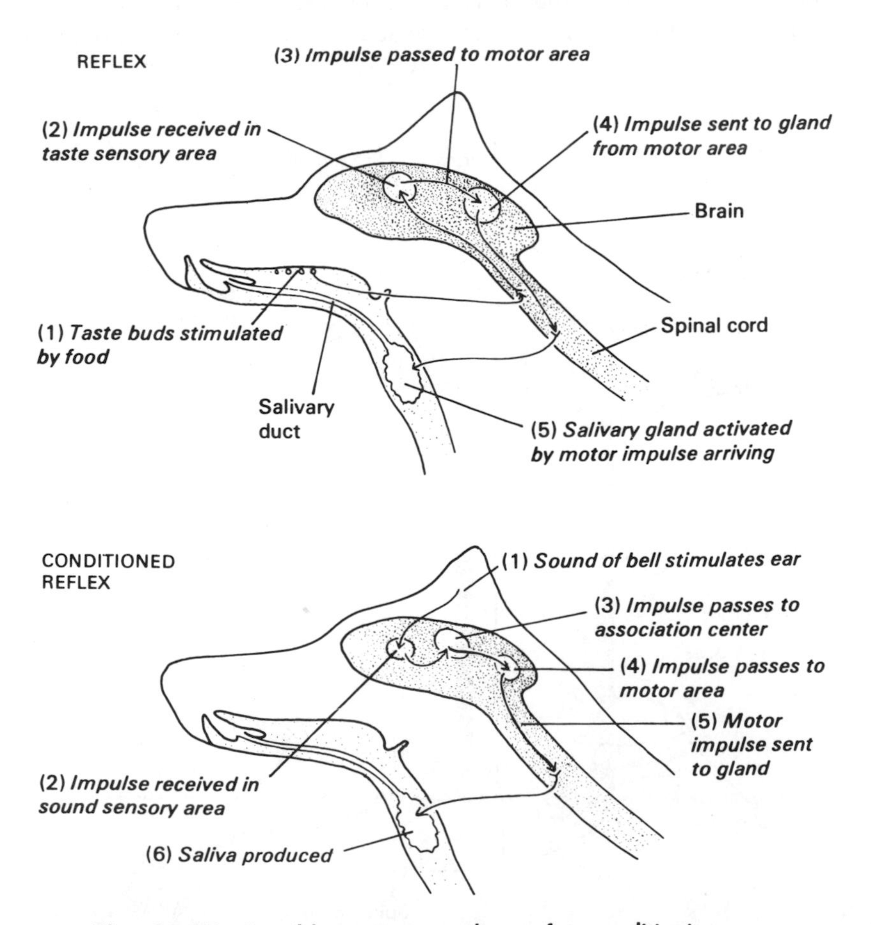

Fig. 29.15 *Possible nervous pathway for conditioning*

A Russian biologist named Pavlov carried out a great many experiments on conditioned reflexes in dogs, one of which has now become known as something of a classic (Fig. 29.15). Pavlov studied the salivation reflex of dogs, produced by the stimuli of taste and smell of food. For several days, he rang a bell at the time the food was given to the dogs. Later, he found that the sound of the bell alone, without the sight and smell of any food, was a sufficient stimulus to initiate salivation. Impulses from the taste and smell centers in the dogs' brains had become correlated with impulses from the ears in the association centers to such an extent that the auditory stimulus eventually could entirely replace the original.

The training of animals is done largely by conditioning them to respond to new stimuli. Many of our own actions, such as walking and riding a bicycle, are complicated sets of conditioned reflexes which we acquired in the first place, perhaps in infancy, by concentration and practice.

29.4 THE ENDOCRINE SYSTEM

The coordination of some of the functions of the body is effected by chemicals called *hormones* secreted by the *endocrine glands.* These are glands which have no ducts or openings (unlike, for example, salivary glands or sweat glands) but their secretions enter the bloodstream as it passes through their tissues, and are carried in solution in the blood all over the body. Each hormone acts on certain tissues in specific ways. Their effects are much slower and more general in character than nervous action: many of the changes they control are rather long-term in their nature, such as the changes taking place during growth or in the development of the characteristics of sexual maturity. Others, such as the changes in heart-rate and breathing brought about by the secretions of the adrenal glands (see below), take place within a few seconds; but this is still slow compared with the fraction of a second needed for a reflex action.

When hormones in the bloodstream reach the liver, they are converted into relatively inactive compounds; a hormonal response is therefore of limited duration and does not persist indefinitely. The inactivated compounds are excreted in due course by the kidneys, and this is why pregnancy can be detected by the identification of its hormonal products in the urine.

(a) **Thyroid gland.** The thyroid gland lies in the neck, in front of the windpipe. It produces several hormones, of which one of the most important is a relatively simple iodine-containing compound called *thyroxine.* Thyroxine controls the rate of chemical activity in man, particularly the rate of tissue respiration. If the thyroid gland produces insufficient thyrox-

ine, the individual tends to sluggishness and overweight; if too much is produced, he will be thin, overactive and anxious. Thyroxine deficiency in infants causes a condition called *cretinism,* in which physical and mental development are both retarded; if identified in its early stages it can be successfully treated by giving just the right amounts of thyroxine.

In amphibia, thyroxine is concerned in metamorphosis, that is, the change of a tadpole into an adult. Feeding tadpoles on thyroxine induces early metamorphosis.

(b) **Adrenal glands.** The two adrenal glands lie just above the kidneys at the back of the abdomen. There are two zones in each gland, an outer *cortex* and an inner *medulla.*

The adrenal cortex produces several hormones including *cortisone,* concerned in the deamination of amino acids to glucose (see Section 20.12(*d*)). The cortex begins to secrete hormones when it is stimulated by the presence of a certain pituitary hormone (see Section 29.4(*f*)) in the blood.

When the brain receives sensory impulses which are associated with danger, or with other situations requiring vigorous action, motor impulses are transmitted to the adrenal medulla, which begins to secrete a hormone called *adrenaline,* the flight or fright hormone. When adrenaline reaches the heart in the bloodstream the heartbeat quickens; when it reaches the blood vessels of the alimentary canal and the skin they contract, diverting blood to the muscles; it dilates the pupils of the eyes, brings about the release of glucose into the blood, and speeds up the rate of breathing and oxidation of carbohydrates. All these changes are advantageous to an animal in a situation requiring greatly increased activity, such as running away or putting up a fight. In man, chemical stimulation by this hormone, combined with nervous stimulation, produces the changes which we recognize as the symptoms of fear: thumping heart, hollow feeling in the stomach, pale face, and so on. The adrenal glands of a human being secrete adrenaline in many situations of anxiety or excitement, and not only in the face of physical danger.

(c) **Pancreas.** While the pancreas is a gland which secretes a digestive juice through a duct into the alimentary canal, it also contains endocrine cells which secrete a hormone into the bloodstream. The hormone, a protein called *insulin,* controls the metabolism of glucose in the body: it determines how much glucose is laid down as insoluble glycogen, and how much is oxidized to release energy.

Insulin accelerates the conversion of glucose in the blood to glycogen in the liver (see Section 20.12(*a*)), promotes the uptake of glucose from the blood by the body cells, and increases the rate of protein synthesis by some cells. Usually the production of insulin by the pancreas is just sufficient to deal with the carbohydrate in the diet; failure to produce enough of the hormone leads to the condition of *diabetes*. The diabetic cannot regulate the concentration of glucose in his blood. It may rise to about 160 mg per 100 cm^3 of blood, so that glucose is excreted in the urine (although even so too much remains in the blood) or fall below 40 mg per 100 cm^3, leading eventually to convulsions and loss of consciousness. Diabetes can be treated by careful diet and regular injections of insulin.

(d) **Reproductive organs.** The ovary produces several hormones, called *estrogens,* of which estradiol and estrone are the most potent. The ovary begins secreting estrogens just before puberty. They are responsible for the development of the feminine secondary sexual characters (see Section 25.8) and later for their maintenance. They cause the lining of the uterus to thicken just before an ovum is released, so that if the ovum is fertilized the uterus is ready to receive it. In some mammals at least, an increase in estradiol secretion brings the female animal into "heat," that is, it makes it receptive to the male. *Progesterone,* the hormone produced by the corpus luteum after ovulation, promotes the further thickening of the uterine wall and the development of its blood supply. If fertilization occurs, the placenta also begins producing estrogens and progesterone, stimulating the further development of the muscular structure and blood vessels in the uterus walls, and perhaps preventing the uterine muscles from contracting until it is time for the baby to be born.

The testes produce the male sex hormone, *testosterone*; it promotes and maintains the development of the masculine secondary sexual characters.

(e) **Duodenum.** The lining of the duodenum is stimulated by the presence of food to produce a hormone called *secretin.* This enters the bloodstream and, on reaching the pancreas, initiates the secretion of pancreatic enzymes. In this way enzyme production is not wasted but only proceeds when food is present in the duodenum.

(f) **Pituitary gland.** The pituitary gland is an outgrowth from the base of the fore-brain (Fig. 29.10). It produces a number of different hormones, the majority of which act upon and regulate the activity of other endocrine glands, to such an extent that the pituitary gland has been called the "master gland." The development of the Graafian follicle, for example, and its secretion of its own estrogen hormone, is controlled by a pituitary hormone, *follicle-stimulating hormone* (FSH). Another pituitary hormone stimulates the thyroid gland to grow and to secrete thyroxine, and a third controls the production of cortisone by the adrenal cortex.

Some pituitary hormones, however, act directly on the body's organs. For example, *antidiuretic hormone* (ADH) controls the amount of water reabsorbed into the blood from the glomerular filtrate in the kidney tubules (see Section 23.5). *Growth hormone* influences the growth of bone and other tissues. If growth hormone is extracted from pituitary glands and injected into experimental animals, they grow larger, and continue growing for a longer period, than usual. However, growth is affected by other hormones as well, including those secreted by the thyroid gland and the pancreas, and the growth hormone may possibly act indirectly by the stimulation of these other endocrine glands, rather than by directly stimulating the tissues.

The Endocrine System in Homeostasis

The endocrine system is important in the regulation of the composition of the internal environment, or homeostasis (see Section 20.13), especially in maintaining the levels of glucose and water in the body fluids. If the concentration of glucose in the blood rises, the pancreas is stimulated to secrete more of the hormone insulin, which brings about the removal of glucose from the blood to be stored as glycogen in the muscles and liver. A fall in the blood glucose level suppresses the production of insulin.

The osmotic potential, and hence the concentration, of the body fluids is also controlled chemically. When the osmotic potential of the blood falls, antidiuretic hormone is released from the pituitary gland into the circulatory system. When it reaches the kidneys, it stimulates the cells of the kidney tubules to reabsorb more of the water in the glomerular filtrate. A rise in the osmotic potential of the blood leads to decreased secretion of antidiuretic hormone and to the production of a dilute urine.

29.5 INTERACTION AND "FEEDBACK"

If a system is to be effectively controlled, two opposing mechanisms are required: a car, for example, needs both an accelerator and a brake. The chemical control of body function is based on this principle, the hormones having effects which are antagonistic to each other. Adrenaline, for example, brings about the release of glucose into the blood, while insulin removes it. Similarly, follicle-stimulating hormone (FSH) brings about the growth of the ovarian Graafian follicle, while progesterone suppresses it. The control of the body's growth, development and activities in constantly changing situations is maintained through the fine adjustment of the balance between antagonistic hormones.

In part, this balance is kept by a "feedback" mechanism, that is, a system in which information about events in the body is "fed back" to a hormone source, enabling it to adjust its output accordingly. Thus a pituitary hormone stimulates the thyroid gland to produce thyroxine, but when the thyroxine in the circulation reaches the pituitary gland, production of the thyroid-stimulating hormone is suppressed: the "feedback" of thyroxine to the pituitary gland regulates the output of the latter (Fig. 29.16(*a*)). The interaction between the two glands keeps the level of thyroxine in the blood, and hence the rate of chemical activities in the tissues, at the optimum level. Similarly, the ovarian follicles are stimulated to produce estrogens by the pituitary hormone, FSH, but when the concentration of estrogens in the blood reaches a certain level, the secretion of FSH is suppressed (Fig. 29.16(*b*)).

"Delayed feedback"—that is, a system in which the information fed back takes some time to have any effect—leads to rhythmic changes. For

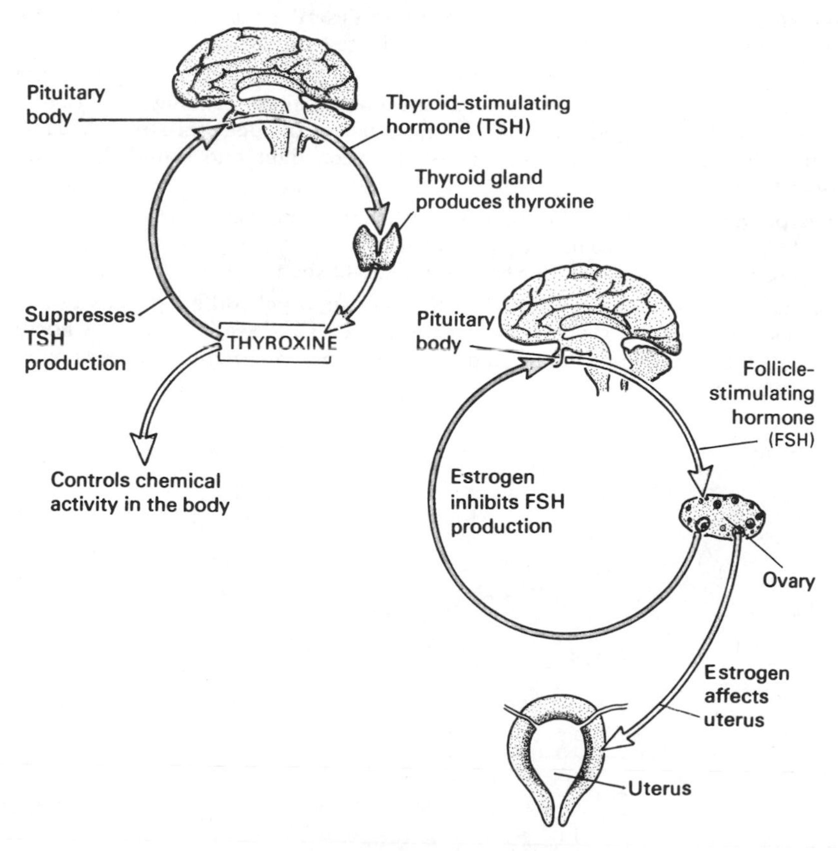

Fig. 29.16 *"Feedback"*

example, the pituitary gland takes about two weeks to respond to a raised level of estrogens in the blood, by which time the uterus lining has thickened and the ovum has been released from the follicle. The response of the pituitary gland is a decrease in FSH secretion, and the follicle's estrogen output consequently falls as well: the reduced estrogen level in the blood leads to the breakdown of the uterine lining characteristic of menstruation. Two weeks later, the pituitary gland's secretion of FSH, no longer inhibited by estrogens in the blood, begins again, and the four-week menstrual cycle repeats itself unless fertilization occurs.

29.6 QUESTIONS

1. List the differences between control by hormones and control by the nervous system.
2. Trace, by diagram or description, the possible reflex arc involved in (*a*) sneezing, (*b*) blinking. (Do not attempt to describe the effector systems in detail and treat the brain as simply an enlarged region of the spinal cord.)
3. All nervous impulses, whether from the eyes, ears, tongue or skin, are basically the same. This implies that the information reaching the brain is little more than a rapid series of electrical pulses of identical strength. How then is it possible for us to distinguish between light and sound, heat and touch?
4. It is possible to train a dog to seek food concealed behind one of several identical doors by flashing a light over the appropriate door. Suggest a nervous pathway by which this behavior is established.
5. Coordination is the term applied to the way in which different activities in the body are made to work efficiently together (see Section 29.1). Choose one of the endocrine glands and describe how its secretion coordinates bodily activities.

UNIT 30
HEREDITY: CHROMOSOMES

30.1 INTRODUCTION

Most living organisms start their existence as a fertilized egg or *zygote* (see Section 25.1). This single cell divides into two cells, which divide again, making four, then eight and so on, to produce eventually the thousands of cells of many different kinds which make up the new organism.

Now if a cell taken from a plant or an animal is placed in a dish with the appropriate nutrients and kept under suitable conditions it will divide and produce hundreds of daughter cells. But these cells are normally all alike; they do not constitute an organism but an undifferentiated mass of tissue. The development of a zygote into an organism must, therefore, be directed in some way to determine that some cells become muscle, some skin, some bone or blood, and that these cells form groups in the right order and in the right place to produce tissues, organs, and ultimately the complete, integrated, coordinated organism.

Moreover, a zygote does not produce just any organism. It will produce one which resembles the parents from whom the zygote was derived. The zygotes of a human and a rabbit may look identical, a nucleus surrounded by a little cytoplasm, but they will develop in quite different ways. The rabbit zygote will produce a rabbit and not a man. An eagle zygote will produce an eagle and not a robin. The study of the mechanism by which the characteristics of the parents are handed on to the offspring is known as *genetics*.

A finch zygote develops into a chick inside the egg shell, isolated from any outside factors other than incubation and the exchange of oxygen and carbon dioxide with the atmosphere. It follows, then, that the "instructions" for the development of the finch chick from the single cell of the zygote must be contained within the zygote itself. Since the zygote was formed by the fusion of the male and female gametes, no other material being involved at all, the "instructions" must have been present in the gametes before their fusion: either in their cytoplasm, or in their nuclei, or in both.

When one examines the gametes of most animals, one finds that the egg has a relatively large volume of cytoplasm associated with its nucleus; on the other hand, the male gamete or sperm consists of little more than a nucleus, the cytoplasm forming a very thin layer around it, extending into a "tail." In spite of this difference in cytoplasm content, all the available evidence suggests that the male and female parents contribute equally to the "instructions" or *genotype* of the zygote. It is therefore likely that the nucleus, rather than the cytoplasm, is of primary importance as the site of the "instructions."

The question next arising is whether there exists any way in which the "instructions" can be seen or studied. Obviously one would not expect to see written directions, but the study of the structure of the nucleus might provide an indication of the nature of the "instructions." We know too that when a cell divides, the "instructions" must be handed on intact and undiminished to each daughter cell; for example, if the two cells resulting from the first division of a frog zygote are separated, both can develop into complete frogs (the same is true of cells separated at the four- and eight-celled stage of embryonic development), showing that the "frog-building instructions" in the zygote are also present in each of the daughter cells. It appears, therefore, that a study of the *dividing* nucleus might be particularly profitable; if we could observe a structure (or structures) in the nucleus which was reproduced exactly, and which on division was shared equally between the two daughter nuclei, we might deduce that this structure was related to the "instructions" for producing a new organism. The sections that follow examine in some detail the events taking place in the nucleus when the cell divides.

30.2 CELL DIVISION

In the early stages of the growth and development of an organism, all its cells are actively dividing to produce new tissues and organs. Most specialized cells subsequently lose this power of division, and it is retained only by a limited number of relatively unspecialized cells such as those in the cambium of plants and in the Malpighian layer of the skin of animals.

The sequence of events taking place during cell division is basically similar in most of those cells that continue to divide. First, the nucleus divides into two; then the whole cell divides, in such a way that one nucleus is associated with each unit of cytoplasm, so that two cells now exist where previously there was only one. Both cells may then enlarge to the size of the parent cell. This process of cell division and enlargement gives rise to growth.

The division of a cell nucleus normally takes place by a series of changes called *mitosis*, the details of which have been elucidated over the last 80 years. There is a second type of cell division, *meiosis,* by which sex cells are formed in the sex organs of plants and animals, which will be presented later (see Section 30.8).

30.3 MITOSIS

Before a cell divides, its nucleus enlarges, and a number of minute thread-like structures called *chromosomes* appear within the nuclear membrane. The number of chromosomes in a nucleus is characteristic of the species (see Section 30.4). The behavior of chromosomes during cell division is usually described in terms of a series of stages, which have been given names such as *prophase, anaphase* and so on. However, the changes take place in a smoothly continuous sequence; the naming of a particular stage does not imply that the process comes to a halt at that juncture, nor that named stages represent equal intervals of time. For example, in the embryonic cells of a particular grasshopper at 38 °C, *prophase* lasts for 100 minutes, *metaphase* 15 minutes, *anaphase* 10 minutes and *telophase* 60 minutes.

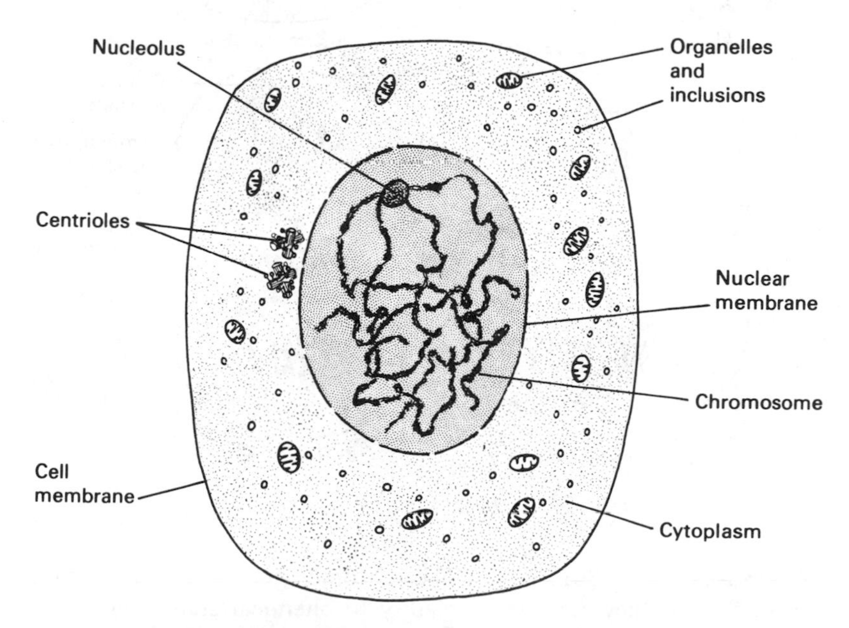

Fig. 30.1 *Animal cell at early prophase*

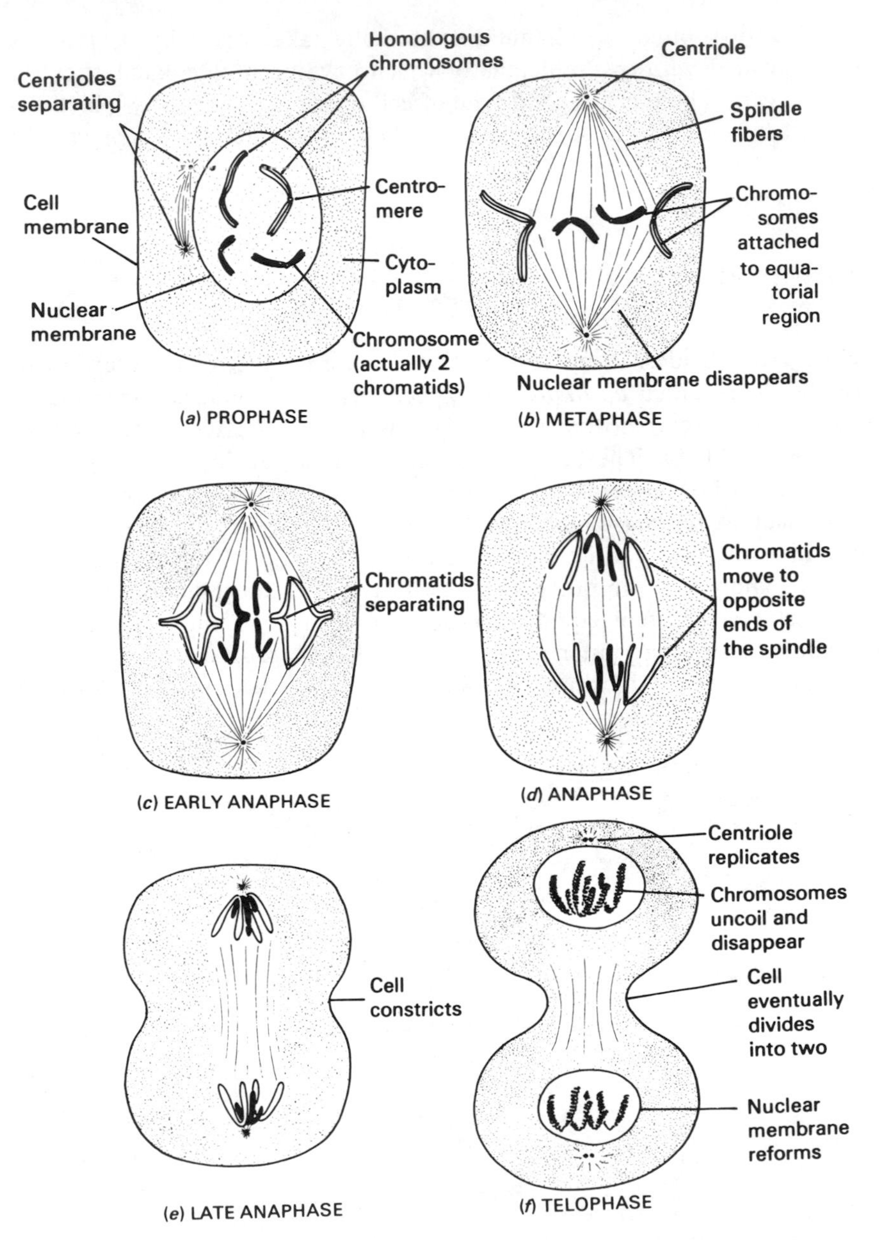

Fig. 30.2 *Mitosis in an animal cell*

(a) **Prophase** (Figs. 30.2(*a*)). The chromosomes become more pronounced, that is, they react more readily to chemical stains and fixatives. Each consists of two strands or *chromatids,* lying close together and joined

in a particular region or *centromere* (Fig. 30.3); these have not been formed by the longitudinal splitting of the chromosome, but by the manufacture of new chromosome material, each chromosome *replicating* itself. Although replication of the chromosomes has occurred before prophase begins, their dual structure is not easily seen until later in mitosis.

The chromosomes shorten and thicken, probably by coiling likc a helical spring, but the coils lie so close to each other that they can only be seen at very high magnifications (Fig. 30.3; Fig. 30.4 shows coiled chromosomes seen at a different stage of cell division). The membrane surrounding the nucleus dissolves, leaving the chromosomes suspended in the cytoplasm, and the one or more *nucleoli* (minute structures normally visible in cell nuclei) also disappear.

(b) **Metaphase** (Fig. 30.2(*b*)). A web-like structure or *spindle* develops in the cytoplasm of the cell; it appears to consist of cytoplasmic fibers. The cells of animals and of some simple plants contain a pair of minute bodies called *centrioles,* lying just outside the nucleus; during metaphase these migrate to opposite ends of the cell, and "fibers" radiating from them meet and join near the center of the cell to form the spindle. Most plant cells have no centrioles, but a spindle is formed nevertheless.

The chromosomes, whose double structure is now clearly evident, attach themselves to the equatorial plane of the spindle, the larger chromosomes at its edge and the smaller ones within the plane.

(c) **Anaphase** (Figs. 30.2(*c*)–(*e*) and 30.5). The two chromatids of each chromosome now separate at the centromere and move away from each other, one to each end of the spindle. There is some experimental evidence

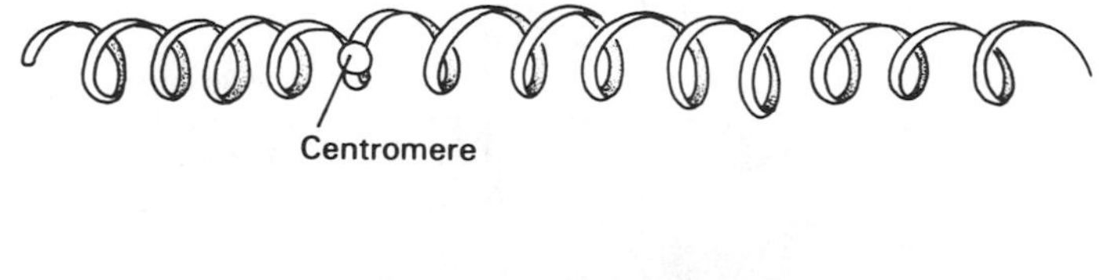

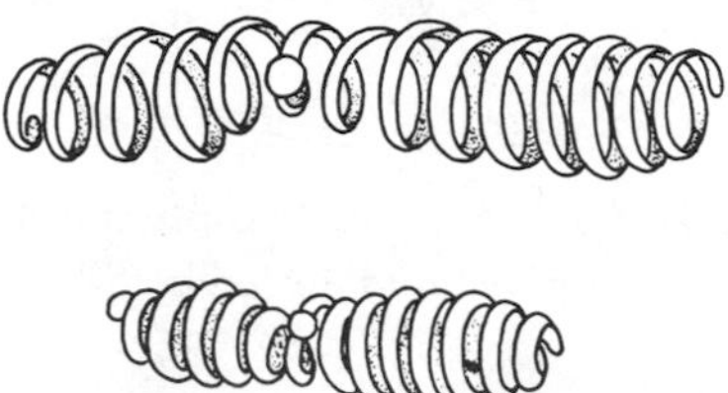

Fig. 30.3 *Diagram illustrating how chromosomes appear to become thicker and shorter during prophase*

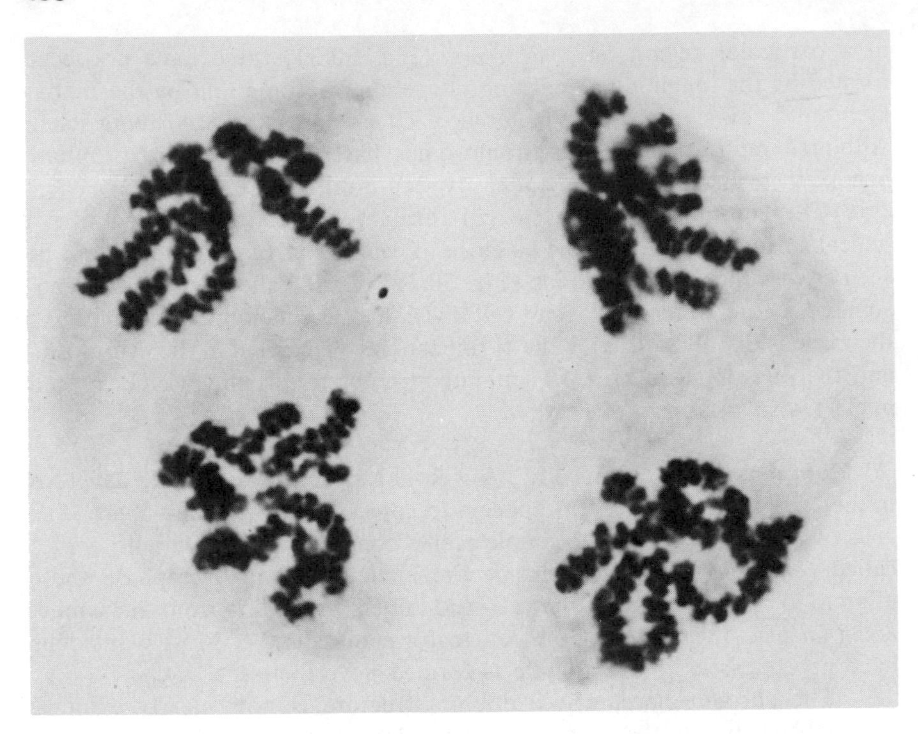

Fig. 30.4 *Chromosomes of* Trillium erectum *at telophase of meiosis*
(A. H. Sparrow, Brookhaven National Laboratory)

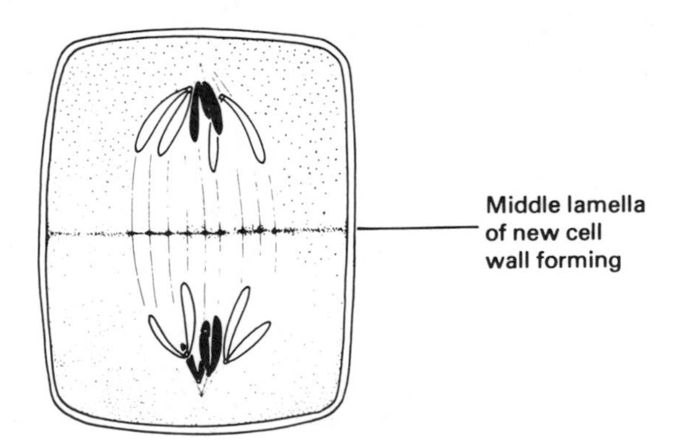

Fig. 30.5 *Late anaphase in a plant cell showing how separation of daughter cells differs from animal cells*

to indicate that the spindle "fibers" are involved in the separation mechanism: the appearance is that of the chromatids first repelling each other at the centromere and then being pulled to opposite ends of the spindle by

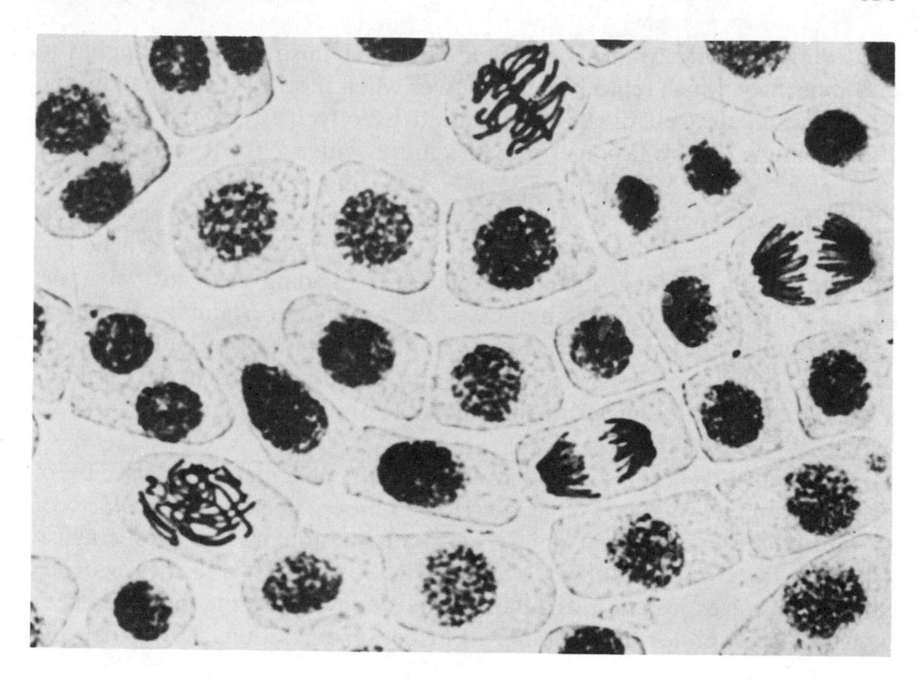

Fig. 30.6 *Cells in a root tip at various stages of mitosis (X500)*
(From McLeish & Snoad, *Looking at Chromosomes*, Macmillan, 1958)

the shortening of the "fibers." However, the existence of such a mechanism has not been verified.

(d) **Telophase** (Fig. 30.2(*f*)). The chromatids, now the chromosomes of the new cells, collect together at the two ends of the spindle and become less distinct, probably because they uncoil and become thinner. The spindle disappears, and a nuclear membrane forms around each group of chromosomes, so that there are now two nuclei in the cell. One or more nucleoli reappear in each daughter nucleus, and in cells containing centrioles each centriole replicates itself. In animal cells, the cytoplasm between the two nuclei constricts and the cell divides into two; in plant cells there is no constriction of the cytoplasm, but a new cell wall forms across the cell in the region originally occupied by the equatorial plane of the spindle (Fig. 30.5).

30.4 CHROMOSOMES

In Section 30.1, we said that the "instructions" for making a new organism might be found in structures in the nucleus which reproduce themselves and which on cell division are shared equally between the daughter nuclei. The preceding account of mitosis makes it clear that this description applies to chromosomes, which reproduce themselves when they form chromatids, and which, when the cell divides, are shared exactly between the new cells. Further study of chromosomes provides more evidence for their importance in transmitting the "instructions."

Chromosomes are so called because they take up certain basic stains very readily (Greek, *chromos* = color, *soma* = body). They can be observed, without staining, in the nucleus of a dividing cell, but when the cell is not dividing they cannot be seen, even after staining. However, isolated flakes and granules of deeply staining material can be seen in the nuclei of cells which are not dividing, and it is thought that the chromosomes persist in the nuclei as fine invisible threads, isolated particles of which still respond to dyes.

Counts of chromosomes show that cells of a particular species of plant or animal contain a characteristic number, the *diploid number,* of chromosomes. For example, human cells contain 46 chromosomes, crayfish 200, fruit fly (*Drosophila*) 8 and rye 14 (Fig. 30.7). This is further evidence to support our expectation that the chromosomes determine the differences between one species and another. Further examination shows that the

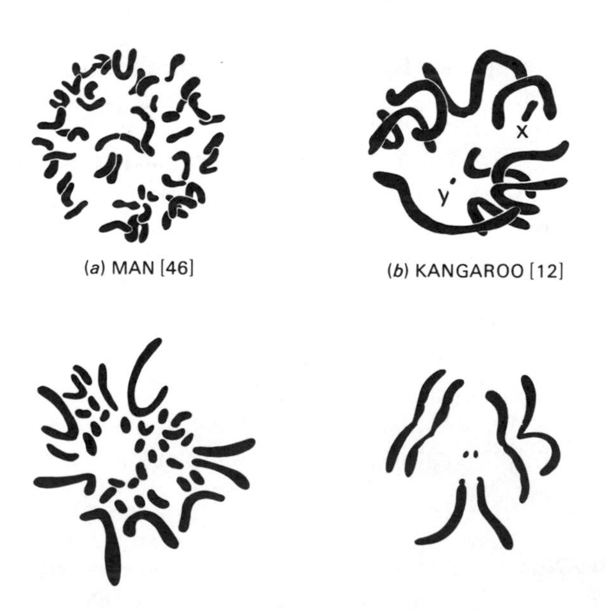

Fig. 30.7 *Chromosomes of different species*
(From Hurst, *The Mechanism of Creative Evolution,* Cambridge University Press, 1933)

chromosomes occur in pairs called *homologous chromosomes*. The members of a pair of homologous chromosomes are not joined together in any way, no more than a pair of shoes are ever joined together, but they have a characteristic length and during anaphase a characteristic shape as well, which depends on the position of the centromere: the chromosome is V-shaped if the centromere is central or a √ shape if it is close to one end. Thus the nuclei of human cells can be regarded as containing 23 pairs of chromosomes, those of crayfish 100 pairs, those of *Drosophila* 4 pairs and so on.

Chromosomes consist of protein associated with a substance called *deoxyribonucleic acid* (DNA); the exact nature of the relation between the two compounds is unknown. Although the constituent chemicals of a cell's cytoplasm are constantly being broken down and rebuilt from fresh material, the two chromosome components remain remarkably stable. Experiments show that during cell division the sharing of chromosome material between the two daughter cells is much more exact than is the case for any other component of protoplasm. While the account given here is necessarily over-simplified, there is in fact a considerable body of experimental evidence pointing to the chromosomes as the main source of the "instructions" which determine that a cell should become like its parent cell, and that as they develop the cells of an organism will endow the animal or plant with the characteristics of its species. Thus the nuclei of the sperm and the ovum each carry a set of chromosomes derived from the male and female parent respectively, and the chromosomes of the zygote determine that it will grow and develop into an individual of the same species as the parents, reproducing in minute detail particular characteristics, like coat or petal color, of both parents.

Fig. 30.8 illustrates onc such example: an alteration in one chromosome is associated with a number of abnormalities in the external appearance of the fly. Since so many different parts of the body are affected, the chromosome aberration must have appeared either at a very early stage of the fly's development from the zygote, or in one of the gametes from which the zygote was formed.

Cells from the salivary glands of *Drosophila* and other flies contain very large chromosomes called giant chromosomes, on which bands can be seen (Fig. 30.9). The size, shape and position of these bands is quite consistent and characteristic for any pair of chromosomes. If, due to some accident in replication, one or other of the bands is lost, there is a corresponding malformation in the adult fly (Fig. 30.10). These bands are thought to represent the site of *genes,* the units of inheritance; although bands have only been observed on these rather unusual giant chromosomes, it is believed that similar structures are carried by all the chromosomes in the body.

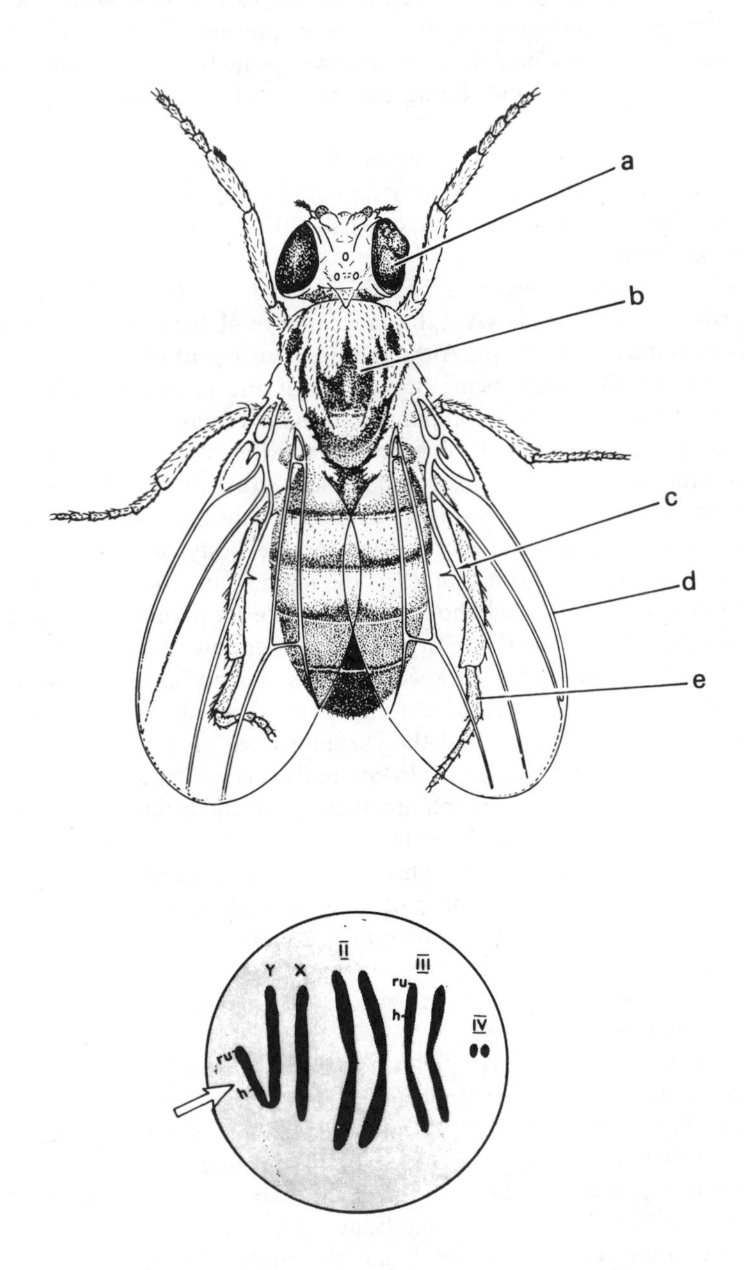

Fig. 30.8 *Effects of a chromosome abnormality in* Drosophila
(After Muller, Journal of Genetics, 1930)

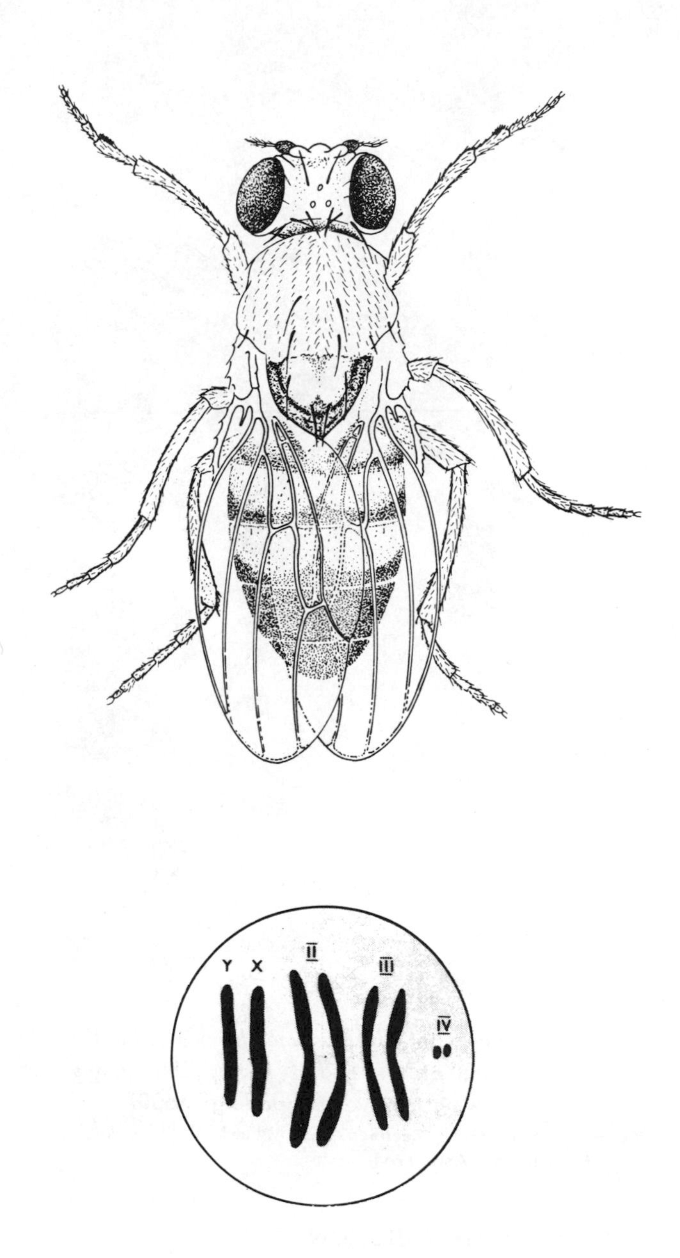

Fig. 30.8 *Effects of a chromosome abnormality in* Drosophila *(cont'd)*

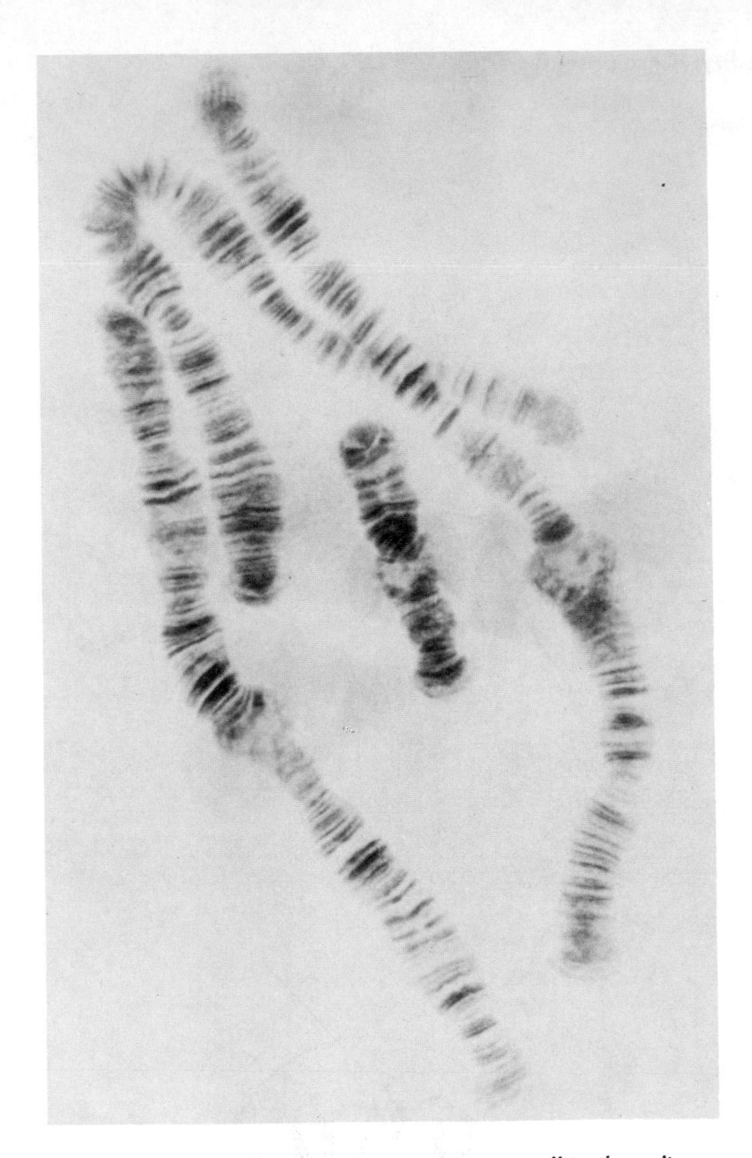

Fig. 30.9 *Four giant chromosomes from a cell in the salivary gland of the midge larva,* Chironomus tentans, *showing transverse banding (X600)*

(Courtesy of Wolfgang Beerman, Max Planck Institute, Tübingen, from *Scientific American*, April 1964)

30.5 GENES AND THEIR FUNCTION

A gene is a theoretical unit of inheritance; the word was coined, and a theory of inheritance developed, long before chromosome structure was investigated in detail or the DNA theory of inheritance was put forward.

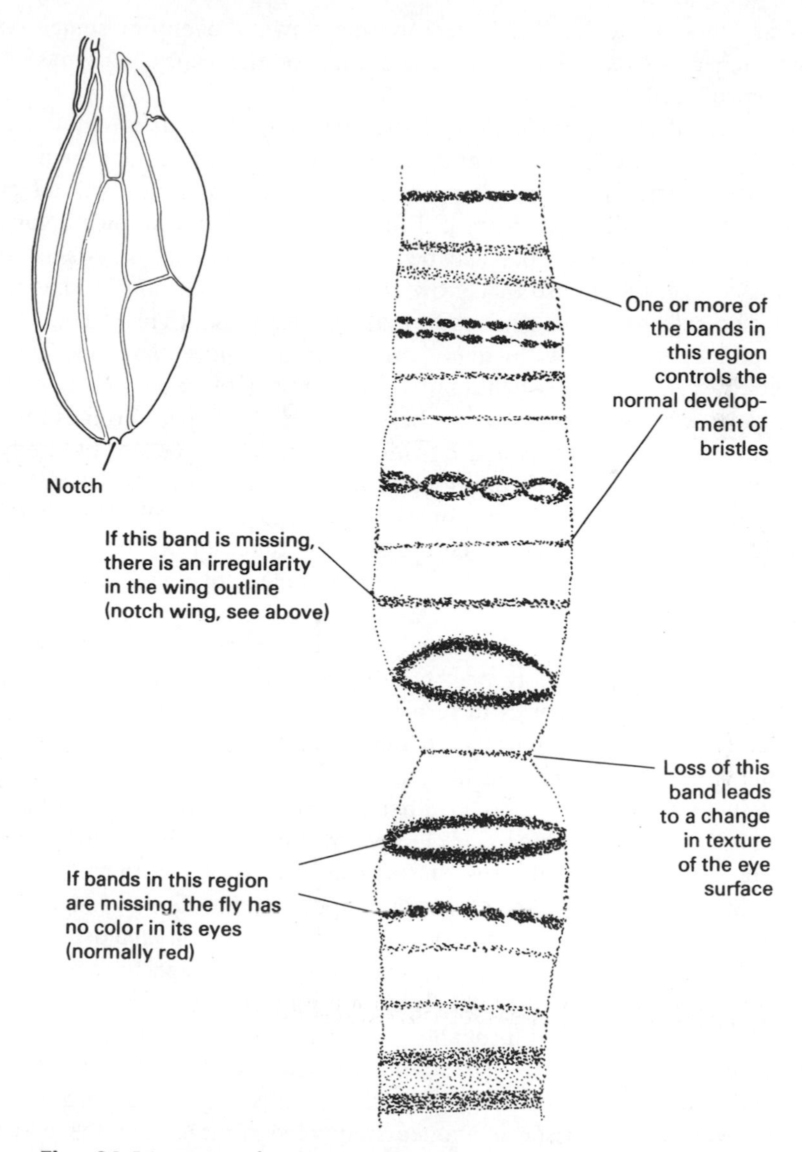

Fig. 30.10 *Part of salivary gland chromosome of* Drosophila, *showing location of four genes*

(From Curt Stein, *Principles of Human Genetics,* 3rd ed., W. H. Freeman, 1973. After Slizynska, *Genetics,* 23, 1938)

A gene is a single piece of information in the "instructions" carried by the zygote; for example, one gene determines whether a rabbit's fur is black or white, and a second gene whether the fur is long or short. Each gene is believed to consist of a group of molecules arranged in a particular manner.

Genes are thought to be distributed in line down the chromosome. At mitosis each chromosome, and therefore each of the genes it carries, is exactly reproduced.

Genes control the production of the enzymes which determine what functions go on in a cell, and eventually in the organ, and in the organism, of which it is a part. If anything happens to a gene in a zygote it will affect the organism which develops from it. For example, there is in mice a gene which determines that the coat will be colored. If this gene is missing in the fertilized egg, the mouse that grows from it will be an *albino,* that is, without pigment: its fur will be white and its eyes pink. (This account is over-simplified; in mice, as in many other species, more than one gene is in fact concerned with pigmentation.) The number of genes in a human cell is not known, but it could be in the region of 1,000 per chromosome.

Two problems are presented by this account. Every cell of the body carries an identical set of chromosomes; if cell characteristics are determined by the genes on the chromosomes, why is not every cell of the body identical? Furthermore, what can be the possible function of, for example, a gene determining eye color when it is in the nucleus of a cell in the wall of the stomach?

It seems that the effect of a gene on the characteristics of the cell of which it is a part depends both on the nature of the gene itself and on the physiology of the cell, which is in turn related to the cell's position in the body. For example, the chemical environment in a certain cell in the scalp allows the gene for blackness in hair to operate in a particular way. Just what the same gene does in cells in other parts of the body is not certain; its action may simply be suppressed. However, most genes have more than one effect, and the most obvious of the characteristics determined by a particular gene is not necessarily the most important one. For instance, the genes responsible for color in the scales of one kind of onion also determine the presence of certain fungicidal chemicals in the plant, and hence confer a resistance to fungal diseases; the color is the more obvious characteristic, but the resistance to disease probably carries a greater advantage to the plant. Similarly, the gene in *Drosophila* which produces the effect of diminutive wings also reduces the expectation of life to half that of normal flies. The wing characteristic is immediately recognizable, but the effect on the life span may be far more important and damaging to the species.

The dependence of a gene's effects on the physiology of the cell which contains it is illustrated by some experiments with certain amphibian embryos. A piece of tissue, which is left undisturbed would develop into skin, was taken from the embryo's abdomen and grafted into a region overlapping the developing eye; the graft eventually became incorporated into the eye as a lens. Since the graft's chromosomes and their genes were unchanged, its new position alone determined the change in the course of its development. This effect has only been observed in a few species; in insects at least the eventual form of individual cells seems to be determined at a very early stage of the embryo's development and is not affected if the cells are moved to a new situation.

30.6 HOW GENES WORK

The nature of the association between the protein and the DNA which constitute chromosomes (see Section 30.4) is unknown, but the structure of the DNA molecule itself has been intensively studied. It is in the shape of a double helix, that is, it is made up of two long spiral strands twisted together, like a double-stranded rope. The helical strands consist of chains of sugar units linked together by phosphate groups, the sugar being *deoxy-ribose,* which has five carbon atoms in its molecule. The two helices are joined by cross-bridges, each sugar unit being linked to its neighbor in the other helix by a pair of organic nitrogen-containing base molecules (Figs. 30.11 and 30.12). In DNA, the bases are of only four principal kinds—*adenine, cytosine, thymine* and *guanine*—and the order in which these base pairs are arranged down the length of the DNA molecule is thought to be the factor which determines the characteristics inherited; a gene is believed to consist of a particular sequence of up to 1,000 base pairs in a DNA molecule.

The sequence in which the bases are arranged along the DNA molecule appears to act as coded instructions for protein manufacture, the order of

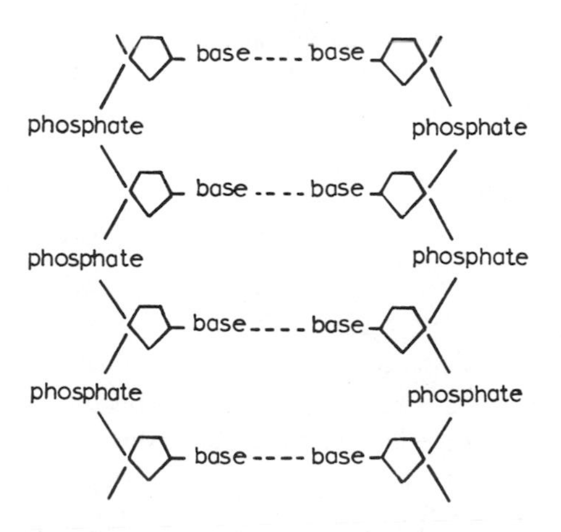

Fig. 30.11 *Part of a DNA molecule (= deoxyribose unit)*

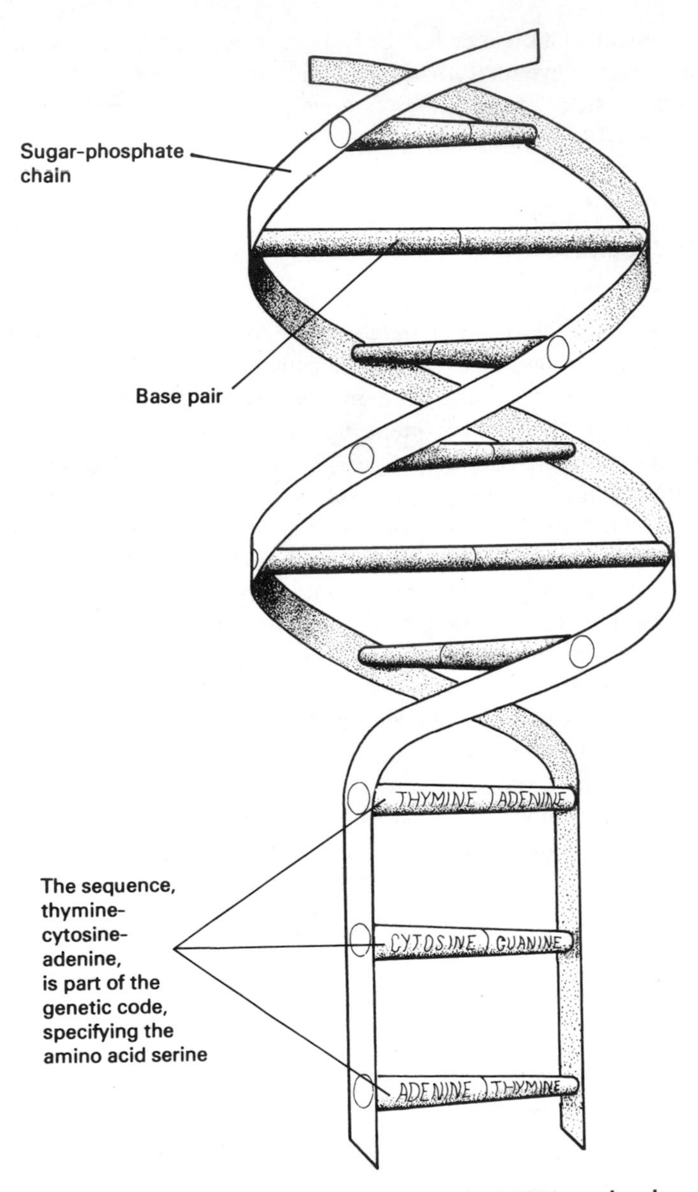

Fig. 30.12 *Diagram showing part of a DNA molecule: the lower part is shown uncoiled to emphasize the position of the base pairs*

bases within the sequence specifying which amino acids are to be linked together to make the protein (see Section 19.3(*b*)). For example, the sequence cytosine–adenine–adenine (CAA) in the DNA molecule specifies the inclusion of the amino acid *valine* in the protein; a row of three thymine units (TTT) specifies *lysine*; the seqeunce adenine–adenine–thymine (AAT) specifies *leucine*. The base sequence CAA–TTT–AAT thus directs the cell to link up a molecule of each of these three amino acids into a peptide, valine–lysine–leucine. More complex proteins are built up in the same way.

Most of the proteins made are enzymes which direct the pattern of chemical activity in the cell. Thus DNA, by determining the kinds of enzyme formed in the cell, will control the cell's activities. A "wrong" sequence of bases in the DNA molecule will result in the "wrong" order of amino acids in the resulting protein and, hence, an ineffective enzyme. This will usually act adversely on the metabolism of the cell. For example, a rabbit with colored fur has a gene which directs the production of the enzyme *tyrosinase,* which catalyses the conversion of the colorless compound tyrosine to the black pigment *melanin.* An albino rabbit has no gene for tyrosinase production, and consequently no pigment is formed from the tyrosine in its body. Similarly, normal humans have a gene which directs the production of an enzyme which catalyses the breakdown in the blood of a chemical called *alcapton.* A person without this gene excretes unchanged alcapton in his urine; its presence is noticeable because it darkens on exposure to air. This relatively harmless effect is associated with pigmentation in certain parts of the body and with the appearance in later life of arthritis.

Clearly, the absence of a gene directing the production of an enzyme essential to the vital chemistry of the cell—rather than enzymes concerned simply with pigmentation or with alcapton breakdown, for instance—could have devastating results leading to the serious malfunction of the cell or even its premature death. Conversely, since the functioning of a normal individual is the result of hundreds of different kinds of chemical changes catalysed by hundreds of enzymes, it is not surprising that such characteristics as intelligence, stature and activity are influenced by the presence or absence of a number of genes.

30.7 VARIATION

The exact replication of chromosomes and genes and their equal distribution between cells at mitosis leads to conformity between parents and offspring in each generation: herring's eggs develop into herring, for example, not into whiting or haddock. Nevertheless, a particular individual may differ in many respects from its brothers and sisters, as well as from its parents. Two black mice may have some white mice among a litter of black ones; in humans, the children of two brown-eyed parents may have either blue eyes or brown. While some variations arise from spontaneous changes or *mutations* in the genes or chromosomes (see Section 30.11), such variations are infrequent and usually harmful. Variations in characteristics like eye or hair color result from the rearrangement in the zygote of the genes and chromosomes derived from the parents. This rearrangement is a direct result of the way in which the chromosomes separate during the processes of cell division which leads to gamete formation, a sequence of events called *meiosis.*

30.8 MEIOSIS

In the reproductive organs, certain cells give rise to the gametes (either sperms or ova) by a special kind of cell division which results in the gametes containing only half the number of chromosomes contained in each of the other cells of the organism. In man, for example, the sperms and the ova each contain only 23 chromosomes; when two gametes fuse, the resulting zygote contains the normal complement of 46 chromosomes, and this number is present in all the cells formed from the zygote by mitotic division. This halving of the chromosome number at gamete formation maintains the number of chromosomes characteristic of the species. If gametes were produced by mitosis, they would contain the full diploid number of chromosomes. Human gametes would each contain 46 chromosomes and the zygote resulting from their fusion would have 92; in the next generation, the offspring would have 184 chromosomes, and so on.

This special process of cell division is called *meiosis*; like mitosis, it is most conveniently described in terms of a series of stages.

(a) **Prophase** (Figs. 30.13(*a*)–(*d*)). At the beginning of meiosis, the chromosomes become apparent in much the same way as in mitosis. However, they still appear as single threads at this stage, although there is reason to believe that two chromatids are present in each chromosome. At the beginning of prophase there is no coiling and consequently no shortening of the chromosomes.

In complete contrast to anything taking place in mitosis, the homologous chromosomes now appear to *attract* each other. The members of each pair come to lie alongside each other so that all their parts correspond exactly. The pairs so formed are called *bivalents*; for example, a human cell with a normal complement of 46 chromosomes would have, at this stage, 23 bivalents. The chromosomes now shorten and thicken by coiling; each can be seen to consist of two chromatids. The homologous chromosomes now seem to repel each other at the centromeres and the pairs begin to separate, except at certain regions called *chiasmata* (singular, *chiasma*)

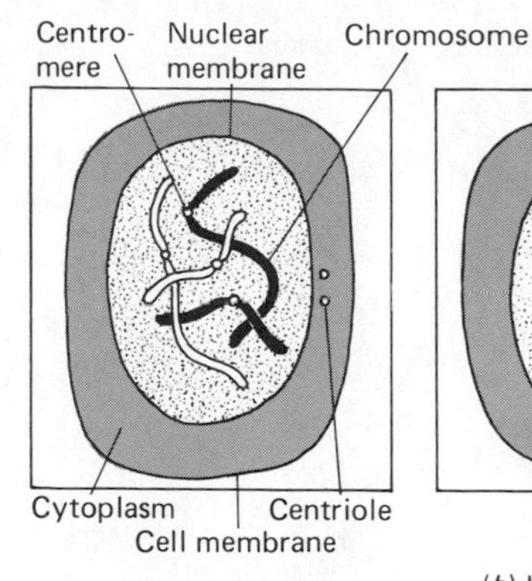

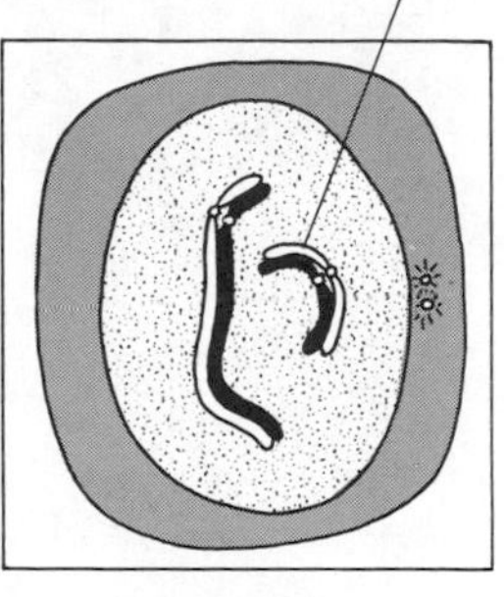

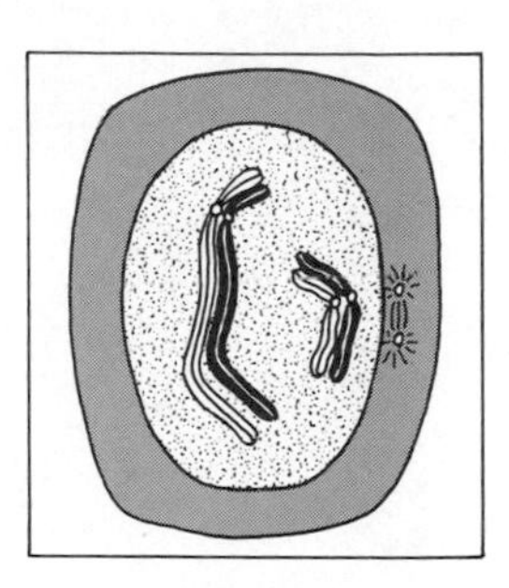

PROPHASE

(*a*) The diploid number of chromosomes appear

(*b*) Homologous chromosomes pair with each other, and then shorten and thicken

(*c*) Replication has occurred and the chromatids become visible

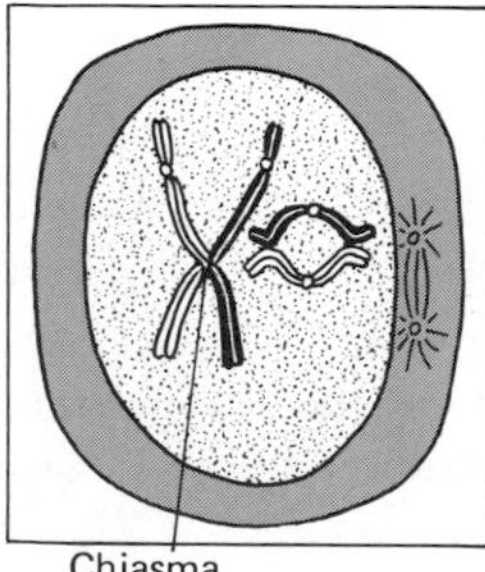

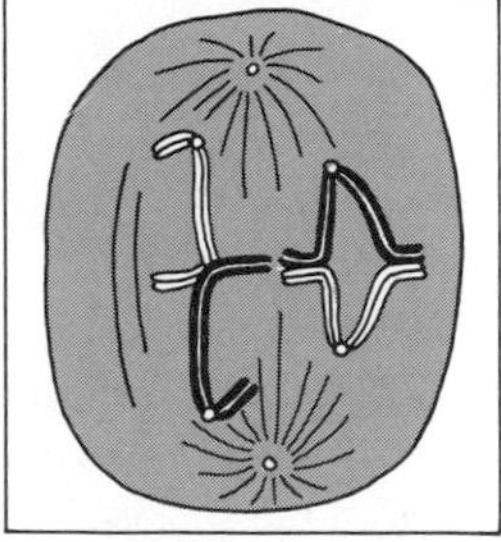

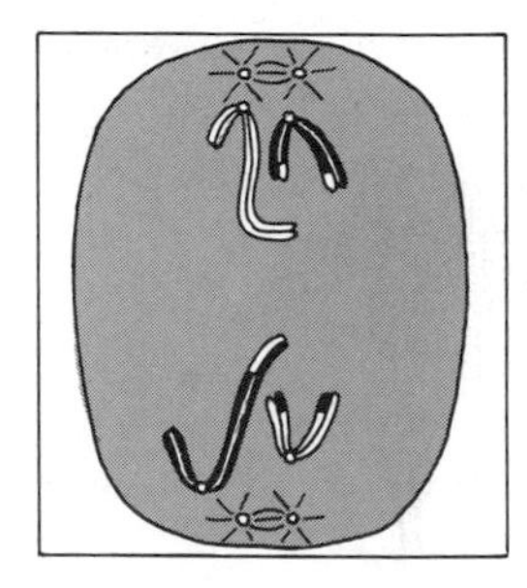

ANAPHASE

(*d*) Homologous chromosomes move apart except at the chiasmata where chromatids have exchanged portions

(*e*) A spindle has formed and homologous chromosomes move to opposite ends taking exchanged portions with them

(*f*) Homologous chromosomes separated but not enclosed in nuclear membranes

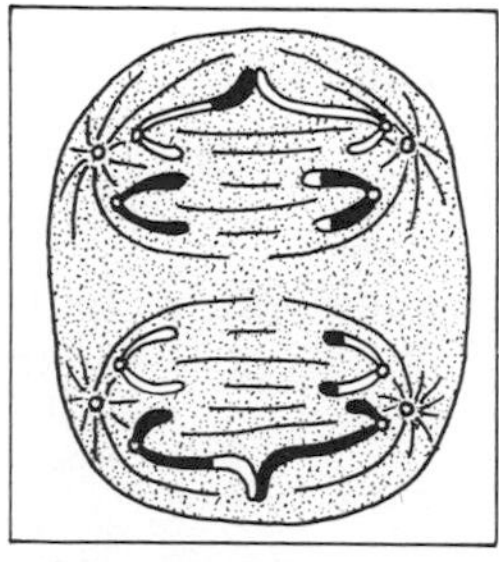

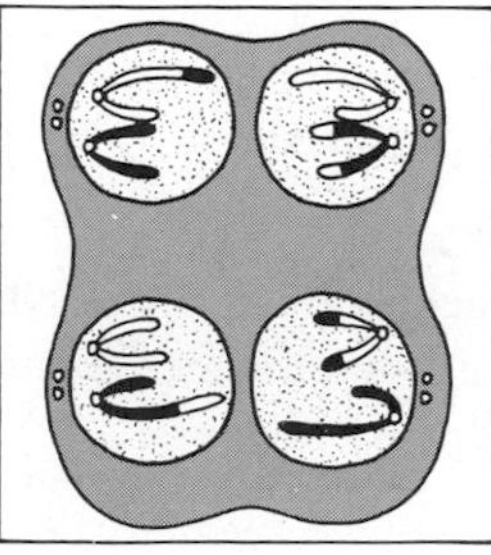

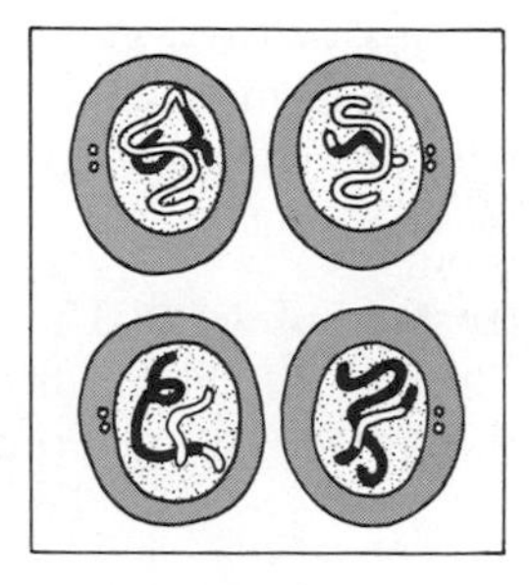

SECOND MEIOTIC DIVISION

(*g*) Spindles form at right angles to the first one and the chromatids separate

TELOPHASE

(*h*) Four nuclei appear, each enclosing the haploid number of chromosomes

(*i*) Cytoplasm divides to form four gametes

Fig. 30.13 *Meiosis in a gamete-forming animal cell*

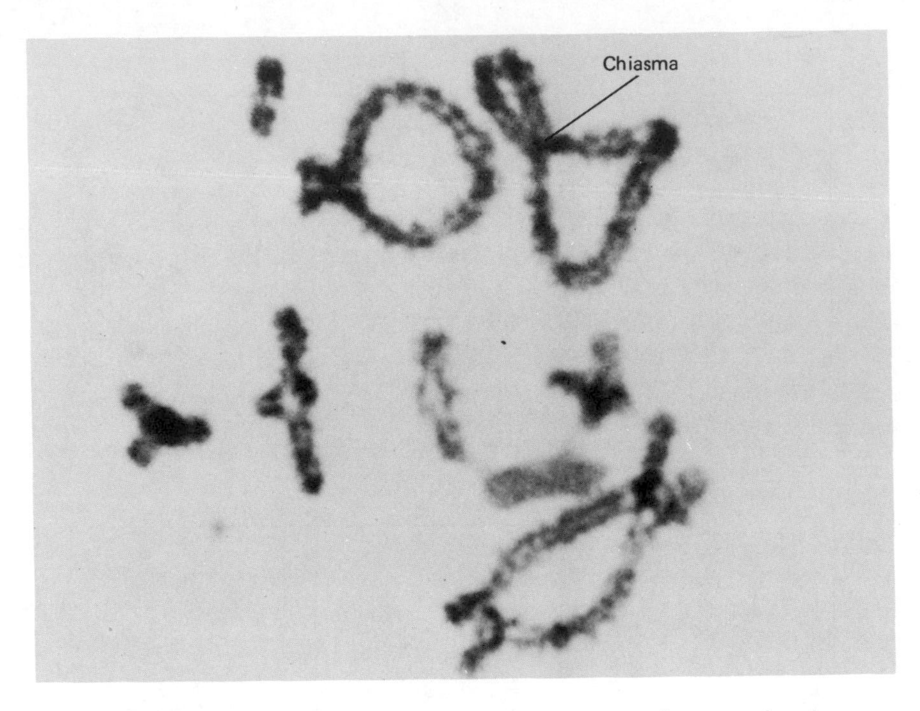

Fig. 30.14 *Meiosis in cell from grasshopper testis* (late prophase)
(From color slide set, *Meiosis in Chorthippus brunneus*, Philip Harris Biological Ltd.)

(Fig. 30.14). In these regions the chromatids appear to have broken and joined up again, in such a way that the fragments have become rearranged. (The significance of this rearrangement or exchange of chromosome material is discussed in Section 30.9.)

The membrane surrounding the nucleus remains intact around the chromosomes throughout prophase.

(b) **Metaphase.** The nuclear membrane disappears, and a spindle is formed. The bivalents, still joined together at the chiasmata, move toward the equatorial region of the spindle.

(c) **Anaphase** (Fig. 30.13(*e*)–(*f*)). The separation of homologous chromosomes, begun in prophase, now continues. The chromosomes, each consisting of two chromatids, move to opposite ends of the spindle in a manner similar to that of single chromatids during mitosis. Consequently *one chromosome of each pair* reaches each end of the spindle (in mitosis *one chromatid of each chromosome* migrates to each end); the resultant

number of chromosomes at each end, half the diploid number, is called the *haploid number* (sometimes the *monoploid number*). This first meiotic division is called the *reduction division.*

(d) **Second meiotic division** (Fig. 30.13(*g*)). The spindle disappears, but a nuclear membrane does not usually form at this stage. Instead, two new spindles appear, lying at right angles to the plane of the first one, each associated with one of the groups of chromosomes resulting from the reduction division. The chromatids now separate and the members of each pair move to opposite ends of the spindle as in mitosis.

(e) **Telophase** (Fig. 30.13(*h*)–(*i*)). A nuclear membrane forms around each of the four groups of chromatids, so that the cell now has four nuclei, each containing the haploid number of chromosomes. Finally the cytoplasm divides so as to separate the nuclei, giving rise to four daughter cells.

Formation of Gametes

(a) **Sperms.** In most animals the four new cells all develop "tails" of cytoplasm to become sperms (*spermatogenesis*).

(b) **Ova.** When ova are formed by meiosis (*oogenesis*) the cytoplasm of the parent cell is not shared equally between the four daughter cells. After the first meiotic division, one of the newly formed nuclei receives the bulk of the cytoplasm; the other nucleus separates from the cell with only a vestige of cytoplasm to form the *first polar body* which, although it may undergo a second meiotic division, cannot function as an ovum and subsequently degenerates. The second meiotic division of the remaining nucleus produces a similar *second polar body* and a mature ovum.

In many vertebrates, the first polar body is not formed until after the potential ovum has been released from the ovary, and the second polar body is only produced after the sperm has actually penetrated the ovum in fertilization.

30.9 NEW COMBINATIONS OF GENES IN THE GAMETES

When an ovum fuses with a sperm, each gamete contributes half the chromosomes in the resulting zygote. In each of the subsequent mitotic divisions of this zygote and of its daughter cells, the full complement of chromosomes derived from both parents is first doubled and then shared equally between the new cells. The result of this replication is that all the new cells contain sets of chromosomes identical in number and parental origin: for example, of the 46 chromosomes in man, 23 are derived originally from the individual's father and 23 from his mother. When gametes are formed, on the other hand, the separation of homologous chromosomes at the first meiotic division is likely to give rise to unequal distribution of paternal and maternal chromosomes in the gametes (Fig. 30.15). At anaphase, it is a matter of chance which a pair of homologous chromosomes migrates to a given end of the spindle. The 23 chromosomes in a human gamete could all be derived from the father, or 10 could be derived from the father and 13 from the mother, or 6 from the father and 17 from the mother, and so on. There are 2^{23} (about 8 million) possible different combinations of individual paternal and maternal chromosomes in the gamete.

If homologous chromosomes were identical in their gene content, this variability in chromosome distribution would have no effect. However, an individual's parents are likely to be genetically dissimilar in many respects: for example, his mother may have blue eyes and fair hair, while his father has brown eyes and black hair. Such a person would be able to produce gametes with combinations of genes quite different from those of either of his parents—such as a combination of genes for blue eyes together with black hair, not present in either parent. (Again, this account is an over-

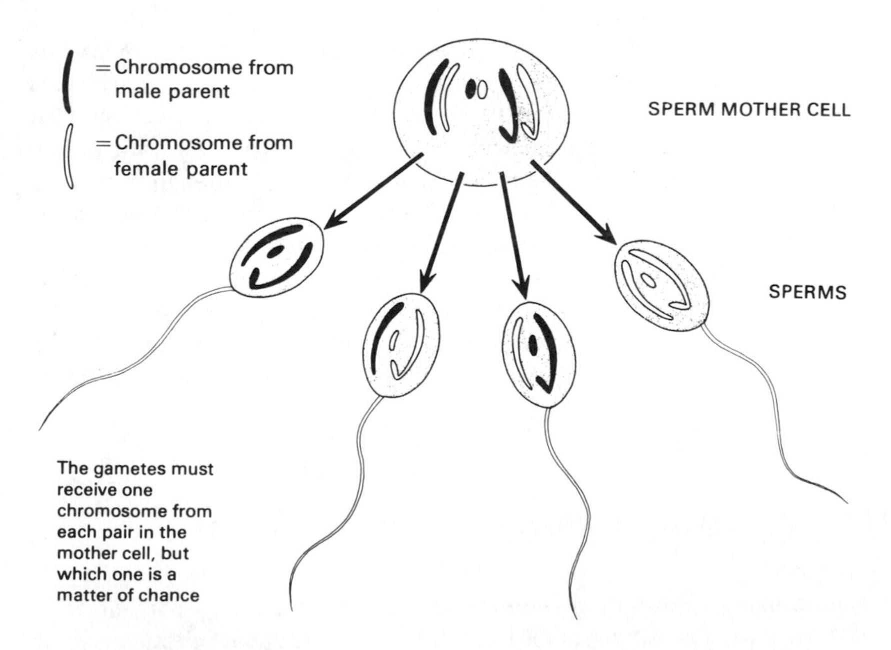

Fig. 30.15 *Some possible combinations of maternal and paternal chromosomes in the gametes as a result of meiosis*

simplification; eye and hair color are usually not controlled by single genes.)

Linkage and crossing over. During the pairing up of homologous chromosomes in the early stages of meiosis, the maternal and paternal chromosomes may exchange portions of chromosome material (see Section 30.8 and Fig. 30.16). This increases the possibility of variations in the gene combinations of the gametes. In the absence of crossing over, the maternal genes carried on the same chromosome, indicated by *A–B–C* in the diagram, would always appear together no matter how the chromosomes became assorted during meiosis. Similarly, the paternal genes represented by *a–b–c* would always remain together. In *Drosophila,* for example, the genes for black body, purple eyes and vestigial wings all occur on the same chromosome; one might therefore expect that a black-bodied *Drosophila* would always have purple eyes and vestigial wings. Crossing over between the chromatids, however, gives the possibility of breaking up these *linkage groups* so that new combinations can arise in the gametes: Fig. 30.16 shows how the *A–b–c* and *a–B–C* groups could arise, and clearly many other groupings are possible. Thus, for example, a black-bodied *Drosophila* could have normal wings, and either red or purple eyes.

30.10 FERTILIZATION

At fertilization, the cytoplasm of the sperm fuses with that of the ovum, and the male nucleus comes to lie alongside the female nucleus: the zygote is formed, but there is often no fusion of nuclear material at this stage. The two nuclei simultaneously undergo mitosis, with the axes of the spindles parallel to each other. At the telophase stage of this division, each of the two groups of adjacent chromatids becomes enclosed in a nuclear membrane; the zygote now contains two nuclei, each having the diploid number of chromosomes. The zygote next divides into two cells, which by subsequent mitotic division give rise to a multicellular organism with the diploid number of chromosomes in all its cells (Fig. 30.17).

(a) Recombination of Genes in the Zygote

The genetic content of the ovum is likely to be different from that of the sperm which fertilizes it. In man, for example, the mother may have curly red hair while the father's hair is straight and black; an ovum carrying genes for curliness and redness of hair could thus be fertilized by a sperm carrying genes for straightness and blackness. The resulting zygote could

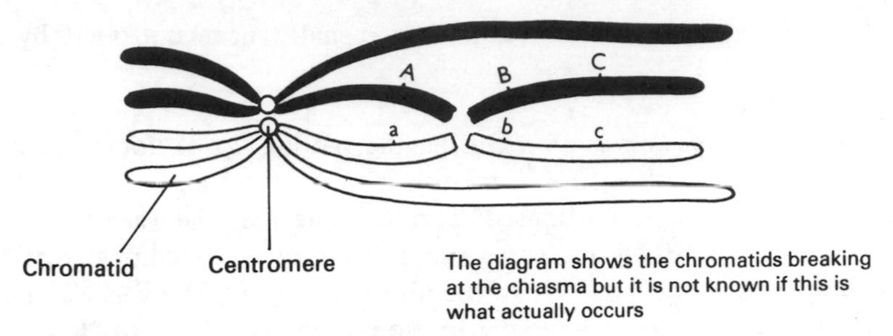

(*a*) PROPHASE Homologous chromosomes have paired up

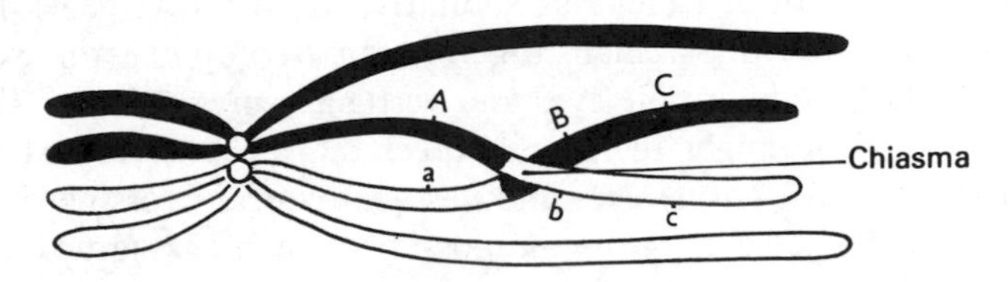

(*b*) PROPHASE The terminal portions of adjacent chromatids have been exchanged

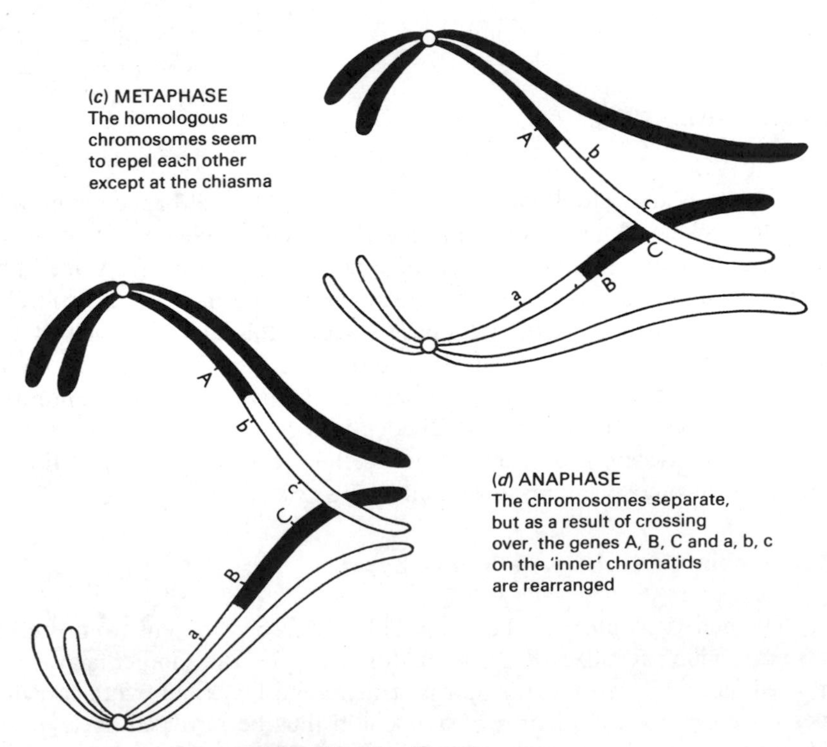

(*c*) METAPHASE
The homologous chromosomes seem to repel each other except at the chiasma

(*d*) ANAPHASE
The chromosomes separate, but as a result of crossing over, the genes A, B, C and a, b, c on the 'inner' chromatids are rearranged

Fig. 30.16 *Crossing over*

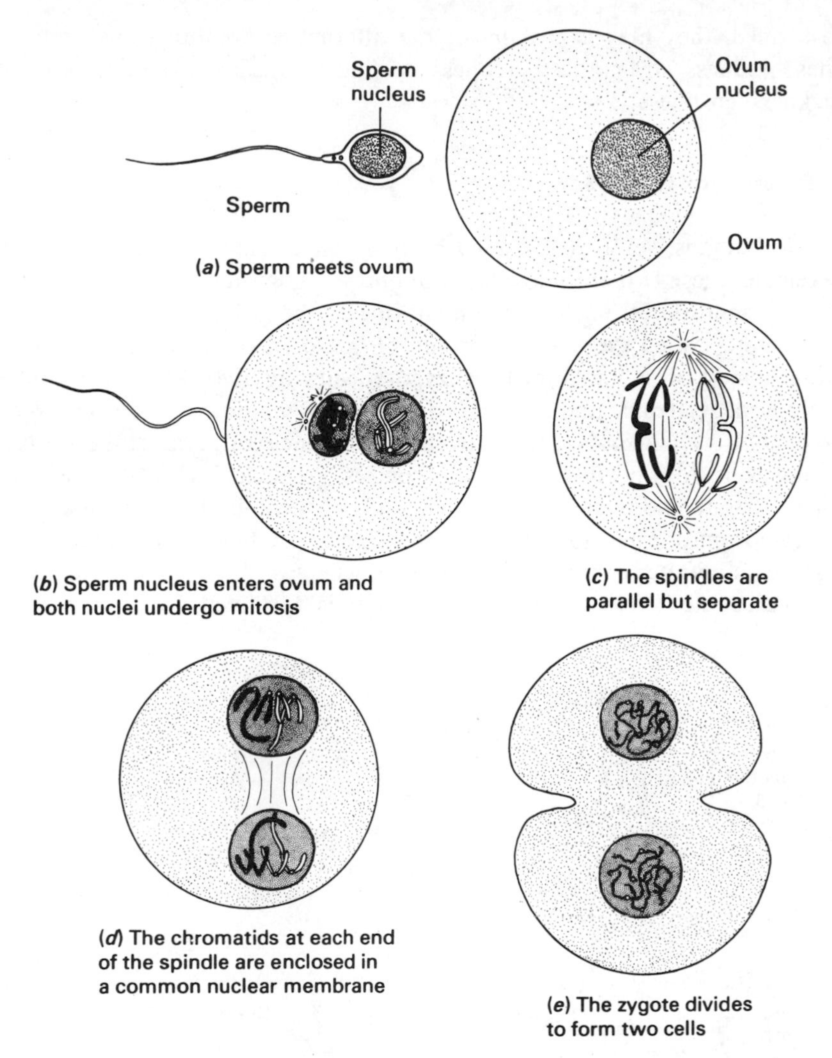

Fig. 30.17 *Fertilization (polar bodies not shown)*

contain genes for redness, blackness, curliness and straightness, and the cells of the child may therefore carry all four genes. However, the genes for curliness and blackness are dominant to those for straightness and redness (a gene is *dominant* if, when a single characteristic is controlled by two contrasting genes, it is the one which is expressed in the visible characteristics of the organism; see also Section 31.1(*a*)). The child's own hair will therefore be curly and black, a new combination of characteristics not represented in either parent. When the child grows up and his own gametes are formed, they may carry any of the alternative combinations of these genes: redness/straightness, redness/curliness, blackness/straightness or blackness/curliness.

(b) Determination of Sex

In humans, sex is determined by one pair of small chromosomes. In the female, these two chromosomes are entirely homologous and are called the X chromosomes, while in the male one of the pair is smaller than the other, and is called the Y chromosome (Fig. 30.7(*b*) shows the X and Y chromosomes of a kangaroo). A zygote carrying two X chromosomes develops into a female and a zygote carrying one X and one Y chromosome into a male; femaleness results from the absence of a Y chromosome, although this does not hold for every species. At meiosis, the sex chromosomes separate and migrate to opposite ends of the spindle, like other pairs of chromosomes. Thus 50 per cent of the sperms produced carry an X chromosome and 50 per cent a Y chromosome, whereas all the ova contain one X chromosome (Fig. 30.18). If an ovum is fertilized by a Y-bearing

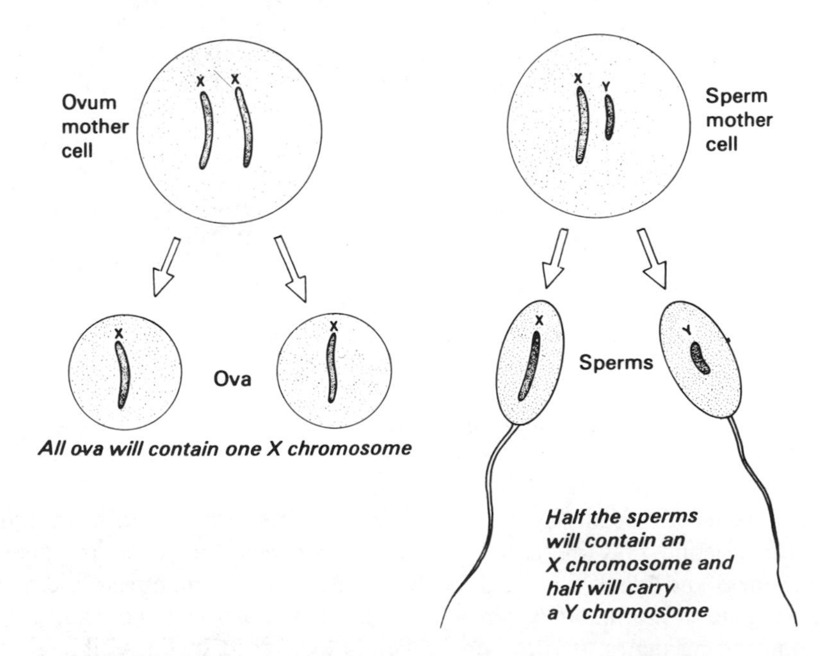

Fig. 30.18 *Diagram illustrating the determination of sex*

Only the X and Y chromosomes are shown; the Y chromosome is not smaller than the X in all animals. Details of meiosis have been omitted; although four gametes are produced, two are sufficient to show the distribution of the X and Y chromosomes.

sperm, the zygote will contain one X and one Y chromosome and will give rise to a boy. Fertilization of an ovum by an X-bearing sperm gives a zygote carrying two X chromosomes, which will develop into a girl.

Since equal numbers of X- and Y-bearing sperms are produced, an ovum is equally likely to be fertilized by either type of sperm. One would therefore expect that equal numbers of boy and girl babies should be born. In fact, slightly more boys than girls are born in most parts of the world. The reason for this is not clear, but it also happens that the mortality rate for boy babies and men is slightly higher than that for girl babies and women, which tends to restore the balance.

(c) Sex Linkage

Certain genes which occur on the X chromosome are more likely to affect males than females. This is best explained in terms of an example (Fig. 30.19). The gene (or group of genes) for a certain form of *color*

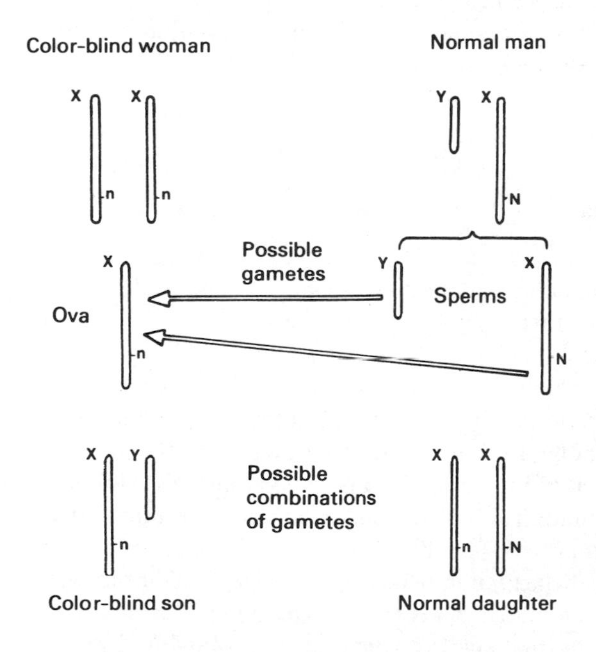

Fig. 30.19 *Sex linkage, showing the possible distribution of X and Y chromosomes between the gametes and the chances of combination in the zygotes*

blindness in humans is carried on the X chromosome; the gene for normal color vision is dominant (see Section 31.1(*a*)) to that for color blindness. A color-blind woman therefore cannot have the gene for normal vision but must carry the genes for color blindness on both her X chromosomes. All her ova therefore carry the gene. If she marries a man with normal color vision, their daughters also have normal vision but their sons are all color blind. This can be accounted for by the fact that the small Y chromosome is homologous with only a part of the X chromosome. The remainder of the X chromosome carries genes (including that for color blindness) which are not represented on the Y chromosome, and it can be assumed that the Y chromosome is not involved at all in the determination of color vision.

Each of the daughters of this marriage carries the gene for color blindness on one of her X chromosomes (the one derived from her color-blind mother) and the gene for normal vision on the other; her own color vision is normal, however, because of the dominance of the gene for normal vision. She is nevertheless a "carrier" of the color-blindness gene; if she marries a man with normal color vision the possible combinations of genes in their children can be calculated as follows:

Parents:	$X_N X_n$ (carrier woman)		$X_N Y$ (normal man)	
Gametes:	X_N	X_n	X_N	Y
Possible combinations of gametes	$X_N X_N$ (normal girl)	$X_n X_N$ (girl carrier)	$X_N Y$ (normal boy)	$X_n Y$ (color-blind boy)

All the daughters of the marriage will therefore have normal color vision, but theoretically it is probable that half of them will be carriers of the color-blindness gene like their mother; half the sons can be expected to have normal vision and half to be color blind. (It should be remembered that this calculation is based on statistical considerations that are only valid when large numbers of individuals can be considered; no human family comprises such large numbers.) The types of children that might be expected from the marriage between a woman carrier and a color-blind man, or between a normal-sighted woman and a color-blind man, can be worked out in a similar way.

Another factor determined by genes on the X chromosome is the disease of *hemophilia,* the characteristic symptom of which is a delay in the clotting time of the blood. Two kinds of sex-linked hemophilia are known, but in addition there are at least three other clotting disorders, which are controlled by genes carried by other chromosomes.

Sexual characteristics like the development of the sexual organs or the secondary sexual characters (see Section 25.8) are not determined by genes on the sex chromosomes; they are the differing expression of genes which are present in both sexes and which are scattered fairly evenly throughout all the chromosomes. Males and females both carry genes controlling factors such as the growth of face and body hair or the development of the mammary glands and of the penis. These genes have different effects according to whether their physiological environment is one of maleness or femaleness, with the result that, for example, the mammary glands in the male are small and functionless, whereas the penis is represented in the female only by a small organ, the *clitoris*.

30.11 MUTATIONS

A *mutation* is a spontaneous change in a gene or a chromosome, which usually produces an alteration in the one or more characteristics under its control. Fig. 30.8 illustrates the effects of a particular chromosome mutation in *Drosophila*. A mutation may not be very important unless the cell in which it occurs is a gamete or a cell giving rise to gametes, or a zygote, when the entire organism arising from this cell may be affected. The mutant form of the gene is inherited in the same way as the normal gene, and is thus passed on to subsequent generations.

On the whole, genes are stable entities because DNA is a stable chemical, but perhaps once in a hundred thousand replications a mutation may occur. The rate of mutation is known for certain genes; in any particular species it is characteristic of a particular gene. The rates are such that in a human ejaculate containing, say, 200 million sperms, there is likely to be a considerable number of nuclei bearing gene and chromosome mutations.

Most mutations that produce an observable effect seem to be harmful if not actually lethal. This is not surprising; any change in a well- but delicately-balanced organism is likely to affect its physiology for the worse. However, the normal gene is usually dominant (see Section 31.1(*a*)) to the mutant gene, so that the effect of the mutation is comparatively seldom seen.

In man, a form of restricted growth called *achondroplastic dwarfism* arises as a result of a dominant mutation; it occurs in about 1 person in every 20,000 of the population but about 80 per cent of affected children die during their first year. *Down's syndrome* (*mongolism*) is another fairly frequent result of a chromosome mutation, in which the ovum carries an extra chromosome, so that the child's cells contain 47 chromosomes instead of 46; it results in mental retardation and certain physical abnormalities, such as slanted eyes, round skull and characteristic short, thick hands and feet.

Radiation and Mutations

It is not known why a particular gene should mutate, but in experimental animals like fruit flies and mice an increase in the mutation rate is produced on exposure to high-energy radiation like X-rays, gamma-rays or ultraviolet light and certain chemical substances. The effects of these artificially induced mutations are similar to those which occur naturally.

All living organisms are exposed to a fairly constant "background" of radiation on the Earth's surface due to cosmic rays and naturally occurring radioactive elements; this has been increased in the last few decades by the presence of radioactive fall-out from nuclear explosions. Individuals also receive radiation from sources such as television tubes and luminous watch dials and from X-rays used in medicine, and people who are concerned with the handling of radioactive materials in nuclear power stations or other industries or in research may receive additional radiation. Obviously it is important to assess both the direct effect of radiation on people's health (for example, prolonged exposure to radiation induces certain cancers, including leukemia) and also its effect on the health of their children as a result of mutations in their reproductive cells.

The maximum safe dose of radiation in respect of direct effects on health is still a matter of controversy; indeed it is still uncertain whether any safe dose exists at all, however small. There is, so far, insufficient information available to determine the relationship between the radiation dose and the mutation rate in man; nevertheless it is probably safe to say that *any* increase in the mutation rate is likely to have harmful effects on the population. The exposure of individuals to the hazards of radiation is therefore limited by law, although to a somewhat arbitrary extent.

30.12 QUESTIONS

1. Sometimes, at meiosis, the bivalent chromosomes fail to separate properly with the result that the gamete so formed contains the diploid number of chromosomes. If such a diploid sperm were to fertilize a normal monoploid ovum, what effect would you expect this to have on the zygote and offspring? Supposing the zygote grew into a normal individual, what might happen when the individual produced gametes by meiosis?

2. A horse and a donkey are related closely enough to be able to reproduce when mated together. The offspring from this mating is a mule and though healthy in all other respects is sterile. From your knowledge of chromosomes and meiosis, suggest a possible explanation.
3. It is possible for a cross between a short-winged, gray-bodied *Drosophila* and a normal-winged, black-bodied *Drosophila,* to produce some short-winged, black-bodied offspring. Explain how this could happen (*a*) if the genes for body color and wing size are on different chromosomes and (*b*) if these genes are on the same chromosome.

Unit 31
Heredity: Genetics

31.1 GENES AND INHERITANCE

An individual inherits the characteristics of the species from its parents; for example, humans inherit highly developed cerebral hemispheres, vocal cords and the nervous coordination necessary for speech, a characteristic arrangement of the teeth and the ability to stand upright, with all its attendant skeletal features. In addition, they inherit certain characteristics peculiar to their parents or their forebears, which are not common to the race as a whole, such as hair and eye color and facial appearance.

Unit Thirty discussed the reasons for believing that the hereditary information is transmitted from the parent organism to its offspring in the chromosomes in the gamete's nucleus, and that *genes,* the units of inheritance, correspond to specific regions of the chromosomes and may consist of large molecules of deoxyribonucleic acid, of a composition which is a characteristic of each gene.

Sometimes the presence of a single gene determines the appearance of a single characteristic, like the eye color of *Drosophila* (Fig. 30.10), but most of the inherited features in humans are controlled by more than one gene (*multifactorial inheritance*); moreover, breeding experiments with humans are not possible. Information on human genetics is therefore often both complex in nature and difficult to collect. Simple examples of inheritance in other animals will therefore be considered first.

(a) Single-factor Inheritance

When a particular characteristic of an organism—coat or petal color, for example—shows no variation among the offspring for an indefinite number of generations, the organism is said to be *pure-breeding* for that characteristic. If a black mouse which is pure-breeding for coat color is mated with a pure-breeding brown mouse, their offspring will not be intermediate in color (dark brown, for example, or patched with brown and black) but will all be black. Each of the baby mice, however, must carry the genes for both blackness and brownness, since each is the product of the fusion of a sperm and an egg; the gene for black fur is said to be *dominant* to that for brown fur, since only that for blackness is expressed in the animal's coat color. The gene for brown fur is said to be *recessive* to that for black fur. The black babies are called the *first filial* (or F_1) *generation*. If, when they are mature, these F_1 black mice are mated among themselves, their offspring, the F_2 generation, will include both black and brown mice; the appearance of brown mice in the F_2 generation is evidence for the theory that the F_1 black mice carried the recessive gene for brown fur, even though it had no effect on their own coat color.

The ratio of black to brown babies in the F_2 generation will be approximately 3:1 when all the litters are considered together, although it must not be assumed that if two black F_1 mice have a litter of four, three of the young ones will be black and one brown. In such a litter of eight baby mice it would not be at all unusual to find that all of them were black, or that five were black and three brown, and so on; the 3:1 ratio can only be expected to appear when large numbers of F_2 individuals are considered.

The gene theory of inheritance accounts for these observations by suggesting that a pure-breeding black mouse carries a pair of genes controlling the production of black pigment in the fur, one of the pair on each of two homologous chromosomes. (These genes are represented by the letters *BB*, the capital letters signifying dominance.) A brown mouse carries the pair of recessive genes for brownness, *bb*, in the same position on the corresponding chromosomes. Genes which, like *B* and *b*, are in the same relative position on homologous chromosomes are called *alleles* or *allelomorphic* genes. During the first meiotic division in gamete formation the homologous chromosomes separate, so that each of the resulting gametes contains only one gene from each pair of alleles. All the gametes from the pure breeding black parent carry the gene for blackness, *B*, and all those from the brown parent carry the gene for brownness, *b*. When these gametes fuse, the zygote will contain both *B* and *b*; *B* is dominant to *b*, and is thus the only gene expressed—that is, all the offspring will have black fur.

When they mature, these black F_1 mice themselves produce gametes; during this meiosis the chromosomes carrying the *B* and *b* genes separate, so that half the gametes produced carry the *B* gene and half the *b* gene. At fertilization, a *B*-carrying sperm is just as likely to fuse with an egg carrying *B* as with one carrying *b*; the result will be either a *BB* or a *bB* zygote. Similarly a *b*-carrying sperm is equally likely to fuse either with an ovum carrying *B* or with one carrying *b*, giving either a *bB* or a *bb* zygote. Thus it is theoretically probable that, of any four of the F_2 offspring, one will be a pure-breeding black *BB* mouse, one a pure-breeding brown *bb* mouse, and two will be black *Bb* mice which, like their parents, carry both the dominant and the recessive genes

The separation at meiosis of the alleles *B* and *b* into different gametes is called *segregation*. The pure-breeding mice, whether black (*BB*) or brown (*bb*), are said to be *homozygous* for coat color. The black mice which will not breed true (*Bb*) are said to be *heterozygous*: if mated with each other their litters are likely to include some brown mice. Individuals which are homozygous for a certain characteristic will, when mated together, produce offspring which are also homozygous for that characteristic; thus two homozygous black *BB* mice will have only black progeny and the offspring of two homozygous brown *bb* mice are all brown.

(b) Genotype and Phenotype

The *BB* mice cannot be distinguished from the *Bb* mice by their appearance: both have black fur. They are said to be the same *phenotypes*; in other words they are identical in appearance for a particular characteristic, in this case blackness. Their genetic constitutions or *genotypes*, however, are different: these black phenotypes have different genotypes.

To decide whether a phenotype is homozygous or heterozygous, it is usual to do a further breeding experiment called a *back-cross*.

(c) The Back-cross

To discover their genotypes, the black F_2 phenotypes are each mated with mice of the same genotype as their brown grandparents, that is, homozygous recessive brown mice, *bb*. The black mice will produce gametes which all carry the *B* gene if they are homozygous; if they are heterozygous, half their gametes carry the *B* gene and the others the *b* gene. The gametes of the brown mice all carry the *b* gene. Thus, when the black parent is heterozygous, the back-cross can be expected to yield both black and brown babies in approximately equal numbers; if the black parent is homozygous, all the babies must also be black since they all receive the dominant gene for blackness, *B*, from this parent.

(d) Incomplete Dominance

If red Shorthorn cows are mated with white Shorthorn bulls the coats of the resulting calves carry both red and white hairs, giving a *red roan*. Neither the red-hair nor the white-hair factor is dominant to the other; they are said to be *codominant*.

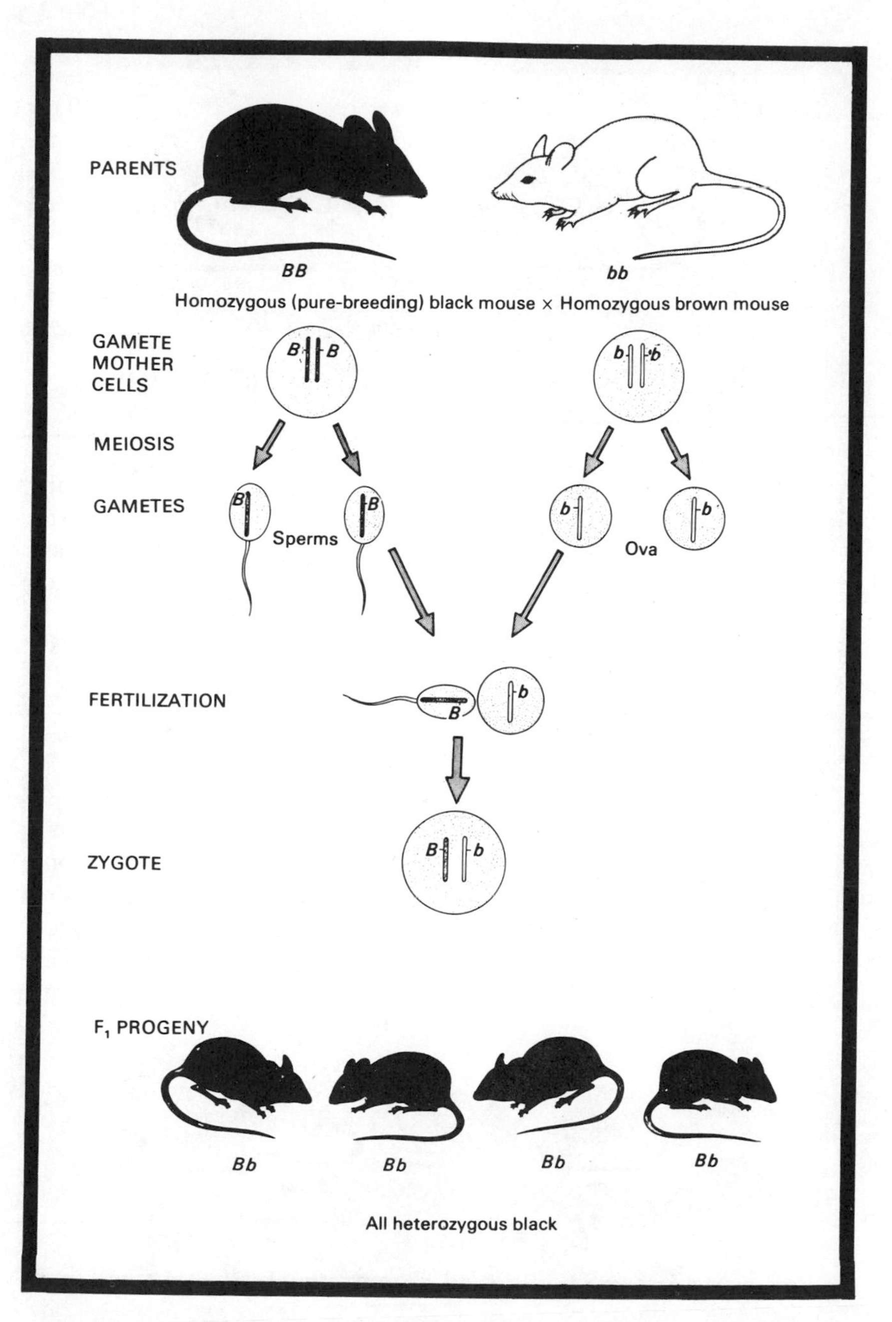

Fig. 31.1 *Inheritance of a single factor for coat color in mice: the F_1 generation.*

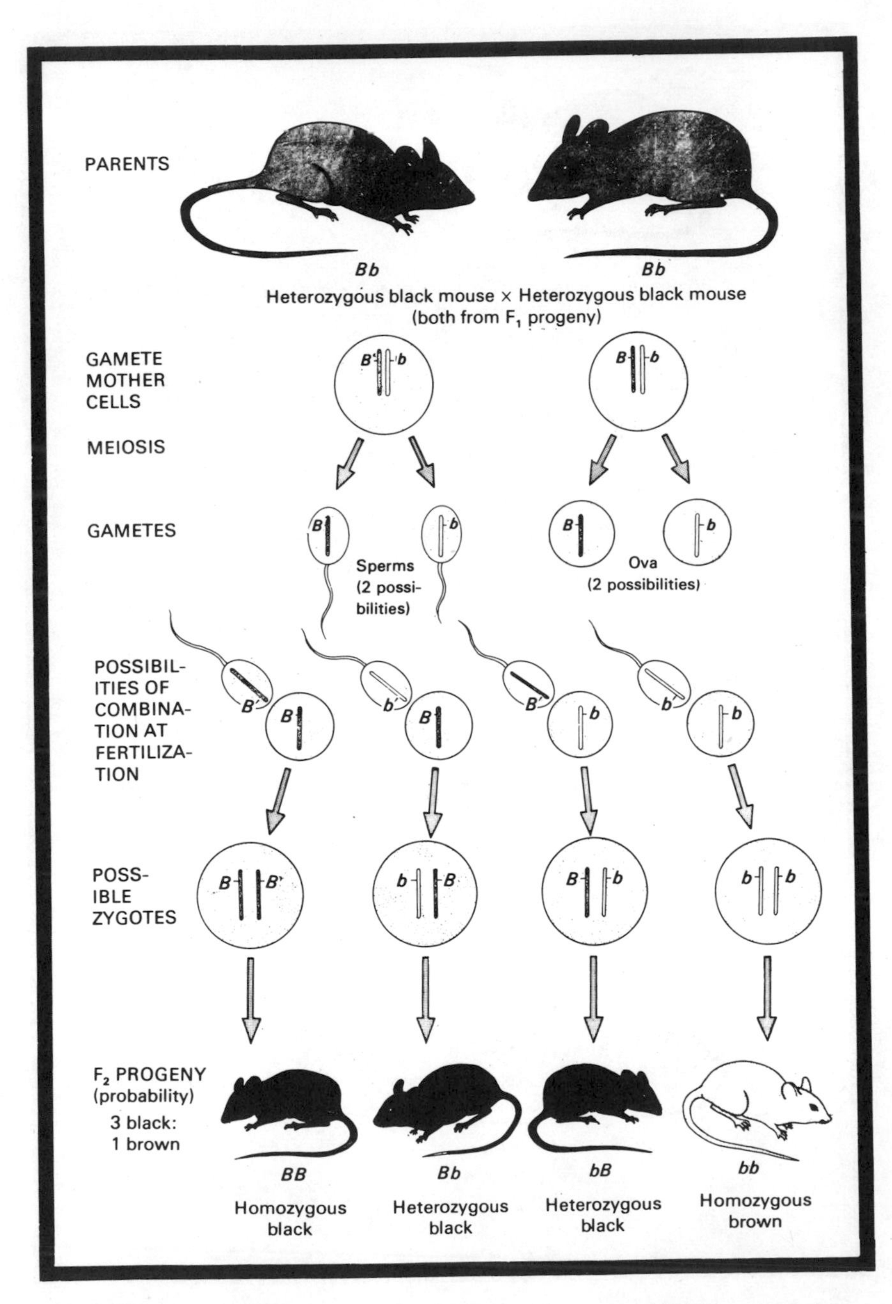

Fig. 31.2 *Inheritance of a single factor for coat color in mice: the F_2 generation*

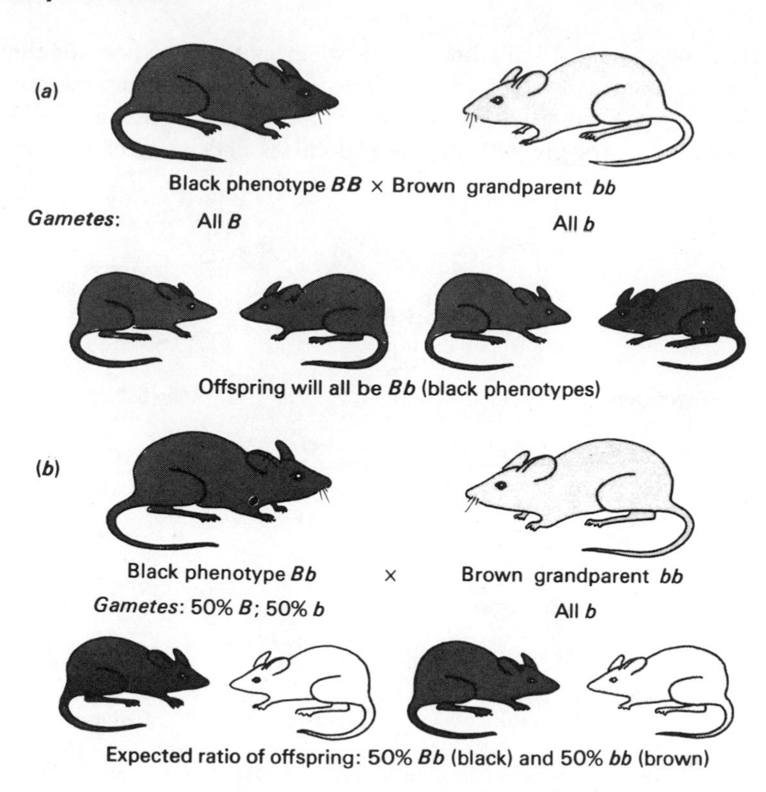

Fig. 31.3 *The back-cross*

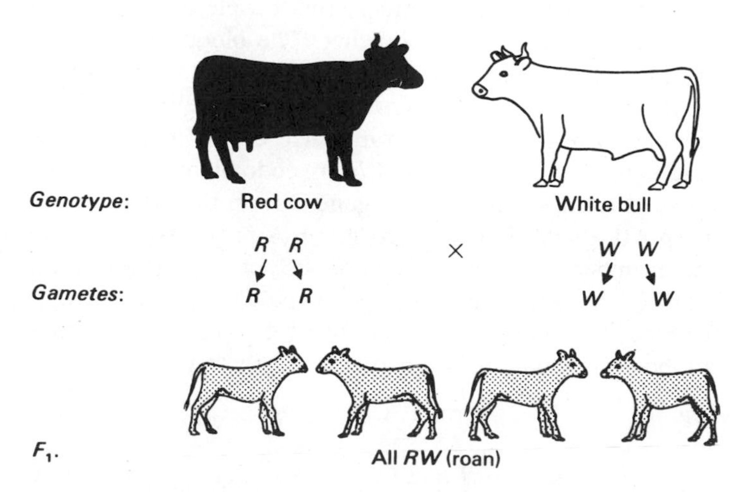

Fig. 31.4 *Inheritance of codominant factors: the F_1 generation*

Red cows and red bulls breed true when mated together: all the offspring of the mating have red coats. White cattle are also homozygous and breed true. The F_1 roan cattle, however, are heterozygous and will not breed true; their progeny will include red calves and white calves, as well as roans.

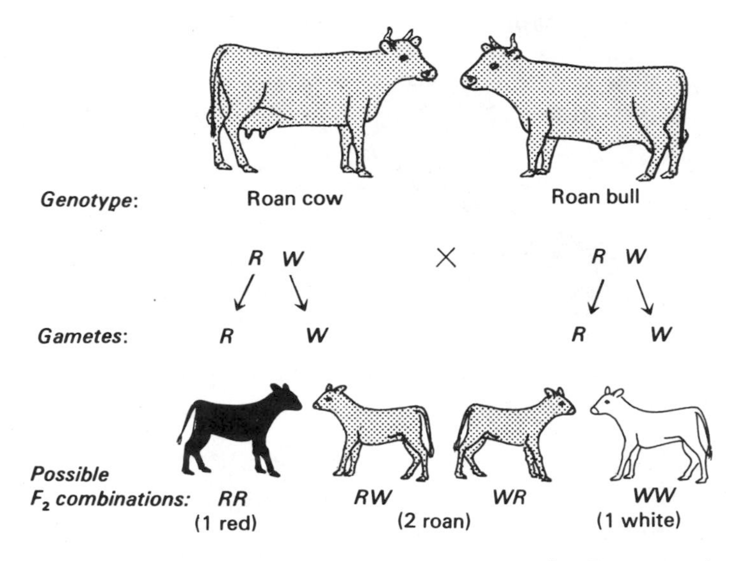

Fig. 31.5 *Inheritance of codominant factors: the F_2 generation*

In humans, the inheritance of *blood groups* is another instance of incomplete dominance. People can be classified into four major groups, A, B, AB and O; when blood samples from people of different groups are mixed, the red cells may clump together. The blood group is determined by a pair of genes, occupying the corresponding sites on homologous chromosomes. Three possible genes *A, B* and *O* can be inherited: each individual inherits two genes, one from each of his parents. Gene *O* is recessive to both *A* and *B,* but *A* and *B* are codominant: that is, if a person inherits gene *A* from one parent and gene *B* from the other, he will belong to the group AB. People belonging to group A may carry either *AA* or *AO* genes and members of group B either *BB* or *BO* genes, while people belonging to group O must have the genotype *OO*. The following example shows the possible blood groups of children of a group A man and a group B woman, both of whom are heterozygous for these genes.

phenotype	group A		group B	
genotype	*AO*		*BO*	
gametes	*A* and *O*		*B* and *O*	
F_1 genotype	*AB*	*AO*	*OB*	*OO*
phenotype	group AB	group A	group B	group O

31.2 HUMAN GENETICS

The "one gene–one character" effects described in Section 31.1 illustrate very clearly the principles of inheritance, first formulated in the 1850s by an Austrian monk, Gregor Mendel. But these are the exceptions rather than the rule; a characteristic is rarely controlled by only one gene. For example, *flower color in sweet peas* is controlled by two pairs of genes. The presence of one gene, *C,* determines the production of the color base and the presence of a second gene, *R,* controls the manufacture of the enzyme which acts on the color base to make a pigment. If the two recessive genes *cc* are present no color base is produced; in the presence of the two genes *rr* no enzyme is formed. Plants containing any of the gene combinations *CCrr, Ccrr, ccRR* and *ccRr* will thus be unable to produce colored flowers. *Coat color in mice* is controlled by at least six genes. In man, most inheritable characteristics are similarly determined by a number of genes. For example, eight of the chemical changes involved in *blood clotting* are known to be under genetic control; several genes are thus concerned in the control of coagulation, and the absence of any one of them may lead to a malfunction in the clotting mechanism, such as hemophilia (see Section 30.10(*c*)).

When one gene only is responsible for an important physiological change, its absence or modification will have serious consequences. Therefore most known instances of single-factor inheritance in humans are associated with rather freakish abnormalities, most of which occur only rarely, perhaps once in many tens of thousands of people. There are, however, a great number of different kinds of genetic abnormality.

Examples of known single-factor inheritance involving a dominant gene in man are white forelock, woolly hair, one form of night-blindness, one form of a deformity called *brachydactyly* in which the fingers are abnormally short owing to the fusion of two phalanges, and *achondroplastic dwarfism* in which the limb bones fail to grow. Single factors controlled by recessive genes include inability to taste the compound phenylthiourea (which to most people tastes bitter), inability to distinguish between the colors red and green, and a particular kind of *albinism,* that is, the absence of pigment from the eyes, hair and skin.

In experimental animals or plants, the type of inheritance and the genetic constitution can often be established by breeding together the brothers and sisters of the F_1 generation, or by back-crossing one or more of the F_1 individuals with the mother or father and producing numbers of offspring large enough to give results that have statistical significance. These methods are obviously not applicable to man and our knowledge of human genetics comes mainly from detailed analyses of the pedigrees of families, particularly those showing abnormal traits such as albinism, from statistical analysis of large numbers of individuals from different families for characteristics such as sex ratio, intelligence or susceptibility to disease, and from individual studies of identical twins (see Section 31.4).

Although single-factor inheritance appears to be relatively rare in man, there is plenty of evidence to suggest genetic control of many physiological, physical and mental characteristics. Body height, eye color, hair color and texture, susceptibility to certain diseases and facial characteristics are all genetically controlled, but the mechanism of control is more complex than the "one gene–one character" effect described in Section 31.1.

31.3 DISCONTINUOUS AND CONTINUOUS VARIATION

The individuals belonging to a species of plants or animals are alike in all major respects; indeed, it is these likenesses which determine that they belong to the same species. Mice, for instance, may be black, brown, white or other colors, and the sizes of their ears and tails may vary, but despite these variations there is no doubt that they are all mice.

(a) Discontinuous Variation

The variations in the coat color of mice are examples of *discontinuous variation,* because there are no intermediate forms. If black and brown mice are bred together they will produce offspring which are either black or brown; there are no mice of intermediate color and none with black and brown mottled coats. No problems therefore arise in deciding into which color category to place the individuals. It is not possible to arrange the mice in a continuous series of colors ranging from brown to black with little difference in color between members of the series.

The way in which sex is inherited is another example of discontinuous variation. With the exception of a small number of abnormalities, an individual is either male or female and there are no intermediates.

In man, blood groups provide an illustration of discontinuous variation (see Section 31.1(*d*)). A person must belong to one or other of the four major groups A, B, AB or O. He cannot, for example, be intermediate between group A and group O: he must be a member of one group or the other.

Eye color, to a certain degree, is inherited in a discontinuous manner; an individual's eyes are either blue or more deeply pigmented. However, there are certain people who would be difficult to classify. Clear-cut examples of discontinuous variation occur among the more serious variants; for example, a person is either an achondroplastic dwarf or he is not, and "mild" cases of this kind of dwarfism do not occur.

The features of discontinuous variation are clearly genetically determined. You cannot alter your blood group by changing your diet; an achondroplastic dwarf cannot grow to full height by eating more food. Moreover, discontinuous variations are likely to be under the control of a small number of genes. The presence of one dominant gene makes an individual an achondroplastic dwarf; the absence of one gene for making pigment causes albinism.

(b) Continuous Variation

When one tries to classify individuals according to their height and weight rather than, say, by their eye color the decisions become more difficult and the classes more arbitrary. People cannot be classified simply as tall or short; there is a whole range of people of intermediate size, with barely measurable differences between them. If necessary, categories can be invented for convenience: we could group together people with heights between 1.4 and 1.6 m, those between 1.6 and 1.8 m, and those between 1.8 and 2.0 m. Such categories do not, however, represent fundamental differences between the groups and they do not indicate discontinuous variations of 0.2 m between the heights of individuals.

There is no obvious reason why continuous variations should not be genetically controlled but they are likely to be under the influence of several genes. Height, for example, is at least partially genetically determined; tall parents tend to have tall children, and vice versa. A person's height might be influenced by, say, 20 genes, each contributing a few centimeters to his stature (the figure of 20 is hypothetical and the number of genes actually involved is unknown). If he inherits most or all of the 20 he will be tall; if only a few, he will be short.

Many continuous variations are the result of two factors: the genotype of the individual and the environment in which he lives. For example, a person may inherit genes for tallness, but if he is undernourished during his years of growth he will not grow as tall as he might have done if he had been well fed. A man will grow fat if he eats more food than he needs; this appears to be an entirely environmental effect, until one realizes that another person may eat just as much food and yet remain slim because of his different inherited constitution. Whether or not one catches a certain disease obviously depends on whether one has been exposed to the disease germs—an exclusively environmental criterion—yet it appears that susceptibility or resistance to some diseases are at least in part under genetic control.

31.4 HEREDITY AND ENVIRONMENT

It is possible to experiment with plants and animals to discover whether an observed variation is due primarily to the genetic constitution or to environmental factors. For instance, plants of a certain species grown in a valley may have larger leaves and taller stems than plants of the same species growing on a mountain side. The two varieties can be collected, planted together in the same situation, and allowed to breed (interbreeding between the varieties being of course avoided). If the differences between the varieties persist in the offspring, they can be assumed to be genetically controlled, but if the daughter plants are indistinguishable, the original variations must have been due solely to the environmental differences.

Similar experiments with man are neither possible nor disirable. Although situations can sometimes be recognized which superficially resemble the experiment, such as the "uniform" environment of an institution for orphaned children, observations in these circumstances are always capable of more than one interpretation. Thus there is usually a great deal of argument about very little evidence when people discuss whether our intelligence, for example, is predominantly determined by the genes we inherit or the conditions of home and school in which we were brought up.

One source of evidence concerning the relative effects of heredity and environment in man is the study of identical twins.

Identical Twins

Twins may be either identical or fraternal (see Section 25.5). Fraternal twins are the result of simultaneous fertilization of two separate ova by two sperms. The chromosomes in the resulting zygotes will thus be quite different, and although the twins develop in the uterus at the same time, they will not necessarily resemble each other any more closely than they resemble other brothers and sisters; for example, they can differ in sex.

Identical twins, on the other hand, are derived from a single fertilized ovum, which separates into two distinct embryos very early in its development. The two embryos thus have identical sets of chromosomes in their cells, since they are derived by mitosis (see Section 30.3) from a single zygote. Identical twins often share a placenta, although they may be enclosed in separate amnions. Such twins, having the same genotypes, are invariably of the same sex, and usually resemble each other very closely, although variations in their position and blood supply while in the uterus may produce differences at birth.

Since identical twins carry exactly the same sets of genetic "instructions," any differences between them must be due to the effects of their environment and not to their genes. The study of identical twins, therefore, can provide valuable evidence for assessing the relative importance of heredity and environment in the development of an individual; for example, one survey found that the average difference in height of identical twins reared together was only 1.7 cm, compared with an average height difference of 4.4 cm for a control group consisting of the same number of fraternal twins (Table 31.1).

Table 31.1 *Average differences in selected physical characteristics between pairs of twins*

Difference in characteristic	*Fraternal twins reared together (50 pairs)*	*Identical twins reared together (50 pairs)*	*Identical twins reared apart (19 pairs)*
Height (cm)	4.4	1.7	1.8
Weight (kg)	4.5	1.9	4.5
Head length (mm)	6.2	2.9	2.2
Head width (mm)	4.2	2.8	2.85

(From Freeman, Newman and Holzinger, *Twins: A study of heredity and environment,* University of Chicago Press, 1937)

Identical twins tend to score very similarly in intelligence tests, even when they have been brought up in different environments; however, educational background can affect the result considerably.

Table 31.2 *Corrected average differences in IQ test scores between pairs of twins*

	Fraternal twins reared together (52 pairs)	*Identical twins reared together (50 pairs)*	*Identical twins reared apart (19 pairs)*
Difference in IQ	8.5	3.1	6.0

(From Freeman, Newman and Holzinger, *Twins: A study of heredity and environment,* University of Chicago Press, 1937)

Detailed histories of identical twins provide impressive illustrations of the importance in their lives of inherited factors. One pair of girl twins was separated soon after birth; one was brought up on a farm and the other in a city, and both contracted tuberculosis at the same age. Another pair of identical twin sisters was separated and adopted shortly after birth but both became schizophrenic within two months of each other during their 16th year. Such individual examples are of interest but are too few to lead to any far-reaching conclusions about the inheritance of human characteristics in general.

31.5 APPLICATIONS OF GENETICS TO HUMAN PROBLEMS

(a) Eugenics

Eugenics is the study of human genetics from the point of view of the improvement of the genetic composition of the human stock (*i*) by encouraging breeding for desirable characteristics, and (*ii*) by eliminating or reducing the incidence of harmful characteristics.

The achievement of this end is a more complex undertaking than it might appear. It might seem obvious that if everyone with a hereditary mental defect were sterilized, or prevented in some other way from having children, then the number of congenital * idiots in the world would be greatly reduced and the condition eventually completely eliminated. However, harmful genes are usually recessive to their normal alleles; this is true of albinism, for example, which appears only in people who carry the two recessive genes, *aa*. These number about one in 20,000 of the population. The marriage of two albinos would give rise to children who were albinos, but such marriages are very rare. However, calculations show that one person in 70 is a carrier for albinism, that is, has the genotype *Aa*. A marriage between two carriers can produce albino (*aa*) children; 99 per cent of all albinos arise from such marriages. There is at present no way in which the carriers can be distinguished from normal (*AA*) individuals, and clearly there could be no question of sterilization until after the first albino child was born into the family, revealing the genetic constitution of the parents. Even if the *aa* albinos did not breed at all, it would still take 22 generations to reduce the frequency of a harmful gene in this way from 1 per cent to 0.1 per cent.

Similarly, if every person with a hereditary mental defect were sterilized, the incidence of the condition would fall by only 8 per cent. This fall, however, represents the elimination of suffering for a large number of people; mentally defective people can hardly provide a suitable home even if they have normal children.

The elimination of a harmful dominant gene can theoretically be accomplished in one generation, provided that the condition appears before the individual reaches the reproductive age.

(b) Genetic Advice

Advice based on a sound knowledge of genetics can sometimes help to avoid unsatisfactory breeding without the adoption of drastic measures like compulsory sterilization. Several countries have *genetic counselling*

* In this context the term *congenital* is used to imply an inherited condition, rather than one resulting from environmental causes during gestation or birth.

services which can advise prospective parents of the chances that they might have a child with an inherited abnormality.

For example, a man with congenital juvenile cataract (an eye defect caused by the presence of a dominant gene), married to a normal woman, would be told that his children would have a 50 per cent chance of inheriting the disease (these are the F_1 offspring of the mating of a *Dd* affective man with a *dd* normal woman). An albino, *aa,* married to a normal person (who might be *AA* or *Aa*), would be told that the chance of any one of his children being an albino is 1 in 140; his brother, showing no sign of albinism, would be advised that he could be an *Aa* carrier, and that he has 1 chance in 420 of having an albino child if he marries a normal person. The chances for other relations, such as an albino's aunts and uncles, can also be calculated in a similar way.

Genetic advice could be far more accurate if heterozygous carriers who show no abnormalities could be distinguished from the apparently identical homozygous people. Differentiation has recently become possible in some instances; for example, people who are heterozygous carriers of a disease called *phenylketonuria,* which can cause severe mental retardation, can be detected with reasonable certainty by a test for the level of a particular amino acid, phenylalanine, in their blood.

(c) Consanguinity

The chances of appearance of harmful characteristics in the children of a marriage between first cousins can be predicted from a knowledge of human genetics.

Fig. 31.6 shows diagrammatically the pedigree of two hypothetical first cousins, Bill and Jane. Suppose cousin Bill is heterozygous for a harmful recessive gene, *n*. Bill could have inherited this gene from any one of his four grandparents. There is a 1 in 2 chance that the gene came from one of the B grandparents, so that there is also a 1 in 4 chance that Jane has inherited the gene. There is thus a 1 in 8 chance ($\frac{1}{2} \times \frac{1}{4}$) that

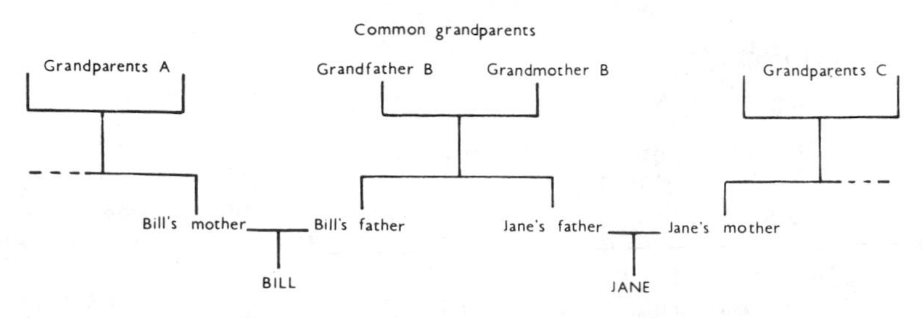

Fig. 31.6 *Lineage of first cousins*

cousin Jane is also *Nn*; if that is the case, the chance of an affected child from their marriage is 1 in 4. The overall chances of an affected child if Bill is *Nn* and marries Jane, are thus 1 in 32 ($\frac{1}{8} \times \frac{1}{4}$).

If the gene is a fairly rare one, occurring in perhaps one in every 100 people, the chance of Bill marrying an *Nn* person from the general population is 1 in 100. The overall chance that one of his children will be affected is then $1/100 \times \frac{1}{4}$, that is, one chance in 400.

Although these considerations would apply equally well to beneficial genes, cousin marriages are not usually encouraged and brother–sister marriages are forbidden by law in nearly all countries. This does not reduce the total number of homozygous recessives in a population, but does reduce the chances of their occurring in a particular family. However, only a few centuries ago the world's population was far below that of today, and the number of our ancestors is thus relatively small; consanguinity is therefore bound to occur in some degree sooner or later.

31.6 INTELLIGENCE

A person's intelligence is the product both of his genetic constitution and of the effects of his environment. He may inherit from his parents the mental equipment for highly intelligent thought, but he needs education and training before his full potential can be realized. Inherited intelligence is almost certainly influenced by a large number of genes and is not capable of simple analysis. When a graph or histogram is plotted to show the different numbers of individuals in a given sample having a particular "intelligence quotient" (Fig. 31.7), it has the same appearance as one drawn for height or skin color: this kind of graph is called a *curve of normal distribution* or *normality curve.* Such curves are characteristic of factors which are influenced by a number of different genes, and in which there is a continuous variation, with every grade of intermediate, rather than the straightforward presence or absence of a condition, as in albinism.

31.7 QUESTIONS

1. Two black guinea pigs are mated together on several occasions and their offspring are invariably black. However, when their black offspring are mated with white guinea pigs, half of the matings result in litters containing black babies only, and the other half produce litters containing equal numbers of black and white babies.

   From these results, deduce the genotypes of the parents and explain the

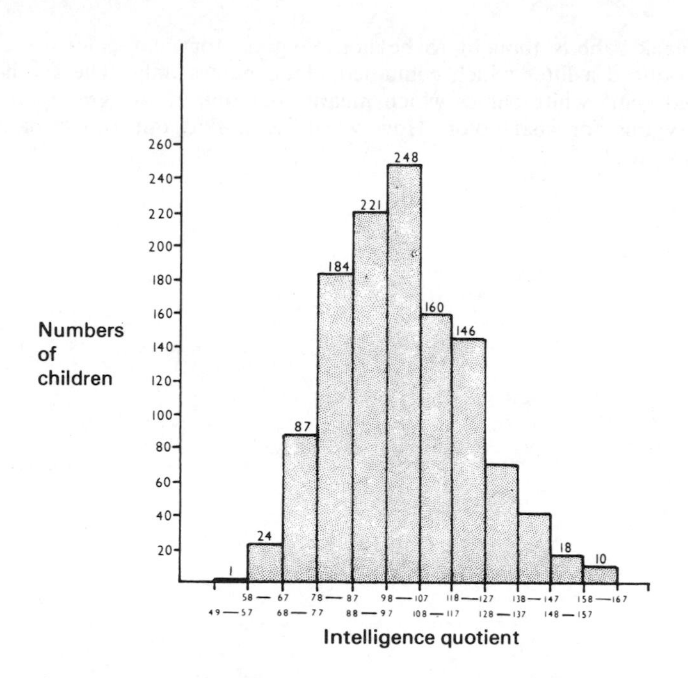

Fig. 31.7 *Distribution of IQ rating in a random sample of 1,207 Scottish eleven-year-old children*

(From C. O. Carter, *Human Heredity*, Penguin Books, 1962)

results of the various matings, assuming that color in this case is determined by a single pair of genes (alleles).

2. The blood groups A, B, AB and O are determined by a pair of allelomorphic genes inherited one from each parent. The gene for group O is recessive to both *A* and *B*. Thus a group A person may have the genotype *AA* or *AO*. Inheritance of *A* from one parent and *B* from the other produces the phenotype AB.

   (*a*) What are the possible blood groups likely to be inherited by children born to a group A mother and group B father? Explain your reasoning.

   (*b*) A woman of blood group A claims that a man of blood group AB is the father of her child. A blood test reveals that the child's blood group is O. Is it possible that the woman's claim is correct? Could the father have been a group B man? Explain your reasoning.

3. A geneticist wishes to find out the color of F_1 flowers from a cross between red- and white-flowered insect-pollinated plants such as snapdragon. Revise, if necessary, Section 7.3 on pollination and describe how he should conduct his experiments, assuming that pollen can effectively be transferred by means of a dry paintbrush.

4. Individuals of a pure-breeding line of *Drosophila* are exposed to X-rays to induce mutations in their gametes. Most mutated genes are recessive to normal genes. How could one find out if mutations had occurred?

5. Two black rabbits thought to be homozygous for coat color were mated and produced a litter which contained black babies only. The F_2, however, included some white babies which meant that one of the grandparents was heterozygous for coat color. How would you find out which parent was heterozygous?

UNIT 32 EVOLUTION AND NATURAL SELECTION

32.1 THE THEORY OF EVOLUTION

The theory of evolution suggests that life on Earth began as relatively simple forms, which over hundreds of millions of years gave rise to succession of living organisms, becoming by a series of small changes more varied and more complex.

While the evolutionary theory offers an explanation of the origin of the great majority of present-day animals and plants, it must be emphasized that it is indeed a theory and not an established fact. Although it is widely regarded as an acceptable hypothesis to account in general terms for the existence of the living organisms we know today, much of the evidence for the theory is either very incomplete or circumstantial.

Some arguments against the theory of evolution are:

(*a*) Genetic changes (mutations) are notoriously deleterious, not beneficial.
(*b*) Evolution is purely by chance selections of genes and gene combinations. Without a "directing force" or "end point" in mind, organisms would become less fit, not better fit. Too much credit is given to the "creative" power of chance.
(*c*) The timetable presented for the development of organs and organisms is too brief to explain the human eye or man himself.
(*d*) Often so-called primitive or ancestral traits of organisms are found to occur in the fossil record much later than so-called advanced traits.
(*e*) If one form gave rise to a more advanced form, then every step in the transition would be better adapted to its environment than the prior steps. Yet "primitive" and "advanced" forms are alive today, but there are none of the transition forms in the fossil record or living today.

Some of the arguments in favor of the theory are outlined in the following paragraphs.

(a) Reproduction and Spontaneous Generation

As far as we know, all living organisms arise by the reproduction of preexisting organisms. The evolutionary theory assumes that when new forms of life appear on Earth they are always derived from organisms that already exist; during the processes of reproduction these have undergone a long series of relatively minor alternations in form until individuals arose which were markedly different from their ancestors. Fossils (see Section 32.1(*b*)) provide some evidence in support of this theory, at least in the case of vertebrates: for example, that mammals were derived from reptiles, reptiles from amphibia, and amphibia from fish.

There is no evidence of any kind for the production of living things in any other way than by reproduction, for example, by the "spontaneous generation" of living organisms from non-living matter. This knowledge weakens any alternatives to the evolutionary theory that may claim that each kind of organism known today arose separately and spontaneously, or was created suddenly at a given point in time—for example, that horses were brought into existence, say, one million years ago and have remained the same ever since, reproducing their kind exactly over thousands of generations.

The question which must arise, however, is "how did the *first* living creatures originate?" Taken to its logical conclusion, the evolutionary theory would suggest that life itself evolved from non-living material. The biologist is obliged to say that, although the spontaneous generation of life is unknown today and is thought to be very improbable throughout the period of time of which we have any detailed knowledge, there was a time, some 500 million years or more ago, when conditions were favorable for such an event or series of events. For example, it is possible that at that time the atmosphere might have been devoid of oxygen but rich in methane and ammonia, and there is some experimental evidence to indicate the feasibility in these conditions of the production of small amounts of amino acids, which might have combined to form proteins.

(b) The Fossil Record

Many of the Earth's rocks have been formed by *sedimentation,* that is, the slow settling out of solid particles in lakes and oceans. *Sedimentary rocks* like sandstone and limestone arise when such particles become cemented together and the layers of sediment are compressed together over millions of years. When the dead remains of animals and plants fall to the bottom of the lake or sea, they become incorporated in the sediment, and may be preserved in a variety of ways as *fossils*. Earth movements over a long period of time may eventually raise the rocks above the water, and the fossils they contain sometimes then become accessible for study (Fig. 32.1).

Where a series of sedimentary rocks is exposed, the lowest layers will be the oldest (if not too contorted by earth movements); scientists can study the fossils in successive layers, and so form some idea of the animals and plants present millions of years ago, and the way in which their characteristics appear to change from one period to another.

When the fossils in the various layers are studied, it appears that (*i*)

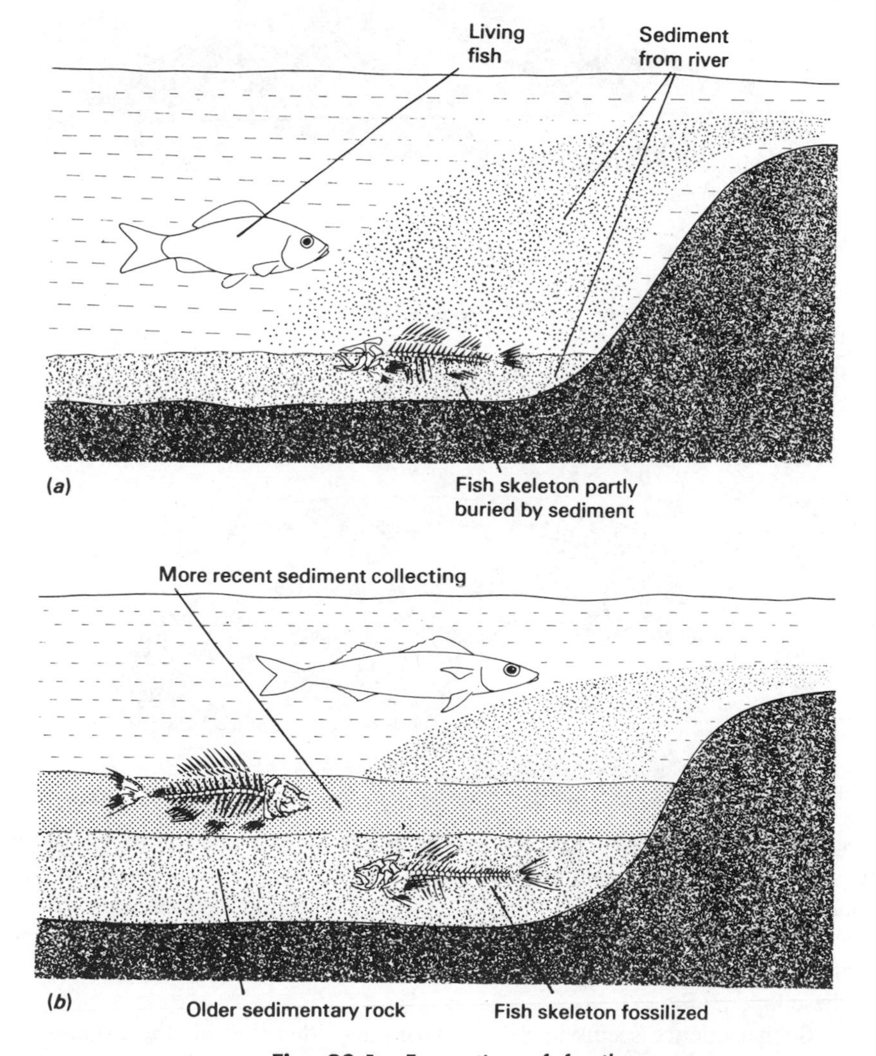

Fig. 32.1 *Formation of fossils*

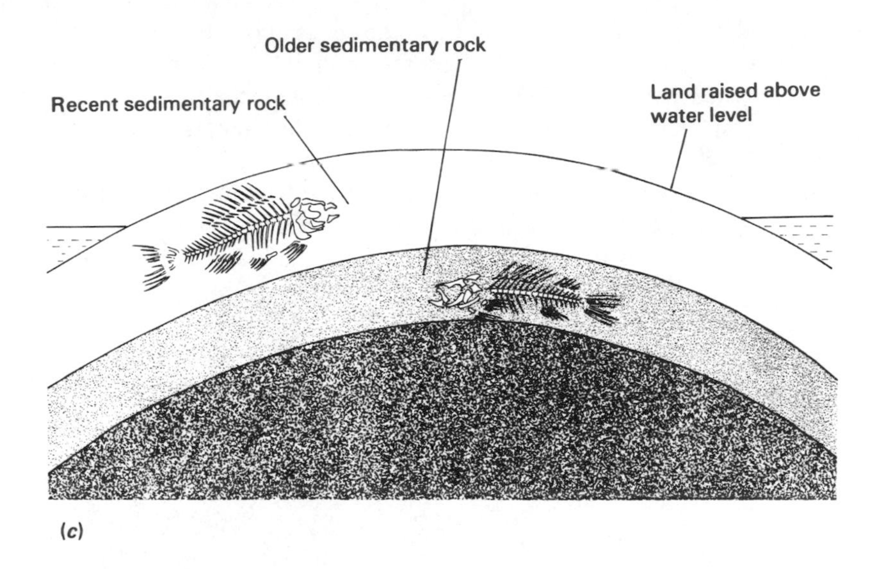

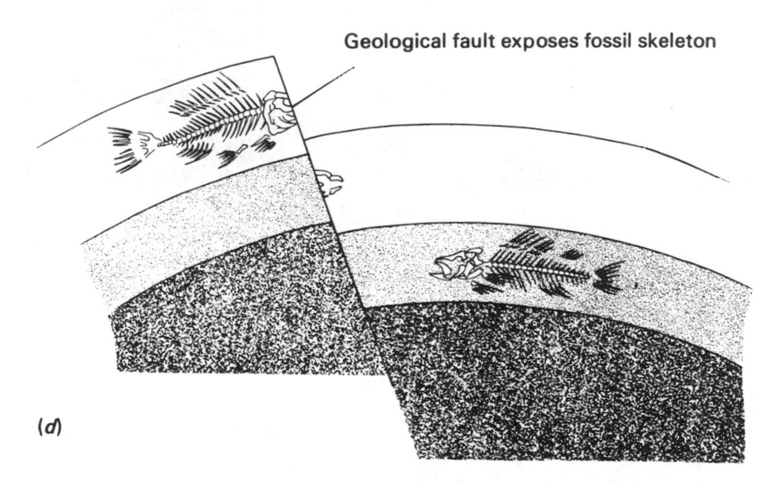

Fig. 32.1 *Formation of fossils (cont'd)*

many species of present-day animals and plants are not found at all in fossil form, and (*ii*) a vast number of species recognizable as fossils in the rocks no longer exist today (Fig. 32.2). For example, sedimentary rocks known to be about 300 million years old contain no traces of fossil mammals, but the remains of "armor-plated" fish, unknown today, are preserved in these layers (Fig. 32.3).

Such evidence seems to detract from any idea that all the organisms existing today have been reproduced exactly since life began. Even if

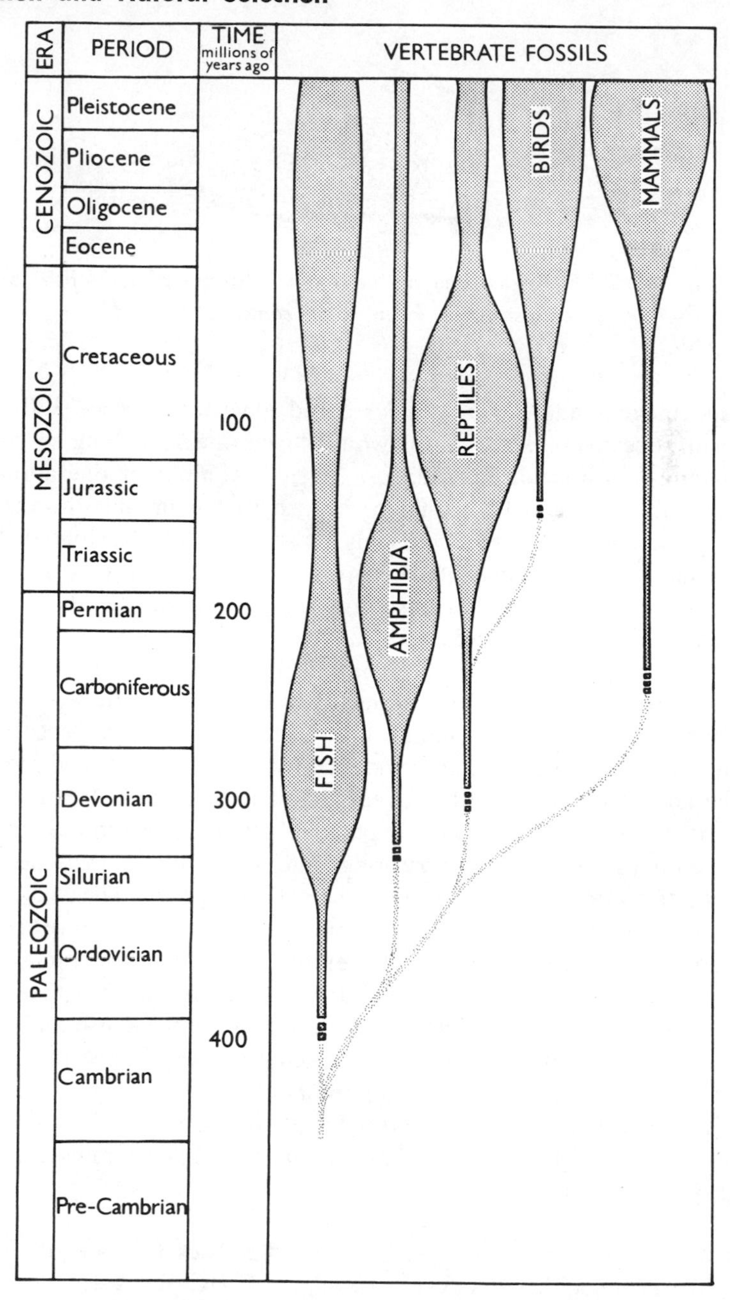

Fig. 32.2 *Chart to show the earliest occurrence and relative abundance of fossil vertebrates (possible evolutionary relationships are shown by faint lines)*

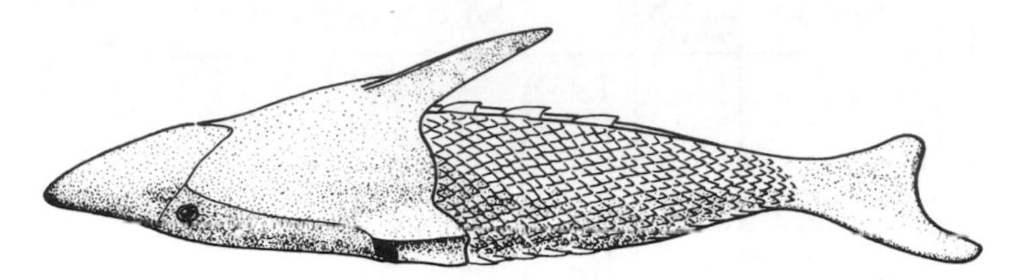

Fig. 32.3 *Pteraspis, one of the extinct "armor-plated" fish (a reconstruction from fossil remains)*

spontaneous generation or *biogenesis* could have taken place 300 million years ago, it seems unlikely that it would "generate" anything so complex as a mammal in a single operation or even a number of operations in a very short time. Although it could be argued that mammals did exist 300 million years ago, but were so sparsely or so unevenly distributed that no fossil remains of them have yet been found, the question of the origin of mammals still remains; such reasoning simply pushes the date of their origin back to an earlier period.

It seems more probable that mammals have been derived from ancestors unlike themselves by a long series of small changes. In fact, scientists have found fossil remains of animals intermediate in many respects between mammals and reptiles, providing evidence for such a theory and indicating that the ancestors may have been reptile-like in character. It must not be supposed, however, that the descendants of present-day reptiles are likely to be mammals, even after a very long time, nor, indeed, that the remote ancestors of mammals closely resemble existing reptiles. The evolutionary theory suggests that present-day mammals and reptiles have common ancestors, which were neither wholly reptilian nor wholly mammalian and which have now become extinct. Similarly the theory suggests that reptiles have fish-like ancestors not represented among present-day fish and that the two groups have continued to evolve in different ways since the first dissimilarities between them became apparent.

The fossilized remains of invertebrates provide little or no evidence that any of the large groups of these animals share a common ancestor. This may be because

(*i*) our knowledge of the fossil record does not go back far enough in time, or

(*ii*) if the common ancestors ever existed, they have not been preserved in fossil form, or

(*iii*) the production of living organisms from non-living matter was not a single unique event but occurred more than once, each time initiating a different primitive form of life which evolved into a distinct group of invertebrate animals.

(c) Circumstantial Evidence

The study of the structure and distribution of modern animals can bring to light features which can be interpreted as evidence for the theory of evolution. Comparison of the structures of the skeletons of vertebrate limbs is one such example (Fig. 32.4). The limbs of birds, bats, whales and lizards look very different from each other, and each is adapted to a particular function—flying, swimming or running—but despite these differences of appearance and function, they all have basically similar skeletal structures. It seems reasonable to regard these limbs as modifications of the limbs of a primitive common ancestor, which have retained during their evolution the same fundamental pattern of bones and joints, while gradually becoming more extensively adapted to each special method of locomotion. However, if these animals did not originate from a common ancestor but came into being independently (perhaps spontaneously) there seems no convincing reason why the bones in limbs performing such different functions should show such striking similarities.

The comparative anatomy of the vertebrates provides many other examples of fundamental likenesses between apparently different species, including features of the skeletal, circulatory and nervous systems; the embryonic forms of different vertebrates also show certain resemblances. These are all instances of circumstantial evidence for the evolutionary theory, that is, they are known facts which do not actually contradict the suggestion that a certain pattern of events took place in the past. However, these events are impossible to reproduce, and are not available for experimental verification; the evidence is therefore often subject to alternative interpretations and further discussion is suited to more advanced study than this book can offer.

32.2 THE THEORY OF NATURAL SELECTION

In 1858 Charles Darwin and Alfred Russel Wallace independently put forward a theoretical explanation of how evolution could have taken place and new species arisen. The theory of evolution by natural selection explains most, although not all, of the observed facts; discoveries made since its publication have provided further support for the theory.

The theory of natural selection is based on observations which can be summarized as follows:

(*a*) *The offspring of animals and plants outnumber their parents.* For example, suppose that a pair of rabbits have 8 offspring which grow up

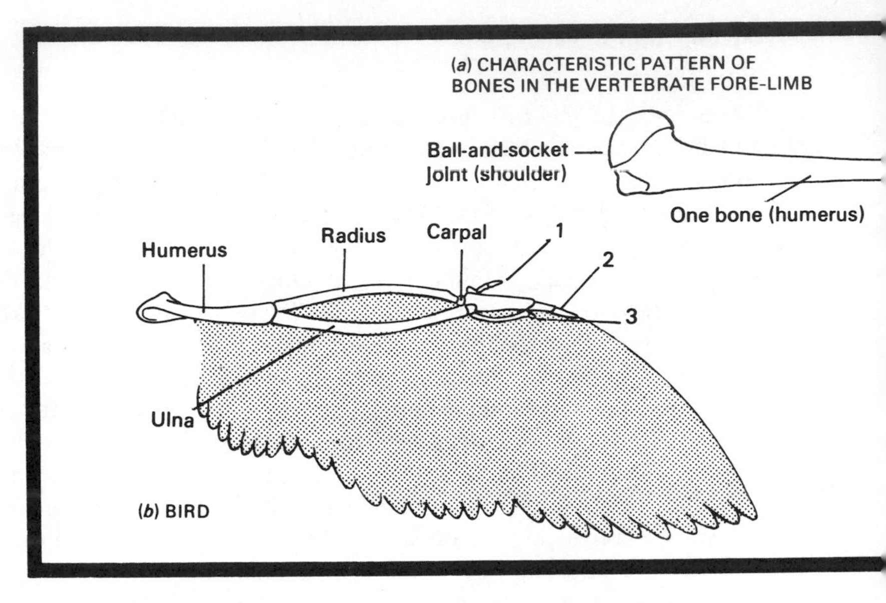

Fig. 32.4 *Comparative anatomy of vertebrate limbs*

and form 4 pairs, each of these pairs also having 8 offspring; if the sequence continued unchecked the number of fourth-generation descendants of the original pair of rabbits would be 512 (that is, 2 → 8 → 32 → 128 → 512).

(*b*) Despite this tendency to increase, *the numbers of any particular species remain more or less constant,* at least in the short term.

It follows from (*a*) and (*b*) that, since fewer organisms survive to maturity than are produced, *there must be a "struggle" for survival.* Of the potential family of 512 fourth-generation rabbits, for instance, only two need to live to maturity if the rabbit population is to remain constant; the other 510 can be expected to die, or not to have been born.

The "struggle" for survival does not, however, imply that individuals actually fight each other; the participants may never meet yet still may compete for food and shelter. Often the competition may be between immature forms such as eggs, larvae or seeds, as well as between the adult organisms; in such a stage of the life history the mortality rate may be very high—for example, very few of the thousands of acorns shed by an oak-tree each autumn normally have any chance of germinating, and still fewer grow into mature trees. The survival "struggle" may be quite passive, and may depend, for example, on the relative resistance of eggs to adverse conditions like cold or drought, or the relative concealment patterns on the body, leading to more or less effective camouflage from predators.

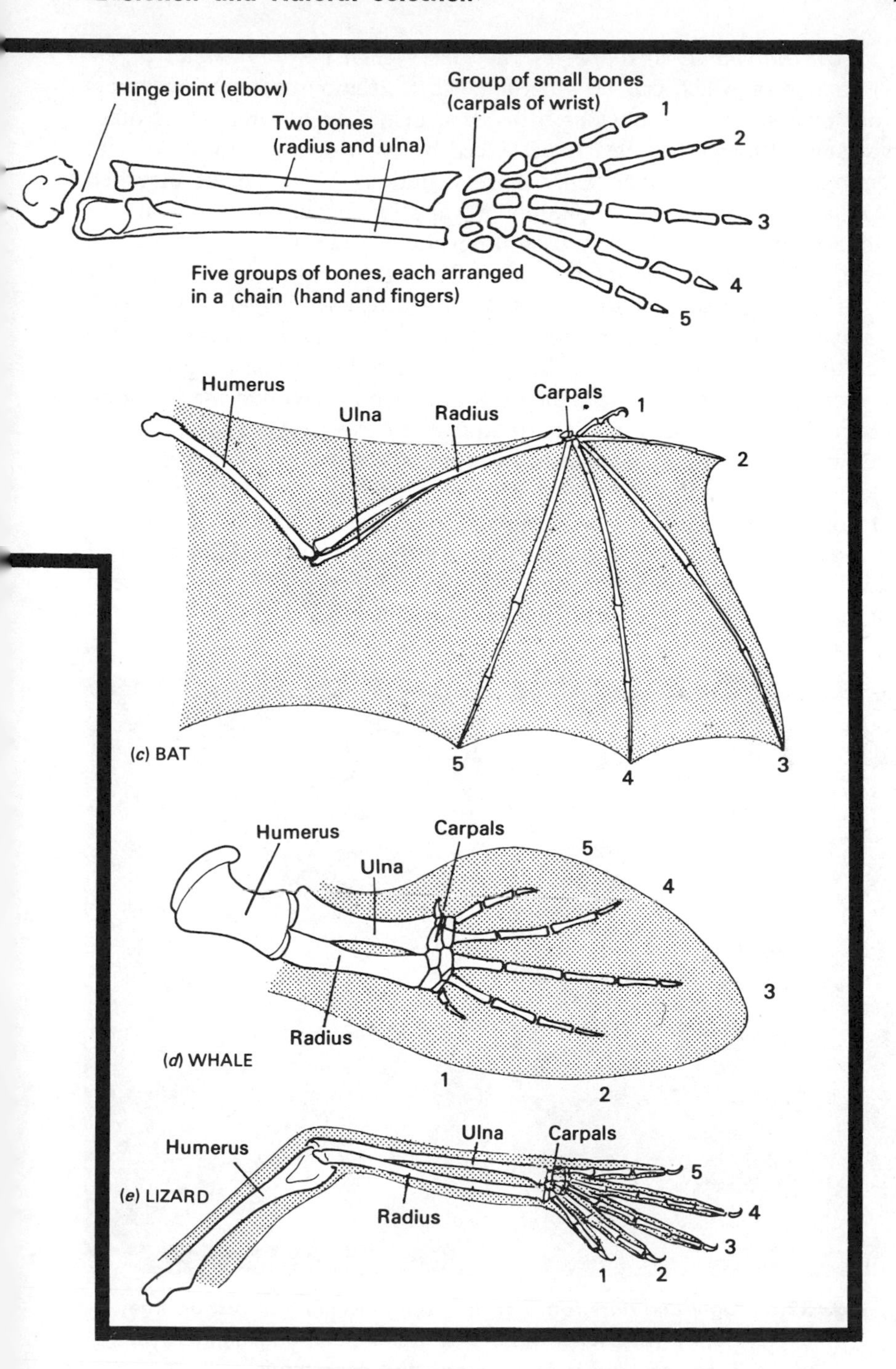
Hinge joint (elbow)
Two bones
(radius and ulna)
Group of small bones
(carpals of wrist)
1
2
3
4
5
Five groups of bones, each arranged
in a chain (hand and fingers)
Humerus
Ulna
Radius
Carpals
1
2
(c) BAT
5
4
3
Humerus
Carpals
Ulna
5
4
3
Radius
(d) WHALE
1
2
Ulna
Carpals
Humerus
(e) LIZARD
Radius
5
4
3
1
2

(*c*) *Individuals of a species vary from each other by small differences,* some of which can be inherited. The variations between members of our own species are obvious to us at a glance, and variation in other organisms, although less clearly perceived by most of us, are very evident to an experienced observer. Only those variations which can be inherited are of any importance in evolution, a variation acquired during an individual's lifetime, such as the well-developed muscular system of an athlete, is not normally passed on to the offspring.

It is suggested that *the varieties which survive longest and which have most offspring tend to be those which are best adapted to the organism's environment or mode of life.* Individuals inheriting harmful variations may die before they reach reproductive age, so that the variation is not passed on. Those which inherit advantageous variations, on the other hand, will like their parents tend to live longer and to have more offspring, some of which may also inherit the variation. Variations which are favorable in a population's particular environment may thus gradually accumulate over many generations, until the members of the population differ so much from those of the original type that they are no longer able to interbreed with them. The "variety" would now be called a new species.

Fig. 32.5 *Light and dark forms of the peppered moth at rest on tree trunks: (a) soot-covered oak trunk near Birmingham; (b) lichen-covered trunk in unpolluted countryside*

(From the experiments of H. B. Kettlewell, University of Oxford, England)

For example, the *peppered moth* is normally light in color, but occasionally a moth with black wings and body occurs. The first recording in Britain of the black variety was made in 1848 in Manchester; by 1895 its numbers had increased so much that it formed 98 per cent of all the peppered moths in the district. Observations showed that the light-colored form was well camouflaged when resting on lichen-covered tree trunks. In industrial areas like Manchester, however, lichens on tree trunks had been reduced or eliminated by the atmospheric pollution which accompanied the Industrial Revolution, and at the same time the tree bark was darkened by soot deposits; in such districts, therefore, the dark moths were better concealed than the light ones when they rested on tree trunks (Fig. 32.5). The moths which were well camouflaged were less likely to be eaten by birds; for example, in Birmingham (another industrial city) a redstart was seen to eat 43 pale moths and only 15 dark ones from equal numbers resting on trees. The "struggle" is a passive competition for concealment, but the outcome is that the dark moths have an improved chance of surviving to lay eggs. Since the dark color is usually due to a single dominant gene (see Section 31.1(*a*)) some of these eggs will give rise to dark-colored offspring. The dark variety cannot yet be called a new species, since dark and pale moths interbreed; however, this account illustrates how a new species might originate by natural selection.

32.3 HERITABLE VARIATION

The sources of heritable variation were not known to Darwin and Wallace, but have since been shown to arise principally in two ways: by mutation and by the recombination of genes.

(*a*) Mutation and Natural Selection

A mutation is a change in a gene or a chromosome (see Section 30.11). Most mutations give rise to harmful effects—a vestigial-winged *Drosophila,* for example, cannot fly and is obviously at a disadvantage compared with normal flies—and are therefore unlikely to be handed on to the offspring. The dark variety of the peppered moth (see Section 32.2(*e*)) results as a rule from a single mutant gene which is dominant to the normal gene, but this is exceptional: usually a mutant gene is recessive to the normal allelomorph. Characteristics controlled by the mutant gene may therefore not be expressed. For example, suppose that a dominant gene *A* mutates to a recessive gene *a,* heterozygous individuals with the genotype *Aa* may be indistinguishable from homozygous *AA* individuals.

If the mutation occurs frequently in the population, however, there is a reasonable chance that eventually two *a* genes may come together in a fertilized ovum, so giving rise to a homozygous *aa* organism in which the new characteristic is fully expressed.

Since so many mutations produce harmful effects in the individuals bearing them, and are therefore likely to be eliminated by natural selection, it might seem that a well-adapted organism which is not subject to mutation would be at an advantage. This may be true so long as the environment does not alter and the organism does not move to a different situation where it is subjected to new selection pressures for which it is poorly adapted. For example, if certain species of bacteria were incapable of mutation, the extensive use of streptomycin could in time bring about the complete elimination of the species. However, gene mutations bring about the appearance of streptomycin-resistant individuals in the population which ensure the survival of the species.

As we have seen, the dark mutant of the peppered moth is at a considerable disadvantage in country districts, but in regions affected by industrial pollution, with the consequent darkening of the tree trunks, the mutant is better protected from predators than the pale form and is thus more likely to survive to breed. The elimination of mutants by natural selection in one environment is the price which populations have to pay if they are to retain the ability to adapt to changes in their surroundings. Mutations should therefore not be regarded merely as accidental mishaps; they are normal events in all populations and are essential for survival and evolution, though they may well be harmful to the individuals affected by them.

(b) Recombination of Genes and Natural Selection

Different populations of the same species often carry different advantageous genes: if the populations interbreed, the two beneficial genes may both be carried by some of the offspring, so conferring a selective advantage on the new phenotype. For example, certain wild grasses carry genes which render them resistant to fungus disease, while some varieties of cultivated wheat, although carrying genes for large grain size, are susceptible to such diseases; crossbreeding can produce a variety of wheat which has a high yield and good resistance to disease. Although the segregation of genes during meiosis (see Section 31.1(*a*)) tends to separate the beneficial genes, their combined selective advantage can enable them to maintain a constant frequency in a population.

Sometimes a genotype has a certain survival value in one environment but not in another. For example, *sickle-cell anemia* occurs when two

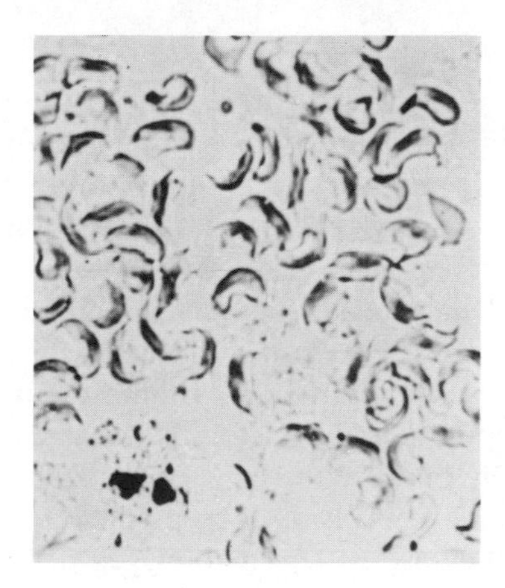

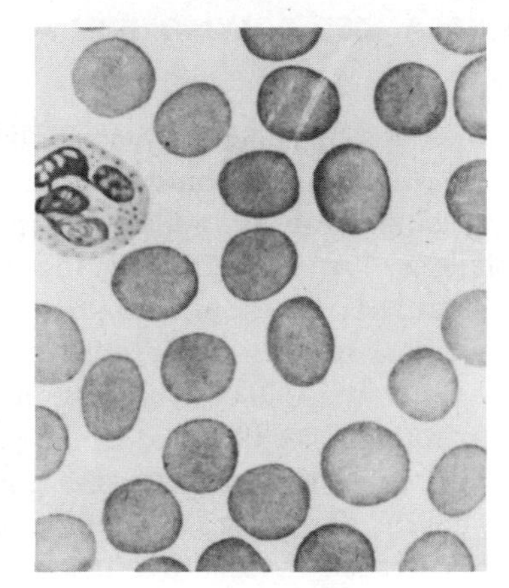

Fig. 32.6 *Sickle-cell anemia: (a) at low oxygen concentrations the red blood cells become distorted; (b) the normal red blood cells for comparison*
(Wellcome Museum of Medical Science)

recessive mutant genes, *h,* combine in an individual; the red blood cells of an affected person become distorted and sickle-shaped when the oxygen concentration in the blood is low (Fig. 32.6). Such people (genotype *hh*) have severe anemia and generally die before reaching reproductive age: that is, natural selection tends to remove the *hh* genotypes from the population. However, the heterozygous individuals (genotype *Hh*) still carry the recessive gene and suffer from anemia to a certain extent; although from one-quarter to one-half of their hemoglobin is affected, usually less than one per cent of their red cells actually show sickling in low oxygen concentrations. Such individuals are said to have the sickle-cell *trait.*

When two *Hh* heterozygotes marry, on average a quarter of their children are likely to have sickle-cell anemia and to die young; in this way the *h* genes could be expected to be gradually eliminated from the population by selection, since *HH* genotypes are likely to have four children capable of breeding for every three produced by the *Hh* genotypes. However, in certain parts of Africa as many as 34 per cent of the population carry the recessive gene, and there is evidence to suggest that these heterozygotes are more resistant to malaria than the *HH* homozygotes. In black populations in America there is an incidence of the sickle-cell trait of only 4–5 per cent, compared with a 15–20 per cent incidence in the African population from which the migrants were originally derived. It therefore seems likely that in non-malarial regions the heterozygotes may lose their selective advantage.

32.4 BALANCED POLYMORPHISM

The existence of genetically controlled varieties within a population provides the raw material on which natural selection acts, with one or other variety being favored or reduced by the selective process. However, in most populations over a short term, the different varieties persist in about the same numbers: the term *balanced polymorphism* is used to describe this phenomenon. For example, although the light forms of moths have a selective advantage in certain environments, a small but constant proportion of dark mutants is still found in the population. This is partly because new mutations continue to appear, and partly because the dark moths still have a certain selective advantage; although more easily seen by predators when resting on lichen-covered tree trunks, they are better camouflaged than the pale moths when in flight. Selection thus operates in their favor in certain circumstances.

Variations in eye and hair color and the existence of different blood groups are examples of genetically controlled variations in human populations. It is not always possible to discern the selective advantage of a particular variant, but this does not mean that the variant is selectively neutral. It would not appear that belonging to a particular blood group could confer a selective advantage, yet there is statistical evidence that individuals of blood group A are less affected than the rest of the population by duodenal ulcers, a characteristic which may enhance their reproductive capacity. This is a further illustration of the fact that the characteristic by which a gene (or group of genes) is most readily recognized is not necessarily the one on which selection is acting (see also Section 30.5).

32.5 ISOLATION AND THE FORMATION OF NEW SPECIES

A mutation or recombination of genes may give rise to a variety with characteristics which enable it to colonize new areas not accessible to the parent stock. In this way the variety may form a breeding population which is isolated from the original population. Further mutations will occur in this isolated group, and these may accumulate until individuals can no longer interbreed with the original stock. In this way a variety will have given rise to a new species.

There are many ways in which a population can become isolated. One obvious example is geographical isolation, where rivers, oceans or mountains separate populations. Another cause could be a difference in breeding season or incompatibility of mating behavior. If variety *A* breeds only in

April while variety *B* breeds in July, the *A* and *B* populations are effectively isolated from each other as far as breeding is concerned, even if they occupy the same area.

Evolution of a new species by natural selection can be regarded as proceeding as follows:

(*a*) mutations and genetic recombinations arise in a population;
(*b*) those which are not so harmful as to be lethal give rise to a balanced polymorphism;
(*c*) environmental change or migration of the population favors certain varieties, which leave more offspring which inherit these same variations;
(*d*) isolation allows other favorable genes to accumulate in the population until it differs so much from the parental stock that interbreeding becomes impossible.

32.6 "THE SURVIVAL OF THE FITTEST"

This is a colloquial expression often used to summarize the theory of evolution by natural selection. However, "fitness" in this context does not refer to the health of an organism but to its production of a large number of offspring which survive to reproductive age. A person may be physiologically "fit"—that is, healthy and strong—but if he fails to have children he is less fit in the evolutionary sense than a poorly-developed, unhealthy person who nevertheless has a large family. Generally speaking, an organism which is healthy and well adapted to its surroundings will survive for a long time and reproduce frequently, leaving behind many offspring which inherit its advantageous characteristics. In human populations, natural selection can only be said to operate on a characteristic if it affects reproductive capacity. For instance, achondroplastic dwarfs are less "fit" than most other people because they have, on average, only 0.25 children per parent compared with 1.27 children born to their normal brothers and sisters. Many dwarfs do not marry, and childbirth is hazardous for the females. Since the gene for dwarfism is dominant, it would tend to disappear from the population were it not maintained by mutation.

From an evolutionary point of view it is more useful to consider the fitness of a population of organisms than the fitness of an individual. Although in the short term fitness depends on the number of viable offspring, in the long term it depends also on the potential for change and adaptation in the population. Within a population, there will be many more variations on which selection may act than in a single creature and its progeny.

32.7 NATURAL SELECTION IN HUMANS

There is no reason apparent at the moment why natural selection should not have been responsible for the evolution of man from an ape-like ancestor, and there is no reason why selection should not still be acting on human populations. The incidence of the sickle-cell trait (see Section 32.3(*b*)) is likely to be the result of selection pressure. Where there is heritable variation and environmental pressure, evolution by natural selection is possible.

However, with the improved efficiency and wider availability of modern medicine, public health measures, education and other benefits of civilization, natural selection has ceased to play so immediate a role in human populations, as it does in other organisms. Sickly babies who would die in "natural" conditions can now be saved, and the congenitally blind, deaf and lame are enabled to live and reproduce. Some people see in this artificial preservation a tendency for mankind to evolve into a race of weaklings. This is to suppose, however, that one day we shall be without modern medicine, good food and so on. It is also wrong to suppose that every weakly child will grow into an enfeebled adult and have unhealthy children. Natural selection is a wasteful process and would have eliminated our diabetics, our short-sighted and our Rhesus babies, many of whom have an important part to play in human affairs.

Man has devoted much thought and effort to minimizing the adverse effects of environmental change by providing himself with, for example, clothing, housing, central heating, air conditioning, and agricultural industry and protection from predators and competitors. This does not mean that selective pressures have disappeared altogether, but they have changed. Resistance to diseases of urban society, indifference to crowding and noise, ability to withstand air pollution and the pace of today's life or assimilate a modern diet without developing ulcers, may all have selective advantages, supposing these characteristics to be controlled to some extent by genes and to become manifest before the end of normal reproductive age.

However, in most advanced countries nowadays selection will operate largely through parental decisions on how many children they will have, since it may be assumed that nearly all the children will reach reproductive age. The "fittest" will still be those with the largest families but this will depend on the conscious decision to procreate rather than on the number left after miscarriage, infant mortality and disease have taken their toll.

Unit 33
Ecology–Interrelationships in Nature

33.1 INTRODUCTION

Ecology is the branch of biology which studies the relationship between organisms and the factors in their surroundings that influence their numbers, distribution, and success. All the millions of life forms on Earth draw upon the physical world for raw materials and energy for life. The physical world also sets conditions suitable or unsuitable for life. And living things have various encounters and make certain adjustments to other life forms. These are all part of ecology.

Because the world has such varied and extreme environments and life has diversified so extensively, there are special branches of ecology. *Synecology* studies groups of organisms or natural communities and how they interrelate. *Autecology* is concerned with the responses of one kind of organism to its environment. There is also desert ecology, marine ecology, island ecology, and alpine ecology, to name a few. As man has altered natural environments by clearing forests and prairies, building highways and dams, and dumping wastes into the atmosphere and waterways, the ecologist is called upon more and more to analyze the problems experienced by individual species and by entire natural communities, and to suggest corrective measures.

33.2 THE ENVIRONMENT

The *environment* includes all of the *abiotic* factors (physical or non-living) and *biotic* factors (living organisms) that make up the surroundings of each living creature (Fig. 33.1). Some of these factors serve as resources, such as water, minerals, light and food. Others may enhance or limit biological activity or success, such as the amount of light, low and high temperatures, and the presence of food or predators. Environmental factors

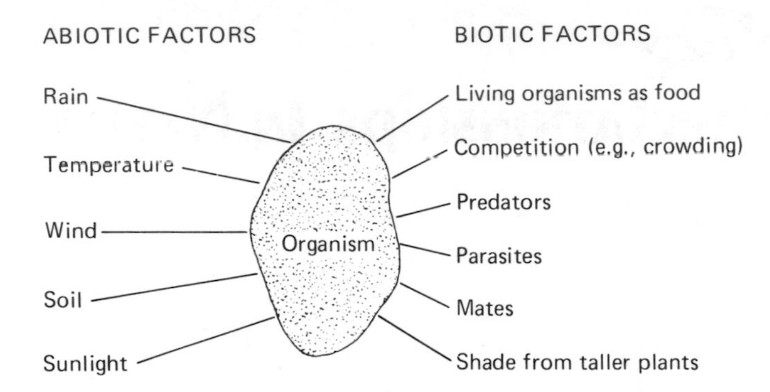

Fig. 33.1 *Environmental factors influence each organism. Some limit and some enhance in a complex of interrelationships, the total environment*

serve as "brakes" upon each species, holding them back so that no one species becomes too numerous or successful. A drought or a severe cold spell limits the whole natural community. Each organism is suited to a particular set of environmental factors called its *habitat* or home. If taken out of its habitat, the organism may die.

(a) Abiotic Factors

The physical or non-living factors of the environment are called *abiotic factors*. The over-riding factor is climate. In turn, climate is the long-term temperature and moisture regime of an area. Other climatic factors that influence the *biota* (life forms) are daily and seasonal weather patterns, the amount of solar energy, wind and fire. The total climate sets the broad limits for life in an area.

The soils (see Unit 34) and geology of an area are abiotic factors that directly affect the *flora* (kinds of plants) that grow there, and in turn, the *fauna* (kinds of animals) that depend upon the plants. The mineral composition, texture, water-holding properties and the pH (alkalinity or acidity) influence the biota on any given site.

The lay-of-the-land (*topography*) and the location on Earth (geography) modify climatic and soil factors and are therefore considered abiotic factors. As the land surface undulates in hills, ridges and mountains, the various surfaces intercept the sun's rays in different ways. A slope that faces the sun is hotter and drier, and hence supports fewer, smaller and different kinds of plants than a slope that faces away from the sun's rays, that is, a shady slope. This is called the *slope effect*.

Another abiotic relationship worldwide exists between *latitude* (distance north and south of the equator) and *altitude* (height above the Earth's surface). Starting at the equator and going northward or southward toward the poles, there is a cooling trend and a reduction in rainfall—one passes from wet tropical, to temperate, and finally into arctic climates. A similar transition in climates occurs as one travels into high mountains, that is, the base of the mountain is hotter and more moist than at the top. This latitude/altitude effect creates natural groupings of plants and animals at a certain altitude that are similar to groupings at lower altitudes but at lower latitudes. An example of this abiotic phenomenon is in the Grand Canyon of Arizona. The climate and biota at the bottom are similar to that in the state of Sonora, Mexico, 1,200 kilometers farther south.

(b) Biotic Factors

Other living organisms in the vicinity of an organism are called the *biotic factors* of its environment. Soil microbes, earthworms, the competition for water and nutrients between the roots of various plants, the shade cast by a tree on some moss or a fern at its base, and a hawk catching a garter snake are all biotic factors. As one life form encounters another, there is some interaction that may range from complete cooperation or dependency to total antagonism, hostility or competition. All these biological interactions further limit the success and numbers of each species of plant and animal within the broad limits set by the abiotic factors of the environment.

33.3 RESOURCE ALLOCATIONS

Energy from the sun drives all meteorological and biological processes on Earth. It is this energy that brings non-living chemicals into forms useful for the assemblage of complex chemicals suitable for living organisms. Ultimately these chemicals are disassembled, energy is lost, and the simple non-living chemicals are returned to the abiotic environment.

Green plants fix carbon dioxide and light energy into simple sugars (see Section 10.2). From simple sugars all other complex organic chemicals are formed by the green plant, including amino acids and proteins. The green plants become a source of nutrients as well as energy for animals that consume them. In turn, animals can be eaten for their nutrient and energy supplies by an animal eater. Thus nutritional and energy relationships are established between specific kinds of organisms within natural communities.

(a) **Food Chains and Food Webs**

All food chains and food webs start with solar energy being captured by green plants. The plants are called *producers* since they create usable food for the other life forms.

All animals obtain their food by taking in complex materials and breaking them down by the processes of digestion into simpler substances that can be absorbed by the body. Some animals, *herbivores* such as rabbits or elephants, do this by eating plants, and others, *carnivores* like lions, seals or owls, by eating other animals; the diet of some species like badgers or man (*omnivores*) includes both plants and animals. All animals, however, derive their food either directly or indirectly from plants. Carnivorous animals feed on other animals, which themselves may feed on smaller animals, but sooner or later in such a series we always come to an animal which feeds on vegetation. For example, pike eat perch; perch eat stickleback which feed on waterfleas; the waterfleas feed on microscopic plants in the pond. This kind of relationship is called a *food chain.* The basis of food chains on land is vegetation in general, but particularly grass and other leaves. In water, the basis is the *phytoplankton* (Fig. 33.2)—the millions of microscopic plants living near the surface of the sea, ponds and lakes. These need only the water around them, the carbon dioxide and salts dissolved in it and sunlight to make all the constitutents of their cells.

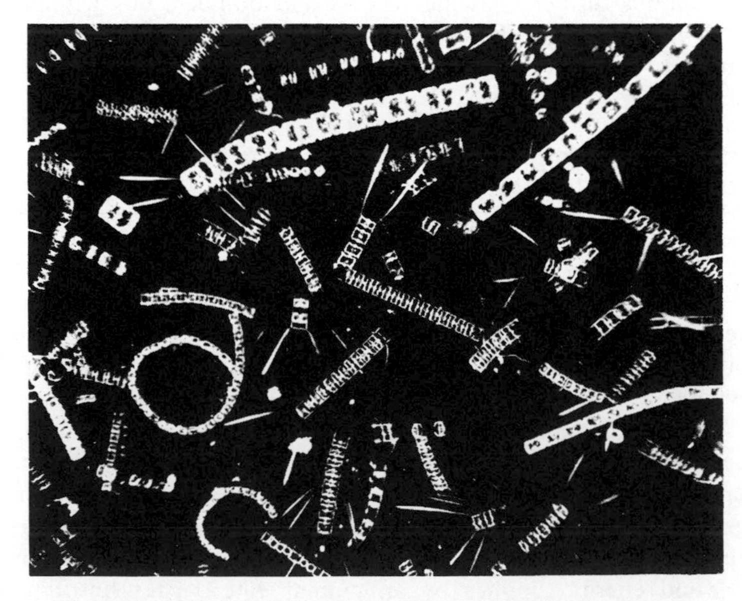

Fig. 33.2 *Phytoplankton (diatoms)*

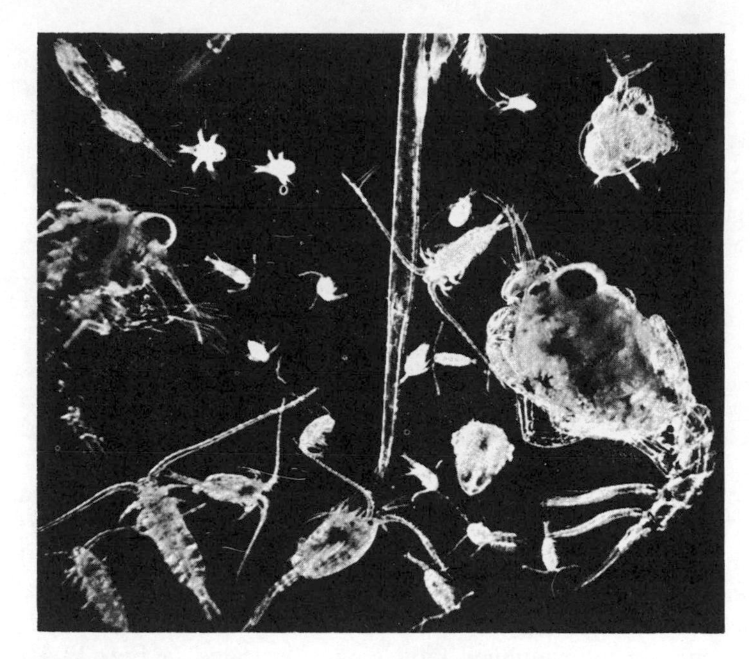

Fig. 33.3 *Zooplankton (mostly adult and larval crustacea from the sea)*
(D. P. Wilson)

Feeding on these microscopic plants are tiny animals, *zooplankton* (Fig. 33.3), such as waterfleas and other crustacea and the larvae (immature forms) of many kinds of animals. The small animals of the zooplankton are eaten by surface-feeding fish such as herring; the herring, in turn, forms part of the diet of man (Fig. 33.4).

In reality, food "chains" are less straightforward than they might appear, since an animal, especially if it is a predator, does not live exclusively on one type of food; a weasel, for example, eats rats, rabbits, birds and other creatures, according to the availability of food at any one time. Similarly, any particular species may be preyed on by several different

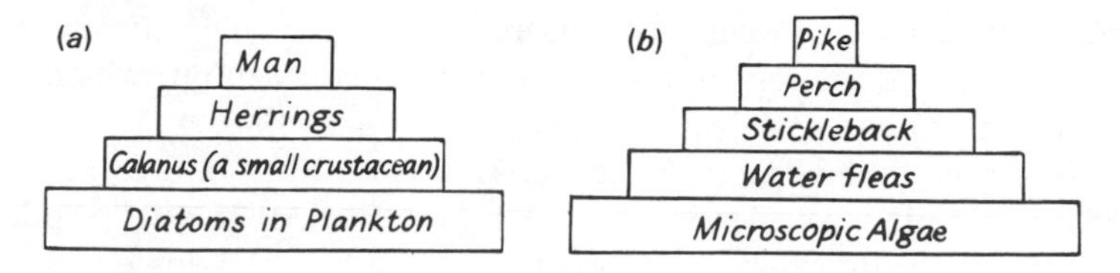

Fig. 33.4 *Examples of food chains*

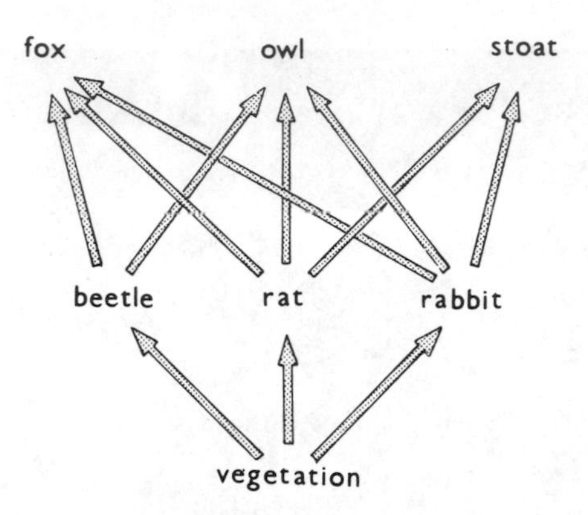

Fig. 33.5 *A food web*

kinds of predator. This more complex relationship can be shown as a "food web"; one such is shown in Fig. 33.5, but even this presents a very over-simplified and generalized picture. It is important to realize also that the links in a food chain do not all involve similiar numbers of individuals. The organisms at the first stage of a chain are usually small and very numerous; for example, in the simplified food chain illustrated in Fig. 33.4(*a*), the diatoms in the phytoplankton are microscopic and very abundant. The crustaceans which feed on the phytoplankton, although also microscopic, are much larger than the diatoms, and each of these small animals eats a great number of diatoms during its lifetime. The herring which eat the zooplankton are larger still, and far less numerous. Finally, large numbers of herring are needed to feed one man.

These numerical relationships are a consequence of the losses involved in the transfer of energy and matter at each stage of a food chain. Most of the food eaten by an animal is either used for its own energy needs or is lost to the body because it has not been wholly digested and absorbed. In general, only about 10 per cent by weight of its food is converted into new living material that can be eaten by another animal. This is why a large mass of living organisms at the beginning of a food chain will support only a small number of carnivorous animals at the end.

The green plants at the base of a food chain are sometimes referred to as the *producers* while the animals which eat them are called *consumers*. A *first-order* consumer eats vegetation; a *second-order* consumer eats animals which feed on vegetation. Consumers such as fungi, which are involved in the processes of decay, are classed as *decomposers*.

If the population of one of the animals or plants in a food chain is altered, all the other species in the chain are affected. If all the pike and perch could be removed from a pond, the number of stickleback in the pond would rise; as the food requirements of the stickleback population increased, the number of water fleas would fall. In the same way, if the diatoms in marine plankton are abundant, the crustaceans which prey on diatoms will flourish, and herring, which eat the crustaceans, may become plentiful. In the 1950s, rabbits in Britain were almost exterminated by the disease of myxomatosis, and the vegetation in formerly rabbit-infested areas changed rapidly. Sheep could graze where the rabbits had previously eaten all the available grass, and trees that were hitherto nipped off as seedlings began to grow to maturity, with the result that chalk downs and other grassland became covered with scrub and eventually with woodland. Foxes ate more voles, beetles and blackberries than before, and attacked more lambs and poultry.

Ultimate Source of Energy

All green plants obtain the energy they need for their metabolism from sunlight; since all animals depend, directly or indirectly, on plant foods, their energy too comes from the Sun. Except for nuclear energy and the energy of the tides, the same is true of all energy released on Earth, whether from the oxidation of foods by plants and animals, or from the combustion of coal (from fossilized forests) or petroleum (formed from deposits of marine organisms which have only partly decomposed during the millions of years since they were buried) or even from hydroelectric power, which depends on the energy of falling water, previously evaporated by the action of the Sun's heat upon the surface of lakes and seas. There is increasing interest in the direct use of the Sun's energy to heat water for domestic use and heating and also to make steam to drive machinery. However, one of the easiest ways to trap sunlight is to grow trees and other plants from which food can be harvested.

Only about 2 per cent of the sunlight falling on a grass pasture is used by the plants for photosynthesis. The rest escapes the leaves, is reflected from their surface or is used to evaporate water from them in transpiration. When the grass is eaten by animals, only about 10 per cent of the food is converted into milk, meat or eggs; the other 90 per cent is used as a source of energy by the animal or lost in feces and urine. It follows that man makes the most efficient use of plants when they are eaten directly, and that their conversion to animal products is a process which is very wasteful of energy.

33.4 THE CARBON CYCLE

Food chains and food webs are but one kind of link in the constant use and reuse of the Earth's chemical resources. The total amount of an element such as, say, carbon, in the Earth's crust does not change: carbon is not manufactured or destroyed. However, the chemical state in which a particular carbon atom exists changes from time to time. Today it may form part of a molecule of carbon dioxide in the air; tomorrow it may be incorporated in a cellulose molecule in a cell wall in a blade of grass. When the grass is eaten by a cow, the carbon atom may become one of many in a protein molecule in the cow's muscle. When the protein molecule is used for respiration the carbon atom will enter the air again as part of a carbon dioxide molecule. Processes like these form the complex series of changes described as the *carbon cycle;* the same kind of cycling applies to nearly all the elements in the Earth's crust. No new matter is created but it is repeatedly rearranged. A great proportion of the atoms of which your body is composed will have been part of many organisms during the millions of years since living things appeared on Earth.

Human activities affect these cycles; for example, the nitrogen excreted from the bodies of humans is not normally returned to the land which produces their food, while the rising demand for energy leads to the burning of carbon fuels like coal or petroleum in ever-increasing quantities, depleting their sources and allowing more and more carbon dioxide to accumulate in the atmosphere.

The carbon cycle may be described in essence in terms of processes which either increase or decrease the amount of carbon dioxide in the environment (Fig. 33.6).

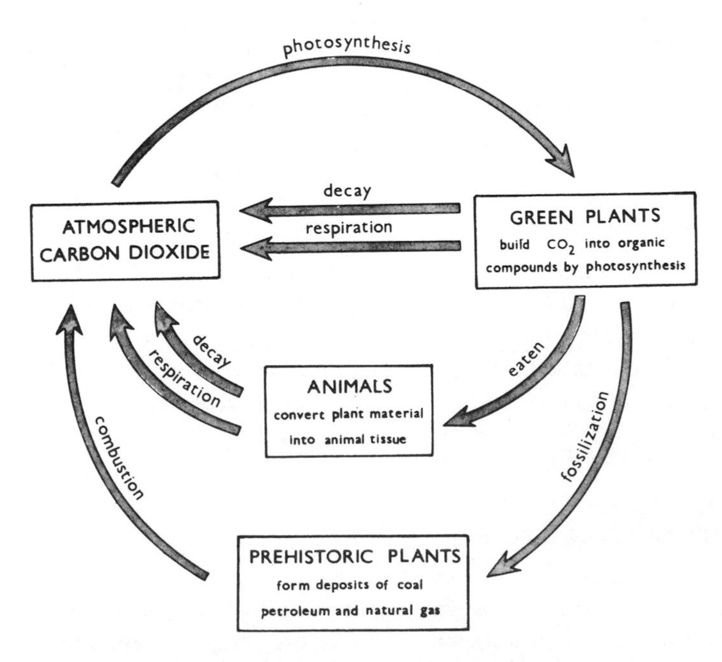

Fig. 33.6 *The carbon cycle*

(a) Removal of Carbon Dioxide from the Atmosphere

Green plants, by their photosynthesis, remove carbon dioxide from the atmosphere or from the water in which they grow. The carbon of the carbon dioxide is incorporated at first into carbohydrates such as sugar or starch and eventually into the cellulose of cell walls, and the proteins, pigments and other organic compounds which constitute living organisms. When the plants are eaten by animals the organic plant matter is digested, absorbed and built into compounds making the animals' tissues. Thus the carbon atoms from the plant become an integral part of the animal.

(b) Addition of Carbon Dioxide to the Atmosphere

(i) **Respiration.** Most living plants and animals obtain energy by oxidizing carbohydrates in their cells to carbon dioxide (see Section 4.2). The microscopic bacteria and fungi that bring about decay oxidize organic material from dead organisms to release energy; again, carbon dioxide is an end-product. In both kinds of respiration carbon dioxide is excreted and returns again to the environment.

(ii) **Combustion.** Most of the familiar fuels are of plant origin: for example wood, coal and its products like coke and coal gas, petroleum and natural gas. All these contain carbon, from carbon dioxide taken up from the air by photosynthesis during the plant's lifetime. When these are burnt their carbon is oxidized to carbon dioxide, which is returned to the atmosphere.

33.5 THE NITROGEN CYCLE

Nitrogen, like carbon, is an essential constitutent of all living material. When a plant or an animal dies and its tissues decompose, one important product is ammonia (NH_3), which is highly water-soluble and therefore quickly washed into the soil where it forms ammonium compounds. The feces of animals contain organic material which is similarly broken down, and their urine is also rich in ammonia and related compounds.

(a) **Nitrifying bacteria.** In the soil are certain species of bacteria, the *nitrite bacteria,* which are so called because they obtain their energy by oxidizing the ammonium compounds to nitrites, that is salts containing the nitrite (NO_2^-) ion. Other bacteria, the nitrate bacteria, further oxidize nitrites to nitrates, salts containing the NO_3^- ion. These bacteria, oxidizing ammonium salts to nitrates, are collectively known as *nitrifying bacteria.*

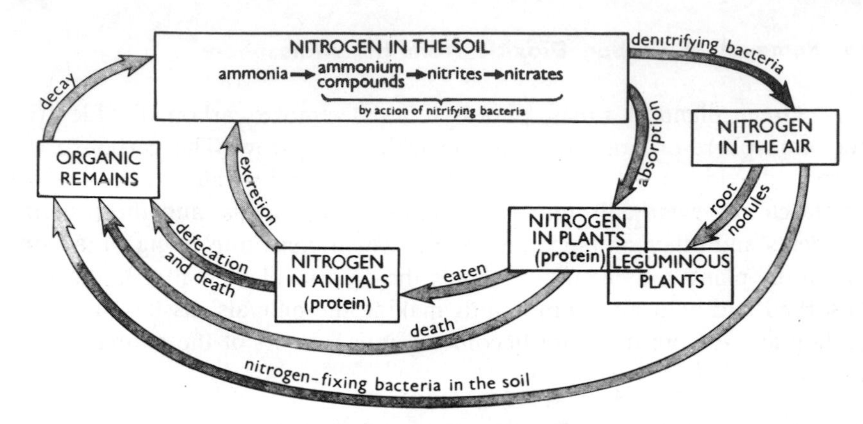

Fig. 33.7 *The nitrogen cycle*

(b) **Nitrogen-fixing bacteria.** Although green plants are surrounded by atmosphere consisting largely of the element nitrogen, they cannot use it in this form to build up food substances. However, there are bacteria in the soil which can take up atmospheric nitrogen and combine it with other elements, or *fix* it. Such nitrogen-fixing bacteria, as well as living free in the soil, are also found in special swellings or *nodules* on the roots of certain plants, especially those of the pea family such as clover, beans or alfalfa (*leguminous* plants) (Fig. 33.8). Since these plants increase the nitrogen content of the soil they are included in the three- or four-year crop rotation used in agricultural practice (see Section 33.5).

(c) **Denitrifying bacteria.** There are also bacteria in the soil that obtain energy by breaking down compounds of nitrogen to gaseous nitrogen which consequently escapes to the atmosphere.

(d) **Lightning.** The high temperature of lightning discharge causes some of the nitrogen and oxygen in the air to combine and form oxides of nitrogen. These dissolve in the rain and are washed into the soil as weak acids, where they form nitrates. Although several million tons of nitrate may reach the earth's surface in this way each year, this forms only a tiny fraction of the total nitrogen being recycled.

33.6 ADAPTATION

An important characteristic of living things is that they can adjust to factors in their environment; we call this adjustment *adaptation*. In nature all living things, plant and animal, are constantly changing to reduce environmental stress whether biotic or abiotic.

Fig. 33.8 *Root nodules on lucerne*
(Rothamsted Experimental Station)

Biological adaptation to the environment is expressed by organisms in two ways, metabolic and genetic. Metabolic (or life) processes of circulation, breathing, photosynthesis, respiration and others are adjusted moment by moment in response to temperature, light, body activity and other factors. Such changes ensure that the organism is "in tune" with its environment. Of course, if conditions are too severe, such as a freeze, no water, a flood, and such, the organism's capacity to change may be exceeded, and it will die. But as long as the organism can adjust to environmental conditions, they are said to be within the organism's *range of tolerance.*

The genetic composition of populations of organisms is also under the influence of the environment. Biotic and abiotic factors can influence the retention or deletion of genes or combinations of genes in a population over many generations. This is called natural selection (see Unit 32). As the genetic makeup of species changes in response to environmental factors, structural and functional features of a species change as a consequence. Such long-term adaptations are inheritable whereas the moment-by-moment metabolic changes are not.

Each species becomes specialized in its activities, responses and form as time goes along. In its own way each species expresses its biotic requirements of protection, nourishment, living place and conditions for reproduction so that it "fits" its environment. The activities or "occupation" of each organism are referred to as its *functional niche*. These are best carried on in the habitat to which the organism is adapted.

33.7 SIGNIFICANCE OF REPRODUCTION

We have learned that all organisms reproduce to carry on their species. Yet there are ecological reasons that make reproduction important. Successful species are those that not only replace their parental members with new individuals but produce large numbers of offspring. That is, reproducing individuals multiply their numbers. Many of these new individuals do not survive because of environmental limiting factors. But the over-production of offspring has the following genetic and ecological benefits:

1. The surplus offspring are food for many other organisms.
2. The offspring are genetically variable so that they can cope with the many environmental factors.
3. The genetically ill-suited individuals are reduced or eliminated.
4. The genetically best-suited individuals will have survival features for themselves and their offspring, especiallv in a changing environment.

Rare and *endangered species* of plants and animals are usually ones with low reproductive capacities and/or where the habitats to which they are specifically adapted have been altered or destroyed. Low reproductive rates do not allow adaptive changes to occur fast enough to cope with the environmental changes.

33.8 SUCCESSION

Just as individuals adapt to their environments, so do natural communities, and this is called ecological *succession*. In its simplest form, community development starts with *pioneer* species that first occupy a site. Their presence and activities modify the environment in such a way that the site becomes suitable for more specialized organisms (successional species). These species in turn further modify the environment by reducing primarily the stress of certain abiotic factors. After a series of such biotic community changes, a *climax* community becomes established which is in harmony with its environment. Individual plants and animals come and go, but the integrity and overall species composition does not change. Most of the original natural vegetation of the world was climax. Most of man's activities have removed or greatly altered these climax communities and replaced them with a few selected domestic species. The Great Plains of the United States were complex climax natural communities of plants and wildlife. They have been replaced by corn, wheat and livestock.

33.9 THE "BALANCE OF NATURE"

In a natural environment, the plants are eaten by the herbivorous animals, which in turn are eaten by the carnivores; the soil is replenished and the plants nourished in their turn by the animals' excreta and by the decay of their tissues when they die, and by the other processes involved in the natural cycles already discussed. If the numbers of predatory animals increase, more herbivores are eaten and the herbivorous population diminishes; the plants, temporarily relieved from grazing, grow more densely. Eventually the predators deplete the herbivores so far that their own numbers decline on account of their restricted food supply; the reduction in the numbers of predators and the plentiful vegetation permit the herbivores to flourish once more. In this way the populations remain basically constant in spite of relatively minor fluctuations. This equilibrium between species is what is meant by the phrase "the balance of nature."

This balance is a *dynamic* equilibrium, and not a static condition. For instance, a given area of woodland may support a population of 50 shrews preying on a population of 10,000 beetles. Five years, or even ten years, later there may still be 50 shrews and 10,000 beetles in the area, but they will not be the same individuals. These numbers will have been maintained against considerable pressures to change them. A pair of shrews may have three litters, each of six young, every year; if all these animals continued to breed, the shrew population could reach 950 after only a year. The beetles' capacity for reproduction is greater still. Obviously there must be a high mortality of both animals, involving other predators and other factors such as disease or food shortages, for the numbers to remain in balance from year to year.

Effect of Man on the Balance of Nature

Civilized man, with his technology and agriculture, interrupts the natural cycles and disturbs the balance of nature to an extent that gives cause for concern.

(a) **Erosion.** Repeated ploughing, overgrazing of pastures and the damage to soil structure by the exclusive use of artificial fertilizers can all lead to soil erosion (see Section 34.4). When this is severe, the land becomes incapable of supporting life of any kind.

When trees are cut down to make way for agriculture or urban development or for use as timber or as a raw material for products like paper or synthetic fibers, the removal of the leaf canopy also exposes the soil below to the erosive forces of wind and rain; this can be particularly serious on thin soils on sloping ground. The topsoil may be washed away into rivers and lakes, silting them up and causing flood damage to the surrounding districts. However, a balanced community can be maintained with care, if tree-felling is limited in extent, and if provision is made for replanting with the appropriate species.

(b) **Eutrophication.** This is the overgrowth of aquatic plants resulting from an excess of nitrogenous salts reaching rivers. The nitrates may come from farmland where heavy application of soluble nitrogenous fertilizers is taking place or from the effluence of treated sewage. Its effects are discussed more fully earlier in this unit.

(c) **Monoculture.** A natural environment usually has a wide variety of vegetation at different levels, flowering and fruiting at different times. This vegetation is exploited in different ways by the animals living there; for example, deer browse on the leafy branches, rabbits crop the turf, squirrels take berries and nuts from the trees, and worms consume the leaves which fall from them. Agricultural practice involves removal of the natural plant and animal community and its replacement with large populations of a single species of plant or animal: arable fields given over to wheat, pastures supporting sheep exclusively. This practice obviously makes the environment unsuitable for the majority of its original inhabitants, as indeed it is meant to; a mixed population of cereal and "weeds" is commercially undesirable. The practice of monoculture, however, has its disadvantages.

Parasites and pests, which in a mixed community find their hosts well spaced, can spread rapidly in a monoculture where suitable hosts are growing closely together. In many parts of Africa, there is evidence to suggest that more protein could be obtained by harvesting the mixed populations of wild animals living in a natural environment than is derived from the herds of sickly cattle which replace them and destroy their habitat.

(d) **Pesticides.** Insect pests can do great damage to food plants in a monoculture because they can spread and multiply easily in the presence of so many suitable hosts. For this reason crops are often sprayed with insecticides such as the chlorinated hydrocarbon DDT. This prevents the

loss of vast quantities of food, but sometimes it upsets the dynamic balance of life in unexpected ways. For example, DDT kills most insects, harmful and beneficial alike, and often affects other animals as well. The codling moth is a common pest of apple trees, for in its larval form it burrows into and ruins the fruit. But when apple orchards were sprayed to eliminate the moth, enormous numbers of red mites appeared because the spray had also killed the spiders which preyed on them.

DDT appeared to be the perfect insecticide for a few years after its introduction. Used against body lice and mosquitoes, which respectively spread typhus and malaria in certain areas and conditions, it has saved thousands of lives. The low concentrations used seemed to be harmless to man and other animals, although larger amounts were known to be poisonous, particularly to fish. Unfortunately, it was not then realized that an animal which takes in DDT with its food or water may not be able to eliminate the insecticide from its body but may retain and accumulate a proportion in the fat deposits, where it may reach a concentration higher than that in the original food. If the fat reserves are mobilized for respiration, or if they are eaten and digested by a predator, the DDT may be released into the blood and may be harmful.

The animals at the end of food chains are particularly vulnerable when DDT is used to kill insect pests on a large scale. In America, DDT was used to kill the beetle which transmits dutch elm disease. The spray and the sprayed leaves reached the soil where the DDT was taken up by worms. When birds, notably the robin, ate the worms, they accumulated lethal doses of DDT and whole populations of birds were wiped out.

A similar event occurred when an insecticide was used to kill gnat larvae in Clear Lake, California. At a concentration of 0.015 part per million of insecticide in the lake water the fish were unharmed. However, the plankton accumulated the compound to a level of 5 ppm. Small fish feeding on the plankton concentrated the insecticide to a level of 10 ppm, and the larger predatory fish contained even more. The western grebes on the lake, which fed on the larger fish, accumulated as much as 1,600 ppm of DDT in their body fat, and five years after treatment of the lake began the birds were dying in large numbers.

DDT is a stable compound and its effects last for a long time, properties which can be disastrous to the balance of nature. When applied to crops it reaches the soil and destroys all insect life there. Eventually it will be washed into the rivers, lakes and oceans; if sufficient accumulates there, it may in time poison fish and other marine life. Meanwhile work continues on the development of pesticides which are selective in their effects, or which break down into harmless substances soon after their application.

There are far too many examples of man's depredation of the

biological resources of the Earth: the excessive killing of animals for food or profit, to the point where they are exterminated; irrigation schemes which make dry areas more productive but which spread the water snails that carry the parasite causing the disease bilharziasis; clearing tropical forests for agriculture and so providing conditions in which the tsetse fly can breed and spread sleeping sickness.

There is a real danger that the ever-increasing human population, demanding constantly improving living standards, might make inroads upon the world's biological systems which could, if allowed to continue uncontrolled, eventually make the Earth's surface uninhabitable by man.

33.10 APPLIED ECOLOGY

Crop Rotation

Different crops make differing demands on the soil, so by changing the crop grown on a particular field from year to year the soil is not depleted of one particular group of minerals. This practice is called *crop rotation;* it has been a part of good farming practice for many centuries. Growing a leguminous crop like clover or beans every few years helps to maintain the nitrogen content of the soil, because of the nitrogen-fixing bacteria in their root nodules. In addition, a year or two of grass improves the soil's crumb structure.

Rotating the crops also reduces the hazards from parasitic infestations. For example, if potatoes are grown in the same field year after year, an increased incidence of the disastrous fungus disease potato blight can be expected. If the field is kept free of potatoes for a few years, so that no host is available to the fungus for several seasons, a healthier crop is likely when potatoes are grown again.

33.11 MANURE AND ARTIFICIAL FERTILIZERS IN AGRICULTURE

The soil's content of nitrogen and other elements is constantly being reduced by washing out of soluble salts by rain, and by the removal of salts by continual cropping. If the minerals are not replaced, either by farmyard manure or by artificial fertilizers such as ammonium nitrate, ammonium sulphate ("sulphate of ammonia") or "superphosphates," the crop yield falls and soil deteriorates.

At Rothamsted Experimental Station, England, wheat has been grown and harvested on an experimental strip of land for over a century without anything being added to the soil. The yield has dropped in this time from 14.7 kg to 6.9 kg per 100 m^2. However, the concentration of nitrogen in the soil has remained steady for the last eighty years, probably because nitrogen-fixing organisms in the soil replace the nitrogen removed in the crop.

Other experimental strips at Rothamsted have been used to study the effects of adding other minerals or mixtures of minerals to the soil (Fig. 33.9); the heaviest yields are obtained on a strip which has been treated year by year with a compound fertilizer containing nitrogen, potassium, magnesium, phosphorus and sodium. Annual applications of farmyard manure produce yields which are nearly as good.

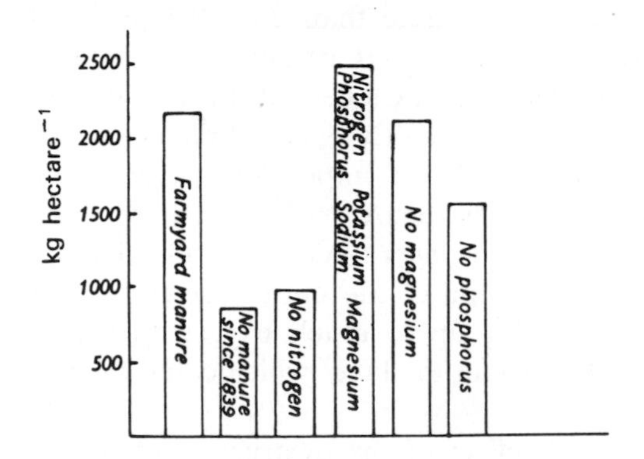

Fig. 33.9 *Rothamsted Experimental Station, Broadbalk Field: average yearly wheat yields, 1852–1925*

Hazards of Using Chemical Fertilizers

Over a long period, say fifty years or so, the yield of crops from a field can be increased by using a program of crop rotation and by regular applications of farmyard manure. Yields are also increased if chemical fertilizers are applied regularly, but the long-term effects of these, in particular the soluble nitrogen compounds such as ammonium salts, are far less satisfactory. Whereas organic manure increases the humus content of the soil (see Section 14.1(*b*)), helping to maintain a satisfactory crumb structure and hence good porosity and permeability to air, the continued exclusive use of chemicals leads to the breakdown and loss of humus, with consequent deterioration of crumb structure and a decrease in porosity.

One result of this is poor aeration of the soil, so that plant roots are deprived of oxygen, without which they cannot absorb the salts effectively.

The unabsorbed nitrates are washed by rain from the soil, and eventually drain into rivers and lakes. The growth of microscopic algae already present in these waters is stimulated by the excess of nitrates; sewage effluents from cities or from intensive animal-rearing units and the phosphates present in certain detergent products can have the same effect. The algae grow quickly and in time die, and are decomposed by bacterial action. The decay bacteria obtain the oxygen they need from the gases dissolved in the water, but in these conditions of exceptional bacterial activity this oxygen supply may become so depleted that fish and other aquatic animals suffocate and die. A foul mud accumulates in the water, consisting of the partially decomposed remains of the algae.

This kind of overgrowth of aquatic plants is called *eutrophication*. One lake already seriously damaged in this way is Lake Erie, which receives water draining from more than 75,000 km^2 of farmland on which artificial fertilizers are regularly used, and also the sewage effluents from large cities such as Detroit and Cleveland. In its waters eutrophication has produced an organic mud of partly decayed plant remains, separated from the waters above by an almost impervious layer of iron oxides. Some scientists think that further removal of oxygen from the upper layers of water might cause this oxide layer to dissolve, allowing the mud to circulate through the lake. The action of the decay bacteria on the organic matter would then resume, using up so much oxygen that anaerobic conditions might be created throughout the lake, which could then support no aquatic life at all.

Another harmful effect of the destruction of the crumb structure of soils by the excessive use of chemical fertilizers is that light soils in particular are much more likely to become dry and powdery, and to be blown away by the wind, especially when not protected by a plant cover or by windbreaks like hedges or tree groves. This leads to loss of valuable topsoil and in extreme cases to dust-bowls and deserts.

Even in the short term, the heavy application of nitrate fertilizers to crops has raised the level of free nitrate in the food plants to a marked degree, and concern has been expressed that these increased concentrations constitute a hazard to health.

However, the increase of crop yields by the application of chemical fertilizers is so valuable to farmers that their continued use appears to be an economic necessity. Two possible solutions to the problems arising from their use are:

(*a*) the development of sparingly soluble fertilizers which release their nutrients into the soil gradually, at a rate to suit the demands of the plants, and

(*b*) the more efficient use of organic manure, of which enormous quantities are wasted at present, to help conserve the soil's humus content and to maintain its structure.

33.12 QUESTIONS

1. Trace the food chains involved in the production of the following articles of human diet: eggs, cheese, bread, meat, wine. In each case show how the energy in the food originates from sunlight.
2. Discuss the advantages and disadvantages of man's attempting to exploit a food chain nearer to its source, e.g. the diatoms of Fig. 33.4.
3. Construct a diagram on the lines of the carbon cycle (Fig. 33.46) to show the cycling process for oxygen.
4. How do you think evidence is acquired to assign animals such as a badger and a wood pigeon to their position in a food web?
5. What would be the desirable qualities of an insecticide to control greenfly on sugar-beet leaves? What might be the disadvantages of total eradication of an insect pest such as the greenfly?
6. Select an organism you have studied in this book and list its adaptations to its environment.
7. Describe some of the abiotic and biotic factors in a natural community near your home.

Unit 34
Soil

34.1 COMPONENTS OF SOIL

The soils on Earth are the source of minerals for life, both plant and animal. Hence they are vital abiotic components in the natural environment. Soil is a complex mixture of (*a*) particles of sand or clay, (*b*) humus, (*c*) water, (*d*) air, (*e*) dissolved salts, and (*f*) micro-organisms.

(a) Inorganic particles. These are formed from rocks which have been *weathered* or broken down. Rock weathering takes place both by physical means, such as seasonal temperature changes producing alternate freezing and thawing of water in crevices, and by chemical action, especially attack by rain-water, which behaves as a very dilute acid because of the traces of dissolved carbon dioxide it contains. The particles constituting *sand* are of 2 to 0.02 mm diameter, those of *silt* 0.02 to 0.002 mm and those of *clay* less than 0.0002 mm diameter. Chemically, sand consists of silica (silicon dioxide), and clay of a mixture of complex compounds of aluminum and silicon oxides; both are often stained red or brown by iron oxides.

Aggregates of these inorganic particles, together with humus, produce *crumbs* up to about 3 mm in diameter, which form the "skeleton" of the soil. The crumb structure of a soil depends on the relative proportions of clay, sand and humus it contains, and also on the activities of plant roots; a good crumb structure is one of the most valuable attributes of an agricultural soil.

(b) Humus. This decomposed organic matter incorporated into the crumbs originates mainly from decaying plant remains. The presence of humus in the crumbs affects the color and physical properties of the soil. Humus is black and structureless; it often forms a coating around sand particles and may be important in "glueing" particles together to form soil crumbs. A sandy soil deficient in humus tends to have a poor crumb structure and is easily blown away if exposed by ploughing. The excessive use of chemical fertilizers on certain soils, especially in dry climates, may destroy part of their humus content, and lead to the formation of "dust-bowls" or the advance of desert margins.

(c) **Water.** The sand particles and clay aggregates in moist soil are covered with a thin film of water, which is held to them by the forces of *capillary attraction,* just as water is held in the pores of blotting paper. Water may also penetrate the aggregates and be held to the clay particles by chemical forces. When a soil contains as much water as it can hold by capillary and chemical attraction (that is, any more would drain away by the force of gravity), it is said to be at *field capacity*. Capillary attraction also tends to distribute water from regions above field capacity, perhaps deep in the soil, to drier regions in the same way that moisture is drawn up a strip of blotting paper when one end is dipped in water. The forces holding water in the soil also set up considerable opposition to the "suction" of plant roots when the soil begins to dry out.

(d) **Air.** The spaces between the aggregates or sand particles are normally filled with air, unless the soil is water-logged, when the air is driven out by water. If the air spaces are so blocked, plant root cells and aerobic micro-organisms in the soil are cut off from the oxygen they need for their respiration.

(e) **Mineral salts.** The soil water contains very small quantities of salts in solution, dissolved out either from the surrounding rock or from the humus in the soil. These salts are vital to plant growth (see Section 10.6).

(f) **Micro-organisms.** Many microscopic plants, fungi and animals live in the soil, but among the most important to plant life are the bacteria which break down the organic matter and humus to form soluble salts which can be taken up in solution by the roots. Other bacteria, the *nitrogen-fixing bacteria,* convert atmospheric nitrogen to organic nitrogen compounds (see Section 33.3 for a more detailed discussion).

34.2 TYPES OF SOIL

(a) Heavy Soils

A soil in which clay particles predominate and which has a poor crumb structure is sticky and difficult to dig or plow. This is partly because the chemical and capillary forces holding the particles together act *at their surface*; the minute clay particles have an enormous total surface area and

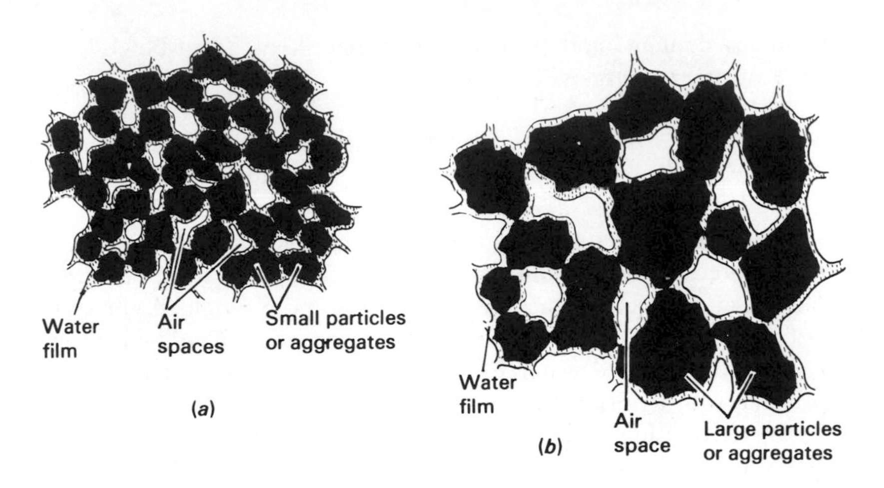

Fig. 34.1 *Two kinds of soil structure: (a) heavy, (b) light*

the forces between them are consequently very large indeed, making them difficult to separate. When dry, heavy soils form hard clods which do not break up readily during cultivation.

The small distances between particles (Fig. 34.1(*a*)) tend to produce poor aeration and drainage, but the large surface presented by the particles retains a high proportion of water, even in conditions of drought. There is also comparatively little loss of minerals in heavy rain, since they are held chemically to the clay particles.

A heavy soil can be made lighter, more easily workable and more permeable to water and air by adding either organic matter or lime. The lime makes the clay particles clump together or *flocculate,* and the clumps of particles then produce a structure resembling that of a lighter soil. The crumb structure of a clay soil can also be improved by the action of plant roots, for example, by growing grass on it for a year or two.

(b) Light Soils

The large inorganic particles of a light, sandy soil give it its gritty, open texture (Fig. 34.1(*b*)). The wider separation of the particles leads to better aeration and drainage, but there is a relatively smaller surface available to hold the water film. This reduced surface means that the surface forces between the particles are smaller than in heavy soils. Moreover, water adheres to sand particles by physical means only, and not by the chemical forces which are significant in clays. Light soils therefore dry out easily; their particles are readily separated in plowing and digging, and the clumps are easily broken up when the soil is dry. Soluble salts are more liable to be washed out by rain from a light soil than from a heavy one, but its content of both minerals and humus, and also its water-holding properties, can be improved by adding farmyard manure or compost.

(c) Loam

A soil with a balanced mixture of particle sizes, a good humus content and a stable crumb structure is called a loam. Loams are the most productive soils in agriculture.

(d) Acid Soils

In some water-logged soils, the lack of air between the soil particles prevents the aerobic decay bacteria from taking their part in the breakdown of dead plants into humus. As a result the partly decomposed remains of vegetation accumulate, sometimes in layers many meters thick, becoming acid as a consequence of the incompleteness of the decomposition. Peat bogs are formed in this way, but soils become acid in other ways as well. For example, repeated applications of ammonium sulphate (widely used as a chemical fertilizer) will make a normal soil acid unless accompanied by lime, which neutralizes the acidity. Most plants cannot grow in acid conditions, and very acid soils are of little value for crop growing.

34.3 EXPERIMENTS ON SOILS

Experiment 34.1 *To find the weight of water in a soil sample*

A sample of soil (about 25 g) is placed in a weighed evaporating basin which is then reweighed. The basin is heated over a water-bath for several hours to drive off the soil water, allowed to cool in a desiccator (so that the sample cannot absorb water from the air), and weighed again. Heating and reweighing are continued until two weighings give identical results, showing that all the soil water has evaporated. The difference between the first and final weighings gives the weight of water that was originally present.

Higher temperatures than that of the water-bath cannot be used as they will cause the decomposition of the humus in the soil and lead to an erroneously high value for the weight loss.

Experiment 34.2 *To find the weight of organic matter in a soil sample*

The sample of dry soil left from Experiment 34.1 is heated in the same evaporating basin, but this time on a sand-tray or gauze over a bunsen burner, until it loses no further weight. All the organic matter has now been completely oxidized to carbon dioxide and water, so that this second loss in weight represents the weight of humus and other organic matter originally present.

Experiment 34.3 *To find the proportion of air (by volume) in a soil sample (Fig. 34.2)*

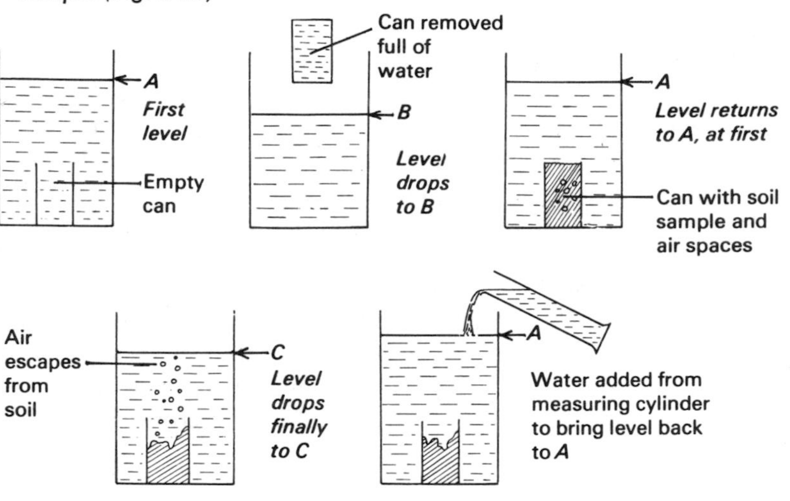

Fig. 34.2

Two juice or soup cans of equal size are required, both with cleanly cut tops. Their capacity is measured, using a measuring cylinder. One can is perforated at the bottom, driven down, open end first, into the soil, and then dug out without disturbing the soil in it; the soil is cut off level at the top, without compressing it. The other can is placed empty in a large vessel of water and the water level *A* marked; the can is now removed full of water and the level drops to *B*. The can full of soil is now placed in the vessel of water. The water level at first returns to *A*, but air bubbles soon begin to rise from the soil and the level falls again to *C*. (There is no need to mark levels *B* and *C*.) All the air must be forced out of the soil by stirring it with a stick under the water. Finally, sufficient water is added from a measuring cup to bring the level up to *A* once more. The volume of water added will be equal to the volume of air that has escaped from the soil.

Note. It is interesting to compare the results of these experiments for several different kinds of soil, or for the same soil under different conditions.

Experiment 34.4 *To make a rough estimate of the proportions of solid components of a soil sample (Fig. 34.3)*

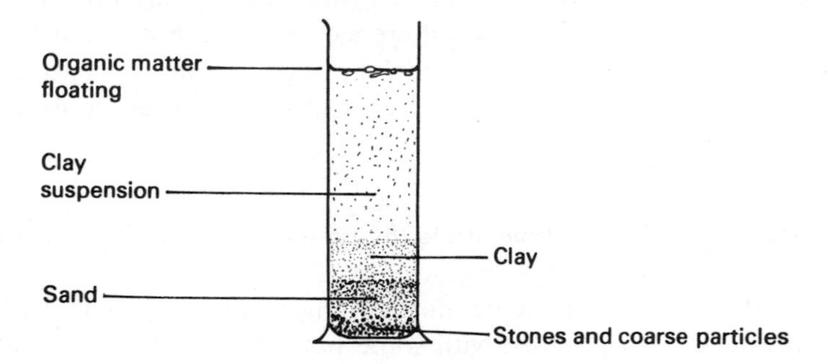

Fig. 34.3

About 20 cm³ of the soil is thoroughly shaken with 200 cm³ of water in a measuring cylinder and the mixture allowed to settle. The particles will settle according to their surface area and density, small stones first, then grains of sand and then clay, so that more or less distinct layers appear. Some fine clay particles will remain suspended in the water even after long standing, and some fragments of organic material will float to the top. The depth of the various layers can be read off and the results compared with those obtained for different soils.

Experiment 34.5 *To compare the permeability of two soil samples to water*

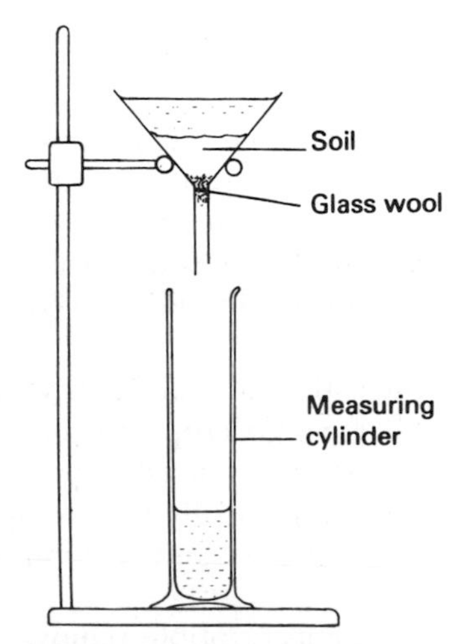

Fig. 34.4 *To examine the permeability to water of a soil sample*

Two glass funnels are plugged with glass wool and half-filled with equal volumes of sandy and clay soils respectively. Both are then filled with water, the level of which is kept constant by topping up throughout the experiment so that there is no difference in pressure between the two. The water running through each funnel in a given time is collected and measured; the ratio of the volumes collected gives an indication of the ratio of the permeability of the two soils.

Experiment 34.6 *To demonstrate the presence of micro-organisms in soil*

Two Petri dishes of sterile nutrient agar are prepared. The agar surface in one dish is sprinkled with some particles of fresh soil, and that in the other dish with some grains of clean sand that have been sterilized by heating. Sterilized spatulas must be used to handle the soil and the sand. If bacteria or fungi are present in the soil they will form colonies on the surface of the agar within two days. The absence of such colonies from the control dish will prove that the bacteria were in the soil and did not come from the air, the dish, the medium or the spatula.

34.4 SOIL EROSION

Soil erosion means the removal of topsoil, usually by the action of wind or rain. It is part of the natural pattern of landscape development and difficult to prevent altogether, but certain kinds of agricultural malpractice can enhance its effects to the point of disaster; these are examples of applied ecology.

(a) **Deforestation.** The soil cover on steep slopes is usually fairly thin but support the growth of trees. If the forests are cut down to make way for agriculture, the soil is no longer protected by a leafy canopy from the driving rain. Consequently, some of the soil is washed away into the rivers, which tend to become choked with silt and overflow their banks.

(b) **Erosion.** Plowing loosens the soil and destroys its natural structure. If humus is not replaced after repeated cropping, or if weeds or the stubble remaining after harvest are destroyed year after year by burning, the water-holding properties of the soil will gradually deteriorate, so that the soil dries out easily and may be blown away as dust. On sloping ground such soil may be eroded by water.

(*i*) *Sheet erosion* is the imperceptible removal of thin layers of soil.

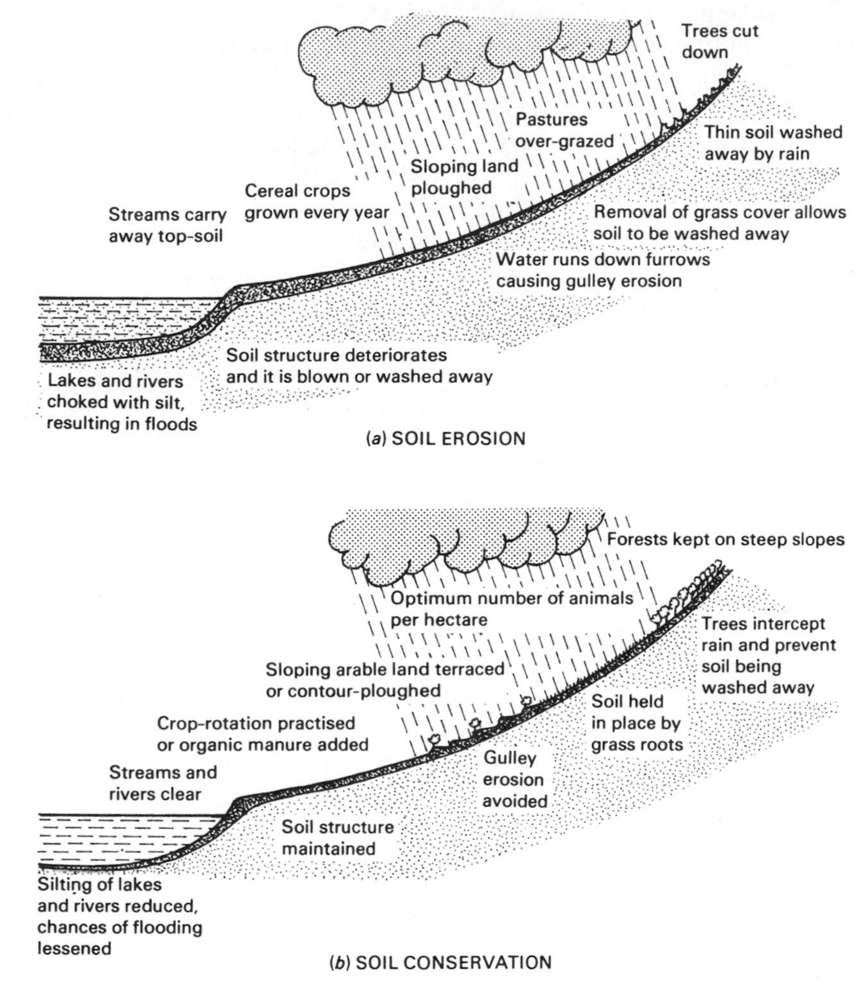

Fig. 34.5 *Comparison of agricultural techniques*
(From W. E. Shewell-Cooper, *The ABC of Soils,* English Universities Press)

(*ii*) *Rill erosion.* Especially on irregular slopes, heavy rain may cut channels into the soil, which deepen as the volume of run-off increases and so form gulleys.

(*iii*) *Gulley erosion.* Gulleys may reach enormous proportions, so that thousands of hectares of topsoil are carried off. Gulleys may follow the tracks made by vehicles, goats, cattle and other farm animals, and can be made much worse by careless plowing. Contour-plowing, in which the furrows follow the ground contours as closely as possible, helps to prevent this kind of erosion, as does the terracing of hillsides.

(c) **Overgrazing.** A given area of farmland can only support a limited number of grazing animals satisfactorily; if this optimum number is exceeded the animals grow more slowly because of the relative scarcity of food. In addition, the stock, especially close-cropping animals like sheep and goats, graze the pastures very closely, leaving little plant cover on the ground, while their hooves trample and compact the soil into a hard layer. Consequently the penetration of the soil by rain-water is reduced; it dries out quickly and may eventually be blown away.

34.5 QUESTIONS

1. What are the conditions in the soil which make it a suitable environment for microscopic animals and plants?
2. What do you suppose is the biological significance of the following agricultural practices: (*a*) plowing farmyard manure (animal feces and straw) into the soil, (*b*) adding lime to the soil, (*c*) spreading ammonium sulphate on the land?
3. Outline the ways in which careless agricultural practices can lead to rapid erosion of a light sandy soil.
4. When the properties of soils are compared in Experiments 34.1, 34.3 and 34.5 the soil samples should all be at field capacity if the comparisons are to be valid. Explain why this is necessary.

Further Reading

Asimov, I.: *The genetic code*. New American Library (1967).

Baron, W. M. M.: *Organization in plants*. Edward Arnold (1963).

Beadle, G. and M. Beadle: *The language of life*. Doubleday (1966).

Bonner, J. and A. W. Galston: *Principles of plant physiology*. W. H. Freeman (1951).

Bracegirdle, B. and P. H. Miles: *An atlas of plant structure,* Volumes 1 and 2. Heinemann (1971, 1973).

Buchsbaum, R.: *Animals without backbones,* Volumes 1 and 2. Penguin (Harmondsworth, 1971).

Carter, C. O.: *Human heredity*. Penguin (Harmondsworth, 1970).

Clegg, P. C. and A. G. Clegg: *Biology of the mammal*. Heinemann (1963).

Coker, E. G.: *Soils and fertilizers*. Macdonald (1971).

Freeman, W. H. and B. Bracegirdle: *An atlas of histology*. Heinemann (1966).

Green, J. H.: *An introduction to human physiology*. Oxford University Press (1963).

Grove, A. J. and G. E. Newell: *Animal biology*. University Tutorial Press (1969).

Hurry, S. W.: *The microstructure of cells*. John Murray (1965).

Imms, A. D.: *Insect natural history*. Collins (1971).

Ingold, C. T.: *The biology of fungi*. Hutchinson (1967).

Mackean, D. G.: *Experimental work in biology*. John Murray (1971). Book 1: *Food tests*. Book 2: *Enzymes*. Book 3: *Soil*. Book 4: *Photosynthesis*. Book 5: *Germination and tropisms*. Book 6: *Diffusion and osmosis*. Book 7: *Respiration and gaseous exchange*.

Mather, K.: *Genetics for schools*. John Murray (1953).

Patten, B. M. and B. M. Carlson: *Foundations of embryology,* 3rd ed. McGraw-Hill (1974).

PEACOCK, H. A.: *Elementary microtechniques.* Edward Arnold (1973).

PELCZAR, M. J., JR.: *Microbiology,* 3rd ed. McGraw-Hill (1972).

RUSSELL, SIR E. J.: *The world of the soil.* Collins (1961).

SINNOTT, E. W., L. C. DUNN and T. DOBZHANSKY: *Principles of genetics.* McGraw-Hill (1958).

SMITH, R. L.: *Ecology and field biology,* 2nd ed. Harper & Row (1974).

STORER, T. I., R. L. USINGER and J. W. NYBAKKEN: *Elements of zoology,* 8th ed. McGraw-Hill (1968).

SUTCLIFFE, J.: *Plants and water.* Edward Arnold (1967).

SWANSON, C. T.: *The cell.* Prentice-Hall (1969).

VICKERMAN, K. and F. E. G. COX: *The protozoa.* John Murray (1967).

WEIER, T. E., C. R. STOCKING and M. G. BARBOUR: *Botany: an introduction to plant biology,* 5th ed. Wiley and Sons (1974).

YAPP, W. B.: *The life and organization of birds.* Edward Arnold (1970).

YOUNG, J. Z.: *The life of vertebrates.* Oxford University Press (1962).

abdomen, insect, 197–98
abiotic, 489–90
abscission, 60–61
absorption, 269, 278–79
accessory food factors, 254 (*See also* Vitamins)
accommodation, 391–92
achondroplastic dwarfism, 443, 463, 464–65, 487
acid soil, 511
actinomycetes, 175
active transport, 20, 23, 133–34, 278, 307
adaptation, 250–54, 498–99
adenine, 439–40
adipose tissue, 283, 331, 334
adrenal gland, 295, 420–21
adrenaline, 295, 420, 422
adventitious roots, 67, 72–73, 74–75, 76–77, 78–79, 81, 111–12
aerobic:
  bacteria, 164
  respiration, 36–42
afterbirth, 353
agglutinins, 297
agriculture, 95–96, 504, 511
agrimony fruit, 103
air:
  exhaled, 41, 314–15, 318–19
  in soil, 509, 512
  passage, 311–12
albinism (albino), 438, 441, 463, 468–69
albumen, 255–56, 293
alcapton, 411
alcohol (ethanol), 42–43, 179, 274
algae, 5, 155, 180–83
algin, 183
alimentary canal, 164–65, 188, 269–71, 279–80
allantois, 255–56
alleles (allelomorphic genes), 457
alligator, 195
altitude, 491
alveolus, 311–12, 313
ameba, 32, 185–88, 287, 293, 379
amino acids, 163, 262, 276, 280, 281, 285, 286, 293, 420
  absorption of, 278, 295
  breakdown products of, 322
  deamination of, 284, 286, 420
  metabolism of, 284, 295, 350
ammonia, 18, 164
amnion:
  bird's egg, 255–56
  man, 349–51, 353
amniotic fluid, 351, 353
amphibia, 5, 194–95, 235–47, 339, 420, 477
ampulla, 400–01
amylases, 271, 274, 281, 289, 290
anaerobic respiration, 36, 42–43, 164 (*See also* Fermentation)
anal fin, 226
anaphase:
  meiosis, 444–45, 446, 447, 448
  mitosis, 427, 428–30
androecium, 85, 87
androgens, 354
animals:
  characteristics of, 2–3, 11–12, 143
  classification, 4
animal dispersal, 103–04
animal kingdom, 184–96
annual plants, 71
anopheles, 216
antagonistic:
  hormones, 422–44
  muscles, 368
antenna, 193, 197, 202–04, 221
anther, 85, 87–89, 90–91, 92–97
antibiotics, 167, 174–75
antibodies, 168, 174, 296–97
antidiuretic hormone (ADH), 295, 328, 422
antigens, 296
antiseptics, 167
antitoxins, 166, 297
anus, 269, 280
aorta, 298, 300, 302, 304, 314, 342
aphid, 210–11
appendix, 164, 270, 279
apple, 60, 99–100
applied ecology, 504
aqueous humor, 388
arachnids, 4, 193
arteriole, 299, 324, 333, 336
artery, 299–301
arthropods, 192–93, 197
artificial:
  fertilizers, 504–55, 511
  propagation, 80–83
ascaris, 192
asexual reproduction:
  bacteria, 161
  flowering plants, 71–83
  fungi, 174–75, 177
  protista, 186
association centers, 412–43
astasia, 189
atmosphere, 497
atrium:
  heart, 302, 304
  tadpole, 244
auditory nerve, 397
autecology, 489
autoclave, 168
autonomic action, 379, 381
autoradiograph, 154
auxins, 122–24, 125
axil, 47, 55
axillary bud, 47, 60, 61, 72
axon, 403–05

bacillus, 163
backbone, 4–5, 225, 227, 235, 248, 253, 361–63
back-cross, 458, 461, 463
bacteria, 5, 14, 161–68, 169, 296–97
  antibiotic action on, 167, 174–75
bacterial disease, 60
balance:
  of nature, 501–04
  sense of, 399–400
balanced polymorphism, 486
ball and socket, 362–63, 480
barb, 249
barbule, 249
bark, 50, 54, 60–61, 81
base, in DNA, 439
bass, 4
bastard wing, 251
bat, 481
beak, bird, 248, 250

beer, 178
begonia, 5
behavior:
  parental, 234, 255–57
  reproductive, 233–34, 240–41, 254–55
Benedict's solution, 268
beri beri, 263, 265
bicarbonate indicator, 151–52
bicuspid valve, 304
biennial plants, 71
bile, 275–76, 281, 285
binary fission, 186
bio-degradable, 164
biogenesis, 478
biota, 490
biotic, 489, 491
birds, 5, 195, 248–58, 339, 477, 480
  reproduction, 254–57
birth:
  control, 355, 357
  human, 352
biuret test, 268
bivalents, 442–43
bladder, urinary, 323, 328, 340–41, 342
blind spot, 389
blue-green algae, 5, 169–70
blood, 165
  circulation, 298–307
  clotting, 463
  composition, 291, 302
  glucose, 285
  groups, 462
  plasma, 286
  pressure, 301, 305
  serum, 293
  system, 200
  transport system, 294
  vessels, 298–302, 312
blowfly, 215
blue-green algae, 5, 159
body temperature, 286
bolus, 272
bomb calorimeter, 260
bone, 359
bottling, 166
Bowman's capsule, 324, 326
bract, 90
brain, 407, 410
breast, 354–55
breathing, 200, 236–38, 310–21, 318–19 (***See** also* respiratory system)
bronchioles, 311
bronchus, 311, 313
brooding pouch, 255
brown algae, 183
bubonic plague, 165, 216
bud, 46–47, 61–63
  growth, 62–63, 64–65
  scales, 61–63, 64–65, 66, 73
budding, 177
bulbs, 47, 72–76
buttercup, 5, 86–87, 94
butterfly, 2, 213, 224
buzzard (hawk), 250

calcium, 263
calorie, 260
calyx, 85, 89
cambium, 48–51, 54, 426
canine, 372, 374
canning, 166
capillary, 278, 299–302, 305–7, 332–33
  attraction in soil, 509
caracoid, 254
caracoid bone, 252–54
carbohydrates, 35, 259, 261, 282
carbon cycle, 495–96
carbon dioxide, 3, 58–59, 145, 149–50, 293, 300, 302, 315, 319, 497
carbonic anhydrase, 314
cardiac muscle, 368
carnassial teeth, 374
carnivore, 492
carnivorous, 238, 374, 375
carpals, 251, 480–81
carpels, 85, 87
carpo-metacarpals, 251, 254
carrier, 452
carrogeenum, 183
cartilage, 194, 359, 361, 363–64
caterpillar, 2, 197, 206, 218–22
catkin, 95–96
caudal fin, 226
cecum, 164, 188, 279
cell, 7, 305–07
  body, 14, 403
  characteristics of, 7
  differences, 11–12
  division of, 10–11, 63, 68–69
  membrane, 9, 11–12, 13, 24, 427–28, 443
  parts of, 7–11
  sap, 11, 23, 25
  specialized, 14
  sting, 15
  wall, 8–10, 25, 27, 59
cellulose, 3, 11, 261
  wall, 23, 164
cement, tooth, 373
censer mechanisms, 102
central nervous system, 407
centriole, 427–29, 443
centromere, 428–29, 443, 448
cereals, 5
cerebellum, 411
cerebral hemispheres, 413
cerebrum, 413
cervix, 340, 353
chalaza, 255
chemical fertilizers, 504–06, 511
chemoreceptors, 203
chewing muscles, 369
chiasma, 442, 444, 448
chick, 256, 257
chironomus, 436
chitin, 172
chlamydomonas, 181
chlorophyll, 3, 47, 147–48, 164–65, 172, 188
chloroplast, 27, 50, 57, 145, 172, 181
cholera, 164, 165, 218
choroid, 388
chromatid, 428, 432, 443, 448–49
chromosome, 427–29, 431–36, 446, 450
  abnormalities, 433–34
  replication, 428–29, 443
chrysalis, 221
chyme, 275
cilia, 14, 185, 187, 312, 344
ciliary body, 388, 391
ciliates, 187
circulation, 237
circulatory system, 291, 293, 298–308
clavicle, 254
claw, 248
clay soil, 508, 510
cleanliness, 168
cleavage, 242
climax community, 500
clinostat, 117–18

clitoris, 453
cloaca, 241
clots, blood, 296
cobalt chloride, 140–41
cobalt, in nutrition, 263
coccus, 163
cochlea, 398–99
codominant, 458, 461
coelenterates, 190–91
cohesion-tension theory, 135
coleoptile, 109, 112–13, 122–23, 125–27
coleorhiza, 109, 112–13
collagen fibers, 361
collecting tubule, 325
colon, 280
colonies, bacteria, 169
color blindness, 451–52
color vision, 206, 393
combustion, 497
compensation point, in photosynthesis, 146
complete metamorphosis, 197, 199
components, of soil, 508
composite flower, 88
composition of blood, 291
compound eye, 197, 198, 204–06
concentration gradient, 19
conception, 349
conditioned reflex, 418–19
conditioning, 418
condom, 355
condoyle, 362
cones, in eye, 388
conjugation, 177
conjunctiva, 387
consanguinity, 469–70
consumers, 494
contraception, 355–56
contraceptive pill, 357
contractile:
  roots, 76
  vacuole, 186
cooling, in plants, 135
coordination, 370, 403–24
copper, in nutrition, 263
copulation, 346
coral, 4, 190
cork, 60
cork cambium, 52, 54
corms, 47, 76–77
cornea, 388
cornified layer, 165, 330–31
corolla, 85, 94
corona radiata, 348
coronary arteries, 302
corpus luteum, 343–44, 356
cortex:
  adrenal, 420–21
  in plants, 48–52, 68
  renal, 323, 325
cortisone, 420
cotyledon, 97, 106–09, 111
courtship display, 254
coverts, 249
Cowper's gland, 342, 346
coxa, 207, 223
crab, 192
crayfish, 33
cremaster, 222
cretinism, 420
crista, 400–401
crocodile, 195
crocus:
  corm, 76–77
  stigma, 97
crop, 198
  rotation, 504
crossing over, 447–48
cross-pollination, 90–91, 94
crumb soil, 508, 510
crustacea, 4, 191–92, 359, 493
crystalline:
  cone, 204, 205, 206, 207
  lens, 388
culture solutions, 157
culturing bacteria, 169
cupula, 400–01
curlew, 250
curve of normal distribution, 470
cuticle:
  invertebrate, 199, 203–04, 220, 359, 362
  leaf, 56
cuttings, 81
cyst, 186
cytoplasm, 7, 8, 9, 10, 12, 13, 14, 23, 25, 27, 291, 427–28, 443
cytosine, 439–40

2, 4-D, 124
daffodil, 5, 73–76
daisy, 5, 88
dandelion, 5, 67, 88–89
Darwin, Charles, 479
DDT, 502–03
deamination, 286, 420
decay bacteria, 164, 511
deciduous plants, 5, 60, 136
decomposers, 163, 494
deep freezing, 166
deficiency diseases, 265–66, 420
deforestation, 514
dendrite, 403–05
dendron, 403
denitrifying bacteria, 498
dentine, 372–73
deoxyribonucleic acid (DNA), 433, 436, 439–41
dermis, 331, 333
destarching, 148
detergents, 164
detoxification, 286
development:
  bird, 255–57
  frog, 241
diabetes, 421
dialysis tubing, 22
diaphragm, 316, 342, 355, 387
diastole, 304
diatomaceous earth, 182
diatoms, 182–83, 494
dicotyledons (dicots), 5, 107–08, 110–11
diet, 259
diffusion, 17–20
  in fish, 230
  in frog, 237–38, 243
  in insects, 200–01
  in leaves, 134, 136, 144
  in lungs, 294, 313–15
  in tissue fluid, 294, 305, 307
diffusion gradient, 19, 144, 315
digestion, 3, 269–90, 492
digestive action, 281
digestive juices, 166, 269
digit, 254
dinosaur, 195
diphtheria, 165
diploid number, 432
direction of sound, 399
discontinuous variation, 464–65

disease, 185, 296
  carriers, 214, 217–18
dispersal, 102–04
  seed, 75
distance judgment, 394
dominant gene, 449, 457
dormancy, of seeds, 108–09
dormant bud, 62, 66
dorsal fin, 226
dorsal root, 410, 415–16
dorsal root ganglion, 410, 417
double vision, 396
down feathers, 249
Down's syndrome, 453
dried food, 166
drosophila, 432–34, 437–38, 447, 453, 483
drugs, 274
dry weight, 159
duodenum, 275, 281, 421
dutch elm disease, 487
dwarfism, 443, 463, 487
dysentery, 185, 218

ear, 397–98
ear drum:
  frog, 235
  man, 397–98
earthworm, 191, 361, 491
ecdysis (molting), 199, 200, 221, 359
echinoderms, 4, 194
ecology, 489–507
ectoplasm, 185
egestion, 2, 185, 273
egg:
  bird, 255
  fish, 232–34
  frog, 241–42
  insect, 218
  membrane, 255
egg white, digestions of, 288
ejaculation, 346
elbow joint, 362
embryo:
  bird, 256
  flowering plant, 97, 106, 109
  frog, 242–43
  man, 348–49, 350–51
  tissue grafts in, 438–39
enamel, dental, 372–73
endangered species, 500
endocrine, glands, 295, 419
  system, 403, 419–24
endolymph, 399
endoplasma, 185
endoskeleton, 194, 359
endosperm, 107, 109
endothelium, 299
energy:
  for photosynthesis, 143–44
  ultimate source, 495
energy requirements, 260–61
energy value, 259–60
environment, 489
environment and heredity, 465–67
enzymes, 35, 111, 163, 165, 347
  digestive, 164, 269, 274–78, 281, 287
  experiments with, 289–90
epicotyl, 106, 108
epidermis:
  man, 330, 333, 336
  plant, 8, 27, 48–52, 56, 58–59, 67
epididymis, 341–43, 346
epigeal germination, 110–11
epiglottis, 273, 313
epithelium, 269, 278, 310, 313–14, 386–87
erectile tissue, 343
erector muscle, 331, 333, 336
erosion, 502, 514–16
erythrocytes (*See* Red blood cells)
esophagus, 269–70, 272–74 (***See also*** Gullett)
essential amino acids, 262
estrogen, 344, 349, 354, 356, 421, 422–23
estuaries, 33
ethanol, 41–42, 178, 274 (*See also* alcohol)
etiolation, 121
eucaryotic, 169
eugenics, 468
euglena, 188–89
Eustachian tube, 272, 397
eutrophication, 164, 502
evaporation, 33, 134–35, 335, 336
evergreen trees, 5, 60, 137
evolution, theory of, 473–79
excretion, 2
  man, 295, 322, 325–28
  segmented worms, 192
excretory organs, 198, 322
excretory products, 322
exoskeleton, 359, 361, 362
expansion, 25
expiration, 315, 317
explosive fruit, 102
extension, 25
extensor muscles, 368
external fertilization, 339
extracellular, 270
eye:
  fish, 226
  frog, 235–36
  insect, 193, 197, 204–06, 221, 223
  man, 165, 387–97
eye color, 447
eyelid, 235, 387, 465

fat cells, 283
fat-soluble vitamins, 264
fatty acids, 261–62, 278, 281
fauna, 490
feathers, 195, 248–51, 336
feces, 167, 218, 269, 280, 495
feedback, 422–44
feeding, 2, 173, 238
feet, bird, 250
female reproductive organs, 340–44
femoral artery, 300
femur, 223, 253–54, 362, 371
fermentation, 43, 177
ferns, 5
fertilization, 84, 96–98, 339, 346–47, 447, 449
  bird, 255
  fish, 232, 234
  flowering plants, 84, 96–98
  frog, 241–42
  man, 346–48
fertilizers, 504–06, 508, 510
fetus, human, 353
fiber cells, 49
fibrin, 296
fibrinogen, 286, 293, 296
fibrous roots, 66–67
fibula, 254, 360–61
field capacity, 509
field of vision, 394–95
filament, 85, 87–89, 91–92
filial generation, 457
filter feeding, 232
fingers, 362, 367
fins, 226–29

fish, 4, 32, 194, 225–34, 339, 475, 477–78
  fossils, 475–76
flaccid, 26
flagellates, 188–89
flagellum, 161, 181, 185, 189
flatworms, 4, 190
flavor, 396
flexor muscles, 368
flight feathers, 248–52
flocculation, of clay, 510
flora, 490
floret, 88–89
flower, 85–97
  bud, 74
  stock, 72, 74–75, 78, 87
flukes, 190
fly, 167
flying:
  bird, 250–54
  insect, 208–09
follicle (*See* graafian follicle)
follicle-stimulating hormone (FSH), 421, 422–24
food, 259–68
  digestion of, 269–90
  preservation and hygiene, 166
food chains, 492, 493
food tests, 267
food vacuole, 185–86
food webs, 492
forearm, 367–68
fore-brain, 411
fossils, 474–78
fovea, 388
foxes, 495
fraternal twins, 352
french bean, 108
frog, 5, 32, 194, 235–47, 417
frog-spawn, 241
frontal lobe, 410, 412
frozen food, 166
fructose, 276, 281
fruit, 85
  dispersal, 102–04
  formation, 95, 97–102
  types, 102–04
fruiting bodies, 173
fry, 234
functional niche, 499
fungi, 3, 5, 60, 81, 172–79
fur, 336

galactose, 278, 281
gall bladder, 271, 276, 285
gametes, 84, 339, 342, 442–46, 459–60 (*See also* ovum; sperm)
ganglion, 407
gaseous exchange, 35, 310, 313–15
gastric glands, 281
gastric juice, 274, 281
gel, 19
genes, 433–41, 447, 456–72, 484–85
genetic advice, 468–69
  counseling, 468–69
genetics, 425, 456–72
genotype, 426, 458–62
geologic fault, 476
geotropism, 119–20, 125, 126
germination, 71, 102, 106, 108–17
gestation period, 352
giant chromosomes, 433, 436–37
gill:
  fish, 225, 230–31
  tadpole, 243, 247
gill bar, 230, 233
gill raker, 231–33
girdle scar, 64–66
girdles, skeletal, 369
gizzard, 198
gland, defined, 269
glomerular filtrate, 325, 422
glomerulus, 324–27
glucose, 3, 35, 261
  in blood, 293, 295, 307, 422
  in plants, 144
  metabolism of, 35–36, 42–43, 280
  product of digestion, 276, 281
  storage of, 282–83
  test for, 267
glycerol, 262, 276, 278, 281
glycogen, 261, 272, 282–83, 285, 309, 337
goose-pimples, 336
graafian follicle, 343–44, 348, 356, 421, 422
grafting, 81–83
granular layer, 330–31, 333
grasses, 5, 90–93
gravity, effect on growth, 118–20
gray matter, 409, 415
growing point, 46, 68, 78, 111
growth, 2, 25
  of birds, 62
growth hormones, 422
grub, 197
guanine, 439–40
guard cells, 27, 57–59
gullet, 187, 188, 198, 270–71
gulley erosion, 515
guttation, 30
gynaecium, 85, 87

habitat, 183, 239, 490
hair color, 447, 449–50
hair follicle, 332–34
hair plexus, 334, 381
half flower, 86–88
haploid number, 445
hard palate, 272
hare, 395
harvestmen, 193
hawkweed, 88–89
hazel catkin, 96
head, insect, 197
hearing, sense of:
  fish, 226
  insect, 203
  man, 398–99
heart, 300, 302–05, 308, 313
heat, 334–35, 337, 415
heat cramps, 335
heat distribution, 295
heat stagnation, 335
heat stroke, 335
heavy soil, 509–10
height, inherited character, 465
hemocoel, 198, 202, 223
hemoglobin, 236, 276, 292, 314
hemophilia, 452
hepatic artery, 285
hepatic portal vein, 278, 285, 302
hepatic vein, 298
herbaceous plants, 5, 71
herbivore, 164, 278, 374, 376–77, 492
heredity, 425–72
  and environment, 465–67
heritable variation, 483–85
hermaphrodite animals, 339
herring, 4, 231, 233, 493
herring gull, 250

heterozygous, 457–60, 462, 468, 471
hibernation, 224, 337
hilum, 106, 108
hind-brain, 411
hinge joints, 365, 481
hip joint, 362, 364
hollyhock, 92
holophytic nutrition, 143–58
holozoic nutrition, 143, 188
homeostasis, 287, 293, 422
homoiothermic animals, 5, 6, 225, 248, 334
homologous chromosomes, 428, 433, 446–47
homozygous, 457–60, 462, 468, 471
honey guides, 87
hookworms, 191
hormones, 293, 295, 419, 422–23
  deactivation of, 286–87, 419
  plant, 122–26
horse chestnut, 64–66
host, 143
housefly, 214–15, 217
human genetics, 463–64
humerus, 251, 254, 360, 362, 367–68, 480–81
humidity, 136
hummingbird, 195
humus, 508, 511
hybrid, 6
hydra, 4, 15, 406–07, 408
hydroid, 191
hydrochloric acid, in stomach, 166, 274, 281
hydrotropism, 121
hypha, 172–73, 174, 175
hypocotyl, 106–08, 110
hypogeal germination, 111, 113
hypothalamus, 328–29

ichneumon fly, 220
identical twins, 352, 466–67
ileum, 277, 295, 308
  absorption in, 278–79
  digestion in, 276, 281
image formation, 390–91
imago, 199, 221
immunity, 166, 296
immunization, 168–69
immunology, 298
impulses, nerve, 379–80
incisor teeth, 372, 374, 376–77
inclusions, 427
incomplete dominance, 458
incomplete metamorphosis, 199
incubation, 255
incus, 398
indoleacetic acid (IAA), 122–24
infection, prevention of, 166–69, 296
inferior vena cava, 308
inflorescence, 66, 85, 90, 91
ingestion, 271
inheritance, 456–64
inhibition, 406
inner ear, 226, 398
inoculation, 297
inorganic particles in soil, 508
insect pollination, 91, 93–95
insecticides, 93–5, 216–18, 502–03
insects, 4, 193, 197–224
inspiration, 315–16
instinct, 413
insulation, 336
insulin, 420–21
integument, 97
intelligence, 470–71
intercostal muscles, 316–17
internal environment, 287
internal fertilization, 339
internal respiration, 35
internal sense organs, 384
internode, 46–47, 72, 78
interrelationships, 459–91
intracellular, 31, 270
intra-uterine device (IUD), 355
invertebrates, 4, 184, 361
  fossil, 478–79
involuntary, 368, 411
iodine, in nutrition, 263
  test for starch, 147, 267–68
iris, 5, 78, 388, 414
iron, in nutrition, 263
  storage, 286
irritability (sensitivity), 2, 3
isolation, 168, 486–87
isotopic labelling, 129–31, 153, 155

jaw, 360, 373, 376
jellyfish, 4, 361
johnson grass, 77–78
joints:
  insect, 206–07
  man, 359, 362–66
joule, 259–60
judgment, of distance, 394

kangaroo, 432
keel, 88–89, 95, 254
keratin, 330–31, 333
kidney, 32, 33, 269–70, 295, 323–27, 422
kilojoule, 260
knee joint, 362, 365
knee jerk, 415, 417

labium, 210–12, 214
labor, 352
labrum, 212
lachrymal duct, 388
lactase, 281
lacteals, 279, 308
lactic acid, 42–43, 283
lactose, 281
Lake Erie, 502
lamina, 55
large intestine, 280 (*See also* Colon; Rectum)
large white butterfly, 218–24
larva, 197, 199, 216–17, 493 (*See also* Caterpillar; Grub; Maggot; Tadpole)
laryngeal cartilage, 273
larynx, 270–71, 311, 319
latent heat, 135
  of vaporization, 335
lateral bud, 47, 61–62, 66, 76–77, 79
lateral line, 227
lateral roots, 47, 53, 110, 112
latitude, 491
leaf, 46–47, 52, 58, 63, 64–65, 72–75, 107
  characteristics of, 137
  distribution of stomata in, 141
  form, 5
  photosyntheses in, 27, 56, 58, 143–56
  transpiration from, 129–41
  venation, 5
  water movement in, 30–31
leaf-fall, 60–61, 136–38
leaf scars, 65–66
leaf stalk (*See* Petiole)
leg, insect, 207–08
leguminous plants, 498, 504
lens:
  human eye, 388, 389, 390

lens (*cont.*)
insect eye, 204–06
lenticel, 47, 50, 52
leucine, 441
leucocytes (*See* White blood cells)
life history:
fish, 232–34
frog, 240–47
insect, 197–99, 218–24
ligament, 251, 363–64
light, effect on growth, 117–18, 120–22
light soil, 510–11
light intensity, 392
lightning, 498
limbs:
comparison, 480–81
insect, 206–07, 223
rabbit, 361, 371
limiting factors, 155–56, 489–91
linkage, 447
lipases, 271, 281
lipid, 172
litmus paper, 18
liver, 276, 279, 285, 292, 300
liverworts, 5
lizards, 195, 481
loam, 511
locomotion (*See* Movement and Locomotion)
locust, 200
long sighted, 395–96
lung:
frog, 195
man, 270, 294–95, 298, 310–21
lupin, 88, 94–95, 98, 104, 107
lymph, 307
lymph nodes, 292, 307
lymphatic duct, 306, 307
lymphatic system, 279, 305–08
lymphatic vessel, 279, 306–08
lymphocytes, 307
lysins, 297, 441
lyrozyme, 388

maggot, 197
magnesium, in nutrition, 263
maize, 109–13
malaria, 188, 214, 216–18, 485
male reproductive organs, 341–46
malleus, 397
malpighian layer, 330–31, 333, 336, 426
malt, 178
maltase, 281
maltose, 274, 276, 281
mammals, 5, 195–96, 339, 477
mammary gland, 196, 339, 354, 455
mandible:
bird, 248, 254
insect, 210–12, 219
manganese, in nutrition, 263
mantle, 193
manure, 504–05
maple fruit, 103
marrow, bone, 292–93
mass flow hypothesis, 131–32
masseter muscle, 369
mastication, 273, 281, 376
mating, 240–41, 255
maxilla, 210–12
mayfly, 199
meat, 495
median fin, 226
medulla:
adrenal gland, 420
medulla (*cont.*)
kidney, 323, 325
medulla oblongata, 410–13
medusae, 191
meiosis, 442–45, 459–60
meissner's corpuscle, 33, 381–82
melanin, 330, 441
membranes: cell, 11, 24
semipermeable, 21, 23
memory, 413
Mendel, Gregor, 463
menstruation, 355–56, 424
mental defectives, 468
mesenteric artery, 298, 300
mesophyll, 60, 134
metabolism, 32, 44–45, 336
of carbohydrates, 282
of proteins, 284
metacarpals, 360, 367
metamorphosis:
frog, 241–47, 420
insect, 199, 218–24
metaphase:
meiosis, 444, 448
mitosis, 427, 429
metatarsals, 248, 254
methylene blue, 19
micropyle, insect egg, 218
seed, 97, 106, 108
mid-brain, 411
middle ear, 397
middle lamella, 8, 9, 430
midrib, 55, 60
milk, 267, 274, 339, 354, 495
milk teeth, 372
mineral salts, 159, 164, 259, 263, 293, 359, 509
mites, 193
mitosis, 427–31, 438
molar teeth, 372, 376, 378
molds, 5, 174
molecules, 17
mollusks, 4, 194
molt, 220
monarch butterfly, 224
monera, 161
mongolism, 453
monocotyledons (monocots), 107, 111
monoculture, 502
monoploid (haploid) number, 445
mosquito, 211–12, 214, 216–17
mosses, 5
moth, 203, 224
motor areas, 412
motor end plate, 403
motor neurone, 403–04, 409, 415
mold, 173–74, 183
mouse, 344
mouth:
digestion in, 272–74, 281
fish, 230–32
frog, 238–39, 243
mouth parts of insects, 210–15
movement and locomotion, 2–3, 184, 362, 369
ameba, 4, 185
bird, 250–54
euglena, 188
fish, 227–29, 362
frog, 236, 362
insect, 206, 361
mammals, 362
Paramecium, 187
movement of food, 271

mucor, 174
mucus, 243, 274, 278, 296, 312, 319
multifactorial inheritance, 456, 463, 465, 470
multipolar neurone, 403–04
muscles:
  insect, 206–09
  vertebrate, 364, 366–71
mushrooms, 5
mutation, 442, 453, 483, 484, 487
mycelium, 173–74
myxomatosis, 495

nasal cavity, 272
natural communities, 489, 491
natural selection, 479–85, 488
navel, 354
neck, 254
nectar, 85
nectary, 85, 87, 94
nematodes, 191
nerve, 407–08
nerve cell, 15, 404–07
nerve ending, 15, 331–32, 380–82, 383
nerve fiber, 15, 373, 379, 386, 406, 409–10, 412
nervous system, 404–19
nest:
  bird, 254, 256
  stickleback, 233–34
neural spine, 363
neurone, 16, 403–05
nictitating membrane, 248
nitrate, 157–58, 164, 497–98, 499, 506
nitrifying bacteria, 497
nitrite bacteria, 497
nitrogen cycle, 497–98
nitrogen-fixing bacteria, 498, 504, 509
node, 46–47, 78
nodules, root, 107, 498–99
normality curve, 470
nose, 319, 386
nostril:
  bird, 248
  fish, 225–26, 227
  frog, 235–56, 238
nourishment, 1
nuclear membrane, 427–28, 449
nucleolus, 427–29, 431
nucleoplasm, 7, 427
nucleus, cell, 7, 8, 9, 12, 13, 14–15, 59, 97, 181, 291, 292, 345, 346–47, 427–29 (*See also* Chromosomes)
nutrients, 261
nutrition, types of, 143
nymph, 199–200

obelia, 191
ocellus, 206
olfactory lobe, 410
olfactory pit, 203
omnivores, 375, 376, 492
oogenesis, 445
operculum:
  fish, 194, 226, 230–32
  tadpole, 244, 245, 246
opsonins, 297
optic nerve, 387–88, 389, 340
oral groove, 188
orbit, 361, 369, 387
organ, 13
organelles, 427
organism, 13
osmoregulation, 32–33
  ameba, 32–33, 186
osmoregulation (*cont.*)
  man, 328–29
osmosis, 17, 20–23, 32–33, 144
  in animals, 32–33
  in capillaries, 307
  in palisade cells, 145
  in plants, 23–32
osmotic potential, 22, 134, 166
osmotic pressure, 22
ossicles, 397
ostrich, 195
otolith, 399–400
outer ear, 397
oval window, 398
ovary:
  animals, man, 198, 340–41, 343–44
  flowering plants, 85, 87–89, 90–91, 97–101, 103
overcooling, 336–37
overgrazing, 516
overheating, 335–36
overweight, 284–85
oviduct, 198, 255, 340–41, 346, 350
oviparous animals, 232, 234
ovipositor, 198, 223
ovulation, 342, 344, 356
ovule, 84–85, 87–89, 90, 97–101
ovum, 97–101, 242–43, 339–40, 445, 449–50, 459–60
  mother cell, 450
owl, 250
oxygen:
  in exhaled air, 315, 319–20
  for germination, 115–16
  product of photosynthesis, 144–46, 150–51, 153
  transport of, 294, 300, 302, 310
  uptake of, 36, 42–43, 201, 230–31, 236–38, 243–44, 292, 313–15
oxygen debt, 43
oxyhemoglobin, 292, 294

pacinian corpuscle, 334, 381–82
pain, 380, 415
pairing in birds, 254
palate, 378
palisade cells, 57, 59–60, 144–45
palp, 210–12
pancreas, 274–75, 281, 420
pancreatic amylase, 281
pancreatic juice, 275, 281
pappus, 89, 103
parachute mechanism, 102
parallax, 394
Paramecium, 187–88, 287
parasites, 143, 185, 220, 502
  bacterial, 165, 167
  fungal, 173, 175–76
  insect-borne, 214–18
  protozoan, 188
parental care:
  bird, 256–57
  mammal, 354
  stickleback, 234
patella, 306–01, 365
Pavlov's experiments, 419
peat, 511
pectoral fin, 226
pectoral girdle, 369
pectoral muscles, bird, 251, 253
peg-and-socket joints, 206–07
Pelargonium, 148
pellegra, 265
pelvic fin, 226

pelvic girdle, 253–54, 277, 355, 362, 369
pelvis:
  kidney, 323–24, 325, 327
  skeleton, 254, 343, 362–63, 371
penicillin, 174–75
penis, 342, 346, 453
peppered moth, 482–83, 484, 485
pepsin, 274, 278, 281, 288
pepsinogen, 278
peptidase, 281
peptides, 274–76, 281, 441
perch, 233, 492–94
perennial plants, 71–72
pericarp, 85, 99–101, 103, 109
perilymph, 398
periosteum, 364
peripheral nervous system, 407
peristalsis, 273
permeable, 7
pesticides, 216, 218, 502–04
petal, 85, 87–89, 97–101, 103
petiole, 52, 55, 60–61, 65
PH, 289, 490
phagocytes, 293
phalanges:
  bird, 254, 361
  man, 367
  rabbit, 361
pharynx, 244, 272–73, 311–12
phenotype, 458–62
phenylalanine, 469
phenylketonuria, 469
phloem, 48–51, 54, 68
  translocation in, 129–32
phosphate, 164, 263, 439–40
photosynthesis, 3, 56, 58, 134, 143–56, 164, 495
phototropism, 118, 121
phylum, 4
phytoplankton, 193, 492, 494
pickled food, 166
pike, 233, 492–94
pine, 92, 137
pinna, 361, 397, 399
pinworms, 191
pioneer species, 500
pith, 48–51
pituitary gland, 328, 411, 421, 423–24
placenta:
  flowering plant, 97, 99
  man, 349–53, 421
Planaria, 190
plants:
  characteristics of, 2–3, 11–12, 143
  classification of, 5
plasma, food, 291, 293, 306
plasma membrane, 11
plasmolysis, 26, 166
platelets, 293
pleural fluid, 306, 311, 318
pleural membrane, 311, 318
Pleurococcus, 188
plumule, 106, 108–09, 111–12
poikilothermic animals, 4, 225, 235, 334
polar bodies, 445, 449
poliomyelitis (polio), 218
pollen, 84–85, 92–97
pollen sac, 85, 92–95
pollen tube, 96–97
pollination, 90–97
polyp, 191
pons, 412
poppy, 99, 101
pore, 101
potassium, in nutrition, 263
potato, 77, 79
potato blight, 175, 504
potometer, 138–40
precursor, 278
predators, 489, 494
preening, 249
pregnancy, 348–52, 419
premolar teeth, 372, 374, 376, 378
prenatal care, 354
preservation techniques, 166
pressure equalization, 398
procaryote, 161
producers, 492, 494
progeny, 459–60, 462, 487
progesterone, 349, 351, 356, 421
prolegs, 219–20, 222
prophase:
  meiosis, 442–41
  mitosis, 427–29
proprioceptors, 203, 370, 384
prop roots, 113
prostate gland, 342, 343
protein, 16–15, 259, 261–62, 296, 358 (***See*** *also* Enzymes)
  digestion of, 262, 281, 284, 288
  plasma, 286, 293
  test for, 267
proteinases, 271, 281
Protista (*See* Single-celled organisms)
protoplasm, 7, 262
protozoa, 4, 167, 184–89, 287
pseudopodium, 185
pseudotrachea, 214, 215, 218
Pteraspis, 478
ptyalin (*See* Salivary amylase)
puberty, 344, 421
pulmonary artery, 302, 303, 304, 312, 314
pulp, dental, 373
pupa, 197, 216, 217, 221
pupation, 197, 216, 221–22, 223
pupil, 388
pure breeding, 456
pus, 296
pygostyle, 254
pyloric sphincter, 274, 275
pyramid, kidney, 323–24
pyrenoid, 181

quill, 248–49

rabbit, 188, 344, 361, 366, 370–71, 495
radiation, 454
radiation sterilization, 166, 168
radicle, 106–10, 119, 124–25
radioisotopes, 129–31
radius, 251, 254, 360, 480–81
range of tolerance, 499
rare species, 500
raspberry, 98
ray, wood, 54
receptacle, 85, 87–89, 98–103
recessive gene, 457
recombination of genes, 447, 484–85
rectum, 280, 343
red blood cells, 291–92, 306, 314
reduction division, 445
reflex action, 379, 414–18
reflex arc, 415–18
refrigeration, 166
region of elongation, 29, 124
relay neurone, 415, 417
renal artery, 298, 300, 323–24, 327, 342
renal tubule, 323, 325

renal vein, 298, 302, 323–24
rennin, 274, 281
reproduction, 2, 474, 500 (*See also* Asexual reproduction; Sexual reproduction)
reproductive organs, 340–46, 421, 442
reptiles, 5, 195, 339, 477
residual air, 317
resistance to disease, 165
resource allocation, 491
respiration, 1–2, 35, 146, 310, 419, 497
respiratory system, 200, 231
retina, 388, 390
rhizome, 46, 77–79
rhythm method, of birth control, 355
ribs, 254, 292, 316–18, 360, 362
rickets, 266
rill erosion, 515
roan coat color, 458, 462
rock weathering, 508
rods, of eye, 388
root, 2, 28–30, 46–47, 64–67, 119
  adventitious, 67
  cap, 29, 69–70, 109
  fibrous, 66–67
  hairs, 28–30, 67–68, 110
  nodule, 107
  pressure, 30, 134
  regions, 68–69, 125
  structure, 67
  system, 58, 66–67
  tap, 47–66
  tip, 68–69
  uptake in, 133
rose, 5, 99, 101
roughage, 259, 267, 280
round window, 398
roundworms, 4, 191
rumen, 164
runners, 46
ryegrass, 90–91

sacculus, 226, 398, 399
salamanders, 194
salivary glands, 281, 419, 436–37
saliva, 273, 281, 288, 289
salivary amylase, 274, 281, 289–90
salt, in diet, 263
  in soil, 159–60, 509
  translocation of, 132–33, 135
  uptake by roots of, 133, 156–57
sand, 508, 510
saprophytes, 143, 162, 173
saprophytic fungi, 173
saprophytic nutrition, 143, 188
scale, fish, 225
scale leaves, 68, 78–79
scale scars, 64–65
scapula:
  bird, 254
  man, 362, 365
  rabbit, 361
scion, 81
sclera, 388
scorpions, 193
scrotum, 341
scurvy, 263, 266
sea star (starfish), 194
sebaceous gland, 331, 334, 536
secondary sex characters, 354–55, 421
secondary thickening, 54
secretin, 421
secretion, 270
sedimentary rock, 474–76
sedimentation, 474
seed, 103
  dispersal of, 75, 102–04
  germination of, 102–06, 108–17
  respiration of, 37–40, 41–42
  structure, 106
  types, 107
seedling, 2
segmented bodies, 197
segmented worms, 4, 191
segregation of genes, 458, 484
self-pollination, 90–91
semen, man, 346
semicircular canals, 226, 398, 399–400
semilunar valves, 303–04
seminal fluid, frog, 241
seminal vesicle, 342–43, 346
seminiferous tubule, 341–42, 345
semipermeable membranes, 21, 23, 24
sensation, intensity of, 300
sense of balance, 399–400
sensitivity, 2–3, 117–22
sensory neurone, 301, 403–04, 409, 411, 415
sensory system:
  fish, 226–27
  insect, 202
  man, 379–402
sepal, 85, 87, 89, 97–101, 103
seta, 223
sewage disposal, 166, 170
sex determination, 450–51
sex hormones, 421
sex linkage, 451
sexual reproduction, 339, 500
  bird, 254–57
  fish, 232–34
  flowering plant, 71, 84–105
  frog, 240–41
  fungi, 174
  insect, 197, 218
  man, 339–58
  protozoa, 186
shaft:
  bone, 366
  feather, 249
sheath (*See also* Diaphragm)
sheep, 347, 348, 377, 495
sheet erosion, 514
shell, 255–56
shivering, 336
shoot, of plants, 2, 46, 79, 117–19
short sight(ed), 395–96
shoulder joint, 365
shrimp, 192
sickle-cell anemia, 484–85
sieve tube, 15, 51–53, 59, 132
sight, sense of:
  fish, 226
  insect, 204–06
  man, 387, 390–97
silica, 182
silkworm, 224
silt, 508
simple eye, 206
single-celled organisms, 184 (*See also* Bacteria; Yeast)
  animals, 4, 184
  plants, 184
single-factor inheritance, 456–58
sinus:
  uterine, 350
  skull, 412
skeleton:
  bird, 251–54
  fossil fish, 475–76

skeleton (*cont.*)
man, 359–70
rabbit, 361
skin:
bird, 248
frog, 235–39
man, 165, 330–38, 381–83
sense, 381, 382–83
skull:
bird, 254
dog, 375
man, 360, 369, 412
sheep, 377
sleeping sickness, 184, 214
"slimming diet," 285
slope effect, 490
small intestine, 270 (*See also* Duodenum; Ileum)
smell, sense of:
fish, 227
insect, 203
man, 386–87
snakebite, 297
snakes, 195
sneezing, 415, 417
snowdrop bulb, 72, 74
sodium, 263
soft palate, 272
soil, 159, 508–16
solar energy, 490, 491, 495
sound, sense of:
insect, 203
man, 398
spawn, 234, 240–41
special senses, 384–86
species, 5, 486–87
spectacles (eye glasses), 395–96
sperm, 241, 339, 341–42, 344–46, 445–46, 449–50, 459–60
spermatic cord, 341
spermatogenesis, 445
sperm duct, 341–43
sperm mother cells, 446, 450
sphincter, 273
sphigmomanometer, 305
spiders, 193
spikelet, 90–91
spinal column, 316–18
spinal nerves, 409, 410, 415
spinal reflex, 417
spindle, 429–31, 443–45, 449
spine, 227, 318, 361, 363, 371
spiracles:
insect, 200–01, 222
tadpole, 244
Spirogyra, 181–82, 188
spleen, 292
sponges, 4, 189–90
spongy layer, 57, 60
spontaneous generation, 474, 478
sporangium, 174
spore:
bacterial, 162–63
fungal, 174–77
springwood, 54
stamen, 85, 87, 91, 97–101, 103
standard, 88
stapes, 398
starch, 172, 261
digestion of, 273–74, 276, 281, 288
test for, 147–48, 288–89
stem, 46–55, 58, 60–61, 63, 72, 90
stem tuber, 77–79
stereoscopic vision, 392–93
sterile, 88
sterilization, 168
sternum, 251–54, 292, 316–17, 360, 361
stickleback, 4, 226, 232–34, 492–94
stigma, 85, 87–89, 90–91, 94, 97–101
sting cell, 15
stock, 81
stoma, 14, 27, 47, 56–59, 134, 136, 141, 175
stomach, 166, 276
digestion in, 274–75, 281
stool, 269
storage:
of digested food, 282
of vitamins, 287
storage organs, 71–80
strawberry, 99–100
streptococcus, 291
streptomycin, 167, 175
stretch receptors, 384
style, 85, 87–89, 91, 97–101
subclavian artery, 300
subclavian vein, 308
subcutaneous fat, 334
succession, 500
successional species, 500
succus entericus, 276, 281
sucrase, 281
sucrose ("sugar"), 261, 281
sugars, 163
summerwood, 54
sunlight, 143, 156, 495
sun stroke, 335
support, skeletal, 361
supporting cell, 385, 386
surface area of organisms, 337
survival of the fittest, 487
suspensory ligament, 388
swallowing, 236, 272
sweat duct, 331
sweat gland, 331–33, 419
sweating, 335, 336
swim bladder, 229–30
swimming:
fish, 227–29
frog, 236
synapse, 405–06, 413, 415
synecology, 489
synovial fluid, 363–65
synovial joint, 363–365
synovial membrane, 363–65
system, 13
systole, 304, 305

tactile bristle, 203
tadpole, 2, 195, 237, 243–47, 420
tail, bird, 226, 227, 244, 245, 247, 253, 254
tapeworm, 190
tap root, 47, 53, 67
tarsals, 360
tarsus, 206, 223, 254
taste, sense of:
insect, 203, 223
man, 385–86
taste bud, 385–86
tear gland, 387–88
teeth, 372–78
telophase:
meiosis, 443, 445
mitosis, 427, 428, 430–31
temperature, 136, 289–90
temperature control, 334
temporalis muscle, 369
tendon, 364, 366
tentacles, 190

terminal bud, 46–47, 62, 64, 66
termites, 188
testa, 97, 106–10
test for:
  fats, 268
  glucose, 268
  proteins, 268
  starch, 267
testis, 341–43, 345
testosterone, 421
tetanus, 165, 297
thirst, 330
thoracic duct, 308
thorax:
  insect, 197
  man, 310–11, 318
thymine, 439–40
thyroid gland, 419–20
thyroid-stimulating hormone, 423
thyroxine, 244, 419–20
tibia, 223, 360, 371
tibio-tarsus, bird, 254
ticks, 193
tidal air, 317
tissue, 13
tissue fluid, 293, 305–07
tissue respiration, 35
toad, 194
tomato, 99
tone, muscle, 368
tongue:
  frog, 239
  man, 272, 385–86
tonsils, 307
topography, 490
touch, sense of:
  insect, 202, 223
  man, 382
toxins, 165, 296
trace elements, 157, 263
trachea:
  insect, 200–02, 216
  man, 270, 273, 311, 313, 318
tracheole, 200–02
translocation, 129–33
transpiration, 133–40
transpiration stream, 28, 134–35
transport, of foods, 295
  of gases, 294
  of salts, 135
  of wastes, 295
trees, 54, 60, 134–35, 136, 137
triceps muscle, 368
trichina worm, 191
tricuspid valve, 303, 304–05
trillium, 430
trochanter, 207, 223
tropisms, 117–22
trout, 4
trypsin, 274, 281
tsetse fly, 214
tube feet, 194
tuber, 77, 79
tuberculosis, 165
turgid, 25
turgor, 23, 25
turgor pressure, 25, 48, 132
twigs, 63–64
twins, 352, 466–67
2, 4-D, 124
typhoid, 164, 165
tyrosinase, 441

ulna, 251, 360, 480–81
umbilical cord, 349, 350, 353–54
umbilical stalk, 256
urea, 286, 307, 322, 327, 350
ureter, 323, 340, 341, 343
urethra, 328, 342, 343, 346
uric acid, 322, 327
urine, 32, 167, 286, 326, 327, 419, 495
uterus, 340, 349, 350–53
utriculus, 226, 398, 399

vaccination, 297
vaccine, 297
vacuole:
  animal, 10, 185–86
  plant, 8, 9, 10, 14, 23, 27, 59, 134, 145
vagina, 340, 346, 350, 353
valine, 441
valve:
  in heart, 303–05
  in lymphatic vessel, 308
  in vein, 301
vane, feather, 249
variation, 441–42, 464–65 (*See also* Mutation)
vascular bundle, 48, 50, 53, 60, 65, 141
vascular tissue, 68
vasoconstriction, 336, 412
vasodilation, 335–36, 412
vectors, 216
vegetative reproduction, 71, 80
vein:
  blood vessel, 299, 301–02
  leaf, 31, 50, 55, 60
vena cava, 303, 304
ventilation:
  insect, 201–02
  man, 310, 315–19
ventral fin, 226
ventral root, 410, 415
ventricle, 302, 314
venule, 301, 306
vertebrae, 254, 292, 360–61, 363
vertebral column, 227, 363
vertebrates, 4, 184, 361–62, 411, 477, 480–81
  fossil, 475–78
vertigo, 401
vessel, plant, 29, 49–51, 59, 66, 135
vibration, sense of, in fish, 227
vibrissae, 334
villi:
  in ileum, 277–79, 295
  in placenta, 349
vinegar, 178
virus, 169–70, 217, 296, 297
vision, 390–91
vitamins, 164, 259, 263–66, 287
vitreous humor, 388
viviparous animals, 232
vocal cords, 319
voice, 319, 355
voluntary muscle, 368
vulva, 340, 355

walking, 206, 253
wall, cell, 11
Wallace, Alfred Russell, 479
warm-blooded animals (*See* Homoiothermic animals)
water (*See also* Osmoregulation; Transpiration)
  absorption of, 281
  essential for germination, 116
  for photosynthesis, 143–44, 153–55
  identification of, 137
  in diet, 259, 267
  in soil, 509–14

water (*cont.*)
  movement in plants, 27–32
  product of respiration, 35–36, 43, 322
  root response, 122
  supplies, 167
water cultures, 157–58
water fleas, 492, 494
water sac, 351, 353
water-soluble vitamins, 265
water supplies, 167
weathering, 508
weedkillers, 124
whale, 481
whiskers, 334
whisky, 178
white blood cells, 14, 165, 291–92, 296, 306, 307
white deadnettle, 86–87, 94
white matter, 409, 410, 415
whorls, 85
wilting, 25–26
wind, 136
  dispersal, 102–03
  pollination, 91–93, 95–96
windpipe, 419
wine, 178
wing:
  bat, 481
  bird, 248–54, 481
wing (*cont.*)
  flower, 88–89, 95
  insect, 197, 208–9, 223
winter annuals, 71
winter twigs, 63–64
wisdom teeth, 378
womb (*See* Uterus)
wood rays, 54
woody plants, 71
world population, 354, 504
worms, 4

X chromosome, 432, 434, 435, 450–52
xeropthalmia, 265
xylem, 30–31, 48–51, 54, 68
  translocation in, 132–33
  water movement in, 29–30

Y chromosome, 432, 434, 435, 450–52
yeast, 43, 177
yellow fever, 214
yolk, 243, 255
yolk sac, 256, 349

zone of elongation, 68
zooplankton, 232, 492
zygote, 84, 339, 346–47, 425–26, 447, 449, 453
  development of, 348, 425